From Stein to Weinstein and Back

Symplectic Geometry of Affine Complex Manifolds

American Mathematical Society

Colloquium Publications
Volume 59

From Stein to Weinstein and Back

Symplectic Geometry of Affine Complex Manifolds

Kai Cieliebak
Yakov Eliashberg

American Mathematical Society
Providence, Rhode Island

2010 *Mathematics Subject Classification*. Primary 32Q28, 53D35.

For additional information and updates on this book, visit
www.ams.org/bookpages/coll-59

Library of Congress Cataloging-in-Publication Data

Cieliebak, Kai, 1966–
From Stein to Weinstein and back : symplectic geometry of affine complex manifolds / Kai Cieliebak, Yakov Eliashberg.
p. cm. — (Colloquium publications ; v. 59)
Includes bibliographical references and index.
ISBN 978-0-8218-8533-8 (alk. paper)
1. Symplectic geometry. 2. Stein manifolds. I. Eliashberg, Y., 1946– II. Title.

QA665.C54 2012
515′.946—dc23

201219063

∞ The paper used in this book is acid-free and falls within the guidelines
established to ensure permanence and durability.
Visit the AMS home page at `http://www.ams.org/`

10 9 8 7 6 5 4 3 2 1 17 16 15 14 13 12

To my parents, Snut and Hinrich. Kai

To Ada. Yasha

Contents

Preface

In Spring 1996 Yasha Eliashberg gave a Nachdiplomvorlesung (a one semester graduate course) "Symplectic geometry of Stein manifolds" at ETH Zürich. Kai Cieliebak, at the time a graduate student at ETH, was assigned the task to take notes for this course, with the goal of having lecture notes ready for publication by the end of the course. At the end of the semester we had some 70 pages of typed notes, but they were nowhere close to being publishable. So we buried the idea of ever turning these notes into a book.

Seven years later Kai spent his first sabbatical at the Mathematical Sciences Research Institute (MSRI) in Berkeley. By that time, through work of Donaldson and others on approximately holomorphic sections on the one hand, and gluing formulas for holomorphic curves on the other hand, Weinstein manifolds had been recognized as fundamental objects in symplectic topology. Encouraged by the increasing interest in the subject, we dug out the old lecture notes and began turning them into a monograph on Stein and Weinstein manifolds.

Work on the book has continued on and off since then, with most progress happening during Kai's numerous visits to Stanford University and another sabbatical 2009 that we both spent at MSRI. Over this period of almost 10 years, the content of the book has been repeatedly changed and its scope significantly extended. Some of these changes and extensions were due to our improved understanding of the subject (e.g., a quantitative version of J-convexity which is preserved under approximately holomorphic diffeomorphisms), others due to new developments such as the construction of exotic Stein structures by Seidel and Smith, McLean, and others since 2005, and Murphy's h-principle for loose Legendrian knots in 2011. In fact, the present formulation of the main theorems in the book only became clear about a year ago. As a result of this process, only a few lines of the original lecture notes have survived in the final text (in Chapters 2–4).

The purpose of the book has also evolved over the past decade. Our original goal was a complete and detailed exposition of the existence theorem for Stein structures in [**42**]. While this remains an important goal, which we try to achieve in Chapters 2–8, the book has evolved around the following two broader themes: The first one, as indicated by the title, is the correspondence between the complex analytic notion of a Stein manifold and the symplectic notion of a Weinstein manifold. The second one is the extent to which these structures are flexible, i.e., satisfy an h-principle. In fact, until recently we believed the border between flexibility and rigidity to run between subcritical and critical structures, but Murphy's h-principle extends flexibility well into the critical range.

The book is roughly divided into "complex" and "symplectic" chapters. Thus Chapters 2–5 and 8–10 can be read as an exposition of the theory of J-convex functions on Stein manifolds, while Chapters 6–7, 9 and 11–14 provide an introduction to Weinstein manifolds and their deformations. However, our selection of material on both the complex and symplectic side is by no means representative for the respective fields. Thus on the complex side we focus only on topological aspects of Stein manifolds, ignoring most of the beautiful subject of several complex variables. On the symplectic side, the most notable omission is the relationship between Weinstein domains and Lefschetz fibrations over the disc.

Over the past 16 years we both gave many lecture courses, seminars, and talks on the subject of this book not only at our home institutions, Ludwig-Maximilians-Universität München and Stanford University, but also at various other places such as the Forschungsinstitut für Mathematik at ETH Zürich, University of Pennsylvania in Philadelphia, Columbia University in New York, the Courant Institute of Mathematical Sciences in New York, University of California in Berkeley, Washington University in St. Louis, the Mathematical Sciences Research Institute in Berkeley, the Institute for Advanced Study in Princeton, and the Alfréd Rényi Institute of Mathematics in Budapest. We thank all these institutions for their support and hospitality.

Many mathematicians and students who attended our lectures and seminars or read parts of preliminary versions of the book provided us with valuable comments and critical remarks. We are very grateful to all of them, and in particular to M. Abouzaid, S. Akbulut, J. Bowden, V. Braungardt, J. Daniel, T. Ekholm, C. Epstein, J. Etnyre, C. Fefferman, F. Forstnerič, U. Frauenfelder, A. Gerstenberger, R. Gompf, A. Huckleberry, P. Landweber, J. Latschev, L. Lempert, R. Lipshitz, C. Llosa Isenrich, D. McDuff, M. McLean, K. Mohnke, J. Morgan, E. Murphy, S. Nemirovski, L. Nirenberg, K. Nguyen, A. Oancea, N. Øvrelid, P. Ozsváth, L. Polterovich, P. Seidel, A. Stadelmaier, A. Stipsicz, D. Thurston, T. Vogel, E. Volkov, J. Wehrheim, and C. Wendl.

We thank G. Herold, T. Müller, and S. Prüfer for creating the figures, and J. Wright Sharp for her help with English and LaTeX.

And most of all, we thank our spouses, Suny and Ada, for their continued support.

Kai Cieliebak
Mathematisches Institut
Ludwig-Maximilians-Universität
Theresienstr. 39
80333 München, Germany

Yakov Eliashberg
Department of Mathematics
Stanford University
Stanford, California 94305, USA

CHAPTER 1

Introduction

1.1. An overview

Stein manifolds. A *Stein manifold* is a properly embedded complex submanifold of some $\mathbb{C}^N$. As we show in this book, Stein manifolds have symplectic geometry built into them, which is responsible for many phenomena in complex geometry and analysis. The goal of this book is a systematic exploration of this symplectic geometry (the "road from Stein to Weinstein") and its applications in the complex geometric world of Stein manifolds (the "road from Weinstein to Stein").

Stein manifolds are necessarily noncompact, and properly embedded complex submanifolds of Stein manifolds are again Stein. Stein manifolds arise, e.g., from closed complex projective manifolds $X \subset \mathbb{C}P^N$: if $H \subset \mathbb{C}P^N$ is any hyperplane, then the affine algebraic manifold $X \setminus H$ is Stein. Using this construction, it is not hard to see that every closed Riemann surface with at least one point removed is Stein. In fact, as we will see below, any open Riemann surface is Stein. Already this example shows that the class of Stein manifolds is much larger than the class of affine algebraic manifolds.

Stein manifolds can also be described intrinsically. The characterization most relevant for us is due to Grauert [**77**]. Let (V, J) be a complex manifold, where J denotes the complex multiplication on tangent spaces. To a smooth real-valued function $\phi : V \to \mathbb{R}$ we can associate the 1-form $d^{\mathbb{C}}\phi := d\phi \circ J$ and the 2-form $\omega_\phi := -dd^{\mathbb{C}}\phi$. The function is called *(strictly) plurisubharmonic* or, as we prefer to say, *J-convex* if $g_\phi(v, w) := \omega_\phi(v, Jw)$ defines a Riemannian metric. Since g_ϕ is symmetric, this is equivalent to saying that ω_ϕ is a symplectic (i.e., closed and nondegenerate) form compatible with J, i.e., $H_\phi = g_\phi - i\omega_\phi$ is a Hermitian metric. A function $\phi : V \to \mathbb{R}$ is called *exhausting* if it is proper (i.e., preimages of compact sets are compact) and bounded from below.

Since the function $\phi_{\rm st}(z) := |z|^2$ on $\mathbb{C}^N$ is exhausting and i-convex with respect to the *standard complex structure* i on $\mathbb{C}^N$, every Stein manifold admits an exhausting J-convex function (namely the restriction of $\phi_{\rm st}$ to V). A combination of theorems of Grauert [**77**] and Bishop and Narasimhan [**18, 144**] asserts that the converse is also true: *a complex manifold which admits an exhausting J-convex function is Stein.*

Note that the space of exhausting J-convex functions on a given Stein manifold (V, J) is convex, and hence contractible. It is also open in $C^2(V)$, so a generic J-convex function is a *Morse function* (i.e., it has only nondegenerate critical points) and a generic path of J-convex functions consists of Morse and *generalized Morse* functions, i.e., functions with only nondegenerate and birth-death type critical points.

Weinstein manifolds. A *Weinstein structure* on a $2n$-dimensional manifold V is a triple (ω, X, ϕ), where ω is a symplectic form, $\phi : V \to \mathbb{R}$ is an exhausting generalized Morse function, and X is a complete Liouville vector field which is gradient-like for ϕ. Here the Liouville condition means that the Lie derivative $L_X\omega$ coincides with ω. The quadruple (V, ω, X, ϕ) is called a *Weinstein manifold.* We will see that homotopic (for an appropriate definition of homotopy, see Section 11.6) Weinstein manifolds are symplectomorphic. This structure was introduced in a slightly different form by A. Weinstein in [**187**] and then formalized in [**49**]. Since then it has become a central object of study in symplectic topology, see e.g. [**32, 169, 24**].

As it was explained above, after fixing an exhausting J-convex generalized Morse function $\phi : V \to \mathbb{R}$ on a Stein manifold (V, J) one can associate with the triple (V, J, ϕ) the symplectic form ω_ϕ. It turns out that the gradient vector field $X_\phi := \nabla_{g_\phi}\phi$ of ϕ, computed with respect to the metric g_ϕ which it generates, is Liouville with respect to the form ω_ϕ. After composing ϕ with a suitable function $\mathbb{R} \to \mathbb{R}$ we may further assume that the vector field X_ϕ is complete. Then the assignment

$$(J, \phi) \mapsto \mathfrak{W}(J, \phi) := (\omega_\phi, X_\phi, \phi)$$

yields a canonical map from Stein to Weinstein structures. A different choice of exhausting J-convex generalized Morse function leads to a homotopic, and hence symplectomorphic, Weinstein manifold. Note that this map forgets the most rigid datum, the integrable complex structure J. A major theme of this book is the reconstruction of Stein structures from Weinstein structures (the "road from Weinstein to Stein").

From Weinstein to Stein. We say that two functions $\phi, \phi' : V \to \mathbb{R}$ are *target equivalent* if there exists an increasing diffeomorphism $g : \mathbb{R} \to \mathbb{R}$ such that $\phi' = g \circ \phi$. In the following theorem we always have to allow for *target reparametrizations,* i.e., replacing functions by target equivalent ones, but we suppress this trivial operation from the notation.

THEOREM 1.1. *(a) Given a Weinstein structure* $\mathfrak{W} = (\omega, X, \phi)$ *on* V*, there exists a Stein structure* (J, ϕ) *on* V *such that* $\mathfrak{W}(J, \phi)$ *is Weinstein homotopic to* $\mathfrak{W}$ *with fixed function* ϕ.

(b) Given a Weinstein homotopy $\mathfrak{W}_t = (\omega_t, X_t, \phi_t)$, $t \in [0, 1]$, *on* V *beginning with* $\mathfrak{W}_0 = \mathfrak{W}(J, \phi)$, *there exists a Stein homotopy* (J_t, ϕ_t) *starting at* $(J_0, \phi_0) = (J, \phi)$ *such that the paths* $\mathfrak{W}(J_t, \phi_t)$ *and* $\mathfrak{W}_t$ *are homotopic with fixed functions* ϕ_t *and fixed at* $t = 0$. *Moreover, there exists a diffeotopy* $h_t : V \to V$ *with* $h_0 = \mathrm{Id}$ *such that* $h_t^* J_t = J$ *for all* $t \in [0, 1]$.

(c) Given a Weinstein homotopy $\mathfrak{W}_t = (\omega_t, X_t, \phi_t)$, $t \in [0, 1]$, *on* V *connecting* $\mathfrak{W}_0 = \mathfrak{W}(J_0, \phi_0)$ *and* $\mathfrak{W}_1 = \mathfrak{W}(J_1, \phi_1)$ *with* $\phi_t = \phi_1$ *for* $t \in [\frac{1}{2}, 1]$, *there exists a Stein homotopy* (J_t, ϕ_t) *connecting* (J_0, ϕ_0) *and* (J_1, ϕ_1) *such that the paths* $\mathfrak{W}(J_t, \phi_t)$ *and* $\mathfrak{W}_t$ *are homotopic with fixed functions* ϕ_t *and fixed at* $t = 0, 1$.

Theorem 1.1 fits in the following more global, partially conjectural picture. To avoid discussing subtleties concerning the appropriate topologies on the spaces of Stein and Weinstein structures, we restrict our attention here to the compact case. Let W be a compact smooth manifold with boundary. In the following discussion we always assume that all considered functions on W have ∂W as their regular level set. A *Stein domain* structure on W is a pair (J, ϕ), where J is a complex structure and $\phi : W \to \mathbb{R}$ is a J-convex generalized Morse function. A *Weinstein domain* structure on W is a triple (ω, X, ϕ) consisting of a symplectic form on W, a generalized Morse function $\phi : W \to \mathbb{R}$, and a Liouville vector field X which is gradient-like for ϕ. Let us denote by $\mathfrak{Stein}$ and $\mathfrak{Weinstein}$ the spaces of Stein and Weinstein domain structures on W, respectively. Let $\mathfrak{Morse}$ be the space of generalized Morse functions on W.

We have the following commutative diagram

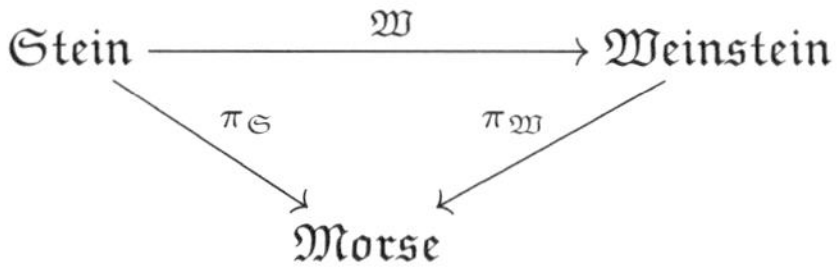

where $\mathfrak{W}(J, \phi) = (\omega_\phi, X_\phi, \phi)$ as above, $\pi_{\mathfrak{W}}(\omega, X, \phi) := \phi$, and $\pi_{\mathfrak{S}}(J, \phi) := \phi$. Consider the fibers $\mathfrak{Stein}(\phi) := \pi_{\mathfrak{S}}^{-1}(\phi)$ and $\mathfrak{Weinstein}(\phi) := \pi_{\mathfrak{W}}^{-1}(\phi)$ of the projections $\pi_{\mathfrak{S}}$ and $\pi_{\mathfrak{W}}$ over $\phi \in \mathfrak{Morse}$.

THEOREM 1.2. *The map* $\mathfrak{W}_\phi := \mathfrak{W}|_{\mathfrak{Stein}(\phi)} : \mathfrak{Stein}(\phi) \to \mathfrak{Weinstein}(\phi)$ *is a weak homotopy equivalence.*

Note that (a compact version of) Theorem 1.1(a) is equivalent to the fact that the map $\mathfrak{W}_\phi$ induces an epimorphism on π_0, while Theorem 1.1(c) implies that the induced homomorphism is injective on π_0 and surjective on π_1. Conversely, it is easy to see that Theorem 1.1(b) and injectivity of $\mathfrak{W}_\phi$ on π_0 imply Theorem 1.1(c).

To put Theorem 1.1(b) into a more global framework, let us denote by $\mathcal{D}$ the identity component of the diffeomorphism group of W. Fix a Stein domain structure (J, ϕ_0) on W (the function ϕ_0 will play no role in what follows; the only important fact is that it exists). For a function $\phi \in \mathfrak{Morse}$, we introduce the spaces

$$\mathcal{D}_J(\phi) := \{h \in \mathcal{D} \mid \phi \text{ is } h^*J\text{-convex}\},$$

$$\mathcal{P}_J(\phi) := \{(h, \gamma) \mid h \in \mathcal{D}_J(\phi),\ \gamma : [0,1] \to \mathfrak{Weinstein}(\phi),\ \gamma(0) = \mathfrak{W}(h^*J, \phi)\},$$

$$\mathcal{P}_J := \bigcup_{\phi \in \mathfrak{Morse}} \mathcal{P}_J(\phi).$$

We denote by $\mathfrak{Weinstein}_J$ the connected component of $\mathfrak{W}(J, \phi_0)$ in $\mathfrak{Weinstein}$ (for any choice of ϕ_0, the component is independent of this choice).

CONJECTURE 1.3. The projection $\pi_{\mathcal{P}} : \mathcal{P}_J \to \mathfrak{Weinstein}_J$, $(h, \gamma) \mapsto \gamma(1)$ is a Serre fibration.

Note that (a compact version of) Theorem 1.1(b) is just the homotopy lifting property of $\pi_{\mathcal{P}}$ for homotopies of *points*, so it is a special case of Conjecture 1.3. We believe that this conjecture can be proven by further developing techniques discussed in this book. By an easy topological argument (see Appendix A.1), Conjecture 1.3 combined with Theorem 1.2 would imply

CONJECTURE 1.4. The map $\mathfrak{W} : \mathfrak{Stein} \to \mathfrak{Weinstein}$ is a weak homotopy equivalence.

Let us emphasize that in this book we are interested in the classification of Stein structures up deformation, and not up to biholomorphism. The classification of Stein complex structures up to biholomorphism is very subtle. For example, C^∞-small deformations of the round ball in $\mathbb{C}^n$, $n \geq 2$, give rise to uncountably many pairwise non-biholomorphic Stein manifolds; see e.g. [**116**] for an exposition of this beautiful subject.

Existence of Stein structures. Theorem 1.1 reduces complex-geometric questions about Stein manifolds to symplecto-geometric questions about Weinstein manifolds. Our next task is to develop techniques for answering those questions.

Let us first analyze necessary conditions for the existence of a Weinstein (or Stein) structure on a given smooth manifold V of real dimension $2n$. Clearly, one necessary condition is the existence of an *almost complex structure* J, i.e., an endomorphism of the tangent bundle with $J^2 = -\mathrm{Id}$. A second necessary condition arises from the following property of Morse functions with gradient-like Liouville fields (see Chapter 2): their Morse indices are $\leq n$. By Morse theory, this implies that V has a handlebody decomposition with handles of index at most n. This observation of Milnor [**139**] was the result of a long development, beginning with Lefschetz [**121**] and followed by Serre [**170**] and Andreotti and Frankel [**7**].

It turns out that for $\dim_{\mathbb{R}} V \neq 4$, these two conditions are sufficient for the existence of a Weinstein structure on V, so in combination with Theorem 1.1(a) we get the following existence theorem which was proved in [**42**]:

THEOREM 1.5 (Existence of Stein structures). *Let (V, J) be an almost complex manifold of dimension $2n \neq 4$ and $\phi : V \to \mathbb{R}$ an exhausting Morse function without critical points of index $> n$. Then there exists an integrable complex structure $\widetilde{J}$ on V homotopic to J for which the function ϕ is target equivalent to a $\widetilde{J}$-convex function. In particular, $(V, \widetilde{J})$ is Stein.*

We prove in this book several refinements and extensions of this result, some of which are due to Gompf [**71, 72, 73**] and Forstnerič and Slapar [**63**].

Theorem 1.5 settles the existence question for Stein structures in dimensions $\neq 4$. In dimension 4 the situation is drastically different. For instance, Lisca and Matič proved in [**125**] that $S^2 \times \mathbb{R}^2$ does not admit any Stein complex structure. On the other hand, Gompf proved the following topological analogue of Theorem 1.5:

THEOREM 1.6 (Gompf [**70**]). *Let V be an oriented open topological 4-manifold which admits a (possibly infinite) handlebody decomposition without handles of index > 2. Then V is homeomorphic to a* Stein surface *(i.e., a Stein manifold of complex dimension 2). Moreover, any homotopy class of almost complex structures on V is induced by an orientation preserving homeomorphism from a Stein surface.*

Let us point out that the Stein surfaces in Theorem 1.6 are usually not of finite type, where a Stein manifold is said to be of *finite type* if it admits an exhausting J-convex function with only finitely many critical points. Gompf's result which uses the technique of Casson handles, as well as Lisca and Matič's theorem which uses Seiberg–Witten theory, are beyond the scope of this book. See however Chapter 16 for some related discussion. For example, we prove that $S^2 \times \mathbb{R}^2$ is not homeomorphic to any Stein surface of finite type.

Deformations of Stein structures. It turns out that the Weinstein problems in parts (b) and (c) of Theorem 1.1 cannot be reduced, in general, to differential topology even when $\dim V > 4$. On the contrary, they are tightly related to the core problems of symplectic topology.

It is easy to see that a Weinstein structure (ω, X, ϕ) on $\mathbb{R}^{2n}$ for which ϕ has no other critical points besides the minimum is symplectomorphic to the standard structure on $\mathbb{R}^{2n}$. On the other hand, as we already pointed out, homotopic Weinstein structures are symplectomorphic. Seidel and Smith [**167**], McLean [**137**], and Abouzaid and Seidel [**3**] have recently constructed for each $n \geq 3$ infinitely many "exotic" Weinstein structures on $\mathbb{R}^{2n}$ which are not symplectomorphic to the standard one and which, moreover, are pairwise non-symplectomorphic. Then Theorem 1.1(a) allows us to transform these Weinstein structures to Stein structures which are not Stein homotopic among each other and to $\mathbb{C}^n$; in particular they do not admit exhausting J-convex functions without critical points of positive index.

One can also reformulate this result as a failure of the following "*J-convex h-cobordism problem*": Let W be a smooth cobordism between manifolds $\partial_- W$ and $\partial_+ W$. A *Stein structure* on W is a complex structure J on W which admits a J-convex function $\phi : W \to \mathbb{R}$ which has $\partial_\pm W$ as its regular level sets. Then the above results by Abouzaid, McLean, Seidel and Smith imply that *for each $n \geq 3$ there exists a Stein cobordism (W, J) diffeomorphic to $S^{2n-1} \times [0, 1]$ for which the corresponding J-convex function ϕ cannot be chosen without critical points.*

By contrast, we prove the following uniqueness theorem in complex dimension two (first sketched in [**47**]; for the diffeomorphism part see [**83, 43, 133**]).

THEOREM 1.7. *Let (W, J) be a minimal compact complex surface with J-convex boundary diffeomorphic to S^3. Suppose that there exists a symplectic form ω taming J, i.e., such that ω is positive on complex directions. Then W is diffeomorphic to the 4-ball and admits a J-convex Morse function $\phi : W \to \mathbb{R}$ which is constant on ∂W and has no other critical points besides the minimum.*

Here a *complex surface* (i.e., a complex manifold of complex dimension 2) is called *minimal* if it contains no embedded holomorphic spheres of self-intersection -1. See Chapter 16 for a discussion of related uniqueness results due to McDuff, Hind, Wendl, and others.

It turns out that the Weinstein problems can be reduced to differential topology in the *subcritical* case when the Morse functions have no middle-dimensional critical points. Moreover, based on work of Murphy [**143**], who discovered that in contact manifolds of dimension > 3 there is a class of Legendrian knots which obey an h-principle, we define a larger class of *flexible* Weinstein manifolds for which problems of symplectic topology can be reduced to differential topology. For example, we have

THEOREM 1.8. *Let V be a manifold of dimension $2n \neq 4$, Ω a homotopy class of nondegenerate (not necessarily closed) 2-forms on V, and $\phi : V \to \mathbb{R}$ an exhausting Morse function without critical points of index $> n$. Then:*

(a) There exists a flexible Weinstein structure (ω, X, ϕ) on V with $\omega \in \Omega$, and this structure is unique up to Weinstein homotopy.

(b) Every diffeomorphism of V preserving the homotopy class Ω is diffeotopic to a symplectomorphism of (V, ω).

Another application of the flexible technique is

THEOREM 1.9. *Let (V, J) be a contractible Stein manifold. Then $V \times \mathbb{C}$ admits an exhausting J-convex Morse function with exactly one critical point, the minimum.*

For our final application, recall that the *pseudo-isotopy problem* in differentiable topology concerns the topology of the space $\mathcal{E}(M)$ of functions on $M \times [0, 1]$ without critical points that are constant on $M \times 0$ and $M \times 1$. Work of Cerf, Hatcher and Wagoner, Igusa, and Waldhausen has led to a description of $\pi_0\mathcal{E}(M)$ for $\dim M \geq 7$ in terms of algebraic K-theory.

Given a topologically trivial Stein cobordism $(M \times [0,1], J)$, one can ask about the topology of the space $\mathcal{E}(M \times [0,1], J)$ of J-convex functions without critical points that are constant on $M \times 0$ and $M \times 1$ (provided that this space is nonempty). Understanding of the topology of the inclusion map $\mathcal{I} : \mathcal{E}(M \times [0,1], J) \to \mathcal{E}(M)$ is the content of the *J-convex pseudo-isotopy problem*. We prove the following result in this direction.

THEOREM 1.10. *If* $\dim M > 3$ *and the Stein structure* J *is flexible, the homomorphism* $\mathcal{I}_* : \pi_0\mathcal{E}(M \times [0,1], J) \to \pi_0\mathcal{E}(M)$ *is surjective.*

We conjecture that $\mathcal{I}_*$ is an isomorphism.

1.2. Plan of the book

This book is organized as follows.

In Chapters 2 and 3 we explore basic properties and examples of J-convex functions and hypersurfaces. In particular, we prove Richberg's theorem on smoothing of J-convex functions and derive several important corollaries.

In Chapter 4 we construct special hypersurfaces that play a crucial role in extending J-convex functions over handles.

The next two chapters contain background material which is standard but sometimes not easy to find in the literature. The necessary complex analytic background is discussed in Chapter 5, and the symplecto-geometric one in Chapter 6.

In Chapter 7 we review several h-principles that we use in this book. We begin with a review of the Smale–Hirsch immersion theory and Whitney's theory of embeddings. We then discuss Gromov's results about symplectic and contact isotropic immersions and embeddings, Murphy's h-principle for loose Legendrian knots, and Gromov's theory of directed embeddings and immersions with applications to totally real embeddings. We finish this chapter with an h-principle for totally real discs with Legendrian boundaries, which we deduce from previously discussed h-principles and which plays an important role in the proofs of the main results of this book.

Theorem 1.5 is proved in Chapter 8. This chapter also contains several new results concerning surrounding of subsets by J-convex hypersurfaces, with applications to holomorphic and polynomial convexity. We also prove here several refinements of Theorem 1.5, some of which are due to Gompf, and Forstnerič and Slapar.

In Chapter 9 we review Morse–Smale theory and the h-cobordism theorem. In particular, we discuss basic facts concerning gradient-like vector fields. We also review the "two-index theorem" of Hatcher and Wagoner and basic notions of pseudo-isotopy theory.

In Chapter 10 we develop a Morse–Smale type theory for J-convex functions. In particular, we show how the Morse-theoretic operations which are used in the proof of the h-cobordism theorem—reordering of critical points, handle-slides, and cancellation of critical points—can be performed in the class of J-convex functions.

In Chapter 11 we introduce Weinstein structures and study their basic properties. We discuss Stein and Weinstein homotopies, and we introduce the classes of subcritical and flexible manifolds which play an important role for the "road from Morse to Weinstein".

In Chapter 12 we discuss modifications of Weinstein structures near critical points and stable manifolds and prove Weinstein analogues of the results proven in Chapter 10 for J-convex functions.

In Chapter 13 we prove a more precise version of Theorem 1.5 by first constructing a Weinstein structure and then proving Theorem 1.1(a).

Chapters 14 and 15 contain our main results about deformations of Weinstein and Stein structures. In Chapter 14 we classify flexible Weinstein structures up to homotopy and show that the problem of simplification of the Morse function corresponding to a flexible Weinstein structure can be reduced to Morse–Smale theory. In particular, we prove Theorem 1.8.

In Chapter 15 we show that every Weinstein homotopy can be transformed to a Stein homotopy. In particular, we prove Theorem 1.1(b) and (c) and deduce various corollaries, including Theorems 1.9 and 1.10.

Chapter 16 concerns the situation in complex dimension 2. In particular, we discuss the method of filling by holomorphic discs and prove Theorem 1.7. We also discuss the classification of Stein fillings of 3-dimensional contact manifolds and review known results about Stein surfaces.

Finally, in Chapter 17 we sketch McLean's construction of exotic Stein structures in higher dimension and explain how they are distinguished by symplectic homology.

Notation. Throughout this book we use the following notation. For a subset $A \subset X$ of a topological space we denote by $\operatorname{Int} A$ and $\bar{A}$ its interior resp. closure, and $A \Subset B$ means that $\bar{A}$ is a compact subset of $\operatorname{Int} B$. For A closed we denote by $\mathcal{O}p\, A$ a sufficiently small (*but not specified*) open neighborhood of A.

Manifolds are always assumed to be smooth and second countable.

Part 1

J-Convexity

CHAPTER 2

J-Convex Functions and Hypersurfaces

In this chapter we introduce the notion of J-convexity for functions and hypersurfaces and discuss their relation. After considering some basic properties and examples, we derive an explicit formula for the normalized Levi form on $\mathbb{C}^n$.

2.1. Linear algebra

A *complex vector space* (V, J) is a real vector space V of dimension $2n$ with an endomorphism J satisfying $J^2 = -\mathrm{Id}$. Scalar multiplication of $v \in V$ with $a + ib \in \mathbb{C}$ is then defined by $(a + ib)v := av + bJv$. A *Hermitian form* on (V, J) is an $\mathbb{R}$-bilinear map $H : V \times V \to \mathbb{C}$ which is $\mathbb{C}$-linear in the first variable and satisfies $H(X, Y) = \overline{H(Y, X)}$. Moreover, if H is positive definite, it is called a *Hermitian metric*. We can write a Hermitian form H uniquely as

$$H = g - i\omega,$$

where g is a symmetric and ω a skew-symmetric bilinear form on the real vector space V. The forms g and ω determine each other,

$$g(X, Y) = \omega(X, JY), \qquad \omega(X, Y) = g(JX, Y),$$

for $X, Y \in V$. Moreover, the forms ω and g are invariant under J, which can be equivalently expressed by the equation

$$\omega(JX, Y) + \omega(X, JY) = 0.$$

Conversely, given a skew-symmetric J-invariant form ω, we can uniquely reconstruct the corresponding Hermitian form H:

$$H(X, Y) := \omega(X, JY) - i\omega(X, Y). \tag{2.1}$$

For example, consider the complex vector space $(\mathbb{C}^n, i)$ with coordinates $z_1 = x_1 + iy_1, \dots, z_n = x_n + iy_n$. It carries the standard Hermitian metric

$$(v, w) := \sum_{j=1}^{n} v_j \bar{w}_j = \langle v, w \rangle - i\omega_{\mathrm{st}}(v, w),$$

where $\langle \cdot, \cdot \rangle$ is the Euclidean metric and $\omega_{\mathrm{st}} = \sum_{j=1}^n dx_j \wedge dy_j$ the *standard symplectic form* on $\mathbb{C}^n$.

For a symmetric bilinear form $Q : V \times V \to \mathbb{R}$, we denote the corresponding quadratic form

$$Q : V \to \mathbb{R}, \qquad Q(v) := Q(v, v)$$

by the same letter. Recall that the symmetric bilinear form can be recovered from the quadratic form by

$$Q(v, w) = \frac{1}{2}\Big(Q(v + w) - Q(v) - Q(w)\Big).$$

We can uniquely decompose a quadratic form on a complex vector space (V, J) into its complex linear and antilinear parts,

$$\begin{aligned} Q(v) &= Q^{\mathbb{C}}(v) + \bar{Q}^{\mathbb{C}}(v), \\ Q^{\mathbb{C}}(v) &= \frac{1}{2}\Big(Q(v) + Q(Jv)\Big) = Q^{\mathbb{C}}(Jv), \\ \bar{Q}^{\mathbb{C}}(v) &= \frac{1}{2}\Big(Q(v) - Q(Jv)\Big) = -\bar{Q}^{\mathbb{C}}(Jv). \end{aligned}$$

The quadratic form Q is called *J-convex* if $Q^{\mathbb{C}}(v) > 0$ for all $v \neq 0$.

EXAMPLE 2.1. For the quadratic form

$$Q(z) = \sum_{j=1}^{n} (\lambda_j x_j^2 + \mu_j y_j^2) \tag{2.2}$$

on $\mathbb{C}^n$ we have $Q^{\mathbb{C}}(z) = \frac{1}{2}\sum_j (\lambda_j + \mu_j)|z_j|^2$. So Q is i-convex if and only if

$$\lambda_j + \mu_j > 0 \text{ for all } j = 1, \ldots, n. \tag{2.3}$$

Note that geometric convexity implies i-convexity but not conversely, since i-convexity only requires that the average of the coefficients over each complex line is positive.

In fact, this example captures all i-convex quadratic forms:

LEMMA 2.2. *Every i-convex quadratic form $Q\colon \mathbb{C}^n \to \mathbb{R}$ can be put in the form (2.2) by a complex linear change of coordinates, where the coefficients λ_j, μ_j satisfy (and are uniquely determined by) the conditions*

$$\lambda_j + \mu_j = 2, \qquad \lambda_j \geq \mu_j, \qquad \mu_1 \leq \mu_2 \leq \cdots \leq \mu_n. \tag{2.4}$$

PROOF. A general quadratic function can be uniquely decomposed into its complex linear and antilinear parts as

$$Q(z) = \sum_{ij} a_{ij} z_i \bar{z}_j + \mathrm{Re}\left(\sum_{ij} b_{ij} z_i z_j\right), \qquad a_{ij} = \bar{a}_{ji},\ b_{ij} = b_{ji}.$$

Under a complex linear coordinate change $z_i \mapsto \sum_k c_{ik} z_k$, the matrices $A = (a_{ij})$ and $B = (b_{ij})$ transform as

$$A \mapsto C^t A \bar{C}, \qquad B \mapsto C^t B C$$

for $C = (c_{ij}) \in GL(n, \mathbb{C})$. Using this, we can first transform the Hermitian matrix A to a diagonal matrix with entries 0 or ± 1. Since A represents the complex linear part $Q^{\mathbb{C}}$, which is positive by hypothesis, we obtain $A = \mathrm{Id}$. Now we can still transform B according to $B \mapsto C^t B C$ for unitary matrices C (thus preserving $A = \mathrm{Id}$). By a lemma of Schur (see e.g. [**183**]), using this we can transform B to a diagonal matrix with nonnegative real entries ν_i and thus Q to

$$Q(z) = \sum_{j=1}^{n} \left(|z_j|^2 + \nu_j \mathrm{Re}\,(z_j^2)\right) = \sum_{j=1}^{n} \left((1 + \nu_j) x_j^2 + (1 - \nu_j) y_j^2\right). \qquad \square$$

Of course, using rescalings $z_j \mapsto r_j z_j$, we can vary the form of Q, e.g., to the following one that will be useful:

$$Q(z) = \sum_{j=1}^{k}(\lambda_j x_j^2 - y_j^2) + \sum_{j=k+1}^{n}(\lambda_j x_j^2 + y_j^2), \qquad \lambda_j > 0. \tag{2.5}$$

2.2. J-convex functions

An *almost complex structure* on a smooth manifold V of real dimension $2n$ is an endomorphism $J : TV \to TV$ satisfying $J^2 = -\mathrm{Id}$ on each fiber. The pair (V, J) is called an *almost complex manifold.* It is called a *complex manifold* if the almost complex structure J is *integrable*, i.e., J is induced by complex coordinates on V. By the theorem of Newlander and Nirenberg [**149**], a (sufficiently smooth) almost complex structure J is integrable if and only if its *Nijenhuis tensor*

$$N(X,Y) := [JX, JY] - [X,Y] - J[X, JY] - J[JX, Y], \qquad X, Y \in TV,$$

vanishes identically. An integrable almost complex structure is called a *complex structure.*

In the following let (V, J) be an almost complex manifold. To a smooth function $\phi : V \to \mathbb{R}$ we associate the 2-form

$$\omega_\phi := -dd^{\mathbb{C}}\phi,$$

where the differential operator $d^{\mathbb{C}}$ is defined by

$$d^{\mathbb{C}}\phi(X) := d\phi(JX)$$

for $X \in TV$. The form ω_ϕ is in general not J-invariant. However, it is J-invariant if J is integrable. To see this, consider the complex vector space $(\mathbb{C}^n, i)$. Given a function $\phi : \mathbb{C}^n \to \mathbb{R}$, define the complex valued $(1,1)$-form

$$\partial\bar\partial\phi := \sum_{i,j=1}^{n} \frac{\partial^2\phi}{\partial z_i \partial \bar z_j} dz_i \wedge d\bar z_j.$$

Using the identities

$$dz_j \circ i = i\, dz_j, \qquad d\bar z_j \circ i = -i\, d\bar z_j,$$

we compute

$$d^{\mathbb{C}}\phi = \sum_j \Big(\frac{\partial\phi}{\partial z_j} dz_j \circ i + \frac{\partial\phi}{\partial \bar z_j} d\bar z_j \circ i\Big) = \sum_j \Big(i\frac{\partial\phi}{\partial z_j} dz_j - i\frac{\partial\phi}{\partial \bar z_j} d\bar z_j\Big),$$

$$dd^{\mathbb{C}}\phi = -2i \sum_{i,j} \frac{\partial^2\phi}{\partial z_i \partial \bar z_j} dz_i \wedge d\bar z_j.$$

Hence

$$\omega_\phi = 2i\partial\bar\partial\phi, \tag{2.6}$$

and the i-invariance of ω_ϕ follows from the invariance of $\partial\bar\partial\phi$.

A function $\phi : V \to \mathbb{R}$ on an almost complex manifold is called *J-convex*[1] if $\omega_\phi(X, JX) > 0$ for all nonzero tangent vectors X. If ω_ϕ is J-invariant, it defines

[1]Throughout this book, by convexity and J-convexity we will always mean *strict* convexity and J-convexity. Nonstrict (J-)convexity will be referred to as *weak* (J-)convexity.

by (2.1) a unique Hermitian form

$$H_\phi := g_\phi - i\omega_\phi, \qquad g_\phi := \omega_\phi(\cdot, J\cdot),$$

and ϕ is J-convex if and only if the Hermitian form H_ϕ is positive definite.

EXAMPLE 2.3. Let $f : \mathbb{C}^n \supset U \to \mathbb{C}$ be holomorphic. Then $|f|^2$ is weakly i-convex. Moreover, outside the zero set of f the function $\log|f|$ satisfies $dd^{\mathbb{C}}(\log|f|) = 0$.

From (2.6) we can derive a simple expression for the form H_ϕ associated to a function $\phi : \mathbb{C}^n \to \mathbb{R}$ in terms of the Hermitian matrix $a_{ij} := \frac{\partial^2\phi}{\partial z_i \partial \bar{z}_j}$. For $v, w \in \mathbb{C}^n$ we have

$$\begin{aligned}\omega_\phi(v,w) &= 2i\sum_{ij} a_{ij}dz_i \wedge d\bar{z}_j(v,w) = 2i\sum_{ij} a_{ij}(v_i\bar{w}_j - w_i\bar{v}_j) \\ &= 2i\sum_{ij}(a_{ij}v_i\bar{w}_j - \bar{a}_{ij}\bar{v}_i w_j) = -4\,\mathrm{Im}\Big(\sum_{ij} a_{ij}v_i\bar{w}_j\Big),\end{aligned}$$

hence

$$H_\phi(v,w) = 4\sum_{i,j=1}^{n} \frac{\partial^2\phi}{\partial z_i \partial \bar{z}_j} v_i \bar{w}_j. \tag{2.7}$$

The Hermitian form H_ϕ is related to the *(real) Hessian* Hess_ϕ (given by the matrix of real second derivatives) as follows:

$$\begin{aligned}\mathrm{Hess}_\phi(v,w) &= \sum_{ij}\Big(\frac{\partial^2\phi}{\partial z_i\partial z_j}v_i w_j + \frac{\partial^2\phi}{\partial \bar{z}_i\partial \bar{z}_j}\bar{v}_i\bar{w}_j + \frac{\partial^2\phi}{\partial z_i\partial \bar{z}_j}v_i\bar{w}_j + \frac{\partial^2\phi}{\partial \bar{z}_i\partial z_j}\bar{v}_i w_j\Big) \\ &= 2\mathrm{Re}\sum_{ij}\Big(\frac{\partial^2\phi}{\partial z_i\partial z_j}v_i w_j + \frac{\partial^2\phi}{\partial z_i\partial \bar{z}_j}v_i\bar{w}_j\Big),\end{aligned}$$

and hence

$$\mathrm{Hess}_\phi(v,w) + \mathrm{Hess}_\phi(iv,iw) = 4\,\mathrm{Re}\sum_{ij}\frac{\partial^2\phi}{\partial z_i\partial \bar{z}_j}v_i\bar{w}_j = g_\phi(v,w).$$

In particular, the corresponding quadratic forms $H_\phi(v) = H_\phi(v,v)$ and $\mathrm{Hess}_\phi(v) = \mathrm{Hess}_\phi(v,v)$ satisfy

$$H_\phi(v) = \mathrm{Hess}_\phi(v) + \mathrm{Hess}_\phi(iv), \tag{2.8}$$

i.e., H_ϕ is twice the complex average of Hess_ϕ.

The *(Morse) index* of a critical point of ϕ is the maximal dimension of a subspace on which the real Hessian Hess_ϕ is negative definite. The normal form (2.5) shows

COROLLARY 2.4. *The Morse index of a critical point of an i-convex function $\phi : \mathbb{C}^n \to \mathbb{R}$ is at most n.*

This corollary is fundamental for the topology of Stein manifolds. We will give an alternative proof in Section 2.8 and generalize it to almost complex manifolds in Section 3.1.

The relation (2.8) generalizes to Kähler manifolds, as we will explain now. We refer to [**12**] for basic facts about Kähler geometry. A *Kähler manifold* is a complex manifold (V, J) with a Hermitian metric $H = g - i\omega$ satisfying the Kähler condition

$d\omega = 0$. Alternatively, the Kähler condition can be expressed as $\nabla J = 0$, i.e., the complex structure J is parallel with respect to the Levi-Civita connection ∇ of the metric g.

Using the metric $g = \langle\ ,\ \rangle$, we associate to a smooth function $\phi : V \to \mathbb{R}$ its *gradient vector field* $\nabla\phi$ defined by $d\phi = \langle \nabla\phi, \cdot\rangle$ and its *(real) Hessian* $\mathrm{Hess}_\phi : T_pV \times T_pV \to \mathbb{R}$ defined by

$$\mathrm{Hess}(X,Y) := \langle \nabla_X \nabla\phi, Y\rangle = X \cdot d\phi(Y) - d\phi(\nabla_X Y).$$

Torsion freeness of the Levi-Civita connection and $d(d\phi) = 0$ yields the symmetry $\mathrm{Hess}(X,Y) = \mathrm{Hess}(Y,X)$. On a Kähler manifold we use $\nabla J = 0$ to compute for $\lambda_\phi := -d^{\mathbb{C}}\phi$:

$$\begin{aligned} d\lambda_\phi(X, JY) &= X \cdot \lambda_\phi(JY) - (JY) \cdot \lambda_\phi(X) - \lambda_\phi([X, JY]) \\ &= X \cdot \lambda_\phi(JY) - (JY) \cdot \lambda_\phi(X) - \lambda_\phi(J\nabla_X Y - \nabla_{JY} X) \\ &= X \cdot d\phi(Y) + (JY) \cdot d\phi(JX) - d\phi(\nabla_X Y) - d\phi(\nabla_{JY} JX) \\ &= \mathrm{Hess}_\phi(X,Y) + \mathrm{Hess}_\phi(JX, JY). \end{aligned}$$

So we have shown

PROPOSITION 2.5. *For a smooth function $\phi : V \to \mathbb{R}$ on a Kähler manifold the Hermitian form H_ϕ is related to the real Hessian Hess_ϕ by*

$$H_\phi(X) = \mathrm{Hess}_\phi(X) + \mathrm{Hess}_\phi(JX).$$ □

2.3. The Levi form of a hypersurface

Let Σ be a smooth (real) hypersurface in an almost complex manifold (V, J). Each tangent space $T_p\Sigma \subset T_pV$, $p \in \Sigma$, contains a unique maximal complex subspace $\xi_p \subset T_p\Sigma$ which is given by

$$\xi_p = T_p\Sigma \cap JT_p\Sigma.$$

These subspaces form a codimension 1 distribution $\xi \subset T\Sigma$ which we will refer to as the *field of complex tangencies*. Suppose that Σ is cooriented by a transverse vector field ν to Σ in V such that $J\nu$ is tangent to Σ. The hyperplane field ξ can be defined by a Pfaffian equation $\{\alpha = 0\}$, where the sign of the 1-form α is fixed by the condition $\alpha(J\nu) > 0$. The 2-form

$$\omega_\Sigma := d\alpha|_\xi$$

is then defined uniquely up to multiplication by a positive function. As in the previous section we may ask whether ω_Σ is J-invariant. The following lemma gives a necessary and sufficient condition in terms of the Nijenhuis tensor.

LEMMA 2.6. *Let (V, J) be an almost complex manifold. The form ω_Σ is J-invariant for a hypersurface $\Sigma \subset V$ if and only if $N|_{\xi\times\xi}$ takes values in ξ. The form ω_Σ is J-invariant for every hypersurface $\Sigma \subset V$ if and only if for all $X, Y \in TV$, $N(X,Y)$ lies in the complex plane spanned by X and Y. In particular, this is the case if J is integrable or if V has complex dimension 2.*

PROOF. Let $\Sigma \subset V$ be a hypersurface and α a defining 1-form for ξ. Extend α to a neighborhood of Σ such that $\alpha(\nu) = 0$. For $X, Y \in \xi$ we have $[X,Y] \in T\Sigma$ and therefore $J[X,Y] = a\nu + Z$ for some $a \in \mathbb{R}$ and $Z \in \xi$. This shows that

$$\alpha(J[X,Y]) = 0$$

for all $X, Y \in \xi$. Applying this to various combinations of X, Y, JX and JY, we obtain

$$\alpha\Big(N(X,Y)\Big) = \alpha([JX,JY]) - \alpha([X,Y]),$$
$$\alpha\Big(JN(X,Y)\Big) = \alpha([X,JY]) + \alpha([JX,Y]).$$

The form ω_Σ is given by

$$\omega_\Sigma(X,Y) = \frac{1}{2}\Big(X\cdot\alpha(Y) - Y\cdot\alpha(X) - \alpha([X,Y]\Big) = -\frac{1}{2}\alpha([X,Y]).$$

Inserting this in the formulae above yields

$$-\frac{1}{2}\alpha\Big(N(X,Y)\Big) = \omega_\Sigma(JX,JY) - \omega_\Sigma(X,Y),$$
$$-\frac{1}{2}\alpha\Big(JN(X,Y)\Big) = \omega_\Sigma(X,JY) + \omega_\Sigma(JX,Y).$$

Hence the J-invariance of ω_Σ is equivalent to

$$\alpha\Big(N(X,Y)\Big) = \alpha\Big(JN(X,Y)\Big) = 0,$$

i.e., $N(X,Y) \in \xi$ for all $X, Y \in \xi$. This proves the first statement and the "if" in the second statement. For the "only if" it suffices to note that if $N(X,Y)$ does not lie in the complex plane spanned by X and Y for some $X, Y \in TV$, then we find a hypersurface Σ such that $X, Y \in \xi$ and $N(X,Y) \notin \xi$. □

A hypersurface Σ is called *Levi-flat* if $\omega_\Sigma \equiv 0$. This is exactly the Frobenius integrability condition for the field of complex tangencies ξ on Σ. Hence, on a Levi-flat hypersurface, ξ integrates to a real codimension 1 foliation.

The hypersurface Σ is called *J-convex* (resp. *weakly J-convex*) if $\omega_\Sigma(X,JX) > 0$ (resp. ≥ 0) for all nonzero $X \in \xi$. If ω_Σ is J-invariant, it defines a Hermitian form L_Σ on ξ by the formula

$$L_\Sigma(X,Y) := \omega_\Sigma(X,JY) - i\omega_\Sigma(X,Y)$$

for $X, Y \in \xi$. The Hermitian form L_Σ is called the *Levi form* of the (cooriented) hypersurface Σ. We will also use the notation $L_\Sigma(X)$ for the quadratic form $L_\Sigma(X,X)$. Note that Σ is Levi-flat if and only if $L_\Sigma \equiv 0$, and J-convex if and only if L_Σ is positive definite. We will sometimes also refer to ω_Σ as the Levi form. As pointed out above, the Levi form is defined uniquely up to multiplication by a positive function.

Given a J-convex hypersurface and a defining 1-form α for the field of complex tangencies ξ, the 2-form $d\alpha$ has rank $\dim_{\mathbb{R}} \Sigma - 1$. Hence there exists a unique vector field R on Σ satisfying the conditions

$$\alpha(R) = 1, \qquad i_R d\alpha = 0.$$

If the hypersurface Σ is given as a regular level set $\{\phi = 0\}$ of a function $\phi : V \to \mathbb{R}$, then we can choose $\alpha = -d^{\mathbb{C}}\phi$ as the 1-form defining ξ (with the coorientation of Σ given by $d\phi$). Thus the Levi form is given by

$$\omega_\Sigma(X,Y) = -dd^{\mathbb{C}}\phi(X,Y) = \omega_\phi(X,Y).$$

This shows that regular level sets of a J-convex function ϕ are J-convex (being cooriented by $d\phi$). It turns out that the converse is also almost true (similarly to the situation for convex functions and hypersurfaces):

LEMMA 2.7. *Let $\phi : V \to \mathbb{R}$ be a smooth function on an almost complex manifold without critical points such that all its level sets are compact and J-convex. Then one can find a convex increasing function $f : \mathbb{R} \to \mathbb{R}$ such that the composition $f \circ \phi$ is J-convex.*

PROOF. For a function $f : \mathbb{R} \to \mathbb{R}$ we have

$$d^{\mathbb{C}}(f \circ \phi) = f' \circ \phi \, d^{\mathbb{C}}\phi,$$
$$\omega_{f\circ\phi} = -dd^{\mathbb{C}}(f \circ \phi) = -f'' \circ \phi \, d\phi \wedge d^{\mathbb{C}}\phi + f' \circ \phi \, \omega_\phi.$$

By J-convexity of the level sets, there is a unique vector field R on V associated as above to the defining 1-form $-d^{\mathbb{C}}\phi$ for the fields of complex tangencies ξ satisfying

$$d\phi(R) = 0, \qquad -d^{\mathbb{C}}\phi(R) = 1, \qquad i_R\omega_\phi|_\xi = 0.$$

Then $TV = \mathbb{R}R \oplus \mathbb{R}JR \oplus \xi$, and for $Y \in \xi$ we have

$$\omega_{f\circ\phi}(R, Y) = \omega_{f\circ\phi}(R, JY) = 0, \qquad \omega_{f\circ\phi}(Y, JY) = f' \circ \phi \, \omega_\phi(Y, JY) > 0.$$

Let us pick a J-invariant metric on $TV = \mathbb{R}R \oplus \mathbb{R}JR \oplus \xi$ satisfying $|R| = |JR| = 1$ and $\omega_\phi(Y, JY) \geq |Y|^2$ for $Y \in \xi$. Using $-d\phi \wedge d^{\mathbb{C}}\phi(R, JR) = 1$, we compute for $X = aR + bJR + Y \in TV = \mathbb{R}R \oplus \mathbb{R}JR \oplus \xi$:

$$\begin{aligned}\omega_{f\circ\phi}(X, JX) = (a^2 + b^2)[f'' + f'\omega_\phi(R, JR)] + bf'\omega_\phi(JR, JY) \\ + af'\omega_\phi(Y, JR) + f'\omega_\phi(Y, JY),\end{aligned}$$

where we have abbreviated $f' = f' \circ \phi$ and $f'' = f'' \circ \phi$. By compactness of the level sets, there exists a smooth function $h : \mathbb{R} \to [1, \infty)$ satisfying $h(y) \geq 2\max_{\{\phi=y\}} |\omega_\phi|$. Abbreviating $h = h \circ \phi$ and using $h^2 \geq h$, we can estimate

$$\begin{aligned}\omega_{f\circ\phi}(X, JX) &\geq (a^2 + b^2)[f'' - h^2 f'] - hf'\sqrt{a^2 + b^2}|Y| + f'|Y|^2 \\ &\geq (a^2 + b^2)[f'' - \frac{3}{2}h^2 f'] + \frac{1}{2}f'|Y|^2.\end{aligned}$$

Now solve the linear differential equation $f''(y) = 2h(y)^2 f'(y)$ with initial condition $f'(y_0) > 0$. The solution exists for all $y \in \mathbb{R}$ and satisfies $f' > 0$, so $f \circ \phi$ is J-convex in view of

$$\omega_{f\circ\phi}(X, JX) \geq \frac{1}{2}f'[(a^2 + b^2)h^2 + |Y|^2] > 0. \qquad \square$$

REMARK 2.8. The proof of the preceding lemma also shows that if $\phi : V \to \mathbb{R}$ is J-convex, then $f \circ \phi$ is J-convex for any function $f : \mathbb{R} \to \mathbb{R}$ with $f' > 0$ and $f'' \geq 0$.

REMARK 2.9. Consider a hypersurface Σ in an almost complex manifold (V, J) and an almost complex submanifold $W \subset V$ transverse to Σ. Then $\Sigma \cap W$ is a hypersurface in (W, J) with field of complex tangencies $\xi \cap TW$ and the Levi form of $\Sigma \cap W$ equals the restriction of L_Σ to $\xi \cap TW$. In particular, Σ is J-convex if and only if $\Sigma \cap W$ is J-convex for all almost complex submanifolds $W \subset V$ transverse to Σ of complex dimension 2.

2.4. Completeness

A vector field is called *complete* if its flow exists for all forward and backward times. For a J-convex function ϕ, we define its *gradient* $\nabla_\phi \phi$ with respect to $g_\phi = \omega_\phi(\cdot, J\cdot)$ by $d\phi = g_\phi(\nabla_\phi \phi, \cdot)$. (Note that g_ϕ is nondegenerate but not necessarily symmetric.) In general, $\nabla_\phi \phi$ need not be complete:

EXAMPLE 2.10. The function $\phi(z) := \sqrt{1+|z|^2}$ on $\mathbb{C}$ satisfies

$$\frac{\partial^2 \phi}{\partial z \partial \bar{z}} = \frac{\partial}{\partial z} \frac{z}{\sqrt{1+|z|^2}} = \frac{1}{\sqrt{1+|z|^2}^3},$$

so $g_\phi = 4(1+|z|^2)^{-3/2}\langle\ ,\ \rangle$, where $\langle\ ,\ \rangle$ is the standard metric. In particular, ϕ is i-convex. Its gradient is determined from

$$d\phi = \frac{x\,dx + y\,dy}{\sqrt{1+|z|^2}} = \frac{4}{\sqrt{1+|z|^2}^3}\langle \nabla_\phi \phi, \cdot \rangle,$$

thus $\nabla_\phi \phi = \frac{1+|z|^2}{4}(x\frac{\partial}{\partial x} + y\frac{\partial}{\partial y})$. A gradient line $\gamma(t)$ with $|\gamma(0)| = 1$ is given by $\gamma(t) = h(t)\gamma(0)$, where $h(t)$ satisfies $h' = \frac{1+h^2}{4}h$. This shows that $\gamma(t)$ tends to infinity in finite time, hence the gradient field $\nabla_\phi \phi$ is not complete.

However, the gradient field $\nabla_\phi \phi$ can always be made complete by composing ϕ with a sufficiently convex function. Recall that a function $\phi : V \to \mathbb{R}$ is called exhausting if it is proper and bounded from below.

PROPOSITION 2.11. *Let $\phi : V \to [a, \infty)$ be an exhausting J-convex function on an almost complex manifold. Then for any diffeomorphism $f : [a,\infty) \to [b,\infty)$ such that $f'' > 0$ and $\lim_{y\to\infty} f'(y) = \infty$, the function $f \circ \phi$ is J-convex and its gradient vector field is complete.*

PROOF. The function $\psi := f \circ \phi$ satisfies

$$dd^{\mathbb{C}}\psi = f'' \circ \phi\, d\phi \wedge d^{\mathbb{C}}\phi + f' \circ \phi\, dd^{\mathbb{C}}\phi\,.$$

In particular, ψ is J-convex if $f' > 0$ and $f'' > 0$. We have

$$\begin{aligned} g_\psi(X,Y) &= -dd^{\mathbb{C}}\psi(X, JY) \\ &= +f'' \circ \phi\, [d\phi(X)d\phi(Y) + d^{\mathbb{C}}\phi(X)d^{\mathbb{C}}\phi(Y)] + f' \circ \phi\, g_\phi(X,Y)\,. \end{aligned}$$

Let us compute the gradient $\nabla_\psi \psi$. We will find it in the form

$$\nabla_\psi \psi = \lambda \nabla_\phi \phi$$

for a function $\lambda : V \to \mathbb{R}$. The gradient is determined by

$$g_\psi(\nabla_\psi \psi, Y) = d\psi(Y) = f' \circ \phi\, d\phi(Y)$$

for any vector $Y \in TV$. Using $d\phi(\nabla_\phi \phi) = g_\phi(\nabla_\phi \phi, \nabla_\phi \phi) =: |\nabla_\phi \phi|^2$ and $d^{\mathbb{C}}\phi(\nabla_\phi \phi) = g_\phi(\nabla_\phi \phi, J\nabla_\phi \phi) = 0$, we compute the left hand side as

$$\begin{aligned} &g_\psi(\nabla_\psi \psi, Y) \\ &= \lambda\Big\{f'' \circ \phi\, [d\phi(\nabla_\phi \phi)d\phi(Y) + d^{\mathbb{C}}\phi(\nabla_\phi \phi)d^{\mathbb{C}}\phi(Y)] + f' \circ \phi\, g_\phi(\nabla_\phi \phi, Y)\Big\} \\ &= \lambda\{f'' \circ \phi\, |\nabla_\phi \phi|^2 d\phi(Y) + f' \circ \phi\, d\phi(Y)\}. \end{aligned}$$

Comparing with the right hand side, we find

$$\lambda = \frac{f' \circ \phi}{f'' \circ \phi \, |\nabla_\phi \phi|^2 + f' \circ \phi}.$$

Since ϕ is proper, we only need to check completeness of the gradient flow for positive times. Consider an unbounded gradient trajectory $\gamma : [0, T) \to V$, i.e., a solution of

$$\frac{d\gamma}{dt}(t) = \nabla_\phi \phi(\gamma(t)), \qquad \lim_{t \to T} \phi(\gamma(t)) = \infty.$$

Here T can be finite or $+\infty$. The function ϕ maps the image of γ diffeomorphically onto some interval $[c, \infty)$. It pushes forward the vector field $\nabla_\phi \phi$ (which is tangent to the image of γ) to the vector field

$$\phi_*(\nabla_\phi \phi) = h(y) \frac{\partial}{\partial y},$$

where t and y are the coordinates on $[0, T)$ and $[c, \infty)$, respectively, and

$$h(y) := |\nabla_\phi \phi|^2 \left(\phi^{-1}(y)\right) > 0.$$

Similarly, ϕ pushes forward $\nabla_\psi \psi = \lambda \nabla_\phi \phi$ to the vector field

$$\phi_*(\nabla_\psi \psi) = \lambda\left(\phi^{-1}(y)\right) h(y) \frac{\partial}{\partial y} = \frac{f'(y)h(y)}{f''(y)h(y) + f'(y)} \frac{\partial}{\partial y} =: v(y).$$

Hence completeness of the vector field $\nabla_\psi \psi$ on the trajectory γ is equivalent to the completeness of the vector field v on $[c, \infty)$. An integral curve of v satisfies $\frac{dy}{ds} = v(y)$, or equivalently,

$$ds = \frac{f''(y)h(y) + f'(y)}{f'(y)h(y)} dy.$$

Thus completeness of the vector field v is equivalent to

$$+\infty = \int_c^\infty \frac{f''(y)h(y) + f'(y)}{f'(y)h(y)} dy = \int_c^\infty \frac{f''(y)dy}{f'(y)} + \int_c^\infty \frac{dy}{h(y)}.$$

The first integral on the right hand side is equal to $\int_c^\infty d\big(\ln f'(y)\big)$, so it diverges if and only if $\lim_{y \to \infty} f'(y) = \infty$. $\square$

We will call an exhausting J-convex function *completely exhausting* if its gradient vector field $\nabla_\phi \phi$ is complete.

2.5. J-convexity and geometric convexity

Next we investigate the relation between i-convexity and geometric convexity. Consider $\mathbb{C}^n = \mathbb{C}^{n-1} \oplus \mathbb{C}$ with coordinates $(z_1, \dots, z_{n-1}, u + iv)$. Let $\Sigma \subset \mathbb{C}^n$ be a hypersurface which is given as a graph $\{u = f(z, v)\}$ for some smooth function $f : \mathbb{C}^{n-1} \oplus \mathbb{R} \to \mathbb{R}$. Assume that $f(0,0) = 0$ and $df(0,0) = 0$. Every hypersurface in a complex manifold can be locally written in this form.

The Taylor polynomial of second order of f around $(0,0)$ can be written as

$$T_2 f(z, v) = \sum_{i,j} a_{ij} z_i \bar{z}_j + 2 \operatorname{Re} \sum_{i,j} b_{ij} z_i z_j + v\, l(z, \bar{z}) + cv^2, \tag{2.9}$$

where l is some linear function of z and $\bar{z}$, and $a_{ij} = \frac{\partial^2 f}{\partial z_i \partial \bar{z}_j}(0,0)$. Let Σ be cooriented by the gradient of the function $f(z,v) - u$. Then the 2-form ω_Σ at the point 0 is given on $X, Y \in \xi_0 = \mathbb{C}^{n-1}$ by

$$
\begin{aligned}
\omega_\Sigma(X,Y) &= 2i\partial\bar{\partial} f(X,Y) = 2i \sum_{i,j} a_{ij} dz_i \wedge d\bar{z}_j(X,Y) \\
&= 2i \sum_{i,j} a_{ij}(X_i \bar{Y}_j - \bar{X}_j Y_i) = 2 \operatorname{Re}\Big(2i \sum_{i,j} a_{ij} X_i \bar{Y}_j\Big) \\
&= -4 \operatorname{Im}(AX, Y),
\end{aligned}
\tag{2.10}
$$

where A is the Hermitian $(n-1) \times (n-1)$ matrix with entries a_{ij}. Hence the Levi form at 0 is

$$L_\Sigma = 4(A\cdot, \cdot).$$

If the function f is (strictly) convex, then

$$T_2 f(z,0) + T_2 f(iz,0) = 2 \sum_{ij} a_{ij} z_i \bar{z}_j$$

is positive for all $z \neq 0$, so the Levi form is positive definite. This shows that geometric convexity of Σ implies i-convexity. The converse is not true; see the first example in Section 2.7 below. It is true, however, locally after a biholomorphic change of coordinates.

PROPOSITION 2.12 (Narasimhan). *A hypersurface $\Sigma \subset \mathbb{C}^n$ is i-convex if and only if it can be made geometrically convex in a neighborhood of each of its points by a biholomorphic change of coordinates.*

PROOF. The "if" follows from the discussion above and the invariance of i-convexity under biholomorphic maps. For the converse write Σ in local coordinates as a graph $\{u = f(z,v)\}$ as above and consider its second Taylor polynomial (2.9). Let $w = u + iv$ and perform in a neighborhood of 0 the biholomorphic change of coordinates $\widetilde{w} - \widetilde{u} + i\widetilde{v} := w - 2\sum_{ij} b_{ij} z_i z_j$. Then $\widetilde{v} = v + O(2)$ and

$$\widetilde{u} = \sum a_{ij} z_i \bar{z}_j + \widetilde{v}\, l(z,\bar{z}) + c\widetilde{v}^2 + O(3).$$

After another local change of coordinates $w' = u' + iv' := \widetilde{w} - \lambda \widetilde{w}^2$, $\lambda \in \mathbb{R}$, we have $v' = \widetilde{v} + O(2)$ and

$$u' = \widetilde{u} + \lambda(v')^2 + O(3) = \sum a_{ij} z_i \bar{z}_j + v' l(z,\bar{z}) + (c + \lambda)(v')^2 + O(3).$$

For λ sufficiently large the quadratic form on the right hand side is positive definite, so the hypersurface Σ is geometrically convex in the coordinates (z, w'). ☐

2.6. Normalized Levi form and mean normal curvature

In a general (almost) complex manifold (V, J), the Levi form L_Σ of a cooriented hypersurface Σ is invariantly defined only up to multiplication by a positive function. However, any Hermitian metric $H = g - i\omega$ on (V, J) provides a canonical choice of defining 1-form $\alpha = i_\nu \omega$, where ν is the unit normal vector field along Σ defining the coorientation. In this case, we will call the form

$$\mathbb{L}_\Sigma(X) := d(i_\nu \omega)(X, JX)$$

the *normalized Levi form* of Σ. Note that if $\Sigma = \phi^{-1}(0)$ for a function $\phi : V \to \mathbb{R}$ with $|\nabla\phi(p)| = 1$, then the 1-forms $i_\nu\omega$ and $-d^{\mathbb{C}}\phi$ coincide at p, and thus

$$\mathbb{L}_\Sigma(X) = -dd^{\mathbb{C}}\phi(X, JX), \qquad X \in T_p\Sigma. \tag{2.11}$$

In certain situations the normalized Levi form can be expressed in terms of curvature, as we will now explain.

Consider first a cooriented hypersurface Σ in $\mathbb{R}^n$ with the Euclidean metric $\langle\ ,\ \rangle$. Its *second fundamental form*

$$II_\Sigma : T\Sigma \to \mathbb{R}$$

can be defined as follows. For $X \in T_x\Sigma$ let $\gamma : (-\epsilon, \epsilon) \to \Sigma$ be a curve with $\gamma(0) = x$ and $\dot\gamma(0) = X$. Then

$$II_\Sigma(X) := -\langle\ddot\gamma(0), \nu\rangle,$$

where ν is the unit normal vector to Σ in x defining the coorientation. The matrix representing the second fundamental form equals the differential of the Gauss map which associates to every point its unit normal vector. Our sign convention is chosen in such a way that the unit sphere in $\mathbb{R}^n$ has positive principal curvatures if it is cooriented by the *outward* pointing normal vector field. The *mean normal curvature* along a k-dimensional subspace $S \subset T_x\Sigma$ is defined as

$$\frac{1}{k}\sum_{i=1}^{k} II_\Sigma(v_i)$$

for some orthonormal basis $v_1, \dots, v_k$ of S. If Σ is given as a graph $\{x_n = f(x_1, \dots, x_{n-1})\}$ with $f(0) = 0$ and $df(0) = 0$, then for $X \in \mathbb{R}^{n-1}$ we can choose the curve

$$\gamma(t) := \Big(tX, f(tX)\Big)$$

in Σ. Taking the second derivative, we obtain

$$II_\Sigma(X) = \sum_{ij} \frac{\partial^2 f}{\partial x_i \partial x_j}(0) X_i X_j = 2T_2 f(X), \tag{2.12}$$

where $T_2 f$ is the second order Taylor polynomial and Σ is cooriented by the gradient of the function $f - x_n$. This leads to the following geometric characterization of i-convexity.

PROPOSITION 2.13. *The normalized Levi form of a cooriented hypersurface $\Sigma \subset \mathbb{C}^n$ with respect to the standard complex structure i and the standard Hermitian metric is given at a point $z \in \Sigma$ by*

$$\mathbb{L}_\Sigma(X) = II_\Sigma(X) + II_\Sigma(iX) \tag{2.13}$$

for $X \in T_z\Sigma$. Thus Σ is i-convex if and only if at every point $z \in \Sigma$ the mean normal curvature along any complex line in $T_z\Sigma$ is positive.

PROOF. Write Σ locally as a graph $\{u = f(z, v)\}$ with $f(0, 0) = 0$ and $df(0, 0) = 0$, and such that the gradient of $\phi = f - u$ defines the coorientation of Σ. Consider the second Taylor polynomial (2.9) of f in $(0, 0)$. In view of (2.12) and (2.10), twice the mean normal curvature along the complex line generated by $X \in \mathbb{C}^{n-1}$ is given by

$$II_\Sigma(X) + II_\Sigma(iX) = 2\Big(T_2 f(X) + T_2 f(iX)\Big) = 4\sum_{ij} a_{ij} X_i \bar{X}_j$$
$$= -dd^{\mathbb{C}} f(X, iX) = -dd^{\mathbb{C}}\phi(X, iX) = \mathbb{L}_\Sigma(X).$$

Here the last equality follows from (2.11), since the gradient of the function $\phi = f - u$ has norm 1 at the point $(0, 0, 0)$. □

Proposition 2.13 generalizes to hypersurfaces in Kähler manifolds as follows. Consider a cooriented hypersurface Σ in a Kähler manifold (V, J, ω). Denote by ν the outward pointing unit normal vector field along Σ, and define the vector field $\tau := J\nu$ tangent to Σ. Then the field of complex tangencies ξ on $T\Sigma$ is the kernel of the 1-form

$$\alpha := g(\tau, \cdot) = i_\nu \omega.$$

Note that the Kähler condition $\nabla J = 0$ implies $\nabla\tau = J\nabla\nu$. Using this together with the metric compatibility $d(g(X, Y)) = g(\nabla X, Y) + g(X, \nabla Y)$ and torsion-freeness $\nabla_X Y - \nabla_Y X = [X, Y]$ of the Levi-Civita connection, we compute for vector fields X, Y tangent to ξ:

$$\begin{aligned} d\alpha(X, Y) &= X \cdot \alpha(Y) - Y \cdot \alpha(X) - \alpha([X, Y]) \\ &= X \cdot g(\tau, Y) - Y \cdot g(\tau, X) - g(\tau, [X, Y]) \\ &= g(\nabla_X \tau, Y) - g(\nabla_Y \tau, X) + g(\tau, \nabla_X Y - \nabla_Y X - [X, Y]) \\ &= g(\nabla_X \tau, Y) - g(\nabla_Y \tau, X) \\ &= g(J\nabla_X \nu, Y) - g(J\nabla_Y \nu, X) \\ &= -g(\nabla_X \nu, JY) + g(\nabla_Y \nu, JX) \\ &= -II_\Sigma(X, JY) + II_\Sigma(Y, JX), \end{aligned}$$

where $II_\Sigma(X, Y) = g(\nabla_X \nu, Y)$ is the second fundamental form of Σ. Inserting $Y = JX$ we obtain

PROPOSITION 2.14. *Let Σ be a cooriented hypersurface in a Kähler manifold (X, J, ω) with second fundamental form II_Σ. Then the normalized Levi form of Σ is given by*

$$\mathbb{L}_\Sigma(X) = II_\Sigma(X) + II_\Sigma(JX). \tag{2.14}$$

In particular, Σ is J-convex if and only if at every point $x \in \Sigma$ the mean normal curvature along any complex line in $T_x\Sigma$ is positive. □

2.7. Examples of J-convex functions and hypersurfaces

An important class of hypersurfaces are boundaries of tubular neighborhoods of submanifolds. In this section we examine their J-convexity for the cases of totally real submanifolds and complex hypersurfaces.

Totally real submanifolds. A submanifold L of an almost complex manifold (V, J) is called *totally real* if it has no complex tangent lines, i.e., $J(TL) \cap TL = \{0\}$ at every point. This condition implies $\dim_{\mathbb{R}} L \le \dim_{\mathbb{C}} V$. For example, the linear subspaces $\mathbb{R}^k := \{(x_1, \ldots, x_k, 0, \ldots, 0) \mid x_i \in \mathbb{R}\} \subset \mathbb{C}^n$ are totally real for all $k = 0, \ldots, n$.

If we are given a Hermitian metric on (V, J), we can define the distance function $\mathrm{dist}_L : V \to \mathbb{R}$,

$$\mathrm{dist}_L(x) := \inf\{\mathrm{dist}(x, y) \mid y \in L\}.$$

PROPOSITION 2.15. *Let L be a properly embedded totally real submanifold of an almost complex manifold (V, J). Then the squared distance function dist_L^2 with respect to any Hermitian metric on V is J-convex in a neighborhood of L. Moreover, L has arbitrarily small neighborhoods with smooth J-convex boundary, and each such neighborhood admits an exhausting J-convex function.*

PROOF. Let $Q : T_pV \to \mathbb{R}$ be the Hessian quadratic form of dist_L^2 at a point $p \in L$. Its value $Q(z)$ equals the squared distance of $z \in T_pV$ from the linear subspace $T_pL \subset T_pV$. (This is most easily seen in geodesic normal coordinates on the normal bundle of L.) Choose an orthonormal basis $e_1, Je_1, \dots, e_n, Je_n$ of T_pV such that $e_1, \dots, e_k$ is a basis of T_pL. In this basis,

$$Q\Big(\sum_{i=1}^n (x_i e_i + y_i Je_i)\Big) = \sum_{j=k+1}^n x_j^2 + \sum_{i=1}^n y_i^2,$$

which is J-convex by (2.3). So dist_L^2 is J-convex on L and therefore by continuity in a neighborhood of L. If L is compact, this concludes the proof because $\{\mathrm{dist}_L \le \varepsilon\}$ is a tubular neighborhood of L with J-convex boundary for each sufficiently small $\varepsilon > 0$, and composition of ϕ with a convex diffeomorphism $f : [0, \varepsilon) \to [0, \infty)$ gives an exhausting J-convex function. For noncompact L we invoke the following argument of Grauert in [**77**].

Pick a locally finite open covering of L by coordinate neighborhoods U_j with smooth local coordinates $z = x + iy \in \mathbb{C}^n$ such that $J = i$ along $L \cap U_j = \{v = 0\}$, where we write $u = (x_1, \dots, x_k)$ and $v = (x_{k+1}, \dots, x_n, y_1, \dots, y_n\}$. Pick a closed covering $W_j \subset U_j$. For each point $p \in W_j \cap L$ consider the function $f_p(u, v) := 2|v|^2 - |u - p|^2$. This function is i-convex and hence J-convex at the critical point p. Pick a closed neighborhood $W \subset \bigcup_j U_j$ of L with smooth boundary such that $2|v| < \mathrm{dist}(W_j, \partial U_j)$ for all $z = (u, v) \in U_j \cap W$. This condition ensures that the boundary of the open cone $K_p := \{z \in U_j \cap W \mid f_p(z) > 0\}$ in W is given by $\{z \in U_j \cap W \mid f_p(z) = 0\}$ for all $p \in W_j \cap L$. To see this, suppose by contradiction that there exists a point $z = (u, v) \in \overline{K}_p \cap \partial U_j$. Then $2|v|^2 - |u - p|^2 = f_p(z) \ge 0$ and the choice of W yield the contradiction

$$\mathrm{dist}(W_j, \partial U_j)^2 \le |u - p|^2 + |v|^2 \le 3|v|^2 < \mathrm{dist}(W_j, \partial U_j)^2.$$

Moreover, we can choose W so small that f_p is J-convex on K_p for all $p \in W_j \cap W$.

Let $\phi_p := g \circ f_p$, where $g(t) = e^{-1/t}$ for $t > 0$ and 0 for $t \le 0$. So the function ϕ_p is positive and J-convex on K_p and can be extended by zero outside to a smooth weakly J-convex function on W. As $\bigcup_{p \in L} K_p = W \setminus L$, there exists a discrete set of points $p_1, p_2, \dots$ such that $\partial W \subset \bigcup_{i=1}^\infty K_{p_i}$. Choose constants $c_i > 0$ so large that the function $\phi := \sum_i c_i \phi_{p_i}$ satisfies $\{\phi < 1\} \subset W$. Hence for every regular value $\varepsilon \in (0, 1)$ of ϕ, the set $W_\varepsilon := \{\phi \le \varepsilon\}$ is a neighborhood of L with smooth J-convex boundary contained in W.

To find an exhausting J-convex function on W_ε, pick an exhausting function $\rho : L \to \mathbb{R}$ and extend it to V. Let dist_L^2 be as above and note that for any function $g : V \to \mathbb{R}_+$ the product $g\,\mathrm{dist}_L^2$ is still J-convex along L. By choosing g appropriately, we can thus ensure that $\psi := \rho + g\,\mathrm{dist}_L^2$ is J-convex on a neighborhood U of L. Let

$\phi : W_\varepsilon \to [0,\varepsilon]$ be as constructed above with $W_\varepsilon \subset U$. Pick a convex diffeomorphism $h : [0,\varepsilon) \to [0,\infty)$. Since ϕ is weakly J-convex, the function $\chi := h\circ\phi+\psi : W \to \mathbb{R}$ (with ψ from above) is J-convex and exhausting. □

Holomorphic line bundles. A complex line bundle $\pi : E \to V$ over a complex manifold V is called a *holomorphic line bundle* if the total space E is a complex manifold and the bundle possesses holomorphic local trivializations. For a Hermitian metric on $E \to V$, consider the hypersurface

$$\Sigma := \{z \in E \;\big|\; |z| = 1\} \subset E.$$

Complex multiplication $U(1) \times \Sigma \to \Sigma$, $(e^{i\theta}, z) \mapsto e^{i\theta} \cdot z$ provides Σ with the structure of a $U(1)$ principal bundle over V. Let α be the 1-form on Σ defined by

$$\alpha\Big(\frac{d}{d\theta}\Big|_0 e^{i\theta}\cdot z\Big) = 1, \qquad \alpha|_\xi = 0,$$

where ξ is the field of complex tangencies on $T\Sigma$. The imaginary valued 1-form $i\alpha$ defines the unique connection on the $U(1)$ principal bundle $\Sigma \to V$ for which all horizontal subspaces are J-invariant. Its curvature is the imaginary valued (1,1)-form Ω on V satisfying $\pi^*\Omega = d(i\alpha)$. On the other hand, α is a defining 1-form for the hyperplane distribution $\xi \subset T\Sigma$, so $\omega_\Sigma = d\alpha|_\xi$ defines the Levi form of Σ. Thus ω_Σ and the curvature form Ω are related by the equation

$$i\omega_\Sigma(X,Y) = \Omega(\pi_*X, \pi_*Y) \tag{2.15}$$

for $X, Y \in \xi$. The complex line bundle $E \to V$ is called *positive* (resp. *negative*) if it admits a Hermitian metric such that the corresponding curvature form Ω satisfies

$$\frac{i}{2\pi}\Omega(X,JX) > 0 \;\big(\text{resp. } < 0\big)$$

for all $0 \neq X \in TV$. Since π is holomorphic, equation (2.15) implies

PROPOSITION 2.16. *Let $E \to V$ be a holomorphic line bundle over a complex manifold. There exists a Hermitian metric on $E \to V$ such that the hypersurface $\{z \in E \;\big|\; |z| = 1\}$ is J-convex if and only if E is a negative line bundle.* □

If V is compact, then the closed 2-form $\frac{i}{2\pi}\Omega$ represents the first Chern class $c_1(E)$,

$$\Big[\frac{i}{2\pi}\Omega\Big] = c_1(E)$$

(see [**113**, Chapter 12]). Conversely, for every closed (1,1)-form $\frac{i}{2\pi}\Omega$ representing $c_1(E)$, Ω is the curvature of some Hermitian connection $i\alpha$ as above [**80**, Chapter 1, Section 2]. So a line bundle over V is positive/negative if and only if its first Chern class can be represented by a positive/negative (1,1)-form. If V has complex dimension 1, we get a very simple criterion.

COROLLARY 2.17. *Let V be a compact Riemann surface and $[V] \in H_2(V,\mathbb{R})$ its fundamental class. A holomorphic line bundle $E \to V$ admits a Hermitian metric such that the hypersurface $\{z \in E \;\big|\; |z| = 1\}$ is J-convex if and only if $c_1(E)\cdot[V] < 0$.*

For example, the corollary applies to the tangent bundle of a Riemann surface of genus ≥ 2.

PROOF. Since $H^2(V, \mathbb{R})$ is 1-dimensional, $c_1(E) \cdot [V] < 0$ if and only if $c_1(E)$ can be represented by a negatively oriented area form. But any negatively oriented area form on V is a negative (1,1)-form. □

REMARK 2.18. If $E \to V$ is just a complex line bundle (i.e., not holomorphic), then the total space E does not carry a natural almost complex structure. Such a structure can be obtained by choosing a Hermitian connection on $E \to V$ and taking the horizontal spaces as complex subspaces with the complex multiplication induced from V via the projection. If we fix an almost complex structure on the total space E such that the projection π is J-holomorphic, then Proposition 2.16 remains valid.

REMARK 2.19. Proposition 2.16 has the following generalization to a holomorphic vector bundle $E \to V$: A Hermitian metric on E determines a unique Hermitian connection with curvature form $\Omega \in \Omega^{1,1}(\mathrm{End}\, E)$. If the curvature is negative in the sense that $i\Omega(X, JX)$ is negative definite for all $0 \neq X \in TV$, then the function $\phi(z) = |z|^2$ on E is J-convex outside the zero section (in particular its level sets $\{|z| = \text{const} > 0\}$ are J-convex).

2.8. Symplectic properties of J-convex functions

In this section we discuss some basic symplectic properties of J-convex functions. The symplectic approach to Stein manifolds will be developed systematically starting from Chapter 11. For more background on symplectic geometry, see Chapter 6.

A *symplectic form* on a manifold V is a 2-form which is closed ($d\omega = 0$) and *nondegenerate* in the sense that $v \mapsto i_v\omega$ defines an isomorphism $T_xV \to T_x^*V$ for each $x \in V$. A 1-form λ such that $d\lambda = \omega$ is symplectic is called a *Liouville form*. The vector field X that is ω-dual to λ, i.e., such that $i_X\omega = \lambda$, is called the *Liouville field*. Note that the equation $d\lambda = \omega$ is equivalent to $L_X\omega = \omega$. If X integrates to a flow $X^t : V \to V$, then $(X^t)^*\omega = e^t\omega$; i.e., the Liouville flow expands the symplectic form. Note that

$$i_X\lambda = 0, \qquad i_X d\lambda = \lambda, \qquad L_X\lambda = \lambda, \tag{2.16}$$

so the flow of X also expands the Liouville form, $(X^t)^*\lambda = e^t\lambda$.

The relevance of these notions comes from the following elementary observation.

LEMMA 2.20. *For a J-convex function ϕ on a complex manifold (V, J) set*

$$\omega_\phi := -dd^{\mathbb{C}}\phi, \qquad \lambda_\phi := -d^{\mathbb{C}}\phi, \qquad X_\phi := \nabla_\phi\phi.$$

Then ω_ϕ is a symplectic form with Liouville field X_ϕ and Liouville form λ_ϕ.

PROOF. By the definition of J-convexity, ω_ϕ is a symplectic form. Since $X_\phi = \nabla_\phi\phi$ is the gradient of ϕ with respect to the metric $g_\phi := \omega_\phi(\cdot, J\cdot)$, for any $Y \in TV$ we have

$$d^{\mathbb{C}}\phi(Y) = g_\phi(\nabla_\phi\phi, JY) = -\omega_\phi(\nabla_\phi\phi, Y) = -i_{X_\phi}\omega_\phi(Y).$$

Hence $i_{X_\phi}\omega_\phi = \lambda_\phi$ and $L_{X_\phi}\omega_\phi = d\lambda_\phi = \omega_\phi$. □

This observation has several easy but important consequences. A zero p of a vector field X is called *hyperbolic* if all eigenvalues of the linearization D_pX have nonzero real parts. In this case p has an injectively immersed *stable manifold*

$$W_p^- = \{x \in V \mid \lim_{t\to\infty} X^t(x) = p\};$$

see Section 9.2.

LEMMA 2.21. *Let (V,ω) be a symplectic manifold with Liouville field X and Liouville form λ, and let p be a hyperbolic zero of X. Then*

$$\lambda|_{W_p^-} \equiv 0.$$

PROOF. Let $x \in W_p^-$ and $v \in T_x W_p^-$. Let $\phi_t : V \to V$ be the flow of X. All eigenvalues of the linearization of X at p have negative real part on $T_p W_p^-$. It follows that the differential $T_x\phi_t : T_xV \to T_{\phi_t(x)}V$ satisfies $\lim_{t\to\infty} T_x\phi_t(v) = 0$. Since $\phi_t(x) \to p$ as $t \to \infty$, this implies

$$e^t\lambda_x(v) = (\phi_t^*\lambda)(v) = \lambda_{\phi_t(x)}(T_x\phi_t \cdot v) \to 0$$

as $t \to \infty$ and hence $\lambda_x(v) = 0$. □

In particular, the lemma implies $\omega|_{W_p^-} \equiv 0$; i.e., the stable manifold $W^-(p)$ is *isotropic* for the symplectic form $\omega = d\lambda$. Since an isotropic submanifold can have at most half the dimension of V (see Section 6.1), it follows that

$$\dim W_p^- \le \frac{1}{2}\dim V.$$

If $X = X_\phi$ is the Liouville field associated to a J-convex function ϕ, then $\dim W_p^-$ equals the Morse index of ϕ at p and we recover Corollary 2.4 (at least for non-degenerate critical points, but the proof easily extends to the degenerate case, see Section 11.4).

Consider next a hypersurface $\Sigma \subset V$ transverse to the Liouville field X of a Liouville form λ and set $\alpha := \lambda|_\Sigma$. Then $\alpha \wedge d\alpha^{n-1}$ is a positive volume form on Σ (where $2n = \dim V$ and Σ is cooriented by X), so α is a *contact form* with *contact structure* $\xi = \ker\alpha$; see Section 6.5. By Lemma 2.21 the stable manifold W_p^- of a hyperbolic zero satisfies $\alpha|_{W_p^-\cap\Sigma} \equiv 0$, so the intersection $W_p^- \cap \Sigma$ is *isotropic* for the contact structure ξ.

Recall from Section 2.3 that on a regular level set $\Sigma = \phi^{-1}(c)$ of a J-convex function ϕ, the contact structure $\xi = \ker\alpha$ defined by the contact form $\alpha = (\lambda_\phi)|_\Sigma$ is just the field of complex tangencies.

We conclude this section with a notion that will play an important role in Part II of this book.

DEFINITION 2.22. We say that a totally real submanifold L in an almost complex manifold (V,J) is *J-orthogonal* to a hypersurface $\Sigma \subset V$ if, for each point $p \in L \cap \Sigma$, $J(T_pL) \subset T_p\Sigma$ and $T_pL \not\subset T_p\Sigma$.

The second condition just means that L is transverse to Σ, so $\Lambda := L \cap \Sigma$ is a submanifold of Σ. The first condition implies that Λ is an integral submanifold for the field of complex tangencies ξ on Σ. If Σ is J-convex and $\dim_{\mathbb{R}} L = \dim_{\mathbb{C}} V = n$, then the second condition $T_pL \not\subset T_p\Sigma$ follows from the first one because integral submanifolds of the contact structure ξ have dimension at most $n-1$ (see Section 6.5).

REMARK 2.23. (a) If L is J-orthogonal to $\Sigma \subset V$ and $\dim_{\mathbb{R}} L = \dim_{\mathbb{C}} V$, then $T_p\Sigma = T_p(\Lambda) \oplus J(T_pL)$ for $p \in \Lambda = L \cap \Sigma$, so the bundle $T\Sigma|_\Lambda$ is uniquely determined by the manifolds $\Lambda \subset L$.

(b) A totally real submanifold L is J-orthogonal to all level sets of a J-convex function ϕ (without critical points) if and only if $d^{\mathbb{C}}\phi|_L \equiv 0$, which implies that L is isotropic for ω_ϕ and intersects each level set $\phi^{-1}(c)$ in an isotropic submanifold.

If L intersects each level set in an isotropic manifold and $\nabla_\phi \phi$ is tangent to L, then L is J-orthogonal to all level sets of ϕ, and the converse holds for $\dim_{\mathbb{R}} L = \dim_{\mathbb{C}} V$ (for the last statement note that $\omega_\phi(\nabla_\phi\phi, v) = d^{\mathbb{C}}\phi(v) = 0$ for all $v \in TL$ implies $\nabla_\phi \phi \in TL$ for L isotropic of half dimension, see Section 6.1). In particular, the stable manifold W_p^- of a nondegenerate critical point of a J-convex function ϕ is J-orthogonal to all regular level sets of ϕ.

Combining the preceding discussion with Morse theory (see Chapter 9), we see that every exhausting J-convex Morse function on a complex manifold (V, J) provides a handlebody decomposition of V whose cells $W_p^- \cap \{\phi \geq c\}$, where p are critical points of ϕ, are attached J-orthogonally along isotropic spheres to regular sublevel sets $\{\phi \leq c\}$. In the proof of the existence theorem in Chapter 8 we will proceed in the reverse direction: starting from a Morse function which is J-convex on $\{\phi \leq c\}$, we will make its attaching spheres in $\{\phi = c\}$ isotropic and the stable manifolds J-orthogonal in order to extend the Stein structure over the next critical level.

2.9. Computations in $\mathbb{C}^n$

In this section we derive some explicit formulas for the normalized Levi form of hypersurfaces in $\mathbb{C}^n$, which will be used in Chapters 4 and 10.

Suppose a hypersurface $\Sigma \subset \mathbb{C}^n$ is given by an implicit equation $\Psi(x) = 0$ and cooriented by the gradient vector field

$$\nabla\Psi = 2\left(\frac{\partial\Psi}{\partial\bar{z}_1}, \dots, \frac{\partial\Psi}{\partial\bar{z}_n}\right) \neq 0$$

along Σ. Recall from (2.7) and (2.8) that the real and complex Hessian forms Hess_Ψ, H_Ψ of Ψ at $p \in \mathbb{C}^n$ are related by

$$H_\Psi(X) = \mathrm{Hess}_\Psi(X) + \mathrm{Hess}_\Psi(iX) = 4\sum_{i,j=1}^n \frac{\partial\Psi}{\partial z_i \partial\bar{z}_j}(p) X_i \bar{X}_j, \tag{2.17}$$

where $X = (X_1, \dots, X_n) \in \mathbb{C}^n$.

LEMMA 2.24. *The second fundamental form and the normalized Levi form of Σ are given for $p \in \Sigma$ and $X \in T_p\Sigma$ (resp. $X \in \xi_p$) by*

$$II_\Sigma(X) = \frac{\mathrm{Hess}_\Psi(X)}{|\nabla\Psi(p)|}, \qquad \mathbb{L}_\Sigma(X) = \frac{H_\Psi(X)}{|\nabla\Psi(p)|}.$$

PROOF. We first prove the formula for the second fundamental form. Consider a curve $\gamma : (-\varepsilon, \varepsilon) \to \Sigma$ with $\gamma(0) = p$ and $\dot\gamma(0) = X$. The t-derivative of the identity $\Psi(\gamma(t)) = 0$ yields $d\Psi(\gamma(t)) \cdot \dot\gamma(t) = 0$, and another derivative at $t = 0$ gives

$$0 = \mathrm{Hess}_\Psi(X) + \langle\nabla\Psi(p), \ddot\gamma(0)\rangle = \mathrm{Hess}_p(X) - |\nabla\Psi(p)|\, II_\Sigma(X).$$

The formula for the normalized Levi form on $X \in \xi_p$ immediately follows from this in view of

$$\mathbb{L}_\Sigma(X) = II_\Sigma(X) + II_\Sigma(iX), \qquad H_\Psi(X) = \mathrm{Hess}_\Psi(X) + \mathrm{Hess}_\Psi(iX). \qquad \square$$

The case $n = 2$. When $n = 2$, we have $\dim_{\mathbb{C}} \xi = 1$ and thus $\mathbb{L}_\Sigma(X)$ has the same value for all unit vectors $X \in \xi$. The complex line ξ is generated by the unit vector $T := \frac{2}{|\nabla\Psi|}\left(-\frac{\partial\Psi}{\partial w}, \frac{\partial\Psi}{\partial\zeta}\right)$. Here we denote the coordinates on $\mathbb{C}^n$ by (ζ, w) instead of (z_1, z_2). Hence equation (2.17) and Lemma 2.24 yield

$$\mathbb{L}_0 := \mathbb{L}_\Sigma(T) = \frac{16}{|\nabla\Psi|^3}\Big(\Psi_{\zeta\bar\zeta}|\Psi_w|^2 - 2\mathrm{Re}\,(\Psi_{\zeta\bar w}\Psi_w\Psi_{\bar\zeta}) + \Psi_{w\bar w}|\Psi_\zeta|^2\Big). \tag{2.18}$$

Next let us write $\zeta = s + it$, $w = u + iv$ and suppose that $\Sigma \subset \mathbb{C}^2$ is given as a graph $\Sigma = \{v = \psi(s,t,u)\}$. Then we obtain

LEMMA 2.25. *The normalized Levi form of the hypersurface* $\Sigma = \{v = \psi(s,t,u)\} \subset \mathbb{C}^2$, *cooriented by the gradient of the function* $\Psi(\zeta, w) = \psi(s,t,u) - v$, *is given by* $\mathbb{L}(X) = \mathbb{L}_0|X|^2$, $X \in \xi$, *where*

$$\mathbb{L}_0 = \frac{(\psi_{ss}+\psi_{tt})(1+\psi_u^2) + \psi_{uu}(\psi_s^2+\psi_t^2) + 2\psi_{su}(\psi_t - \psi_u\psi_s) - 2\psi_{tu}(\psi_s + \psi_u\psi_t)}{(1+\psi_s^2+\psi_t^2+\psi_u^2)^{\frac{3}{2}}}.$$

In particular, the hypersurface Σ *is i-convex if and only if*

$$(\psi_{ss}+\psi_{tt})(1+\psi_u^2) + \psi_{uu}(\psi_s^2+\psi_t^2) + 2\psi_{su}(\psi_t - \psi_u\psi_s) - 2\psi_{tu}(\psi_s + \psi_u\psi_t) > 0.$$

PROOF. The expression for $\mathbb{L}_0$ follows from (2.18) in view of

$$\begin{aligned} 2\Psi_{\bar\zeta} &= \psi_s + i\psi_t, & 4\Psi_{\zeta\bar\zeta} &= \psi_{ss} + \psi_{tt}, & 4|\Psi_\zeta|^2 &= \psi_s^2 + \psi_t^2, \\ 2\Psi_w &= \psi_u + i, & 4\Psi_{w\bar w} &= \psi_{uu}, & 4|\Psi_w|^2 &= 1 + \psi_u^2, \\ 4\Psi_{\zeta\bar w} &= \psi_{su} - i\psi_{tu}, & 4\Psi_w\Psi_{\bar\zeta} &= (\psi_u\psi_s - \psi_t) + i(\psi_s + \psi_u\psi_t), \end{aligned}$$
$$16\,\mathrm{Re}\,(\Psi_{\zeta\bar w}\Psi_w\Psi_{\bar\zeta}) = \psi_{su}(\psi_u\psi_s - \psi_t) + \psi_{tu}(\psi_s + \psi_u\psi_t). \qquad \square$$

The case of general n. Now we return to the case of general n. Suppose that $\Sigma \subset \mathbb{C}^n$ is given as a graph $\Sigma = \{v = \psi(z,u)\}$, where we denote coordinates on $\mathbb{C}^n$ by (z, w) with $z = (z_1, \dots, z_{n-1}) \in \mathbb{C}^{n-1}$ and $w = u + iv \in \mathbb{C}$. Then Lemma 2.25 generalizes in a weaker form to

LEMMA 2.26. *Suppose that for each fixed u the function* $\psi(\cdot, u)$ *is i-convex and denote by* $H_\psi^{\min} > 0$ *the minimum of its complex Hessian form on the unit sphere in* $\mathbb{C}^{n-1}$. *Then the normalized Levi form of the hypersurface* $\Sigma = \{v = \psi(z,u)\} \subset \mathbb{C}^n$, *cooriented by the gradient of the function* $\Psi(z,w) := \psi(z,u) - v$, *is bounded below by*

$$\min_{|X|=1}\mathbb{L}_\Sigma(X) \geq \frac{H_\psi^{\min}(1+\psi_u^2) - |\psi_{uu}|\,|d_z\psi|^2 - 2|d_z\psi_u|\,|d_z\psi|\sqrt{1+\psi_u^2}}{(1+\psi_u^2+|d_z\psi|^2)^{3/2}}.$$

PROOF. Consider a unit vector $X = (Z, W) \in \xi_{(z,w)} \subset \mathbb{C}^n$, where $Z = (Z_1, \dots, Z_{n-1}) \in \mathbb{C}^{n-1}$ and $W \in \mathbb{C}$. Set $\Psi_z := (\Psi_{z_1}, \dots, \Psi_{z_{n-1}}) \in \mathbb{C}^{n-1}$. Then X satisfies $1 = |X|^2 = |Z|^2 + |W|^2$ and

$$0 = (X, \frac{\nabla\Psi}{2}) = \sum_{j=1}^{n-1} Z_j\Psi_{z_j} + W\Psi_w.$$

This implies

$$|W|\,|\Psi_w| \leq |Z|\,|\Psi_z|, \tag{2.19}$$

which via $1 - |Z|^2 = |W|^2 \leq |Z|^2|\Psi_z|^2/|\Psi_w|^2$ yields

$$|Z|^2 \geq \frac{|\Psi_w|^2}{|\Psi_w|^2 + |\Psi_z|^2} = 4\frac{|\Psi_w|^2}{|\nabla\Psi|^2}. \tag{2.20}$$

We further have the relations

$$\Psi_{\bar z_j} = \psi_{\bar z_j}, \qquad \Psi_{z_i\bar z_j} = \psi_{z_i\bar z_j}, \qquad 4|\Psi_z|^2 = |d_z\psi|^2,$$
$$2\Psi_w = \psi_u + i, \qquad 4\Psi_{w\bar w} = \psi_{uu}, \qquad 4|\Psi_w|^2 = 1+\psi_u^2,$$
$$4\Psi_{z_i\bar w} = \psi_{s_iu} - i\psi_{t_iu}, \qquad \sum_{i=1}^{n-1}|4\Psi_{z_i\bar w}|^2 = \sum_{i=1}^{n-1}(\psi_{s_iu}^2+\psi_{t_iu}^2) = |d_z\psi_u|^2,$$

$$|4\,\mathrm{Re}\sum_{i=1}^{n-1}\Psi_{z_i\bar w}Z_i\bar W| \le |d_z\psi_u|\,|Z|\,|W|. \tag{2.21}$$

Combining all these relations we estimate

$$\begin{aligned}
\mathbb{L}_\Sigma(X) &= \frac{4}{|\nabla\Psi|}\left(\sum_{i,j=1}^{n-1}\Psi_{z_i\bar z_j}Z_i\bar Z_j + 2\,\mathrm{Re}\sum_{i=1}^{n-1}\Psi_{z_i\bar w}Z_i\bar W + \Psi_{w\bar w}|W|^2\right)\\
&\ge \frac{1}{|\nabla\Psi|}\left(H_\psi^{\min}|Z|^2 - 2|d_z\psi_u|\,|Z|\,|W| - |\psi_{uu}|\,|W|^2\right)\\
&\ge \frac{|Z|^2}{|\nabla\Psi|\,|\Psi_w|^2}\left(H_\psi^{\min}|\Psi_w|^2 - 2|d_z\psi_u|\,|\Psi_z|\,|\Psi_w| - |\psi_{uu}|\,|\Psi_z|^2\right)\\
&\ge \frac{1}{|\nabla\Psi|^3}\left(H_\psi^{\min}(1+\psi_u^2) - 2|d_z\psi_u|\,|d_z\psi|\sqrt{1+\psi_u^2} - |\psi_{uu}|\,|d_z\psi|^2\right).
\end{aligned}$$

Here in the first line we have used equation (2.17) and Lemma 2.24, in the second line (2.17) and (2.21), in the third line (2.19), and in the last line (2.20) and (2.21). Since $|\nabla\Psi|^2 = 1+\psi_u^2+|d_z\psi|^2$, this concludes the proof. □

CHAPTER 3

Smoothing

In this chapter we develop some techniques for constructing J-convex functions. We begin with the well-known fact that the maximum of two J-convex functions can be uniformly approximated by J-convex functions. For this, we extend the notion of J-convexity to continuous functions such that it is preserved under the maximum construction and show that any continuous J-convex function can be smoothed (Richberg's theorem).

In Section 3.3 we derive a condition under which the maximum and smoothing constructions do not lead to new critical points. In Section 3.4 we show how to deform a family of (possible intersecting) J-convex hypersurfaces into level sets of a J-convex function, and in Section 3.5 we modify J-convex functions near totally real submanifolds.

3.1. J-convexity and plurisubharmonicity

A C^2-function $\phi : U \to \mathbb{R}$ on an open domain $U \subset \mathbb{C}$ is i-convex if and only if it is *subharmonic*,[1] i.e.,

$$\Delta\phi = \frac{\partial^2\phi}{\partial x^2} + \frac{\partial^2\phi}{\partial y^2} = 4\frac{\partial\phi}{\partial z\partial\bar{z}} > 0.$$

A continuous function $\phi : U \to \mathbb{R}$ is called subharmonic if it satisfies

$$\Delta\phi \geq m$$

for a positive continuous function $m : U \to \mathbb{R}$, where the Laplacian and the inequality are understood in the distributional sense, i.e.,

$$\int_U \phi\,\Delta\delta\;dx\,dy \geq \int_U m\,\delta\;dx\,dy \tag{3.1}$$

for any nonnegative smooth function $\delta : U \to \mathbb{R}$ with compact support. The function

$$m_\phi := \sup\{m \mid \text{ inequality (3.1) holds}\}$$

is called the *modulus of subharmonicity* of the function ϕ. Note that to find $m_\phi(z)$ at a point $z \in U$, we only need to test (3.1) for functions δ supported near z. If ϕ is a C^2-function satisfying (3.1), then choosing a sequence of functions δ_n converging to the Dirac measure of a point $z \in U$ and integrating by parts shows $\Delta\phi(z) \geq m(z)$, so for a C^2-function the two definitions agree and $m_\phi = \Delta\phi$.

If $z = x + iy \to w = u + iv$ is a biholomorphic change of coordinates on U, then

$$\Delta_z\delta\;dx \wedge dy = 2i\frac{\partial^2\delta}{\partial z\partial\bar{z}}dz \wedge d\bar{z} = -dd^{\mathbb{C}}\delta = \Delta_w\delta\;du \wedge dv, \tag{3.2}$$

[1] By "subharmonic" we will always mean "strictly subharmonic". Nonstrict subharmonicity will be referred to as "weak subharmonicity". The same applies to plurisubharmonicity discussed below.

so inequality (3.1) transforms into

$$\int_U \phi(w)\Delta\delta(w)du\,dv \geq \int_U m(w)\delta(w)\left|\frac{dz}{dw}\right|^2 du\,dv.$$

This shows that subharmonicity is invariant under biholomorphic coordinate changes and therefore can be defined for continuous functions on Riemann surfaces. Note, however, that the modulus of subharmonicity is not invariant under biholomorphic coordinate changes; it depends on the additional choice of a Riemannian metric.

The following lemma gives a useful criterion for subharmonicity of continuous functions.

LEMMA 3.1. *A continuous function $\phi : U \to \mathbb{R}$ on a domain $U \subset \mathbb{C}$ satisfies $\Delta\phi \geq m$ for a positive continuous function $m : U \to \mathbb{R}$ if and only if*

$$\phi(z) + \frac{m(z)r^2}{4} \leq \frac{1}{2\pi}\int_0^{2\pi} \phi(z + re^{i\theta})d\theta \tag{3.3}$$

for all $z \in U$ and sufficiently small $r > 0$ (depending on z).

PROOF. Fix a point $z \in U$ and consider the function

$$\psi(w) := \phi(w) - \frac{1}{4}m(z)|w - z|^2.$$

For $r > 0$ sufficiently small, (3.3) is equivalent to

$$\psi(z) = \phi(z) \leq \frac{1}{2\pi}\int_0^{2\pi} \phi(z + re^{i\theta})d\theta - \frac{m(z)r^2}{4} = \frac{1}{2\pi}\int_0^{2\pi} \psi(z + re^{i\theta})d\theta,$$

and thus to

$$\psi(z) \leq \frac{1}{2\pi}\int_0^{2\pi} \psi(z + re^{i\theta})d\theta.$$

By a standard result (see e.g. [**103**, Section 1.6]), this inequality is equivalent to $\Delta\psi(z) \geq 0$ in the distributional sense, and therefore to $\Delta\phi(z) \geq \frac{1}{4}m(z)\Delta_w|w-z|^2 = m(z)$. □

Now let (V, J) be an almost complex manifold. A *complex curve* in V is a 1-dimensional complex submanifold of (V, J). Note that the restriction of the almost complex structure J to a complex curve is always integrable.

LEMMA 3.2. *A C^2-function ϕ on an almost complex manifold (V, J) is J-convex if and only if its restriction to every complex curve is subharmonic.*

PROOF. By definition, ϕ is J-convex if and only if $-dd^{\mathbb{C}}\phi(X, JX) > 0$ for all $0 \neq X \in T_xV$, $x \in V$. Now for every such $X \neq 0$ there exists a complex curve $C \subset V$ passing through x with $T_xC = \mathrm{span}_{\mathbb{R}}\{X, JX\}$ (see [**152**]). By formula (3.2) above, $-dd^{\mathbb{C}}\phi(X, JX) > 0$ precisely if $\phi|_C$ is subharmonic in x. □

REMARK 3.3. In the proof we have used the fact that the differential operator $dd^{\mathbb{C}}$ commutes with restrictions to complex submanifolds. This is true because the exterior derivative and the composition with J both commute with restrictions to complex submanifolds.

As a consequence, we obtain the following generalization of Corollary 2.4 to the almost complex case.

COROLLARY 3.4. *The Morse index (i.e., the maximal dimension of a subspace on which the real Hessian is negative definite) of a critical point of a J-convex function on a $2n$-dimensional almost complex manifold is at most n.*

PROOF. Let p be a critical point of a J-convex function $\phi : V \to \mathbb{R}$ and suppose that its Morse index is $> n$. Then there exists a subspace $W \subset T_pV$ of dimension $> n$ on which the Hessian of ϕ is negative definite. Since $W \cap JW \neq \{0\}$, W contains a complex line L. Let C be a complex curve through p tangent to L. Then $\phi|_C$ attains a local maximum at p. But this contradicts the maximum principle because $\phi|_C$ is subharmonic by Lemma 3.2. □

In view of Lemma 3.2, we can speak about *continuous J-convex functions* on almost complex manifolds as functions whose restrictions to all complex curves are subharmonic. Such functions are also called *(strictly) plurisubharmonic*. For functions on $\mathbb{C}^n$, Lemma 3.1 and the proof of Lemma 3.2 show

LEMMA 3.5. *A continuous function $\phi : \mathbb{C}^n \supset U \to \mathbb{R}$ is i-convex if and only if its restriction to every complex line is subharmonic. This means that there exists a positive continuous function $m : U \to \mathbb{R}$ such that*

$$\phi(z) + \frac{1}{4}m(z)|w|^2 \leq \frac{1}{2\pi}\int_0^{2\pi} \phi(z + we^{i\theta})d\theta \tag{3.4}$$

for all $z \in U$ and sufficiently small $w \in \mathbb{C}^n$ (depending on z).

As in the 1-dimensional case, we call the supremum of all functions m satisfying (3.4) the *modulus of i-convexity* of the function $\phi : \mathbb{C}^n \supset U \to \mathbb{R}$ and denote it by m_ϕ. Thus ϕ is i-convex if and only if $m_\phi > 0$. If ϕ is of class C^2, then Lemma 3.1 and the discussion following inequality (3.1) shows

$$m_\phi(x) = \min\{-dd^{\mathbb{C}}\phi(v, iv) \mid v \in \mathbb{C}^n,\ |v| = 1\}.$$

More generally, for a continuous function ϕ on a complex manifold (V, J) equipped with a Hermitian metric, we define the *modulus of J-convexity* m_ϕ via formula (3.4) in holomorphic coordinates for which the Hermitian metric is standard at the point z. Note that m_ϕ depends only on J, ϕ, and the Hermitian metric.

REMARK 3.6. We will need the modulus of convexity only for integrable J. In the case of an almost complex manifold (V, J), we can define the modulus of J-convexity as follows. We fix a Hermitian metric and a locally finite covering of V by coordinate neighborhoods U_i. According to [**152**], there exists for each $p \in U_i$ and unit vector $v \in T_pV$ a holomorphic disc $f_{i,p,v} : \mathbb{C} \supset D \to V$ with $f_{i,p,v}(0) = 0$ and $df_{i,p.v}(1) = v$. Moreover, $f_{i,p,v}$ can be chosen to depend continuously on (p, v) in the C^2-topology. Now we define $m_\phi(p) := \max_{i,v} m_{\phi \circ f_{i,p,v}}(0)$, where the maximum is taken over all unit vectors v and all i such that $p \in U_i$. Remark 3.3 shows that for a C^2-function $\phi : V \to \mathbb{R}$, the modulus of J-convexity is given by

$$m_\phi(x) = \min\{-dd^{\mathbb{C}}\phi(X, JX) \mid X \in T_xV,\ |X| = 1\}. \tag{3.5}$$

We do not know whether the definition of the modulus of J-convexity on an almost complex manifold depends on the chosen holomorphic discs $f_{i,p,v}$. According to Corollary 3.16 below, in the integrable case the definition coincides with the previous one and hence does not depend on the $f_{i,p,v}$.

The following lemma follows from equation (3.1) via integration by parts.

LEMMA 3.7. *If ϕ is a J-convex function on an almost complex manifold (V,J), then $\phi+\psi$ is J-convex for every sufficiently C^2-small C^2-function $\psi : V \to \mathbb{R}$.*

Our interest in continuous J-convex functions is motivated by the following

PROPOSITION 3.8. *If ϕ and ψ are continuous J-convex functions on an almost complex manifold (V,J), then $\max(\phi,\psi)$ is again J-convex.*

More generally, let $(\phi_\lambda)_{\lambda\in\Lambda}$ be a continuous family of continuous J-convex functions, parametrized by a compact space Λ. Then $\phi := \max_{\lambda\in\Lambda}\phi_\lambda$ is a continuous function whose modulus of J-convexity satisfies

$$m_\phi \geq \min_{\lambda\in\Lambda} m_{\phi_\lambda}.$$

PROOF. Continuity of $\phi = \max_{\lambda\in\Lambda}\phi_\lambda$ is an easy exercise. For J-convexity we use the criterion from Lemma 3.1. Let $U \subset V$ be a complex disc and choose a local coordinate z on U. By hypothesis, condition (3.3) holds for all ϕ_λ with functions $m_\lambda = m_{\phi_\lambda}$. Set $m(z) := \min_{\lambda\in\Lambda} m_\lambda$. At any point $z \in U$ we have $\phi = \phi_\lambda$ for some $\lambda \in \Lambda$ (depending on z). Now the lemma follows from

$$\begin{aligned}\phi(z) + \frac{1}{4}m(z)r^2 \leq \phi_\lambda(z) + \frac{1}{4}m_\lambda(z)r^2 &\leq \frac{1}{2\pi}\int \phi_\lambda(z+re^{i\theta})d\theta \\ &\leq \frac{1}{2\pi}\int \phi(z+re^{i\theta})d\theta.\end{aligned}$$

□

REMARK 3.9. If the family $m_{\phi_\lambda} > 0$ is continuous in λ, then $\min_\lambda m_{\phi_\lambda} > 0$ and thus $\max_\lambda \phi$ is J-convex. For example, by equation (3.5), this is the case if all the J-convex functions ϕ_λ are C^2 and their first two derivatives depend continuously on λ.

3.2. Smoothing of J-convex functions

For integrable J, continuous J-convex functions can be approximated by smooth ones. The following proposition was proved by Richberg [**161**]. We give below a proof following [**59**].

PROPOSITION 3.10 (Richberg [**161**]). *Let ϕ be a continuous J-convex function on a* complex *manifold (V,J). Then for every positive function $h : V \to \mathbb{R}_+$ there exists a smooth J-convex function $\psi : V \to \mathbb{R}$ such that $|\phi(x) - \psi(x)| < h(x)$ for all $x \in V$. If ϕ is already smooth on a neighborhood of a compact subset K, then we can achieve $\psi = \phi$ on K.*

REMARK 3.11. (i) A continuous weakly J-convex function (i.e., one whose restriction to each complex curve is weakly subharmonic) cannot in general be approximated by smooth weakly J-convex functions; see [**59**] for a counterexample.

(ii) We do not know whether Proposition 3.10 remains true for almost complex manifolds.

The proof is based on an explicit smoothing procedure for functions on $\mathbb{R}^m$. Pick a smooth nonnegative function $\rho : \mathbb{R}^m \to \mathbb{R}$ with support in the unit ball and $\int_{\mathbb{R}^m} \rho = 1$. For $\delta > 0$ set $\rho_\delta(x) := \delta^{-m}\rho(x/\delta)$. Let $U \subset \mathbb{R}^m$ be an open subset and set

$$U_\delta := \{x \in U \mid \bar{B}_\delta(x) \subset U\}.$$

For a continuous function $\phi : \mathbb{R}^m \supset U \to \mathbb{R}$ define the *mollified function* $\phi_\delta : U_\delta \to \mathbb{R}$,

$$\phi_\delta(x) := \int_{\mathbb{R}^m} \phi(x-y)\rho_\delta(y)d^m y = \int_{\mathbb{R}^m} \phi(y)\rho_\delta(x-y)d^m y. \tag{3.6}$$

The last expression shows that the functions ϕ_δ are smooth for every $\delta > 0$. The first expression shows that if ϕ is of class C^k for some $k \geq 0$, then $\phi_\delta \to \phi$ as $\delta \to 0$ in C^k uniformly on compact subsets of U.

Proposition 3.10 is an immediate consequence of the following lemma, via induction over a countable coordinate covering.

LEMMA 3.12. *Let ϕ be a continuous J-convex function on a complex manifold (V, J). Let $A, B \subset V$ be compact subsets such that ϕ is smooth on a neighborhood of A and B is contained in a holomorphic coordinate neighborhood. Then for every $\varepsilon > 0$ and every neighborhood W of $A \cup B$, there exists a continuous J-convex function $\psi : V \to \mathbb{R}$ with the following properties:*

- *ψ is smooth on a neighborhood of $A \cup B$;*
- *$|\psi(x) - \phi(x)| < \varepsilon$ for all $x \in V$;*
- *$\psi = \phi$ on A and outside W.*

PROOF. The proof follows [**59**]. First suppose that ϕ is i-convex on an open set $U \subset \mathbb{C}^n$. By Lemma 3.5, there exists a positive continuous function $m : U \to \mathbb{R}$ such that (3.4) holds for all $z \in U_{2\delta}$ and $w \in \mathbb{C}^n$ with $|w| \leq \delta$. Hence the mollified function ϕ_δ satisfies

$$\begin{aligned}\phi_\delta(x) + \frac{1}{4}m_\delta(x)|w|^2 &= \int_{\mathbb{C}^n} \Big(\phi(x-y) + \frac{1}{4}h(x-y)|w|^2\Big)\rho_\delta(y)d^{2n}y \\ &\leq \int_{\mathbb{C}^n}\int_0^{2\pi} \phi(x-y+we^{i\theta})d\theta\rho_\delta(y)d^{2n}y \\ &= \int_0^{2\pi} \phi_\delta(x+we^{i\theta})d\theta,\end{aligned}$$

so ϕ_δ is i-convex on $U_{2\delta}$.

Now let $\phi : V \to \mathbb{R}$ be as in the proposition. Pick a holomorphic coordinate neighborhood U and compact neighborhoods $A' \subset W$ of A and $B' \subset B'' \subset W \cap U$ of B with $A \subset \operatorname{int} A' \subset A' \subset W$, such that ϕ is smooth on A'. By the preceding discussion, there exists a smooth J-convex function $\phi_\delta : B'' \to \mathbb{R}$ with $|\phi_\delta(x) - \phi(x)| < \varepsilon/2$ for all $x \in B''$. Pick smooth cutoff functions $g, h : V \to [0,1]$ such that $g = 1$ on A, $g = 0$ outside A', $h = 1$ on B', and $h = 0$ outside B''. Define a continuous function $\widetilde{\phi} : V \to \mathbb{R}$,

$$\widetilde{\phi} := \phi + (1-g)h(\phi_\delta - \phi).$$

The function $\widetilde{\phi}$ is smooth on $A' \cup B'$, $|\widetilde{\phi}(x) - \phi(x)| < \varepsilon/2$ for all $x \in V$, $\widetilde{\phi} = \phi_\delta$ on $B' \setminus A'$, and $\widetilde{\phi} = \phi$ on A and outside B''. Since ϕ is C^2 on $A' \cap B''$, the function $(1-g)h(\phi_\delta - \phi)$ becomes arbitrarily C^2-small on this set for δ small. Hence by Lemma 3.7, $\widetilde{\phi}$ is J-convex on $A' \cap B''$ for δ sufficiently small. So we can make $\widetilde{\phi}$ J-convex on $A' \cup B'$. However, $\widetilde{\phi}$ need not be J-convex on $B'' \setminus (A' \cup B')$.

Pick a compact neighborhood $W' \subset W$ of $A' \cup B''$. Without loss of generality we may assume that ε was arbitrarily small. Then by Lemma 3.7 there exists a continuous J-convex function $\widetilde{\psi} : V \to \mathbb{R}$ (which differs from ϕ by a C^2-small function) satisfying $\widetilde{\psi} = \phi - \varepsilon$ on $A \cup B$, $\widetilde{\psi} = \phi + \varepsilon$ on $W' \setminus (A' \cup B')$, and $\widetilde{\psi} = \phi$ outside W. Now the function $\psi := \max(\widetilde{\phi}, \widetilde{\psi})$ has the desired properties. □

REMARK 3.13. The proof of Lemma 3.12 shows the following additional properties in Proposition 3.10:

(1) If ϕ_λ is a continuous family of J-convex functions parametrized by a compact space Λ, then the family ϕ_λ can be uniformly approximated by a continuous family of smooth J-convex functions ψ_λ.

(2) If $\phi_0 \leq \phi_1$, then the smoothed functions also satisfy $\psi_0 \leq \psi_1$. This holds because the proof only uses mollification $\phi \mapsto \phi_\delta$, interpolation, and taking the maximum of two functions, all of which are monotone operations.

REMARK 3.14. For a Stein manifold (V, J), Proposition 3.10 can alternatively be proved as follows. Embed V as a proper submanifold in some $\mathbb{C}^N$. By Corollary 5.27 below, there exists a neighborhood U of V in $\mathbb{C}^N$ with a holomorphic submersion $\pi : U \to V$ fixed on V. After shrinking U, we may assume that the squared Euclidean distance function dist_V^2 from V is smooth on U. Given a continuous J-convex function $\phi : V \to \mathbb{R}$, the function $\Phi := \phi \circ \pi + \mathrm{dist}_V^2 : U \to \mathbb{R}$ is i-convex and agrees with ϕ on V.

For a compact subset $W \subset V$, pick $\delta > 0$ such that the δ-ball around each point of W is contained in U and define the smooth i-convex function Φ_δ on a neighborhood of W by convolution in $\mathbb{C}^N$. The restriction ϕ_δ of Φ_δ to W is smooth, J-convex, and close to ϕ in $C^0(W)$. If ϕ was already smooth near some compact subset $K \subset \mathrm{Int}\, W$, then ϕ_δ is close to ϕ in $C^2(K)$, and we can interpolate by a cutoff function to achieve $\phi_\delta = \phi$ on K.

To approximate ϕ on the whole manifold V, pick an exhaustion $W_0 \subset W_1 \subset \cdots$ of V by compact subsets with $W_k \subset \mathrm{Int}\, W_{k+1}$ and $\bigcup_{k \in \mathbb{N}} W_k = V$. Using the previous paragraph, we inductively find smoothings ϕ_k of ϕ on W_k with $\phi_k = \phi_{k-1}$ on W_{k-2}. Thus the construction stabilizes and yields a smoothing of ϕ on V.

Proposition 3.8, Remark 3.9, and Proposition 3.10 imply

COROLLARY 3.15. *The maximum of two smooth J-convex functions ϕ, ψ on a complex manifold (V, J) can be C^0-approximated by smooth J-convex functions. If* $\max(\phi, \psi)$ *is smooth on a neighborhood of a compact subset K, then we can choose the smoothings to be equal to* $\max(\phi, \psi)$ *on K.*

More generally, let $(\phi_\lambda)_{\lambda \in \Lambda}$ be a continuous family of J-convex C^2-functions whose first two derivatives depend continuously on a parameter λ varying in a compact metric space Λ. Then $\max_{\lambda \in \Lambda} \phi_\lambda$ *can be C^0-approximated by smooth J-convex functions. If* $\max_{\lambda \in \Lambda} \phi_\lambda$ *is smooth on a neighborhood of a compact subset K, then we can choose the smoothings to be equal to* $\max_{\lambda \in \Lambda} \phi_\lambda$ *on K.* □

We will denote the smoothing of a continuous function $\phi : V \to \mathbb{R}$ by

$$\mathrm{smooth}(\phi).$$

In particular, the smoothing of the maximum of ϕ and ψ will be written as

$$\mathrm{smooth}\max(\phi, \psi).$$

This is a slight abuse of notation because the smoothing of a function depends on various choices. However, the notation is justified by the fact (Remark 3.13) that the smoothing can be done continuously in families.

Using Proposition 3.10, we can now justify the earlier Remark 3.6 that our definitions of the modulus of J-convexity coincide in the integrable case:

COROLLARY 3.16. *Consider two holomorphic discs $f, g : \mathbb{C} \supset D \to V$ in a complex manifold (V, J) with $f(0) = g(0)$ and $df(0) = dg(0)$. Then for any continuous J-convex function $\phi : V \to \mathbb{R}$ the compositions $\phi \circ f$ and $\phi \circ g$ have the same modulus of i-convexity at the origin.*

PROOF. Note first that by Remark 3.3 the statement holds if ϕ is C^2. In the continuous case, let $m = m(0) > 0$ be a constant such that (3.3) holds for $\phi \circ f$ at $z = 0$ for all $r \in [0, 1]$. For given $\varepsilon > 0$, we pick a J-convex smoothing ψ of ϕ such that $|\phi - \psi| < \varepsilon/4$ on the images of f and g. Then (3.3) holds for $\psi \circ f$, and hence also for $\psi \circ g$, with the same constant m and up to an error $\varepsilon/2$. Thus (3.3) holds for $\phi \circ g$ with the constant m up to an error ε. Letting $\varepsilon \to 0$ this shows that $m_{\phi\circ g}(0) \geq m_{\phi\circ f}(0)$, and the converse inequality follows similarly. □

3.3. Critical points of J-convex functions

We wish to control the creation of new critical points under the construction of taking the maximum of two J-convex functions and then smoothing. This is based on the following trivial observation: A smooth function $\phi : M \to \mathbb{R}$ on a manifold has no critical points if and only if there exist a vector field X and a positive function h with $X \cdot \phi \geq h$. Multiplying by a nonnegative volume form Ω on M with compact support, we obtain

$$\int_M (X \cdot \phi)\Omega \geq \int_M h\Omega.$$

Using $(X \cdot \phi)\Omega + \phi L_X \Omega = L_X(\phi\Omega) = d(\phi i_X \Omega)$ and Stokes' theorem (assuming M is orientable over $\operatorname{supp}\Omega$), we can rewrite the left hand side as

$$\int_M (X \cdot \phi)\Omega = -\int_M \phi L_X \Omega.$$

So we have shown: A smooth function $\phi : M \to \mathbb{R}$ on a manifold has no critical points if and only if there exist a vector field X and a positive function h such that

$$-\int_M \phi L_X \Omega \geq \int_M h\Omega$$

for all nonnegative volume forms Ω on M with sufficiently small compact support. This criterion obviously still makes sense if ϕ is merely continuous. However, for technical reasons we will slightly modify it as follows.

We say that a continuous function $\phi : M \to \mathbb{R}$ satisfies $X \cdot \phi \geq h$ (in the distributional sense) if around each $p \in M$ there exists a coordinate chart $U \subset \mathbb{R}^m$ on which X corresponds to a constant vector field such that

$$-\int_U \phi L_X \Omega \geq h(p) \int_U \Omega$$

for all nonnegative volume forms Ω with support in U. Writing $\Omega = g(x)d^m x$ for a nonnegative function g, this is equivalent to

$$-\int_U \phi(x)(X \cdot g)(x)d^m x \geq h(p) \int_U g(x)d^m x. \tag{3.7}$$

This condition ensures that smoothing does not create new critical points:

LEMMA 3.17. *If a continuous function $\phi : \mathbb{R}^m \supset U \to \mathbb{R}$ satisfies (3.7) for a constant vector field X and a constant $h = h(p) > 0$, then each mollified function ϕ_δ defined by equation (3.6) also satisfies (3.7) with the same X, h.*

PROOF. Let g be a nonnegative test function with support in U and $0 < \delta < \operatorname{dist}(\operatorname{supp} g, \partial U)$. Let $y \in \mathbb{R}^m$ with $|y| < \delta$. Applying (3.7) to the function $x \mapsto g(x+y)$ and using translation invariance of X, h and the Lebesgue measure $dx := d^m x$, we find

$$\begin{aligned} -\int_U \phi(x-y) X \cdot g(x) dx &= -\int_U \phi(x) X \cdot g(x+y) dx \\ &\geq h \int_U g(x+y) dx = h \int_U g(x) dx. \end{aligned}$$

Multiplying by the nonnegative function ρ_δ and integrating yields

$$\begin{aligned} -\int_U \phi_\delta(x) X \cdot g(x) dx &= -\int_U \int_{B_\delta} \phi(x-y) \rho_\delta(y) X \cdot g(x) dy\, dx \\ &\geq h \int_U \int_{B_\delta} g(x) \rho_\delta(y) dy\, dx = h \int_U g(x) dx. \end{aligned}$$ □

The next proposition shows that the condition $X \cdot \phi \geq h$ is preserved under taking the maximum of functions.

PROPOSITION 3.18. *Suppose the continuous functions $\phi, \psi : M \to \mathbb{R}$ satisfy $X \cdot \phi \geq h$, $X \cdot \psi \geq h$ with the same X, h. Then $X \cdot \max(\phi, \psi) \geq h$.*

More generally, suppose $(\phi_\lambda)_{\lambda \in \Lambda}$ is a continuous family of functions $\phi_\lambda : M \to \mathbb{R}$, parametrized by a compact metric space Λ, such that all ϕ_λ satisfy $X \cdot \phi_\lambda \geq h$ with the same X, h. Then $X \cdot \max_{\lambda \in \Lambda} \phi_\lambda \geq h$.

PROOF. Let $U \subset \mathbb{R}^m$ be a coordinate chart and $X, h := h(p)$ be as in (3.7). After a rotation and rescaling, we may assume that $X = \frac{\partial}{\partial x_1}$. Suppose first that ϕ, ψ are smooth and 0 is a regular value of $\phi - \psi$. Then $\vartheta := \max(\phi, \psi)$ is a continuous function which is smooth outside the smooth hypersurface $\Sigma := \{x \in U \mid \phi(x) = \psi(x)\}$. Define the function $\frac{\partial \vartheta}{\partial x_1}$ as $\frac{\partial \phi(x)}{\partial x_1}$ if $\phi(x) \geq \psi(x)$ and $\frac{\partial \psi(x)}{\partial x_1}$ otherwise. We claim that $\frac{\partial \vartheta}{\partial x_1}$ is the weak x_1-derivative of ϑ. Indeed, for any test function g supported in U we have (orienting Σ as the boundary of $\{\phi \geq \psi\}$)

$$\begin{aligned} \int_U \frac{\partial \vartheta}{\partial x_1} g\, d^m x &= \int_{\{\phi \geq \psi\}} \frac{\partial \phi}{\partial x_1} g\, d^m x + \int_{\{\phi < \psi\}} \frac{\partial \psi}{\partial x_1} g\, d^m x \\ &= \int_\Sigma \phi g\, dx_2 \cdots dx_m - \int_{\{\phi \geq \psi\}} \phi \frac{\partial g}{\partial x_1} d^m x \\ &\quad - \int_\Sigma \psi g\, dx_2 \cdots dx_m - \int_{\{\phi < \psi\}} \psi \frac{\partial g}{\partial x_1} d^m x \\ &= -\int_U \vartheta \frac{\partial g}{\partial x_1} d^m x, \end{aligned}$$

since $\phi = \psi$ on Σ. This proves the claim. By hypothesis we have $\frac{\partial \vartheta}{\partial x} \geq h$, so the conclusion of the lemma follows via

$$-\int_U \vartheta \frac{\partial g}{\partial x_1} d^m x = \int_U \frac{\partial \vartheta}{\partial x_1} g\, d^m x \geq h \int_U g\, d^m x.$$

Next let $\phi, \psi : U \to \mathbb{R}$ be continuous functions satisfying (3.7). By Lemma 3.17, there exist sequences ϕ_k, ψ_k of smooth functions, converging locally uniformly to ϕ, ψ, such that $X \cdot \phi_k \geq h$ and $X \cdot \psi_k \geq h$ for all k. Perturb the ϕ_k to smooth functions $\widetilde{\phi}_k$ such that 0 is a regular value of $\widetilde{\phi}_k - \psi_k$, $\widetilde{\phi}_k \to \phi$ locally uniformly, and $X \cdot \widetilde{\phi}_k \geq h - 1/k$ for all k. By the smooth case above, the function $\max(\widetilde{\phi}_k, \psi_k)$ satisfies

$$- \int_U \max(\widetilde{\phi}_k, \psi_k) X \cdot g \, d^m x \geq (h - 1/k) \int_U g \, d^m x$$

for any nonnegative test function g supported in U. Since $\max(\widetilde{\phi}_k, \psi_k) \to \max(\phi, \psi)$ locally uniformly, the limit $k \to \infty$ yields the conclusion of the lemma for the case of two functions ϕ, ψ.

Finally, let $(\phi_\lambda)_{\lambda \in \Lambda}$ be a continuous family as in the lemma. Pick a dense sequence $\lambda_1, \lambda_2, \ldots$ in Λ. Set $\psi_k := \max\{\phi_{\lambda_1}, \ldots, \phi_{\lambda_k}\}$ and $\psi := \max_{\lambda \in \Lambda} \phi_\lambda$. By the lemma for two functions and induction, the functions ψ_k satisfy (3.7) with the same X, h for all k. Thus the lemma follows in the limit $k \to \infty$ if we can show locally uniform convergence $\psi_k \to \psi$.

We first prove pointwise convergence $\psi_k \to \psi$. So let $x \in U$. Then $\psi(x) = \phi_\lambda(x)$ for some $\lambda \in \Lambda$. Pick a sequence k_ℓ such that $\lambda_{k_\ell} \to \lambda$ as $\ell \to \infty$. Then $\phi_{\lambda_{k_\ell}}(x) \to \phi_\lambda(x) = \psi(x)$ as $\ell \to \infty$. Since $\phi_{\lambda_{k_\ell}}(x) \leq \psi_{k_\ell}(x) \leq \psi(x)$, this implies $\psi_{k_\ell}(x) \to \psi(x)$ as $\ell \to \infty$. Now the convergence $\psi_k(x) \to \psi(x)$ follows from monotonicity of the sequence $\psi_k(x)$.

So we have an increasing sequence of continuous functions ψ_k that converges pointwise to a continuous limit function ψ. By a simple argument this implies locally uniform convergence $\psi_k \to \psi$: Let $\varepsilon > 0$ and $x \in U$ be given. By pointwise convergence there exists k such that $\psi_k(x) \geq \psi(x) - \varepsilon$. By continuity of ϕ_k and ψ, there exists $\delta > 0$ such that $|\psi_k(y) - \psi_k(x)| < \varepsilon$ and $|\psi(y) - \psi(x)| < \varepsilon$ for all y with $|y - x| < \delta$. This implies $\psi_k(y) \geq \psi(y) - 3\varepsilon$ for all y with $|y - x| < \delta$. In view of monotonicity, this establishes locally uniform convergence $\psi_k \to \psi$, and hence concludes the proof of the proposition. □

Finally, we show that J-convex functions can be smoothed without creating critical points.

PROPOSITION 3.19. *Let $\phi : V \to \mathbb{R}$ be a continuous J-convex function on a complex manifold satisfying $X \cdot \phi \geq h$ for a vector field X and a positive function $h : V \to \mathbb{R}$. Then the J-convex smoothing $\psi : V \to \mathbb{R}$ in Proposition 3.10 can be constructed so that it satisfies $X \cdot \psi \geq \widetilde{h}$ for any given function $\widetilde{h} < h$.*

PROOF. The function ψ is constructed from ϕ in Lemma 3.12 by repeated application of the following three constructions:

(1) Mollification $\phi \mapsto \phi_\delta$. This operation preserves the condition $X \cdot \phi \geq h$ by Lemma 3.17.

(2) Taking the maximum of two functions. This operation preserves the condition $X \cdot \phi \geq h$ by Proposition 3.18.

(3) Adding a C^2-small function f to ϕ. Let $k : V \to \mathbb{R}$ be a small positive function such that $\sup_U (X \cdot f)(x) \geq -k(p)$ for each coordinate chart U around p as in condition (3.7) (for this it suffices that f is sufficiently C^1-small). Then we find

$$- \int_U f(x)(X \cdot g)(x) dx = \int_U (X \cdot f)(x) g(x) dx \geq -k(p) \int_U g(x) dx,$$

so the function $\phi + f$ satisfies $X \cdot (\phi + f) \geq h - k$. In the proof of Lemma 3.12, this operation is applied finitely many times on each compact subset of V, so by choosing the function k sufficiently small, we can achieve that $X \cdot \psi \geq \widetilde{h}$. □

Propositions 3.18 and 3.19 together imply

COROLLARY 3.20. *If two smooth J-convex functions ϕ, ψ on a complex manifold V satisfy $X \cdot \phi > 0$ and $X \cdot \psi > 0$ for a vector field X, then the smoothing ϑ of* $\max(\phi, \psi)$ *can also be arranged to satisfy $X \cdot \vartheta > 0$.* □

REMARK 3.21. (a) Clearly, Proposition 3.19 and Corollary 3.20 also hold without the J-convexity condition, for functions and vector fields on a smooth manifold.

(b) Inspection of the proofs shows that Propositions 3.18 and 3.19 remain valid if all inequalities are replaced by the reverse inequalities.

COROLLARY 3.22. *If two smooth J-convex functions ϕ, ψ on a complex manifold V are C^1-close, then the smoothing of* $\max(\phi, \psi)$ *is C^1-close to ϕ.*

PROOF. Let X be a vector field and $h_\pm : V \to \mathbb{R}$ functions such that $h_- \leq X \cdot \phi, X \cdot \psi \leq h_+$. By the preceding remark, the smoothing ϑ of $\max(\phi, \psi)$ can be constructed such that $\widetilde{h}_- \leq X \cdot \vartheta \leq \widetilde{h}_+$ for any given functions $\widetilde{h}_- < h_-$ and $\widetilde{h}_+ > h_+$. Since X, h_-, h_+ were arbitrary, this proves C^1-closeness of ϑ to ϕ. □

Finally, we apply the preceding result to smoothing of J-convex hypersurfaces.

COROLLARY 3.23. *Let $(M \times \mathbb{R}, J)$ be a compact complex manifold and $\phi, \psi : M \to \mathbb{R}$ two functions whose graphs are J-convex cooriented by ∂_r, where r is the coordinate on $\mathbb{R}$. Then there exists a smooth function $\vartheta : M \to \mathbb{R}$ with J-convex graph which is C^0-close to* $\min(\phi, \psi)$ *and coincides with* $\min(\phi, \psi)$ *outside a neighborhood of the set $\{\phi = \psi\}$.*

PROOF. For a convex increasing function $f : \mathbb{R} \to \mathbb{R}$ with $f(0) = 0$ consider the functions

$$\Phi(x, r) := f\big(r - \phi(x)\big), \qquad \Psi(x, r) := f\big(r - \psi(x)\big).$$

For f sufficiently convex, Φ and Ψ are J-convex and satisfy $\partial_r \Phi > 0, \partial_r \Psi > 0$ near their zero level sets. Thus by Propositions 3.18 and 3.19 the function $\max(\Phi, \Psi)$ can be smoothed, keeping it fixed outside a neighborhood U of the set $\{\max(\Phi, \Psi) = 0\}$, to a function Θ which is J-convex and satisfies $\partial_r \Theta > 0$ near its zero level set. The last condition implies that the smooth J-convex hypersurface $\Theta^{-1}(0)$ is the graph of a smooth function $\vartheta : M \to \mathbb{R}$. Now note that the zero level set $\{\max(\Phi, \Psi) = 0\}$ is the graph of the function $\min(\phi, \psi)$. This implies that ϑ is C^0-close to $\min(\phi, \psi)$ and coincides with $\min(\phi, \psi)$ outside U. □

REMARK 3.24. Note that convex functions on $\mathbb{C}^n$ are also J-convex. On the other hand for any two convex functions ϕ, ψ the function smooth $\max(\phi, \psi)$ is also convex, and therefore has a unique critical point, the minimum.

3.4. From families of hypersurfaces to J-convex functions

The following result shows that a continuous family of J-convex hypersurfaces transverse to the same vector field gives rise to a smooth function with regular J-convex level sets. This will be extremely useful for the construction of J-convex functions with prescribed critical points.

PROPOSITION 3.25. *Let $(M\times[0,1],J)$ be a compact complex manifold such that $M\times\{0\}$ and $M\times\{1\}$ are J-convex cooriented by ∂_r, where r is the coordinate on $[0,1]$. Suppose there exists a smooth family $(\Sigma_\lambda)_{\lambda\in[0,1]}$ of J-convex hypersurfaces transverse to ∂_r with $\Sigma_0=M\times\{0\}$ and $\Sigma_1=M\times\{1\}$. Then there exists a smooth foliation $(\widetilde\Sigma_\lambda)_{\lambda\in[0,1]}$ of $M\times[0,1]$ by J-convex hypersurfaces transverse to ∂_r with $\widetilde\Sigma_\lambda=M\times\{\lambda\}$ for λ near 0 or 1.*

PROOF. The proof has two steps. In the first step we use the maximum construction to make the family of hypersurfaces weakly monotone in the parameter λ, and in the second step we perturb it to make it strictly monotone and thus obtain a foliation.

Step 1. Let $\varepsilon>0$ be so small that the hypersurfaces $M\times\{\lambda\}$ are J-convex for $\lambda\le\varepsilon$ and $\lambda\ge 1-\varepsilon$. We first modify the family such that $\Sigma_\lambda=M\times\{\lambda\}$ for $\lambda\le\varepsilon$ and $\lambda\ge1-\varepsilon$. After a C^2-small perturbation and decreasing ε, we may further assume that $\Sigma_\lambda\subset M\times(\varepsilon,1-\varepsilon)$ for all $\lambda\in(\varepsilon,1-\varepsilon)$. Pick a smooth family of surjective J-convex functions $\phi_\lambda:\mathcal{O}p\,\Sigma_\lambda\to[-1,1]$ [2] with regular level sets $\phi_\lambda^{-1}(0)=\Sigma_\lambda$, and extend ϕ_λ by the values ±1 to a continuous function $M\times[0,1]\to[-1,1]$ (which is not J-convex outside $\mathcal{O}p\,\Sigma_\lambda$). After composing each ϕ_λ with a suitable convex function $\mathbb{R}\to\mathbb{R}$, shrinking the neighborhoods $\mathcal{O}p\,\Sigma_\lambda$ and extending as before by ±1, we may assume that $\phi_\lambda\ge\phi_\mu$ for all $\lambda\le\mu$ with either $\lambda\le\varepsilon$ or $\mu\ge1-\varepsilon$.

By Proposition 3.8, the continuous functions

$$\psi_\lambda:=\max{}_{\nu\ge\lambda}\phi_\nu$$

are J-convex on $U_\lambda:=\psi_\lambda^{-1}([-\frac12,\frac12])$. By construction, they satisfy

$$\psi_\lambda\ge\psi_\mu \text{ for } \lambda\le\mu, \tag{3.8}$$

and $\psi_\lambda=\phi_\lambda$ for $\lambda\le\varepsilon$ and $\lambda\ge1-\varepsilon$. By Proposition 3.18, the ψ_λ satisfy $\partial_r\cdot\psi_\lambda\ge h$ (in the distributional sense) on U_λ for a positive function $h:M\times[0,1]\to\mathbb{R}$.

Next use Proposition 3.10 to approximate the ψ_λ by smooth functions $\widehat\psi_\lambda$ that are J-convex on $\widehat U_\lambda:=\widehat\psi_\lambda^{-1}([-\frac14,\frac14])$. By Remark 3.13, the resulting family $\widehat\psi_\lambda$ is continuous in λ and still satisfies (3.8). By Proposition 3.19, the smoothed functions satisfy $\partial_r\cdot\widehat\psi_\lambda\ge h/2>0$ on $\widehat U_\lambda$, hence the level sets $\widehat\Sigma_\lambda:=\widehat\psi_\lambda^{-1}(0)$ are regular and transverse to ∂_r. We can modify the smoothing construction to achieve $\widehat\psi_\lambda=\phi_\lambda$ near $\lambda=0$ and 1, still satisfying J-convexity, transversality of the zero level to ∂_r, and (3.8). Note that as a result of the smoothing construction, the functions $\widehat\psi_\lambda$, and hence their level sets $\widehat\Sigma_\lambda$, depend continuously on the parameter λ with respect to the C^2-topology.

Step 2. Since $\widehat\Sigma_\lambda$ is transverse to ∂_r, we can write it as the graph $\{r=f_\lambda(x)\}$ of a smooth function $f_\lambda:M\to[0,1]$. By construction, the functions f_λ depend continuously on λ with respect to the C^2-topology, $f_\lambda\le f_\mu$ for $\lambda\le\mu$, and $f_\lambda(x)=\lambda$ for $\lambda\le\varepsilon$ and $\lambda\ge1-\varepsilon$, with some $\varepsilon>0$ (possibly smaller than the one above). Note that $f_\mu(x)-f_\lambda(x)\ge\mu-\lambda$ for $\lambda\le\mu\le\varepsilon$ and $1-\varepsilon\le\lambda\le\mu$. Pick a function $g:[0,1]\to[0,1]$ satisfying $g(\lambda)=0$ for $\lambda\le\varepsilon/2$ and $\lambda\ge1-\varepsilon/2$, $g'(\lambda)\ge-1+\gamma$ for $\varepsilon/2\le\lambda\le\varepsilon$ and $1-\varepsilon\le\lambda\le1-\varepsilon/2$, and $g'(\lambda)\ge\gamma$ for $\varepsilon\le\lambda\le1-\varepsilon$, with some $\gamma>0$; see Figure 3.1. For g sufficiently small, the graphs of the functions

[2] Recall that for a closed subset $A\subset X$ of a topological space, $\mathcal{O}p\,A$ denotes a sufficiently small but not specified open neighborhood of A.

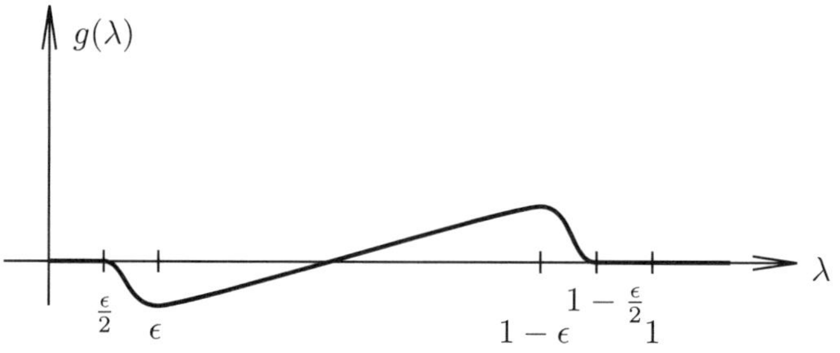

FIGURE 3.1. The function g.

$\widehat{f}_\lambda(x) := f_\lambda(x) + g(\lambda)$ are still J-convex, $\widehat{f}_\lambda(x) = \lambda$ for $\lambda \le \varepsilon/2$ and $\lambda \ge 1-\varepsilon/2$, and

$$\widehat{f}_\mu(x) - \widehat{f}_\lambda(x) \ge \gamma(\mu - \lambda)$$

for all $\lambda \le \mu$. Now mollify the functions $\widehat{f}_\lambda(x)$ in the parameter λ to

$$\widetilde{f}_\lambda(x) := \int_{\mathbb{R}} \widehat{f}_{\lambda-\mu}(x)\rho_\delta(\mu)d\mu,$$

with a cutoff function $\rho : \mathbb{R} \to \mathbb{R}$ as in equation (3.6). Since the functions $f_{\lambda-\mu}$ are C^2-close to f_λ for $\mu \in \operatorname{supp}\rho_\delta$ and δ small, the graph of $\widetilde{f}_\lambda$ is C^2-close to the graph of f_λ and hence J-convex. Moreover, for $\lambda' \ge \lambda$ the $\widetilde{f}_\lambda$ still satisfy

$$\widetilde{f}_{\lambda'}(x) = \int_{\mathbb{R}} \widehat{f}_{\lambda'-\mu}(x)\rho_\delta(\mu)d\mu \ge \int_{\mathbb{R}} \widehat{f}_{\lambda-\mu}(x)\rho_\delta(\mu)d\mu + \gamma(\lambda'-\lambda) = \widetilde{f}_\lambda(x) + \gamma(\lambda'-\lambda).$$

Modify the $\widetilde{f}_\lambda$ such that $\widetilde{f}_\lambda(x) = \lambda$ for $\lambda \le \varepsilon/2$ and $\lambda \ge 1-\varepsilon/2$, and so that their graphs are still J-convex and $\widetilde{f}_\mu(x) - \widetilde{f}_\lambda(x) \ge \gamma(\mu-\lambda)$ for all $\lambda \le \mu$. The last inequality implies that the map $(x,\lambda) \mapsto \big(x, f_\lambda(x)\big)$ is an embedding, thus the graphs of $\widetilde{f}_\lambda$ form the desired foliation $\widetilde{\Sigma}_\lambda$. □

3.5. *J*-convex functions near totally real submanifolds

In this section we discuss modifications of J-convex functions near totally real submanifolds. The following is the main result.

PROPOSITION 3.26. *Let L be a totally real submanifold of a complex manifold (V,J) and $K \subset L$ a compact subset. Suppose that two smooth J-convex functions ϕ, ψ coincide along L together with their differentials, i.e., $\phi(x) = \psi(x)$ and $d\phi(x) = d\psi(x)$ for all $x \in L$. Then, given any neighborhood U of K in V, there exists a J-convex function ϑ with the following properties:*

(a) ϑ coincides with ϕ outside U and with ψ in a smaller neighborhood $U' \subset U$ of K.

(b) ϑ and ϕ coincide along L together with their differentials.

(c) ϑ can be chosen arbitrarily C^1-close to ϕ, with modulus of J-convexity uniformly bounded from below.

(d) Assume in addition that ϕ, ψ are Morse, and at each critical point p on K the stable and unstable spaces for $\nabla_\phi\phi$ and $\nabla_\psi\psi$ satisfy $E_p^-(\phi) = E_p^-(\psi) \subset T_pL$ and $E_p^+(\phi) = E_p^+(\psi)$, and T_pL is isotropic with respect to the symplectic form $\omega_\phi = -dd^{\mathbb{C}}\phi$. Then there exists a vector field X on V which is gradient-like for both ϕ and ψ.

(e) Assume in addition that on some neighborhood N *of* K *in* L *we have* $\nabla_\psi \psi = \lambda \nabla_\phi \phi$ *for a positive function* $\lambda : N \to \mathbb{R}_+$. *Then* $\nabla_\vartheta \vartheta = \mu \nabla_\phi \phi$ *on* L *for a positive function* $\mu : L \to \mathbb{R}_+$

(f) Assume in addition that $r\partial_r \phi \geq \mu r^2$ *and* $r\partial_r \psi \geq \mu r^2$, *where* r *is the distance from* L *with respect to some Hermitian metric and* $\mu > 0$ *a constant. Then we can arrange that* $r\partial_r \vartheta \geq \mu r^2/2$.

REMARK 3.27. (i) In the notation of Proposition 3.26, ϕ and ϑ can be connected by the family of J-convex functions $\phi_t := (1-t)\phi + t\vartheta$, $t \in [0,1]$, satisfying properties (b)–(f).

(ii) If L is Lagrangian and $\nabla_\phi \phi$ and $\nabla_\psi \psi$ are tangent to L, then so is $\nabla_\vartheta \vartheta$. This follows from the observation that tangency of $\nabla_\phi \phi$ to L for L Lagrangian is equivalent to vanishing of $d\phi \circ J$ on L, which is preserved under convex combinations.

The proof of Proposition 3.26 is based on three lemmas.

LEMMA 3.28. *Let* $\phi, \psi : V \to \mathbb{R}$ *be smooth* J*-convex functions on an almost complex manifold* (V, J), *and set* $\vartheta := (1-\beta)\phi + \beta\psi$ *for a smooth function* $\beta : V \to [0,1]$.

(a) Suppose that

$$|\phi(x) - \psi(x)|\, |dd_x^{\mathbb{C}}\beta| + 2|d_x\beta|\, |d_x(\phi - \psi)| < \min\big(m_\phi(x), m_\psi(x)\big)$$

for all $x \in V$ *(with respect to some Hermitian metric). Then* ϑ *is* J*-convex.*

(b) Suppose that at some point $x \in V$ *we have* $\phi(x) = \psi(x)$, $d\phi(x) = d\psi(x)$, *and* $\nabla_\phi \phi(x) = \lambda \nabla_\psi \psi(x)$ *for some* $\lambda > 0$. *Then* $\nabla_\vartheta \vartheta(x) = \mu \nabla_\phi \phi(x)$ *for some* $\mu > 0$.

PROOF. (a) Adding up

$$dd^{\mathbb{C}}(\beta\psi) = \beta\, dd^{\mathbb{C}}\psi + d\beta \wedge d^{\mathbb{C}}\psi + d\psi \wedge d^{\mathbb{C}}\beta + \psi\, dd^{\mathbb{C}}\beta$$

and the corresponding equation for $(1-\beta)\phi$ at any point $x \in V$, we find

$$\begin{aligned} -dd^{\mathbb{C}}\vartheta &= -(1-\beta)\, dd^{\mathbb{C}}\phi - \beta dd^{\mathbb{C}}\psi + d\beta \wedge d^{\mathbb{C}}(\phi - \psi) \\ &\quad + d(\phi - \psi) \wedge d^{\mathbb{C}}\beta + (\phi - \psi) dd^{\mathbb{C}}\beta \\ &\geq \min(m_\phi, m_\psi) - 2|d\beta|\, |d(\phi - \psi)| - |\phi - \psi|\, |dd^{\mathbb{C}}\beta| \\ &> 0. \end{aligned}$$

(b) At the point x the terms $\phi - \psi$ and $d\phi - d\psi$ vanish, so the computation in (a) shows $-dd^{\mathbb{C}}\vartheta = -(1-\beta)\, dd^{\mathbb{C}}\phi - \beta dd^{\mathbb{C}}\psi$. Hence at the point x we have $d\vartheta = d\phi = d\psi$ and the associated metrics satisfy

$$g_\vartheta = (1-\beta)g_\phi + \beta g_\psi.$$

Now at x we make the ansatz $\nabla_\vartheta \vartheta = \mu \nabla_\phi \phi = \mu\lambda \nabla_\psi \psi$ and compute

$$\begin{aligned} g_\vartheta(\nabla_\vartheta \vartheta, \cdot) &= (1-\beta)\mu g_\phi(\nabla_\phi \phi, \cdot) + \beta\mu\lambda g_\psi(\nabla_\psi \psi, \cdot) \\ &= (1-\beta)\mu\, d\phi + \beta\mu\lambda\, d\psi = \mu(1 - \beta + \beta\lambda) d\vartheta, \end{aligned}$$

which yields the correct equation if $\mu = (1 - \beta + \beta\lambda)^{-1}$. □

LEMMA 3.29. *For any constants* $a > 0$ *and* $0 < \delta < \varepsilon$ *there exists a smooth function* $f : [0, \varepsilon] \to \mathbb{R}_{\geq 0}$ *with the following properties (see Figure 3.2):*

(i) $f(x) = ax$ *near* 0 *and* $f(x) = 0$ *near* ε;

(ii) $-\delta \leq f'(x) \leq a$ *and* $xf''(x) \geq -\delta$ *for all* $x \in [0, \varepsilon]$.

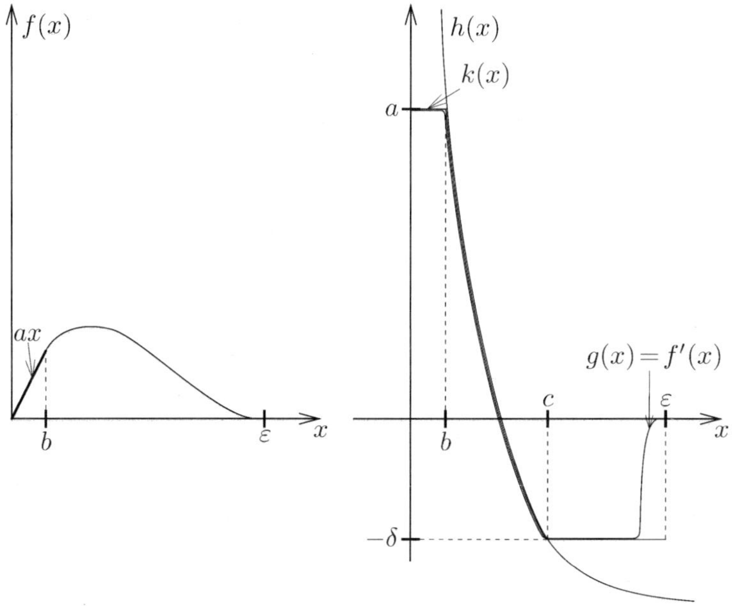

FIGURE 3.2. Construction of the function f.

PROOF. We need to find a function g $(= f')$ satisfying

(i) $g(x) = a$ near 0 and $g(x) = 0$ near ε;
(ii) $-\delta \leq g(x) \leq a$ and $xg'(x) \geq -\delta$ for all $x \in [0, \varepsilon]$;
(iii) $\int_0^\varepsilon g(x)dx = 0$ and $\int_0^y g(x)dx \geq 0$ for all $y \in [0, \varepsilon]$.

For a constant $c \in (0, \varepsilon)$ (which will be determined later) consider the function $h(x) := -\delta \ln(x/c) - \delta$. It satisfies $xh'(x) = -\delta$ and $h(c) = -\delta$. Let $b \in (0, c)$ be determined by $h(b) = a$. Let $k : [0, \varepsilon] \to \mathbb{R}$ be the continuous function which agrees with h on $[b, c]$, equals a on $[0, b]$ and $-\delta$ on $[c, \varepsilon]$; see Figure 3.2. We estimate its integral by

$$\int_0^\varepsilon k(x)dx = \int_0^c k(x)dx + \int_c^\varepsilon k(x)dx \leq ac - \delta(\varepsilon - c) < 0$$

for c sufficiently small. Now the function g is obtained by smoothing k, connecting it to 0 near ε, and increasing its integral to make it zero. □

Next we prove Proposition 3.26 in the special case that $\psi = \phi + a\,\mathrm{dist}_L^2$, where $a > 0$ and dist_L is the distance from L with respect to some Hermitian metric. Note that according to Proposition 2.15, this function is J-convex near L.

LEMMA 3.30. *Let L be a totally real submanifold of a complex manifold (V, J) and $K \subset L$ a compact subset. Let ϕ be a smooth J-convex function and U a neighborhood of K in V. Then there exists a Hermitian metric on V such that for any constant $a > 0$ there exist a J-convex function $\bar{\phi}$ with the following properties:*

(a) $\bar{\phi}$ coincides with ϕ outside U and with $\phi + a\,\mathrm{dist}_L^2$ in a smaller neighborhood $U' \subset U$ of K.

(b) $\bar{\phi}$ and ϕ coincide along L together with their differentials.

(c) $\bar{\phi}$ can be chosen arbitrarily C^1-close to ϕ, with modulus of J-convexity uniformly bounded from below.

(d) Assume in addition that ϕ is Morse and at each critical point p on K the stable space E_p^- for $\nabla\phi$ satisfies $E_p^- \subset T_pL$, and T_pL is isotropic with respect to the symplectic form $\omega_\phi = -dd^{\mathbb{C}}\phi$. Then there exists a vector field X on V which is gradient-like for both ϕ and $\bar{\phi}$.

(e) Assume in addition that $r\partial_r\phi \geq \mu r^2$, where r is the distance from L with respect to the Hermitian metric and $\mu > 0$ a constant. Then we can arrange that $r\partial_r\bar{\phi} \geq \mu r^2/2$.

PROOF. Fix a compact neighborhood W of K in V on which dist_L^2 is J-convex. Set

$$\widetilde{\phi} := \phi + f(\rho), \qquad \rho := \mathrm{dist}_L^2,$$

where $f : [0, \varepsilon] \to \mathbb{R}$ is the function from Lemma 3.29, with constants $0 < \delta < \varepsilon$ to be determined later. Then $\widetilde{\phi}$ coincides with ϕ on $\{\rho \geq \varepsilon\}$ and with $\phi + a\rho$ near L. Let us show that $\widetilde{\phi}$ is J-convex on W. Indeed,

$$dd^{\mathbb{C}}\widetilde{\phi} = dd^{\mathbb{C}}\phi + f''(\rho)d\rho \wedge d^{\mathbb{C}}\rho + f'(\rho)dd^{\mathbb{C}}\rho.$$

By Proposition 2.15, there exist constants m_L, M_L such that

$$m_L|v|^2 \leq -dd^{\mathbb{C}}\rho(v, Jv) \leq M_L|v|^2$$

for $v \in T_xW$, $x \in W$. Moreover, on W we have $|d\rho| \leq C_L\sqrt{\rho}$, where the constant C_L depends only on the geometry of $L \cap W$. Thus for $v \in T_xW$, $x \in W$, $|v| = 1$, we have

$$\begin{aligned}
-dd^{\mathbb{C}}\widetilde{\phi}(v, Jv) &= -dd^{\mathbb{C}}\phi(v, Jv) + f''(\rho)\Big(d\rho(v)^2 + d\rho(Jv)^2\Big) - f'(\rho)dd^{\mathbb{C}}\rho(v, Jv) \\
&\geq m_\phi - \max\{0, -f''(\rho)\}C_L\rho - \max\{0, -f'(\rho)\}M_L \\
&\geq m_\phi - C_L\delta - M_L\delta \geq \mu/2
\end{aligned}$$

for δ sufficiently small, where $\mu := \min_W m_\phi$.

Note that $\widetilde{\phi}$ is arbitrarily C^1-close to ϕ for ε small. Fix a cutoff function β with support in W and equal to 1 on a neighborhood $W' \subset W$ of K. The function

$$\bar{\phi} := (1 - \beta)\phi + \beta\widetilde{\phi}$$

satisfies $\bar{\phi} = \phi$ outside W and $\bar{\phi} = \phi + a\rho$ near K. Moreover, since the estimates $m_\phi \geq \mu$ and $m_{\widetilde{\phi}} \geq \mu/2$ are independent of ε and δ (provided δ is sufficiently small), Lemma 3.28 implies that $\bar{\phi}$ is J-convex if ε and δ are sufficiently small. By construction, $\bar{\phi}$ has properties (a)–(c).

Suppose now that ϕ satisfies the assumptions of (d). Consider a critical point $p \in K$ of index $\ell \leq k = \dim L$. We first construct nice coordinates near p. Consider the Hermitian vector space (T_pV, J, ω_ϕ). Since by assumption $E_p^- \subset T_pL$ are isotropic subspaces of T_pV, we find a unitary isomorphism $\Phi : (\mathbb{C}^n, i, \omega_{\mathrm{st}}) \to (T_pV, J, \omega_\phi)$ mapping $\mathbb{R}^\ell$ to E_p^- and $\mathbb{R}^k$ to T_pL. Let $F : \mathbb{R}^k \supset \mathcal{O}p\,(0) \to L$ be a smooth embedding with $d_0F = \Phi|_{\mathbb{R}^k}$. If $k < n$, extend F to a smooth embedding $\mathbb{R}^n \supset \mathcal{O}p\,(0) \to L$ with $d_0F = \Phi|_{\mathbb{R}^n}$. Using Proposition 5.55, we can extend F to a smooth embedding $F : \mathbb{C}^n \supset \mathcal{O}p\,(0) \to V$ such that F^*J agrees with i to second order along $\mathbb{R}^n$. In particular, it satisfies $d_0F = \Phi$. We pull back all data under F

and denote them by the same letters. Consider the standard complex coordinates $z_j = x_j + iy_j$ on $\mathbb{C}^n$ and write $z = (u, v, w)$ with

$$u := (x_1, \dots, x_\ell), \quad v = (x_{\ell+1}, \dots, x_k), \quad w := (x_{k+1}, \dots, x_n, y_1, \dots, y_n).$$

By construction, u are coordinates on E_p^- and (u, v) are coordinates on L. Moreover, the metric g_ϕ coincides with the standard metric on $\mathbb{C}^n$ in these coordinates at the point $p = 0$, and thus (v, w) are coordinates on E_p^+. We choose the Hermitian metric near $p = 0$ so that it coincides with the standard metric on $\mathbb{C}^n$ to second order along $\mathbb{R}^k$ (this is possible because J and i agree to second order along $\mathbb{R}^k$), and thus $\rho(z) = |w|^2 + O(|w|^3)$. We define the vector field

$$X(u, v, w) := (-u, v, w)$$

near $p = 0$. Since the splitting $T_pV = E_p^- \oplus E_p^+$ is orthogonal with respect to the Hessian $H_p\phi$, and the Hessian is positive (resp. negative) definite on E_p^+ (resp. E_p^-), we have

$$\begin{aligned} d\phi \cdot X(z) &= H_p\phi(z, X(z)) + O(|z|^3) \\ &= -H_p\phi(u, 0, 0) + H_p\phi(0, v, w) + O(|z|^3) \geq \gamma|z|^2 \end{aligned}$$

for some constant $\gamma > 0$. On the other hand, $d\rho \cdot X(z) = 2|w|^2 + O(|w|^3)$ implies $|w|^2 \leq d\rho \cdot X(z) \leq 3|w|^2$ and hence

$$d\widetilde{\phi} \cdot X(z) \geq \gamma|z|^2 - 3\max\{0, -f'(\rho)\}|w|^2 \geq \gamma(|u|^2 + |v|^2) + (\gamma - 3\delta)|w|^2 \geq \frac{\gamma}{2}|z|^2$$

provided that $\delta \leq \gamma/6$. This shows that X is gradient-like for ϕ and $\widetilde{\phi}$ (and hence for $\bar{\phi}$) near $p = 0$. Outside a neighborhood of the critical points of ϕ, its gradient vector field $\nabla_\phi \phi$ is also gradient-like for $\bar{\phi}$ if the functions are sufficiently C^1-close (which can be arranged by making ε, δ small). The gradient-like vector field for ϕ and $\bar{\phi}$ is now obtained by interpolation between X near the critical points and $\nabla_\phi \phi$ outside.

Finally, we prove (e). Using $f' \geq -\delta$ we estimate for $\widetilde{\phi} = \phi + f(\rho)$:

$$r\partial_r \widetilde{\phi} = r\partial_r \phi + 2f'(\rho)r^2 \geq (\mu - 2\delta)r^2 \geq \frac{3}{4}\mu r^2$$

for δ sufficiently small. Next, using $|\phi - \widetilde{\phi}| \leq Br^2$ for some constant B, we estimate for $\bar{\phi} = (1 - \beta)\phi + \beta\widetilde{\phi}$:

$$r\partial_r \bar{\phi} = (1 - \beta)r\partial_r \phi + \beta r\partial_r \widetilde{\phi} + (\widetilde{\phi} - \phi)r\partial_r \beta \geq \frac{3}{4}\mu r^2 - Br^2|r\partial_r \beta| \geq \mu r^2/2,$$

provided we can make $|r\partial_r \beta|$ arbitrarily small. For this, we write β as a product $\beta_1\beta_2$. Here $\beta_1 = h \circ \pi$ for the projection $\pi : V \to L$ along $r\partial_r$ and h a cutoff function on L, so $r\partial_r \beta_1 = 0$. The second function is of the form $\beta_2 = g(r)$ for a function $g : [0, \varepsilon] \to [0, 1]$ which equals 1 near 0 and 0 near 1. The proof of Lemma 3.29 shows that we can find such a function g with $|\partial_r \beta_1| = |rg'(r)| \leq \delta$ for arbitrarily small δ. □

Proof of Proposition 3.26. Fix a compact neighborhood $\widetilde{K} \subset L \cap U$ of K in L containing no critical points outside K. Take a Hermitian metric on (V, J) as in Lemma 3.30 and consider the function $\rho := \mathrm{dist}_L^2$, square of the distance to

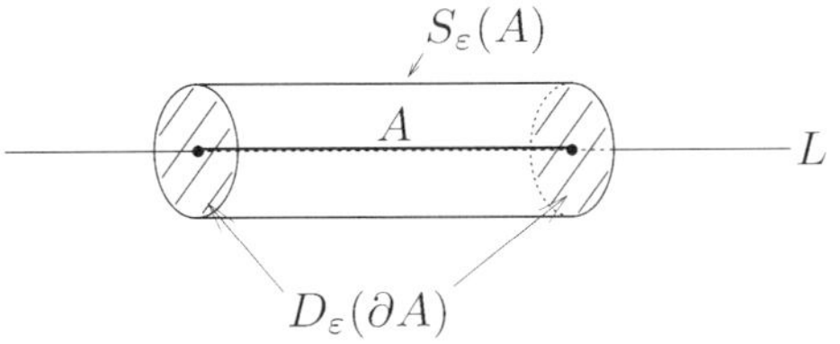

FIGURE 3.3. The tube $D_\varepsilon(A)$ around A.

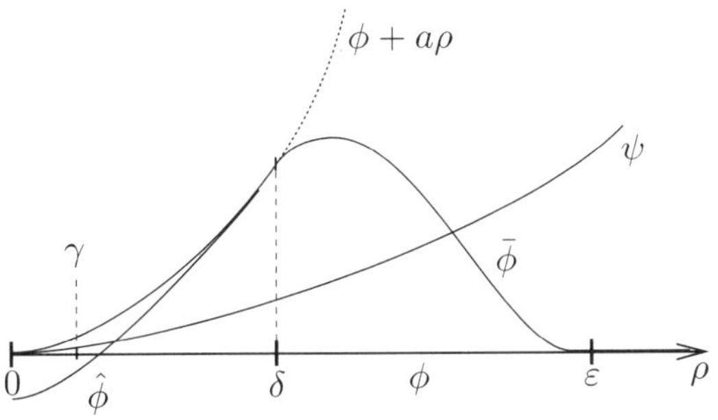

FIGURE 3.4. Construction of the function $\widehat{\vartheta}$.

L, defined on a tubular neighborhood of L. According to Proposition 2.15, this function is J-convex near L. Hence, after shrinking U,

$$\phi_a := \phi + a\rho$$

is J-convex on U for any $a \geq 0$.

For small $\varepsilon > 0$ denote by $\pi_\varepsilon : D_\varepsilon(L) := \{\rho \leq \varepsilon\} \to L$ the projection onto the nearest point. For a subset $A \subset L$, set

$$D_\varepsilon(A) := \pi_\varepsilon^{-1}(A), \qquad S_\varepsilon(A) := D_\varepsilon(A) \cap \partial D_\varepsilon(L);$$

see Figure 3.3.

Since ϕ and ψ agree up to first order along L, there exists an $a > 0$ and $\varepsilon > 0$ such that

$$\phi_a > \psi \text{ on } D_\varepsilon(\widetilde{K}) \setminus L.$$

By Lemma 3.30 we find a J-convex function $\bar{\phi} : D_\varepsilon(\widetilde{K}) \to \mathbb{R}$ which agrees with ϕ near $S_\varepsilon(\widetilde{K})$ and with ϕ_a on $D_\delta(\widetilde{K})$ for some $\delta \in (0, \varepsilon)$, so $\bar{\phi} > \psi$ on $D_\delta(\widetilde{K}) \setminus L$. Next pick a cutoff function $\alpha(\rho)$ which equals 0 for $\rho \geq \delta$ and 1 for $\rho \leq \delta/2$. The function

$$\widehat{\phi} := \bar{\phi} - \mu\alpha$$

is J-convex for $\mu > 0$ sufficiently small. Moreover, it satisfies

$$\widehat{\phi} < \psi \text{ on } D_\gamma(\widetilde{K}), \qquad \widehat{\phi} > \psi \text{ near } S_\delta(\widetilde{K}), \qquad \widehat{\phi} = \phi \text{ near } S_\varepsilon(\widetilde{K}),$$

for some $\gamma \in (0, \delta)$; see Figure 3.4. So the function

$$\widehat{\vartheta} := \begin{cases} \text{smooth}\max(\psi, \widehat{\phi}) & \text{on } D_\delta(\widetilde{K}), \\ \widehat{\phi} & \text{on } D_\varepsilon(\widetilde{K}) \setminus D_\delta(\widetilde{K}) \end{cases}$$

coincides with ψ on $D_\gamma(\widetilde{K})$ and with ϕ near $S_\varepsilon(\widetilde{K})$. Moreover, since $\widehat{\phi}$ is C^1-close to ϕ, by construction and Lemma 3.30, $\widehat{\vartheta}$ is C^1-close to ϕ by Corollary 3.22.

It remains to interpolate between $\widehat{\vartheta}$ and ϕ near $D_\varepsilon(\partial\widetilde{K})$. For this, we fix a cutoff function $\beta : L \to \mathbb{R}$ which equals 1 near K and 0 on $L \setminus \widetilde{K}$ and extend it to $D_\varepsilon(\widetilde{K})$ via the projection π_ε. By Lemma 3.28, the function

$$\vartheta := (1-\beta)\phi + \beta\widehat{\vartheta}$$

is J-convex if we choose $\widehat{\vartheta}$ sufficiently C^1-close to ϕ. Since ψ agrees with ϕ together with their differentials along L, the same holds for ϑ and ϕ. So ϑ has properties (a)–(c).

For property (d), let X be the gradient-like vector field for ϕ and $\bar{\phi}$ from Lemma 3.30. The assumptions on ψ ensure that X is also gradient-like for ψ near K. The function $\widehat{\phi}$ is C^1-close to $\bar{\phi}$ and differs from $\bar{\phi}$ only by a constant near the critical points, so X is gradient-like for $\widehat{\phi}$. Since the function $\widehat{\vartheta}$ is obtained by the maximum construction and smoothing and agrees with ψ near the critical points, X is gradient-like for $\widehat{\vartheta}$ by Corollary 3.20. Finally, X is gradient-like for ϑ because ϑ is C^1-close to $\widehat{\vartheta}$ and equals $\widehat{\vartheta}$ near the critical points.

Property (e) follows from Lemma 3.28, assuming that $\nabla_\phi\phi = \lambda\nabla_\psi\psi$ at the points of L where $d\beta \neq 0$. Property (f) follows from Lemma 3.28 and the fact that this property is preserved under the maximum construction (which is obvious) and under the interpolation $\vartheta := (1-\beta)\phi + \beta\widehat{\vartheta}$ (which was shown in the proof of Lemma 3.28). This concludes the proof of Proposition 3.26. □

The corresponding result for J-convex hypersurfaces is

COROLLARY 3.31. *Let Σ, Σ' be J-convex hypersurfaces in a complex manifold (V, J) that are tangent to each other along a totally real submanifold L. Then for any compact subset $K \subset L$ and neighborhood U of K, there exists a J-convex hypersurface Σ'' that agrees with Σ outside U and with Σ' near K. Moreover, Σ'' can be chosen C^1-close to Σ and tangent to Σ along L.*

PROOF. Pick smooth functions ϕ, ψ with regular level sets $\Sigma = \phi^{-1}(0)$ and $\Sigma' = \psi^{-1}(0)$ such that $d\phi = d\psi$ along L. By Lemma 2.7, after composing ϕ and ψ with the same convex function, we may assume that ϕ, ψ are J-convex on a neighborhood $W \subset U$ of K. Let $\vartheta : W \to \mathbb{R}$ be the J-convex function from Proposition 3.26 which coincides with ψ near K and with ϕ outside a compact subset $W' \subset W$. Since ϑ is C^1-close to ϕ, it has 0 as a regular value and $\Sigma'' := \vartheta^{-1}(0)$ is the desired J-convex hypersurface. □

3.6. Functions with J-convex level sets

According to Lemma 2.7, a function ϕ with compact regular J-convex level sets can be made J-convex by composing it with a sufficiently convex function $f : \mathbb{R} \to \mathbb{R}$. Motivated by this, we now introduce a class of functions from which we can recover J-convex functions, but which gives us greater flexibility. This class of functions will be used throughout the remainder of this book.

Let (V, J) be a complex manifold. We call a continuous function $\phi : V \to \mathbb{R}$ a *function with J-convex level sets*, or *J-lc function* (where "lc" stands for "level-convex") if $g \circ \phi$ is J-convex for some smooth function $g : \mathbb{R} \to \mathbb{R}$ with $g' > 0$ and $g'' \geq 0$.

The following proposition shows that the main properties of J-convex functions carry over to J-lc functions.

PROPOSITION 3.32. *Let (V, J) be a complex manifold.*

(a) If $\phi : V \to \mathbb{R}$ is J-lc, then so is $h \circ \phi$ for every smooth function $h : \mathbb{R} \to \mathbb{R}$ with $h' > 0$.

(b) A proper C^2-function $\phi : V \to \mathbb{R}$ is J-lc if and only if it is J-convex near its critical points and its level sets are J-convex outside the critical points.

(c) Let $\phi_\lambda : V \to \mathbb{R}$, $\lambda \in \Lambda$ be a compact continuous family of J-lc functions such that $g_\lambda \circ \phi_\lambda$ are J-convex for a continuous family of convex functions $g_\lambda : \mathbb{R} \to \mathbb{R}$. Then $\max_{\lambda\in\Lambda}\phi_\lambda$ is J-lc.

(d) Every continuous J-lc function can be C^0-approximated by smooth J-lc functions.

(e) Let $\phi : V \to \mathbb{R}$ be a smooth J-lc function and $g_1, g_2 : \mathbb{R} \to \mathbb{R}$ be weakly convex functions such that $\psi_i = g_i \circ \phi$ are J-convex for $i = 1, 2$. Then $\nabla_{\psi_1}\psi_1 = h\nabla_{\psi_2}\psi_2$ for a positive function $h : V \to \mathbb{R}$.

Here we call a function $f : \mathbb{R} \to \mathbb{R}$ *convex* (resp. *weakly convex*) if f is an increasing diffeomorphism and $f'' > 0$ (resp. $f'' \geq 0$).

REMARK 3.33. For a J-lc function ϕ we will write $\nabla_\phi\phi$ for the gradient $\nabla_\psi\psi$ of some J-convex function $\psi = g\circ\phi$. In view of Proposition 3.32(e), this is well-defined up to multiplication by a positive function. In particular, we can unambiguously speak of the stable manifold of a critical point of a J-lc function.

The proof of Proposition 3.32 uses the following lemma.

LEMMA 3.34. *For every increasing diffeomorphism $g : \mathbb{R} \to \mathbb{R}$, there exists a smooth convex function $f : \mathbb{R} \to \mathbb{R}$ such that $f \circ g$ is convex. More generally, for every compact smooth family of increasing diffeomorphisms $g_\lambda : \mathbb{R} \to \mathbb{R}$, $\lambda \in \Lambda$, there exists a smooth family of convex functions $f_\lambda : \mathbb{R} \to \mathbb{R}$, $\lambda \in \Lambda$, such that the functions $f_\lambda \circ g_\lambda$, $\lambda \in \Lambda$, are all equal and convex.*

PROOF. We have

$$(f \circ g)' = f' \circ g \cdot g', \qquad (f \circ g)'' = f'' \circ g \cdot g'^2 + f' \circ g \cdot g''.$$

So $f \circ g$ is convex if and only if

$$f''(y) > h(y)f'(y), \qquad h(y) := \max\left\{0, -\frac{g'' \circ g^{-1}(y)}{\left(g' \circ g^{-1}(y)\right)^2}\right\}.$$

This inequality can obviously be solved by making h slightly larger and integrating $f'(y) := e^{\int_0^y h(x)dx}$ to obtain f.

In the case of a family $g_\lambda : \mathbb{R} \to \mathbb{R}$, $\lambda \in \Lambda$, fix some $\lambda_0 \in \Lambda$. We need to find a convex function f_{λ_0} such that $f_\lambda := f_{\lambda_0} \circ g_{\lambda_0} \circ g_\lambda^{-1}$ is convex for all λ. By the computation above, with $f := f_{\lambda_0}$ and $\widetilde{g}_\lambda := g_{\lambda_0} \circ g_\lambda^{-1}$ this is equivalent to

$$f''(y) > h(y)f'(y), \qquad h(y) := \max\left\{0, \max_{\lambda\in\Lambda}\frac{-\widetilde{g}''_\lambda \circ \widetilde{g}_\lambda^{-1}(y)}{\left(\widetilde{g}'_\lambda \circ \widetilde{g}_\lambda^{-1}(y)\right)^2}\right\},$$

which again has a smooth solution f. □

PROOF OF PROPOSITION 3.32. (a) Let ϕ be J-lc and $g : \mathbb{R} \to \mathbb{R}$ be a convex function such that $g \circ \phi$ is J-convex. Then $f \circ g \circ \phi = (f \circ g \circ h^{-1}) \circ (h \circ \phi)$ is J-convex for every convex function $f : \mathbb{R} \to \mathbb{R}$. By Lemma 3.34, we find f such that $f \circ g \circ h^{-1}$ is convex, which shows that $h \circ \phi$ is J-lc.

(b) Assume first that ϕ is J-convex near its critical points and its level sets are J-convex outside the critical points. Note that the same properties then hold for $g \circ \phi$ for any smooth function $g : \mathbb{R} \to \mathbb{R}$ with $g' > 0$: J-convexity of the level sets is clearly preserved, and at a critical point we have $dd^{\mathbb{C}}(g \circ \phi) = f' \circ \phi \cdot dd^{\mathbb{C}}\phi$, which shows J-convexity of $g \circ \phi$ near critical points. Now by Lemma 2.7 (applied outside a neighborhood of the critical points), we find g such that $g'' > 0$ and $g \circ \phi$ is J-convex.

Conversely, let ϕ be J-lc and $g : \mathbb{R} \to \mathbb{R}$ be a convex function such that $g \circ \phi$ is J-convex. Then by the preceding observation, $\phi = g^{-1} \circ (g \circ \phi)$ is J-convex near its critical points and its level sets are J-convex outside the critical points.

(c) Let $\phi_\lambda : V \to \mathbb{R}$ and $g_\lambda : \mathbb{R} \to \mathbb{R}$ be as in the proposition such that $g_\lambda \circ \phi_\lambda$ are J-convex. By Lemma 3.34, there exists a smooth family of convex functions $f_\lambda : \mathbb{R} \to \mathbb{R}$ such that the functions $f_\lambda \circ g_\lambda$ are all equal to the same convex function g. Thus $g \circ \phi_\lambda = f_\lambda \circ g_\lambda \circ \phi_\lambda$ is J-convex for all λ, and by Proposition 3.8 so is $\max_\lambda g \circ \phi_\lambda$. Then (a) implies that $g^{-1} \circ \max_\lambda g \circ \phi_\lambda = \max_\lambda \phi_\lambda$ is J-lc.

(d) Let ϕ be J-lc and $g : \mathbb{R} \to \mathbb{R}$ be a smooth convex function such that $g \circ \phi$ is J-convex. Let smooth$(g \circ \phi)$ be a J-convex smoothing as in Section 3.2. Then by (a), a smooth J-lc function approximating ϕ is given by $g^{-1}\text{smooth}(g \circ \phi)$.

(e) This follows from the proof of Proposition 2.11, writing $\psi_1 = (g_1 \circ g_2^{-1}) \circ \psi_2$. □

3.7. Normalized modulus of J-convexity

Consider a Kähler manifold (V, J, ω). In this subsection we will derive conditions on the modulus of J-convexity (of a function or a hypersurface) that ensure K-convexity for complex structures K that are C^2-close to J.

Given a quadratic form Q and a metric on a vector space we define

$$M(Q) := \max_{||T||=1} |Q(T)|, \qquad m(Q) := \min_{||T||=1} Q(T),$$

where we will consider the second quantity only if Q is positive definite.

We begin with the case of a smooth J-convex function $\phi : V \to \mathbb{R}$. Recall from Proposition 2.5 that the Hermitian form $H_\phi = -dd^{\mathbb{C}}\phi$ is related to the real Hessian Hess_ϕ by

$$H_\phi(X) = \text{Hess}_\phi(X) + \text{Hess}_\phi(JX). \tag{3.9}$$

Recall also that $m(H_\phi) : V \to \mathbb{R}_+$ is the modulus of J-convexity. We define the *normalized modulus of J-convexity* by

$$\mu(\phi) := \frac{m(H_\phi)}{\max\{M(\text{Hess}_\phi), |\nabla\phi|\}} : V \to \mathbb{R}_+. \tag{3.10}$$

Next we consider a J-convex hypersurface Σ in a Kähler manifold. Recall that its normalized Levi form is related to the second fundamental form by $\mathbb{L}_\Sigma(X) = II_\Sigma(X) + II_\Sigma(JX)$. We call the ratio

$$\mu(\Sigma) := \frac{m(\mathbb{L}_\Sigma)}{\max\{M(II_\Sigma), 1\}} : \Sigma \to \mathbb{R}_+ \tag{3.11}$$

the *normalized modulus of J-convexity* of the hypersurface Σ.

In the following we will sometimes write $M(\phi)$, $M(\Sigma)$, $m(\phi)$, and $m(\Sigma)$ instead of $M(\mathrm{Hess}_\phi)$, $M(II_\Sigma)$, $m(H_\phi)$, and $m(\mathbb{L}_\Sigma)$, respectively.

The main motivation for these definitions is the fact that a lower bound on $\mu(\phi)$ ensures J-convexity of $\phi\circ f$ for diffeomorphisms $f:V\to V$ sufficiently C^2-close to the identity:

LEMMA 3.35. *(a) Let ϕ be a function on a Kähler manifold (V,J,ω) with $\mu(\phi)\ge\varepsilon>0$. Then $m(\phi\circ f)\ge m(\phi)/2$ (in particular, $\phi\circ f$ is J-convex) for any diffeomorphism $f:V\to V$ with $\|f-\mathrm{Id}\|_{C^2(V)}\le\varepsilon/20$.*

(b) Let Σ be a cooriented hypersurface in a Kähler manifold (V,J,ω) with $\mu(\Sigma)\ge\varepsilon>0$. Then $m\big(f(\Sigma)\big)\ge m(\Sigma)/2$ (in particular, $f(\Sigma)$ is J-convex) for any diffeomorphism $f:V\to V$ with $\|f-\mathrm{Id}\|_{C^2(V)}\le\varepsilon/20$.

PROOF. (a) By the chain rule, the Hessian Hess_ϕ of the function ϕ changes under composition with f by

$$\begin{aligned}\mathrm{Hess}_{\phi\circ f}-\mathrm{Hess}_\phi &= Df\cdot\mathrm{Hess}_\phi\cdot Df^t+D^2f\cdot\nabla\phi-\mathrm{Hess}_\phi\\ &= Df\cdot\mathrm{Hess}_\phi\cdot(Df-\mathrm{Id})^t+(Df-\mathrm{Id})\cdot\mathrm{Hess}_\phi+D^2f\cdot\nabla\phi.\end{aligned}$$

For $\delta:=\|f-\mathrm{Id}\|_{C^2}\le 1/2$, this implies the estimate

$$\|\mathrm{Hess}_{\phi\circ f}-\mathrm{Hess}_\phi\|\le 5\delta\max\{M(\mathrm{Hess}_\phi),|\nabla\phi|\}.$$

For $|X|=1$ and $\delta\le\mu(\phi)/20$, using (3.9) we obtain

$$\begin{aligned}H_{\phi\circ f}(X)&\ge H_\phi(X)-10\delta\max\{M(\mathrm{Hess}_\phi),|\nabla\phi|\}\\ &\ge m(H_\phi)\left(1-\frac{10\delta}{\mu(\phi)}\right)\ge\frac12 m(H_\phi).\end{aligned}$$

(b) Let us first compute how the second fundamental form at a point p on a hypersurface Σ in $\mathbb{R}^m$ (with respect to the standard Euclidean metric) changes under a diffeomorphism $f:\mathbb{R}^m\to\mathbb{R}^m$. After applying a rigid motion, we may assume that $p=0$ and Σ is the graph $y=g(x)$ of a function $g:\mathbb{R}^{m-1}\to\mathbb{R}$ with $g(0)=0$ and $dg(0)=0$. After composing f with a rigid motion, we may assume that $\widetilde\Sigma:=f(\Sigma)$ is again the graph $\widetilde y=\widetilde g(\widetilde x)$ of a function $\widetilde g:\mathbb{R}^{m-1}\to\mathbb{R}$ with $\widetilde g(0)=0$ and $d\widetilde g(0)=0$. Writing $\widetilde z=(\widetilde x,\widetilde y)=f(x,y)=f(z)$, the last condition is equivalent to $\widetilde x_j(0)=\widetilde y(0)=\frac{\partial\widetilde y}{\partial x_i}(0)=0$. So the Taylor expansion of f at the origin is

$$\begin{aligned}\widetilde y&=by+\frac12\langle x,Cx\rangle+O(y|z|+|x|^3),\\ \widetilde x&=Bx+\frac{\partial\widetilde x}{\partial y}y+O(|z|^2),\end{aligned}$$

where

$$B=\left(\frac{\partial\widetilde x_j}{\partial x_i}\right),\qquad b=\frac{\partial\widetilde y}{\partial y},\qquad C=\left(\frac{\partial\widetilde y}{\partial x_i\partial x_j}\right).$$

The equation $y=g(x)=\frac12\langle x,Ax\rangle+O(|x|^3)$ for Σ is thus equivalent to

$$\begin{aligned}\widetilde y&=\frac12 b\langle x,Ax\rangle+\frac12\langle x,Cx\rangle+O(y|z|+|x|^3)\\ &=\frac12 b\langle B^{-1}\widetilde x,AB^{-1}\widetilde x\rangle+\frac12\langle B^{-1}\widetilde x,CB^{-1}\widetilde x\rangle+O(\widetilde y|\widetilde z|+|\widetilde x|^3),\end{aligned}$$

and therefore the equation for $\widetilde{\Sigma} = f(\Sigma)$ is $\widetilde{y} = \widetilde{g}(\widetilde{x}) = \frac{1}{2}\langle \widetilde{x}, \widetilde{A}\widetilde{x}\rangle + O(|\widetilde{x}|^3)$ with

$$\widetilde{A} = b(B^{-1})^t AB^{-1} + (B^{-1})^t CB^{-1}.$$

Since $A, \widetilde{A}$ are the matrices of the second fundamental forms of $\Sigma, \widetilde{\Sigma}$ at the origin, this yields as in (a) the estimate

$$\| II_{f(\Sigma)} - II_\Sigma \| = \|\widetilde{A} - A\| \leq 5\delta \max\{M(II_\Sigma), 1\}.$$

The estimate $m\bigl(f(\Sigma)\bigr) \geq m(\Sigma)/2$ if $\|f - \mathrm{Id}\|_{C^2(V)} \leq \mu(\Sigma)/20$ follows from this as in (a). □

The preceding lemma implies persistence of J-convexity under perturbations of the complex structure in view of

LEMMA 3.36. *There exist constants C_n depending only on $n \in \mathbb{N}$ with the following property. If J, K are two integrable complex structures on the unit ball $B \subset \mathbb{C}^n$, then there exists a biholomorphism $h : (U, J) \to (V, K)$ between neighborhoods of 0 fixing 0 and satisfying*

$$\|h - \mathrm{Id}\|_{C^2(U)} \leq C_n \|J - K\|_{C^2(B)}.$$

PROOF. Suppose $\|J - K\|_{C^2(B)} = \delta$. There exist linear isomorphisms S, T with $\|S - T\| \leq C_n\delta$, for a constant C_n depending only on n, such that $S^*J = T^*K = i$ at 0. The proof of the Newlander–Nirenberg Theorem 5.7.4 in [**103**] yields a biholomorphism $f : (U, S^*J) \to (U', i)$ between neighborhoods of 0, fixing 0, with a bound $\|f - \mathrm{Id}\|_{C^2(U)} \leq C\|J - K\|_{C^2(B)}$. Let $g : (V, T^*K) \to (V', i)$ be the corresponding map for T^*K. Then $h := T \circ g^{-1} \circ f \circ S^{-1}$ (between suitable neighborhoods) has the desired properties. □

COROLLARY 3.37. *Let ϕ be a function on an n-dimensional Kähler manifold (V, J, ω) with $\mu(\phi) \geq \varepsilon > 0$. Then the moduli of convexity (measured with respect to the same reference metric) satisfy $m(\phi, K) \geq m(\phi, J)/4 > 0$ (in particular ϕ is K-convex) for any complex structure K on V with $\|J - K\|_{C^2(V)} \leq \frac{\varepsilon}{20\,C_n}$, where C_n is the constant from Lemma 3.36.*

PROOF. Let $\delta := \|J - K\|_{C^2(V)}$. By Lemma 3.36, in a neighborhood of any point $p \in \Sigma$ there exists a biholomorphism $h : (\mathcal{O}p\, p, J) \to (\mathcal{O}p\, p, K)$ fixing p with $\|h - \mathrm{Id}\|_{C^2} \leq C_n\delta$. Then $m(\phi, K) = m(\phi \circ h, J)$ if measured with respect to metrics g and h^*g, and $m(\phi, K) \geq \frac{1}{2}m(\phi \circ h, J)$ if both are measured with respect to the same reference metric and δ is sufficiently small. On the other hand, by Lemma 3.35 we have $m(\phi \circ h, J) \geq \frac{1}{2}m(\phi, J)$ if $\|h - \mathrm{Id}\| \leq C_n\delta \leq \varepsilon/20$, from which the corollary follows. □

The following result relates the moduli of convexity of functions and hypersurfaces.

PROPOSITION 3.38. *Let $\Sigma \subset \mathbb{C}^n$ be a compact J-convex hypersurface (possibly with boundary). Then there exists a J-convex function $\psi : \mathcal{O}p\,\Sigma \to \mathbb{R}$ such that $\Sigma \subset \{\psi = 0\}$, and at every point of Σ we have*

$$|\nabla\psi| = 1, \qquad m(\psi) \geq \frac{m(\Sigma)}{2}, \qquad M(\psi) \leq \frac{6M(\Sigma)^2}{m(\Sigma)}, \qquad \mu(\psi) \geq \frac{\mu(\Sigma)^2}{12}.$$

The proof is based on the following linear algebra lemma. Consider $\mathbb{C}^n \cong \mathbb{C}^{n-1}\oplus\mathbb{R}\oplus i\mathbb{R}$ with coordinates $(z, u+iv)$. For a quadratic form $Q : \mathbb{C}^{n-1}\times\mathbb{R}\to\mathbb{R}$ define $Q^{\mathbb{C}} : \mathbb{C}^{n-1}\to\mathbb{R}$, $Q^{\mathbb{C}}(z) = Q(z) + Q(iz)$, and for $\lambda > 0$ define

$$Q_\lambda : \mathbb{C}^n \to \mathbb{R}, \qquad (z, u+iv) \mapsto Q(z,u) + \lambda v^2.$$

LEMMA 3.39. *Suppose $Q^{\mathbb{C}}$ is positive definite. Then there exists $\lambda > 0$ such that*

$$m(Q_\lambda^{\mathbb{C}}) \geq \frac{1}{2}m(Q^{\mathbb{C}}), \qquad M(Q_\lambda) \leq \frac{6M(Q)^2}{m(Q^{\mathbb{C}})}.$$

PROOF. Set $M := M(Q)$ and $m := m(Q^{\mathbb{C}})$. Then

$$|Q_\lambda(z,u+iv)| \leq |Q(z,u)| + \lambda v^2 \leq \max\{M,\lambda\}(|z|^2+u^2+v^2),$$

and hence $M(Q_\lambda) \leq \max\{M,\lambda\}$. To estimate $Q_\lambda^{\mathbb{C}}$ we write $Q(z,u) = P(z) + \ell(z)u + au^2$ for a linear form $\ell : \mathbb{C}^{n-1}\to\mathbb{R}$ and $a\in\mathbb{R}$. By definition of M we have $|a| \leq M$ and $\ell(z)u| \leq 2M|z|\,|u|$. Thus

$$\begin{aligned} Q_\lambda^{\mathbb{C}}(z,u+iv) &= Q_\lambda(z,u+iv) + Q_\lambda(iz,-v+iu) \\ &= P(z) + P(iz) + (a+\lambda)(u^2+v^2) + \ell(z)u - \ell(iz)v \\ &\geq m|z|^2 + (\lambda - M)(u^2+v^2) - 2M|z|(|u|+|v|). \end{aligned}$$

The last line is $\geq \frac{m}{2}(|z|^2+u^2+v^2)$ if and only if

$$\frac{m}{2}|z|^2 + (\lambda - M - \frac{m}{2})(u^2+v^2) - 2M|z|(|u|+|v|) \geq 0,$$

which is easily seen to be the case for $\lambda = 6M^2/m$. As noted above, for this choice of λ we have $M(Q_\lambda) \leq 6M^2/m$ (recall that $m \leq M$ by definition). □

PROOF OF PROPOSITION 3.38. Consider the metric decomposition $U = \Sigma\times[-\varepsilon,\varepsilon]$ of a tubular neighborhood $U \supset \Sigma$, so that the coordinate t corresponding to the second factor is the Euclidean distance from a point to $\Sigma = \{t=0\}$. We assume that Σ is cooriented by $-\partial_t$. We will find the desired function ψ of the form

$$\psi(x,t) = -t + \frac{1}{2}\lambda(x)t^2$$

for a suitable function $\lambda : \Sigma\to\mathbb{R}_+$. Then clearly $|\nabla\psi| = 1$ along Σ. Let us compute explicitly the function ψ. Take a point $p\in\Sigma$ and choose unitary coordinates $(x_1+ix_2,\dots,x_{2n-3}+ix_{2n-2},x_{2n-1}+iy)$ centered at p such that the hypersurface Σ is given by an equation $y = f(x)$, $x = (x_1,\dots,x_{2n-1})$, and $df(0) = 0$. Because the computation involves only the 2-jet of f we can assume that f is a quadratic form. Let us introduce new coordinates $u = (u_1,\dots,u_{2n-1}), t$, where t is the signed distance to Σ (assuming that Σ is cooriented by $\frac{\partial}{\partial y}$) and u is the x-coordinate of the projection of the point to Σ. The coordinates are related by the formula

$$(x,y) = (u,f(u)) + \frac{t}{\sqrt{1+|\nabla f(u)|^2}}(-\nabla f(u),1)$$

(see Figure 3.5), i.e.,

$$x = u - \frac{t}{\sqrt{1+|\nabla f(u)|^2}}\nabla f(u), \qquad y = f(u) + \frac{t}{\sqrt{1+|\nabla f(u)|^2}}.$$

The implicit function formulas for the first and second derivatives of t with respect to x and y involve only first and second derivatives of the right hand side

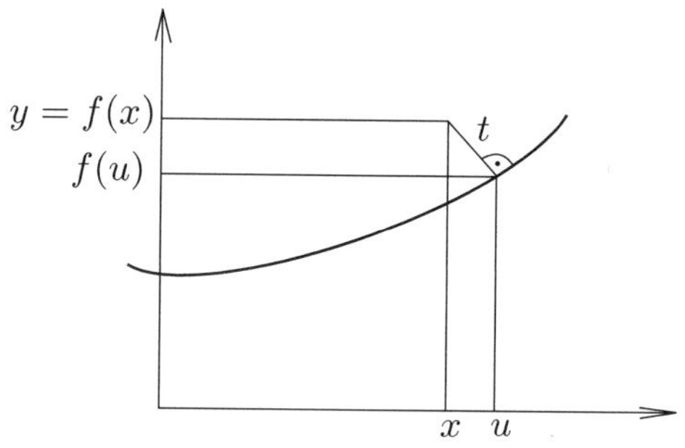

FIGURE 3.5. The coordinate change from (x, y) to (u, t).

with respect to (x, y, u, t), and hence to compute the derivatives at the origin we can ignore all the terms of higher degree than 2 with respect to these variables. We will continue the computation systematically dropping terms of higher order than 2. Thus, writing $f(u) = \frac{1}{2}\langle u, Au\rangle$, we have

$$x = u - t\nabla f(u) = u - tAu, \qquad y = f(u) + t = \frac{1}{2}\langle u, Au\rangle + t.$$

Solving for t and u (and again ignoring higher order terms), we get $u = (1 + tA)x$ and $t = y - f(x)$, hence

$$\psi(x, y) = -y + f(x) + \frac{1}{2}\lambda(p)y^2.$$

Note that $Q = 2f$ is the second fundamental form of Σ and $Q(x) + \lambda(p)y^2$ is the Hessian of ψ at p. So the estimates for $m(\psi)$ and $M(\psi)$ follow from the preceding lemma, and they easily imply the estimate for $\mu(\psi)$ (distinguish the cases $M \geq 1$ and $M < 1$). This proves Proposition 3.38. $\square$

For a Kähler manifold (V, J, ω) and a continuous function $\phi : V \to \mathbb{R}$, we introduce the quantity

$$m(\phi; \varepsilon) := \inf_{\|f - \mathrm{Id}\|_{C^2} \leq \varepsilon} m(\phi \circ f).$$

LEMMA 3.40. *(a) For a continuous family ϕ_λ, $\lambda \in \Lambda$, over a compact parameter space Λ, we have*

$$m(\max{}_{\lambda\in\Lambda}\phi_\lambda; \varepsilon) \geq \min{}_{\lambda\in\Lambda} m(\phi_\lambda; \varepsilon).$$

(b) If $m(\phi; \varepsilon) > 0$, the function ϕ can be smoothed to a function $\widetilde{\phi}$ satisfying

$$m(\widetilde{\phi}; \varepsilon/2) \geq m(\phi; \varepsilon)/2.$$

PROOF. (a) By Proposition 3.8 we have $m(\max{}_\lambda\phi_\lambda) \geq \min{}_\lambda m(\phi_\lambda)$. Using this and $(\max{}_\lambda\phi_\lambda) \circ f = \max{}_\lambda(\phi_\lambda \circ f)$ we deduce

$$m(\max{}_\lambda\phi_\lambda; \varepsilon) = \inf_f m\big(\max{}_\lambda(\phi_\lambda \circ f)\big) \geq \inf_f \min{}_\lambda m(\phi_\lambda \circ f) = \min{}_\lambda m(\phi_\lambda; \varepsilon).$$

(b) Consider first the smoothing ϕ_δ of a function $\phi : \mathbb{C}^n \to \mathbb{R}$ defined by the convolution formula (3.6). For a diffeomorphism f with $\|f - \mathrm{Id}\| \leq \varepsilon/2$ and $y \in \mathbb{C}^n$

with $|y| \leq \delta \leq \varepsilon/2$, the function $f_y(x) := f(x) - y$ satisfies $\|f_y - \mathrm{Id}\| \leq \varepsilon$, and we obtain

$$\begin{aligned}\frac{1}{2\pi}\int_0^{2\pi} \phi_\delta \circ f(z + w\, e^{i\theta}) d\theta &= \int_{\mathbb{C}^n} \frac{1}{2\pi}\int_0^{2\pi} \phi \circ f_y(z + w\, e^{i\theta}) d\theta \rho_\delta(y) dy \\ &\geq \int_{\mathbb{C}^n} \Big[\phi \circ f_y(z) + m(\phi;\varepsilon)(z)|w|^2\Big] \rho_\delta(y) dy \\ &= \phi_\delta \circ f(z) + m(\phi;\varepsilon)(z)|w|^2.\end{aligned}$$

This shows that $m(\phi_\delta; \varepsilon/2) \geq m(\phi;\varepsilon)$ for $\delta \leq \varepsilon/2$, from which the result for the smoothing $\widetilde{\phi}$ easily follows. □

Now we can prove the main result of this section.

PROPOSITION 3.41. *There exist constants c_n depending only on the dimension n with the following property. Let M be a compact manifold (possibly with boundary) of dimension $2n-1$ and (J,ω) a Kähler structure on $V = M \times \mathbb{R}$. Let $g_\lambda : M \to \mathbb{R}$, $\lambda \in \Lambda$, be a C^2-family of functions, parametrized over a compact manifold (possibly with boundary) Λ. Denote by Σ_λ the graph of g_λ, cooriented from below. Suppose that the normalized moduli of convexity with respect to (J,ω) satisfy $\mu(\Sigma_\lambda) \geq \varepsilon > 0$ for all $\lambda \in \Lambda$. Then there exists a smoothing g of the function $\min_{\lambda\in\Lambda} g_\lambda$ whose graph Σ is K-convex for every complex structure K on V with $\|K - J\|_{C^2} \leq c_n\varepsilon^2$.*

PROOF. In the proof, all moduli of convexity are with respect to J.

By Proposition 3.38 there exists a smooth family of J-convex functions $\psi_\lambda : \mathcal{O}p\,\Sigma_\lambda \to \mathbb{R}$ such that $\Sigma_\lambda = \{\psi_\lambda = 0\}$ and along Σ_λ we have $|\nabla\psi_\lambda| = 1$ and $\mu(\psi_\lambda) \geq \varepsilon^2/12$.

By Lemma 3.35, we have $m(\psi_\lambda; \varepsilon^2/240) \geq \varepsilon^2/24$ for all λ. By Lemma 3.40, the function $\max_\lambda \psi_\lambda$ can be smoothed to a function ψ satisfying $m(\psi;\varepsilon/480) \geq \varepsilon^2/48$. Thus it follows from Lemma 3.36 and the definition of $m(\psi;\varepsilon)$ that ψ is K-convex for every complex structure K with $\|K - J\|_{C^2} \leq \frac{\varepsilon^2}{480\, C_n}$.

By Propositions 3.18 and 3.19, we have $\partial_r \cdot \psi > 0$. Hence $\Sigma = \psi^{-1}(0)$ is the graph of a smooth function $g : M \to \mathbb{R}$, and it is K-convex for every complex structure K with $\|K - J\|_{C^2} \leq c_n\varepsilon^2$, where $c_n := \frac{1}{480\, C_n}$ and C_n is the constant from Lemma 3.36. □

CHAPTER 4

Shapes for i-Convex Hypersurfaces

4.1. Main models

A crucial ingredient in the proof of the Existence Theorem 1.5 is the bending of a J-convex hypersurface such that it "surrounds" the core disc of a handle as shown in Figure 8.1. The main goal of this chapter is the proof of the following two theorems which assert the existence of the necessary models in $\mathbb{C}^n$.

Let us fix integers $1 \leq k \leq n$. Viewing $\mathbb{C}^n$ as a real vector space with coordinates $(x_1, \dots, x_n, y_1, \dots, y_n)$, let us define

$$R := \sqrt{\sum_{j=1}^{k} y_j^2}, \qquad r := \sqrt{\sum_{j=1}^{n} x_j^2 + \sum_{j=k+1}^{n} y_j^2}.$$

THEOREM 4.1. *For any $a > 1$ and $\gamma \in (0,1)$ there exists an i-convex hypersurface $\Sigma \subset \mathbb{C}^n$ with the following properties (see Figure 4.1):*

(i) *Σ is given by an equation $\Psi(r,R) = -1$, and cooriented by $\nabla\Psi$, for a function $\Psi(r,R)$ satisfying $\frac{\partial\Psi}{\partial r} > 0$ and $\frac{\partial\Psi}{\partial R} \leq 0$;*

(ii) *in the domain $\{r \geq \gamma\}$ the hypersurface Σ coincides with $\{ar^2 - R^2 = -1\}$;*

(iii) *in the domain $\{R \leq 1\}$ the hypersurface Σ coincides with $\{r = \delta\}$ for some $\delta \in (0,\gamma)$;*

(iv) *Σ is J-convex for any complex structure J which satisfies the estimate*

$$\|J - i\|_{C^2} \leq c(a,n)\gamma^{12}, \tag{4.1}$$

where $c(a,n)$ is a positive constant depending only on a and the dimension n.

The hypersurface Σ divides $\mathbb{C}^n$ into two domains:

$$\mathbb{C}^n = \Omega^{\mathrm{ext}} \cup \Omega^{\mathrm{int}}, \qquad \Omega^{\mathrm{ext}} \cap \Omega^{\mathrm{int}} = \Sigma,$$

where Ω^{int} is the domain which contains the subspace $\{r = 0\}$ which will later correspond to the core disc of a handle. The second theorem provides an i-lc function Ψ on the exterior domain Ω^{ext} which agrees with the standard function $\Psi_{\mathrm{st}}(r,R) = ar^2 - R^2$ for $r \geq \gamma$ and has Σ as a level set. Taking the maximum of a given i-lc function with Ψ and extending it by Ψ_{st} for $r \geq \gamma$ will later allow us to implant given functions near $\{r = 0\}$ into a complex manifold and thus deform J-lc functions near totally real discs. A first example of this construction appears in the proof of Corollary 4.4 below.

THEOREM 4.2. *For any $a > 1$ and $\gamma \in (0,1)$ there exists an i-lc function $\Psi : \Omega^{\mathrm{ext}} \to \mathbb{R}$ with the following properties (see Figure 4.1):*

(i) *Ψ is of the form $\Psi(r,R)$ with $\frac{\partial\Psi}{\partial r} > 0$ and $\frac{\partial\Psi}{\partial R} \leq 0$;*

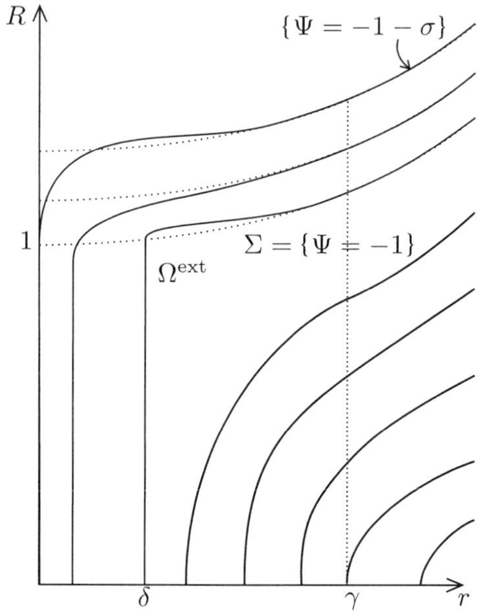

FIGURE 4.1. The hypersurface Σ and the function Ψ.

(ii) $\Psi(r,R) = ar^2 - R^2$ *in* $\Omega^{\text{ext}} \cap \{r \geq \gamma\}$;
(iii) $\Psi \equiv -1$ *on* Σ;
(iv) Ψ *is* J*-lc for any complex structure* J *which satisfies estimate* (4.1) *from Theorem* 4.1.

The proof of these two theorems will occupy the remainder of this chapter. Properties (i)–(iii) of Theorem 4.1 will be proved at the end of Section 4.5, properties (i)–(iii) of Theorem 4.2 at the end of Section 4.6, and property (iv) for both theorems at the end of Section 4.7.

Let us formulate two corollaries of Theorem 4.2 that will be useful in later chapters.

COROLLARY 4.3. *For* $a > 1$, $\gamma \in (0,1)$ *and* J *as in Theorem* 4.1 *and any sufficiently small* $\sigma > 0$ *there exists an open subset* $\Omega \subset \mathbb{C}^n$ *and a* J*-lc function* $\Psi : \Omega \to (-1-\sigma, \infty)$ *with the following properties (see Figure* 4.1*):*
(i) Ψ *is of the form* $\Psi(r,R)$ *with* $\frac{\partial \Psi}{\partial r} > 0$ *and* $\frac{\partial \Psi}{\partial R} \leq 0$;
(ii) $\Psi(r,R) = ar^2 - R^2$ *on* $\Omega \cap \{r \geq \gamma\}$;
(iii) *there exists a diffeomorphism* $f : (-1-\sigma, -1] \to (0, \delta]$ *such that* $f \circ \Psi(r,R) = r$ *on the set* $\{r \leq \delta,\ R \leq 1\}$.

PROOF. Let us fix $a > 1$ and $\gamma \in (0,1)$. We apply Theorem 4.1 with parameters a and $t\gamma$, $t \in (0,1]$. The corresponding hypersurfaces Σ_t can be chosen to depend smoothly on t so that on $\{R \leq 1\}$ they coincide with $\{r = \rho(t)\}$ for a diffeomorphism $\rho : (0,1] \to (0,\delta]$. We perturb the Σ_t such that on $\{r \geq \gamma\}$ they coincide with $\{ar^2 - R^2 = -1 - (1-t)\sigma\}$. Using Proposition 3.25, we can then modify the Σ_t on the set $\{r \leq \gamma,\ R \geq 1\}$ to a foliation. Now define $\Psi|_{\Sigma_t} := -1 - (1-t)\sigma$ and extend it over the domain Ω^{ext} bounded by $\Sigma = \Sigma_1$ by the function in Theorem 4.2. □

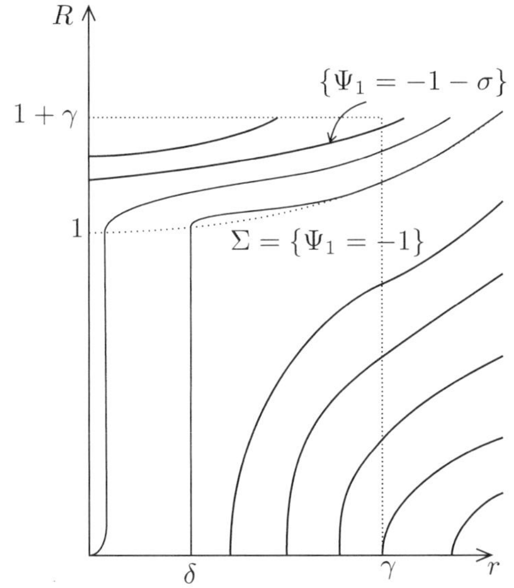

FIGURE 4.2. The function Ψ_1.

COROLLARY 4.4. *For $a > 1$, $\gamma \in (0,1)$, and J as in Theorem 4.1 there exists a smooth family of J-lc functions $\Psi_t : \mathbb{C}^n \to \mathbb{R}$, $t \in [0,1]$, with the following properties (see Figure 4.2):*

(i) *Ψ_t is of the form $\Psi_t(r,R)$ with $\frac{\partial \Psi_t}{\partial r} > 0$ and $\frac{\partial \Psi_t}{\partial R} \leq 0$;*
(ii) *$\Psi_0(r,R) = ar^2 - R^2$, and $\Psi_t(r,R) = ar^2 - R^2$ on $\{r \geq \gamma\} \cup \{R \geq 1+\gamma\}$;*
(iii) *Ψ_t is target equivalent to $ar^2 - R^2$ near $\{r = 0\}$;*
(iv) *$\Psi_1 \equiv -1$ on Σ from Theorem 4.1.*

PROOF. Let $\Psi : \mathbb{C}^n \supset \Omega \to (-1-\sigma, \infty)$ be the function from Corollary 4.3. After repeating the construction for smaller γ, we may assume that $\Omega \cap \{r \leq \gamma\} \subset \{R < 1+\gamma\}$. In particular, we then have $1+\sigma < (1+\gamma)^2$. Set $\Psi_{\rm st}(r,R) := ar^2 - R^2$ and note that $\max_{\{r\leq\gamma,\, R\leq 1+\gamma\}} \Psi_{\rm st} = a\gamma^2$ and $\min_{\{r\leq\gamma,\, R\leq 1+\gamma\}} \Psi_{\rm st} = -(1+\gamma)^2$. Pick numbers $-1-\sigma < b < c < -1$ and smooth increasing functions $g, h : \mathbb{R} \to \mathbb{R}$ with the following properties:

- $g(x) \leq x$, $g(x) = x$ for $x \leq c$ and $g(a\gamma^2) < -1$;
- $h(x) \leq x$, $h(x) = x$ for $x \geq b$ and $h(-1-\sigma) < -(1+\gamma)^2$.

Then the function $\Psi_1 := \text{smooth}\max(g \circ \Psi_{\rm st}, h \circ \Psi)$ has the desired properties for $t = 1$. Now the homotopy $\text{smooth}\max(ar^2 - R^2, \Psi_1 + t)$ connects $ar^2 - R^2$ for very negative t to $\text{smooth}\max(ar^2 - R^2, \Psi_1)$ at $t = 0$, and the homotopy $\text{smooth}\max(ar^2 - R^2 - t, \Psi_1)$ connects the latter to Ψ_1 for large t. □

4.2. Shapes for i-convex hypersurfaces

We now derive the conditions under which a "shape function" $R = \phi(r)$ defines an i-convex hypersurface in $\mathbb{C}^n$. In this and the following section we first consider the *critical* case $k = n$. In Section 4.4 we will see that the same shapes also work for the *subcritical* case $k < n$.

Consider the map

$$\pi : \mathbb{C}^n \to \mathbb{R}^2, \qquad z \mapsto (r, R) := (|x|, |y|)$$

for $z = x + iy$, $x, y \in \mathbb{R}^n$. The image of the map π is the quadrant

$$Q := \{(r, R) \mid r, R \geq 0\} \subset \mathbb{R}^2.$$

A curve $C \subset Q$ defines a hypersurface $\Sigma := \pi^{-1}(C)$ in $\mathbb{C}^n$. We call C the *shape* of Σ. Our goal in this section is to determine conditions on C which guarantee i-convexity of Σ.

As a preliminary, let us compute the second fundamental form of a surface of revolution. Consider $\mathbb{R}^k \oplus \mathbb{R}^l$ with coordinates (x, y) and $\mathbb{R}^k \oplus \mathbb{R}$ with coordinates $(x, R = |y|)$. To a function $\Phi : \mathbb{R}^k \oplus \mathbb{R} \to \mathbb{R}$ we associate the *surface of revolution*

$$\Sigma_\Phi := \{(x, y) \in \mathbb{R}^k \oplus \mathbb{R}^l \mid \Phi(x, |y|) = 0\}.$$

We coorient Σ_Φ by the gradient $\nabla\Phi$ of Φ (with respect to all variables). Denote by $\Phi_R = \frac{\partial \Phi}{\partial R}$ the partial derivative.

LEMMA 4.5. *At every $z = (x, y) \in \Sigma_\Phi$ the splitting*

$$T_z\Sigma_\Phi = \Bigl(T_z\Sigma_\Phi \cap (\mathbb{R}^k \oplus \mathbb{R}y)\Bigr) \oplus \Bigl(T_z\Sigma_\Phi \cap (\mathbb{R}^k \oplus \mathbb{R}y)^\perp\Bigr)$$

is orthogonal with respect to the second fundamental form II. The second subspace is an eigenspace of II with eigenvalue $\Phi_R/|\nabla\Phi|R$.

PROOF. The unit normal vector to Σ_Φ at $z = (x, y)$ is

$$\nu(z) = \frac{1}{|\nabla\Phi|}(\nabla_x\Phi, \frac{\Phi_R}{R}y),$$

where $\nabla_x\Phi$ denotes the gradient with respect to the x-variables. For $Y \perp y$ we get

$$D\nu(z) \cdot (0, Y) = \frac{1}{|\nabla\Phi|}(0, \frac{\Phi_R}{R}Y) + \mu\nu$$

for some $\mu \in \mathbb{R}$. From $\langle \nu(z), D\nu(z) \cdot (0, Y)\rangle = 0$ we deduce $\mu = 0$, so $T_z\Sigma_\Phi \cap (\mathbb{R}^k \oplus \mathbb{R}y)^\perp$ is an eigenspace of II with eigenvalue $\Phi_R/|\nabla\Phi|R$. From this it follows that

$$II\Bigl((0, Y), (X, \lambda y)\Bigr) = \langle D\nu \cdot (0, Y), (X, \lambda y)\rangle = 0$$

for $(X, \lambda y) \in T_z\Sigma_\Phi \cap (\mathbb{R}^k \oplus \mathbb{R}y)$. □

Reduction to the case $n = 2$. Now let $C \subset Q$ be a curve. At a point $z = x + iy \in \Sigma = \pi^{-1}(C)$ consider the subspace $\Lambda_{xy} \subset \mathbb{R}^n$ generated by the vectors $x, y \in \mathbb{R}^n$ and its complexification

$$\Lambda_{xy}^{\mathbb{C}} := \Lambda_{xy} + i\Lambda_{xy}.$$

Let $\Lambda_\perp$ be the orthogonal complement of Λ_{xy} in $\mathbb{R}^n$ and $\Lambda_\perp^{\mathbb{C}}$ its complexification (which is the orthogonal complement of $\Lambda_{xy}^{\mathbb{C}}$ in $\mathbb{C}^n$). Note that $\Lambda_\perp^{\mathbb{C}}$ is contained in $T_z\Sigma$ and thus in the maximal complex subspace ξ_z. So the maximal complex subspace splits into the orthogonal sum (with respect to the metric)

$$\xi_z = \widetilde{\Lambda} \oplus \Lambda_\perp^{\mathbb{C}} = \widetilde{\Lambda} \oplus \Lambda_\perp \oplus i\Lambda_\perp, \tag{4.2}$$

where $\widetilde{\Lambda} = \xi_z \cap \Lambda_{xy}^{\mathbb{C}}$.

LEMMA 4.6. *The splitting* (4.2) *is orthogonal with respect to the second fundamental form II, and $\Lambda_\perp$ and $i\Lambda_\perp$ are eigenspaces with eigenvalues $\lambda_r = \Phi_r/|\nabla\Phi|r$ and $\lambda_R = \Phi_R/|\nabla\Phi|R$, respectively.*

PROOF. Note that Σ can be viewed as a surface of revolution in two ways, either rotating in the x- or the y-variables. So by Lemma 4.5, the splittings

$$\Big(\xi_z \cap (\mathbb{R}x \oplus i\mathbb{R}^n)\Big) \oplus \Big(\xi_z \cap (\mathbb{R}x \oplus i\mathbb{R}^n)^\perp\Big),$$
$$\Big(\xi_z \cap (\mathbb{R}^n \oplus i\mathbb{R}y)\Big) \oplus \Big(\xi_z \cap (\mathbb{R}^n \oplus i\mathbb{R}y)^\perp\Big)$$

are both orthogonal with respect to II and the right hand spaces are eigenspaces. In particular, $\Lambda_\perp = \xi_z \cap (\mathbb{R}x \oplus i\mathbb{R}^n)^\perp$ and $i\Lambda_\perp = \xi_z \cap (\mathbb{R}^n \oplus i\mathbb{R}y)^\perp$ are eigenspaces orthogonal to each other with eigenvalues $\Phi_r/|\nabla\Phi|r$ and $\Phi_R/|\nabla\Phi|R$. Since $\Lambda_{xy}^\mathbb{C}$ is the orthogonal complement of $\Lambda_\perp \oplus i\Lambda_\perp$ in $\mathbb{C}^n$, the lemma follows. □

It follows that the restriction of II to $\Lambda_\perp^\mathbb{C} = \Lambda_\perp \oplus i\Lambda_\perp$ has matrix

$$\begin{pmatrix} \lambda_r & 0 \\ 0 & \lambda_R \end{pmatrix},$$

and by Proposition 2.13 the restriction of the normalized Levi form $\mathbb{L}_\Sigma$ to $\Lambda_\perp^\mathbb{C}$ is given by

$$\mathbb{L}_\Sigma(X) = (\lambda_r + \lambda_R)|X|^2.$$

Now suppose that C is given near the point $\pi(z)$ by the equation $R = \phi(r)$, and the curve is cooriented by the gradient of the function $\Phi(r,R) = \phi(r) - R$. Since $|\nabla\Phi| = \sqrt{\Phi_r^2 + \Phi_R^2} = \sqrt{1+\phi'(r)^2}$, the eigenvalues λ_r on $\Lambda_\perp$ and λ_R on $i\Lambda_\perp$ equal

$$\lambda_r = \frac{\Phi_r}{|\nabla\Phi|r} = \frac{\phi'(r)}{r\sqrt{1+\phi'(r)^2}},$$
$$\lambda_R = \frac{\Phi_R}{|\nabla\Phi|R} = -\frac{1}{\phi(r)\sqrt{1+\phi'(r)^2}}.$$

Hence the preceding discussions shows

LEMMA 4.7. *Let $\Sigma = \pi^{-1}(C)$ be the hypersurface given by the curve $C = \{\phi(r) - R = 0\}$, cooriented by the gradient of $\phi(r) - R$. Then the restriction of the normalized Levi form $\mathbb{L}_\Sigma$ to $\Lambda_\perp^\mathbb{C}$ is given by*

$$\mathbb{L}_\Sigma(X) = \frac{1}{\sqrt{1+\phi'(r)^2}}\Big(\frac{\phi'(r)}{r} - \frac{1}{\phi(r)}\Big)|X|^2.$$

This restriction is positive definite if and only if

$$\mathcal{L}^\perp(\phi) := \frac{\phi'(r)}{r} - \frac{1}{\phi(r)} > 0.$$

In particular, if $\phi'(r) \le 0$ the restriction is always negative definite.

Lemma 4.7 reduces the question about i-convexity of Σ to positivity of $\mathcal{L}^\perp(\phi)$ and the corresponding question about the intersection $\Sigma \cap \Lambda_{xy}^\mathbb{C}$. When $\dim_\mathbb{C} \Lambda_{xy}^\mathbb{C} = 1$, this intersection is a curve which is trivially i-convex, hence Σ is i-convex if and only if $\mathcal{L}^\perp(\phi) > 0$. The remaining case $\dim_\mathbb{C} \Lambda_{xy}^\mathbb{C} = 2$ just means that we have reduced the original question to the case $n = 2$, which we will now consider.

The case $n = 2$. We denote complex coordinates in $\mathbb{C}^2$ by $z = (\zeta, w)$ with $\zeta = s + it$, $w = u + iv$. The hypersurface $\Sigma \subset \mathbb{C}^2$ is given by the equation

$$\sqrt{t^2 + v^2} = R = \phi(r) = \phi(\sqrt{s^2 + u^2}).$$

We want to express the coefficient $\mathbb{L}_0$ of the normalized Levi form $\mathbb{L}_\Sigma(X) = \mathbb{L}_0|X|^2$ at a point $z \in \Sigma$ in terms of ϕ. Suppose that $r, R > 0$ at the point z. After a unitary transformation

$$\zeta \mapsto \zeta \cos\alpha + w \sin\alpha, \qquad w \mapsto -\zeta \sin\alpha + w \cos\alpha,$$

which leaves Σ invariant, we may assume $t = 0$ and $v > 0$. Then near z we can solve the equation $R = \phi(r)$ for v,

$$v = \sqrt{\phi(\sqrt{s^2 + u^2})^2 - t^2} =: \psi(s, t, u).$$

According to Lemma 2.25, the coefficient of the normalized Levi form of the hypersurface $\Sigma = \{v = \psi(s, t, u)\}$ is given by

$$\mathbb{L}_0 = \frac{1}{(1 + \psi_s^2 + \psi_t^2 + \psi_u^2)^{\frac{3}{2}}} \big((\psi_{ss} + \psi_{tt})(1 + \psi_u^2) + \psi_{uu}(\psi_s^2 + \psi_t^2) + 2\psi_{su}(\psi_t - \psi_u\psi_s) - 2\psi_{tu}(\psi_s + \psi_u\psi_t)\big).$$

Note that at the point z we have $t = 0$ and $\psi(s, 0, u) = \phi(r) = \phi(\sqrt{u^2 + s^2})$. Using this, we compute the derivatives at z:

$$\psi_s = \frac{\phi' s}{r}, \qquad \psi_{ss} = \frac{\phi'' s^2}{r^2} + \frac{\phi' u^2}{r^3}, \qquad \psi_u = \frac{\phi' u}{r}, \qquad \psi_{uu} = \frac{\phi'' u^2}{r^2} + \frac{\phi' s^2}{r^3},$$

$$\psi_{su} = \frac{\phi'' su}{r^2} - \frac{\phi' su}{r^3}, \qquad \psi_t = 0, \qquad \psi_{tt} = -\frac{1}{\phi}, \qquad \psi_{tu} = 0.$$

Inserting this in the above expression for $\mathbb{L}_0$, we obtain

$$\begin{aligned}(1 + \phi'^2)^{3/2}\mathbb{L}_0 &= \Big(\frac{\phi'' s^2}{r^2} + \frac{\phi' u^2}{r^3} - \frac{1}{\phi}\Big)\Big(1 + \frac{\phi'^2 u^2}{r^2}\Big) \\ &\quad + \Big(\frac{\phi'' u^2}{r^2} + \frac{\phi' s^2}{r^3}\Big)\frac{\phi'^2 s^2}{r^2} - 2\Big(\frac{\phi'' su}{r^2} - \frac{\phi' su}{r^3}\Big)\frac{\phi'^2 su}{r^2} \\ &= \frac{\phi'' s^2}{r^2} + \frac{\phi' u^2}{r^3} + \frac{\phi'^3}{r} - \frac{1}{\phi}\Big(1 + \frac{\phi'^2 u^2}{r^2}\Big).\end{aligned}$$

We say that the curve C is *cooriented from above* if it is cooriented by the gradient of the function $\phi(r) - R$. Equivalently (since $t = 0$ at z), the hypersurface $\Sigma = \pi^{-1}(C)$ is cooriented by the gradient of $\sqrt{\phi(\sqrt{s^2 + u^2})^2 - t^2} - v$, which is the coorientation we have chosen above. The opposite coorientation will be called *coorientation from below.* The preceding discussion leads to

PROPOSITION 4.8. *Let $\mathbb{L}_\Sigma$ be the normalized Levi form of the hypersurface $\Sigma = \{R = \phi(r)\}$, cooriented from above, and suppose $r > 0$.*

(a) The restriction of $\mathbb{L}_\Sigma$ to $\Lambda_\perp^{\mathbb{C}}$ is given by

$$\mathbb{L}_\Sigma(X) = \frac{1}{\sqrt{1 + \phi'(r)^2}} \Big(\frac{\phi'(r)}{r} - \frac{1}{\phi(r)}\Big)|X|^2.$$

(b) The coefficient $\mathbb{L}_0$ of the restriction of $\mathbb{L}_\Sigma$ to $\Lambda^{\mathbb{C}}_{xy}$ is given in suitable unitary coordinates $\zeta = s + it, w = u + iv$ with $r^2 = s^2 + u^2$ and $R^2 = t^2 + v^2$ by

$$\mathbb{L}_0 = \frac{1}{(1+\phi'^2)^{3/2}}\left(\frac{\phi'' s^2}{r^2} + \frac{\phi' u^2}{r^3} + \frac{\phi'^3}{r} - \frac{1}{\phi}\Big(1 + \frac{\phi'^2 u^2}{r^2}\Big)\right).$$

(c) The maximal absolute value $M(II) := \max\{|II(X)|; X \in T\Sigma, |X| = 1\}$ of the normal curvature of Σ equals

$$M(II) = \max\left(\frac{|\phi''|}{(1+\phi'^2)^{\frac{3}{2}}}, \frac{|\phi'|}{r\sqrt{1+\phi'^2}}, \frac{1}{\phi\sqrt{1+\phi'^2}}\right).$$

If Σ is J-convex, then $\phi' > 0$ and

$$M(II) = \max\left(\frac{|\phi''|}{(1+\phi'^2)^{\frac{3}{2}}}, \frac{\phi'}{r\sqrt{1+\phi'^2}}\right).$$

PROOF. Parts (a) and (b) follow from Lemma 4.7 and the preceding discussion.

For (c), we write Σ as the zero set of the function $\Phi(r, R) = \phi(r) - R$. Assume first that x, y are linearly independent. Recall from Lemma 4.6 that the splitting at $z = (x, y) \in \Sigma$,

$$T_z\Sigma = (T_z\Sigma \cap \Lambda^{\mathbb{C}}_{xy}) \oplus \Lambda_\perp \oplus i\Lambda_\perp,$$

is orthogonal with respect to II and $\Lambda_\perp$, $i\Lambda_\perp$ are eigenspaces with eigenvalues

$$\lambda_r = \frac{\Phi_r}{|\nabla\Phi| r} = \frac{\phi'(r)}{r\sqrt{1+\phi'(r)^2}}, \qquad \lambda_R = \frac{\Phi_R}{|\nabla\Phi| R} = -\frac{1}{\phi(r)\sqrt{1+\phi'(r)^2}}.$$

It remains to compute the eigenvalues on the 3-dimensional space $T_z\Sigma \cap \Lambda^{\mathbb{C}}_{xy}$. Denote by $x^\perp, y^\perp$ the unit vectors in Λ_{xy} orthogonal to x, y such that $(x/r, x^\perp)$ and $(y/R, y^\perp)$ are orthonormal bases defining the same orientation as (x, y). Note that the circle actions by rotation of the x, resp. y, coordinates leave Σ invariant and are generated by $x^\perp$, resp. $y^\perp$. So by Lemma 4.5, these circle actions lead to a II-orthogonal splitting

$$T_z\Sigma \cap \Lambda^{\mathbb{C}}_{xy} = \mathbb{R}x^\perp \oplus T_z\Sigma \cap (\mathbb{R}x \oplus i\mathbb{R}y) \oplus i\mathbb{R}y^\perp.$$

The eigenvalue on $T_z\Sigma \cap (\mathbb{R}x \oplus i\mathbb{R}y)$ equals the curvature of the curve $\gamma(r) = (r, \phi(r))$, which is given (with the correct sign) by

$$\lambda = \frac{\sqrt{|\gamma'|^2|\gamma''|^2 - \langle\gamma'', \gamma'\rangle^2}}{|\gamma'|^3} = \frac{\sqrt{(1+\phi'^2)\phi''^2 - \phi'^2\phi''^2}}{(1+\phi'^2)^{3/2}} = \frac{\phi''}{(1+\phi'^2)^{3/2}}.$$

The eigenvalue on $\mathbb{R}x^\perp$ equals the normal curvature of the circle of radius r in the (x, y)-plane. This circle has curvature $1/r$ and normal projection amounts to multiplication by the factor $\phi'/\sqrt{1+\phi'^2}$ (see Figure 4.3), so the normal curvature equals $\phi'/r\sqrt{1+\phi'^2} = \lambda_r$. Similarly, the eigenvalue on $i\mathbb{R}y^\perp$ equals $-1/\phi\sqrt{1+\phi'^2} = \lambda_R$ and the formula for $M(II)$ follows.

If x, y are linearly dependent, then $T_z\Sigma \cap \Lambda^{\mathbb{C}}_{xy} = T_z\Sigma \cap (\mathbb{R}x \oplus i\mathbb{R}y)$ is 1-dimensional with eigenvalue λ and we obtain the same formula for $M(II)$.

If Σ is i-convex when cooriented from above, then the positivity of the expression in (a) implies that $\frac{\phi'}{r} > \frac{1}{\phi} > 0$, so the third term in the formula for $M(II)$ is dominated by the second term. □

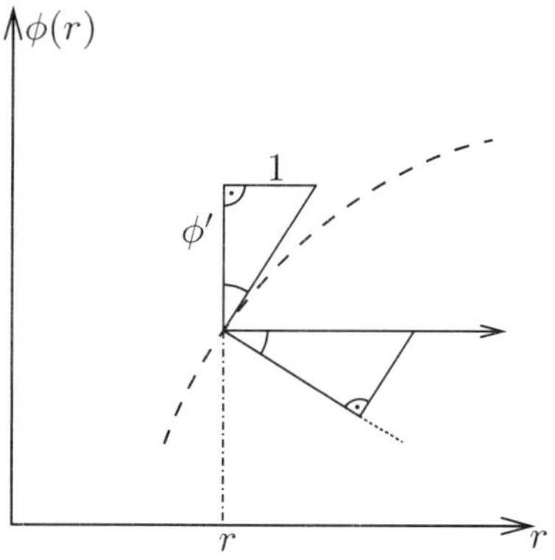

FIGURE 4.3. The normal curvature of the circle of radius r.

REMARK 4.9. The discussion in the preceding proof leads to an alternative derivation of the normalized Levi form by computing the mean normal curvature. The field of complex tangencies $\xi_z \subset T_z\Sigma \cap \Lambda^{\mathbb{C}}_{xy}$ is spanned by the vectors

$$\phi' x^\perp + iy^\perp, \qquad i(\phi' x^\perp + iy^\perp) = -y^\perp + i\phi' x^\perp.$$

The space $T_z\Sigma \cap (\mathbb{R}x \oplus i\mathbb{R}y)$ is spanned by the vector $v := x/r + i\phi'(r)y/R$. Denote by τ the oriented angle from x to y in Λ_{xy}. Then

$$-y^\perp + i\phi' x^\perp = -\cos\tau x^\perp + i\phi'\cos\tau y^\perp + \sin\tau v,$$

and since $x^\perp, iy^\perp, v$ are eigenvectors of II and $|v|^2 = 1+\phi'^2$, we obtain

$$\begin{aligned} &II(\phi' x^\perp + iy^\perp) + II(-y^\perp + i\phi' x^\perp) \\ &= \phi'^2\lambda_r + \lambda_R + \cos^2\tau\lambda_r + \phi'^2\cos^2\tau\lambda_R + \sin^2\tau(1+\phi'^2)\lambda \\ &= \frac{1}{\sqrt{1+\phi'^2}}\left(\frac{\phi'^3}{r} + \frac{\cos^2\tau\phi'}{r} - \frac{1}{\phi}(1+\phi'^2\cos^2\tau) + \phi''\sin^2\tau\right). \end{aligned}$$

Dividing by $|\phi' x^\perp + iy^\perp|^2 = 1+\phi'^2$ and setting $s = \sin\tau r$, $u = \cos\tau r$, this yields the coefficient $\mathbb{L}_0$ in Proposition 4.8(b) for the normalized Levi form $\mathbb{L}_\Sigma(X) = II(X) + II(iX)$ on $\Lambda^{\mathbb{C}}_{xy}$.

4.3. Properties of i-convex shapes

The precise expressions for the normalized Levi form $\mathbb{L}_\Sigma$ in Proposition 4.8 will become important in Section 4.7. For now, we will only be interested in the conditions for positivity $\mathbb{L}_\Sigma$, which we restate in the following proposition.

PROPOSITION 4.10. *The hypersurface $\Sigma = \{R = \phi(r)\}$ is i-convex cooriented from above at $r > 0$ if and only if ϕ satisfies the following two conditions:*

$$\mathcal{L}^\perp(\phi) := \frac{\phi'(r)}{r} - \frac{1}{\phi(r)} > 0, \tag{4.3}$$

$$\mathcal{L}^2(\phi) := \frac{\phi'' s^2}{r^2} + \frac{\phi' u^2}{r^3} + \frac{\phi'^3}{r} - \frac{1}{\phi}\Big(1 + \frac{\phi'^2 u^2}{r^2}\Big) > 0 \tag{4.4}$$

for all (s,u) with $s^2 + u^2 = r^2$. It is i-convex cooriented from below if and only if the reverse inequalities hold.

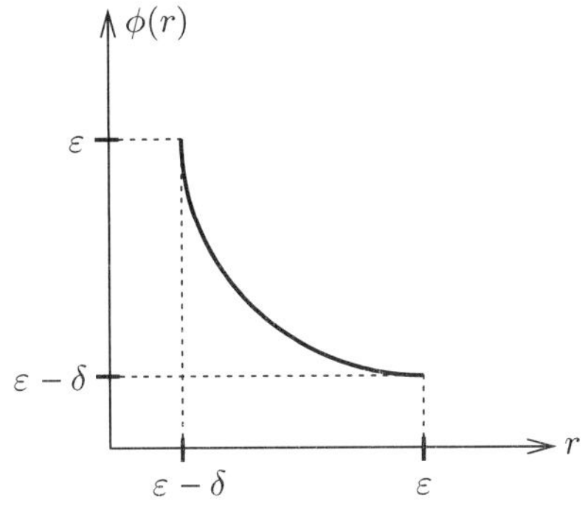

FIGURE 4.4. A quarter circle is an i-convex shape cooriented from below.

The following corollary gives some useful sufficient conditions for i-convexity.

COROLLARY 4.11. *(a) If $\phi > 0$, $\phi' > 0$, $\phi'' \leq 0$, and*

$$\phi'' + \frac{\phi'^3}{r} - \frac{1}{\phi}(1 + \phi'^2) > 0, \tag{4.5}$$

then Σ is i-convex cooriented from above.

(b) If $\phi > 0$, $\phi' \leq 0$, $\phi'' \geq 0$, and

$$\phi'' + \frac{\phi'^3}{r} - \frac{1}{\phi} < 0,$$

then Σ is i-convex cooriented from below.

PROOF. (a) If $\phi' > 0$ and $\phi'' \leq 0$, we get

$$\mathcal{L}^2(\phi) \geq \phi'' + \frac{\phi'^3}{r} - \frac{1}{\phi}(1 + \phi'^2).$$

So positivity of the right hand side implies condition (4.4). Condition (4.3) is also a consequence of $\phi'' + \frac{\phi'^3}{r} - \frac{1}{\phi}(1 + \phi'^2) > 0$.

(b) If $\phi' \leq 0$ and $\phi'' \geq 0$, we get

$$\mathcal{L}^2(\phi) \leq \phi'' + \frac{\phi'^3}{r} - \frac{1}{\phi}.$$

So negativity of the right hand side implies the reverse inequality (4.4). The reverse inequality (4.3) is automatically satisfied. □

As a first application of Corollary 4.11 we have

LEMMA 4.12. *For any $0 < \delta < \varepsilon < \sqrt{2}\delta$ the quarter circle*

$$\phi(r) := \varepsilon - \sqrt{\delta^2 - (\varepsilon - r)^2}, \qquad r \in [\varepsilon - \delta, \varepsilon],$$

defines an i-convex hypersurface $\{R = \phi(r)\}$ cooriented from below (see Figure 4.4).

PROOF. Fix $0 < \delta < \varepsilon < \sqrt{2}\delta$. For $r \in [\varepsilon-\delta, \varepsilon]$ set $s := \sqrt{\delta^2 - (\varepsilon - r)^2} \in [0, \delta]$. We have

$$\phi'(r) = -\frac{\varepsilon - r}{s}, \qquad \phi''(r) = \frac{\delta^2}{s^3},$$

$$\phi'' + \frac{\phi'^3}{r} - \frac{1}{\phi} = \frac{1}{s^3}\left(\delta^2 - \frac{(\varepsilon - r)^3}{r} - \frac{s^3}{\varepsilon - s}\right).$$

Set $t := \varepsilon - r$. Then we need to prove that

$$F(t) := \frac{t^3}{\varepsilon - t} + \frac{s^3}{\varepsilon - s} > \delta^2 \tag{4.6}$$

for all $t \in [0, \delta]$, where $s = \sqrt{\delta^2 - t^2}$. We have

$$F'(t) = t\left(\frac{t(3\varepsilon - 2t)}{(\varepsilon - t)^2} - \frac{s(3\varepsilon - 2s)}{(\varepsilon - s)^2}\right).$$

A short computation shows that the function $G(t) := \frac{t(3\varepsilon - 2t)}{(\varepsilon - t)^2}$ is strictly increasing on $[0, \delta]$. It follows that the function $\frac{F'(t)}{t} = G(t) - G(s)$ has a unique zero when $t = s$, i.e., $t = \frac{\delta}{\sqrt{2}}$, is negative on $[0, \frac{\delta}{\sqrt{2}})$ and positive on $(\frac{\delta}{\sqrt{2}}, \delta]$. Hence the function $F(t)$ attains its minimum at the point $\frac{\delta}{\sqrt{2}}$. We compute $F(\frac{\delta}{\sqrt{2}}) = \frac{\delta^3}{\sqrt{2}\varepsilon - \delta}$, so the condition $\varepsilon < \sqrt{2}\delta$ implies $F(\frac{\delta}{\sqrt{2}}) > \delta^2$ and hence inequality (4.6). □

For the remainder of this chapter we will only be interested in hypersurfaces $\{R = \phi(r)\}$ that are i-convex cooriented from *above*. We will call the corresponding function ϕ satisfying the conditions of Proposition 4.10 an *i-convex shape*. The following lemma lists some elementary properties of i-convex shapes.

LEMMA 4.13 (Properties of i-convex shapes). *(a) If ϕ is an i-convex shape, then so is $\phi + c$ for any constant $c \geq 0$ (i-convexity from above is preserved under upwards shifting).*

(b) If ϕ is an i-convex shape at $r > 0$, then the function $\phi_\lambda(r) := \lambda\phi(r/\lambda)$ is an i-convex shape at λr for each $\lambda > 0$.

(c) If ϕ, ψ are i-convex shapes for $r \leq r_0$, resp. $r \geq r_0$, such that $\phi(r_0) = \psi(r_0)$ and $\phi'(r_0) = \psi'(r_0)$, then the function

$$\vartheta(r) := \begin{cases} \phi(r) & \text{for } r \leq r_0, \\ \psi(r) & \text{for } r \geq r_0 \end{cases}$$

can be C^1-perturbed to a smooth i-convex shape which agrees with ϑ outside a neighborhood of r_0.

(d) If ϕ, ψ are i-convex shapes, then the function

$$\vartheta := \max(\phi, \psi)$$

can be C^0-perturbed to a smooth i-convex shape which agrees with ϑ outside a neighborhood of the set $\{\phi = \psi\}$.

PROOF. (a) If ϕ satisfies one of the inequalities (4.3), (4.4), and (4.5), then $\phi + c$ satisfies the same inequality for any constant $c \geq 0$.

(b) This can be seen by applying the biholomorphism $z \mapsto \lambda z$ on $\mathbb{C}^n$, or from Proposition 4.10 as follows: the function ϕ_λ has derivatives $\phi_\lambda(\lambda r) = \lambda\phi(r)$, $\phi'_\lambda(\lambda r) = \phi'(r)$, $\phi''_\lambda(\lambda r) = \phi''(r)/\lambda$, and the replacement $r \mapsto \lambda r$, $\phi \mapsto \lambda\phi$, $\phi' \mapsto \phi'$, $\phi'' \mapsto \phi''/\lambda$ leaves both conditions in Proposition 4.10 unchanged.

(c) This follows from the fact that for given r, ϕ, ϕ', the set of ϕ'' such that condition (4.4) holds is convex.

(d) After C^2-perturbing ϕ, we may assume that the graphs of ϕ and ψ intersect transversely. Consider an intersection point r_0 such that $\phi(r_0) = \psi(r_0)$ and $\phi'(r_0) < \psi'(r_0)$, so near r_0 we have

$$\vartheta(r) = \begin{cases} \phi(r) & \text{for } r \le r_0, \\ \psi(r) & \text{for } r \ge r_0. \end{cases}$$

We claim that for any $\delta, M > 0$ there exist $r_- < r_0 < r_+$ with $|r_+ - r_-| < \delta$ and a quadratic function $\chi : [r_-, r_+] \to \mathbb{R}$ with the following properties:

- $\chi'' \equiv m \ge M$;
- $\chi(r_-) = \phi(r_-)$, $\chi'(r_-) = \phi'(r_-)$;
- $\chi(r_+) = \psi(r_+)$, $\chi'(r_+) = \psi'(r_+)$.

To see this, take for every sufficiently close $r_- < r_0$ a linear function $a + br$ tangent to ϕ at r_- and add a quadratic term $m(r - r_-)^2/2$ to make it tangent to ψ at some $r_+ > r_0$, and note that $r_+ \to r_0$ and $m \to \infty$ as $r_- \to r_0$.

We make χ smooth by decreasing χ'' from m to $\phi''(r_-)$ near r_- and from m to $\psi''(r_+)$ near r_+.

It remains to show i-convexity of χ. Condition (4.3) holds for χ because it holds for ϕ, ψ and up to an error of order δ for $r \in [r_-, r_+]$ we have $r \approx r_0$, $\chi(r) \approx \phi(r_0) = \psi(r_0)$, and $\chi'(r) \in [\phi'(r_0), \psi'(r_0)]$. Next note that condition (4.4) for $s = 0$ becomes

$$\left(\frac{\chi'}{r} - \frac{1}{\chi}\right)(1 + \chi'^2) > 0,$$

which is satisfied in view of condition (4.3). Since $\chi''(r)$ is uniformly bounded from below independently of δ, there exists a constant $\sigma > 0$ independent of δ such that χ satisfies condition (4.4) for all $|s| \le \sigma$. Moreover, near r_-, resp. r_+, condition (4.4) holds for χ because it holds for ϕ, ψ and χ'' is larger than ϕ'', resp. ψ''. So it remains to consider the region where $\chi'' \equiv m$ in the case $|s| \ge \sigma$. In this region r, χ, χ' are bounded independently of m. On the other hand, the term $\chi'' s^2/r^2$ becomes arbitrarily large as $m \to \infty$, so condition (4.4) holds for m sufficiently large. □

4.4. Shapes in the subcritical case

The following lemma extends i-convex shapes to the subcritical case. For $k \le n$ we set

$$r := \sqrt{x_1^2 + \cdots + x_n^2 + y_{k+1}^2 + \cdots + y_n^2}, \qquad R := \sqrt{y_1^2 + \cdots + y_k^2}.$$

LEMMA 4.14. *Let $\phi(r)$ be an i-convex shape with $\phi' > 0$. Then:*

(a) $\Sigma := \{R = \phi(r)\}$ is an i-convex hypersurface cooriented from above.

(b) Σ intersects the subspace $i\mathbb{R}^n$ i-orthogonally in the sense that $i(i\mathbb{R}^n) = \mathbb{R}^n \subset T_{iy}\Sigma$ for any $iy \in i\mathbb{R}^n \cap \Sigma$.

(c) Σ is transverse to the vector field

$$X = \sum_{i=1}^{k}\left(x_i \frac{\partial}{\partial x_i} - y_i \frac{\partial}{\partial y_i}\right) + \sum_{j=k+1}^{n}\left(x_j \frac{\partial}{\partial x_j} + y_j \frac{\partial}{\partial y_j}\right). \tag{4.7}$$

PROOF. (a) Set $\bar{r} := \sqrt{x_1^2 + \cdots + x_n^2}$ and $\bar{R} := \sqrt{y_1^2 + \cdots + y_n^2}$. By assumption, the hypersurface $\bar{\Sigma} := \{\bar{R} = \phi(\bar{r})\}$ is i-convex cooriented from above. By Lemma 2.7 there exists a convex increasing function $f : \mathbb{R} \to \mathbb{R}$ with $f(0) = 0$ such that the function

$$\bar{\psi} : \mathbb{C}^n \to \mathbb{R}, \quad \bar{\psi}(z) := f\big(\phi(\bar{r}) - \bar{R}\big)$$

is i-convex on the neighborhood $\bar{U} := \bar{\psi}^{-1}\big((-1,1)\big)$ of $\bar{\Sigma}$.

Let us write $z = (z', z'') \in \mathbb{C}^n = \mathbb{C}^k \oplus \mathbb{C}^{n-k}$ with $z' := (z_1, \dots, z_k)$ and $z'' := (z_{k+1}, \dots, z_n)$. The unitary group $U(n-k)$ acts on $\mathbb{C}^n$ by rotation in the second factor, $gz = (z', gz'')$ for $g \in U(n-k)$. Note that the functions $\psi_g(z) := \bar{\psi}(gz)$ are i-convex on the sets $U_g := \psi_g^{-1}\big((-1,1)\big) = g^{-1}(\bar{U})$. Define

$$\psi : \mathbb{C}^n \to \mathbb{R}, \quad \psi(z) := \max{}_{g \in U(n-k)} \bar{\psi}(gz).$$

Since ϕ is increasing, the function

$$g \mapsto \phi\big(\sqrt{|\mathrm{Re}\, z'|^2 + |\mathrm{Re}\,(gz'')|^2}\big) - \sqrt{|\mathrm{Im}\, z'|^2 + |\mathrm{Im}\,(gz'')|^2}$$

for fixed (z', z'') attains its maximum for $\mathrm{Im}\,(gz'') = 0$, so $\mathrm{Re}\,(gz'') = z''$ and

$$\psi(z) = f\Big(\phi\big(\sqrt{|\mathrm{Re}\, z'|^2 + |z''|^2}\big) - |\mathrm{Im}\, z'|\Big) = f\big(\phi(r) - R\big).$$

In particular, ψ is smooth with regular level set $\psi^{-1}(0) = \Sigma$.

We claim that ψ is i-convex near Σ. To see this, consider a point $z \in \Sigma$. By definition we have $\psi_g(z) \le 0$ for all $g \in U(n-k)$. Set

$$A := \{g \in U(n-k) \mid \psi_g(z) \ge -1/2\}.$$

Since A and $\overline{U(n-k) \setminus A}$ are compact, there exists a neighborhood B of z on which $\psi_g > -1$ for all $g \in A$ and $\psi_g \le -1/4$ for all $g \in U(n-k) \setminus A$. Thus $B' := B \cap \psi^{-1}\big((-1/4, 1/4)\big)$ is a neighborhood of z on which $\psi = \max{}_{g \in A} \psi_g \in (-1/4, 1/4)$ and $\psi_g \in (-1, 1/4)$ for all $g \in A$. Since ψ_g is i-convex on $B' \subset U_g$ for all $g \in A$, Proposition 3.8 implies i-convexity of ψ on B'. This shows that ψ is i-convex near Σ, hence its level set Σ is also i-convex.

(b) This is clear from the definition of Σ, and (c) follows from the computation

$$X \cdot \Big(\phi(r) - R\Big) = \phi'(r)\left(\sum_{i=1}^{n} x_i \frac{\partial r}{\partial x_i} + \sum_{i=k+1}^{n} y_i \frac{\partial r}{\partial y_i}\right) + \sum_{j=1}^{k} y_j \frac{\partial R}{\partial y_j} = \phi'(r) r + R > 0.$$

□

4.5. Construction of special shapes

We will now construct special i-convex shapes satisfying the differential inequality in Corollary 4.11(a). One such solution with the desired properties has been constructed in [**42**]. The following simplified construction was pointed out to us by M. Struwe. We will find the function ϕ as a solution of *Struwe's differential equation*

$$\phi'' + \frac{\phi'^3}{2r} = 0 \tag{4.8}$$

with $\phi' > 0$ and hence $\phi'' < 0$. Then the inequality in Corollary 4.11(a) reduces to

$$\frac{\phi'^3}{2r} - \frac{1}{\phi}(1 + \phi'^2) > 0. \tag{4.9}$$

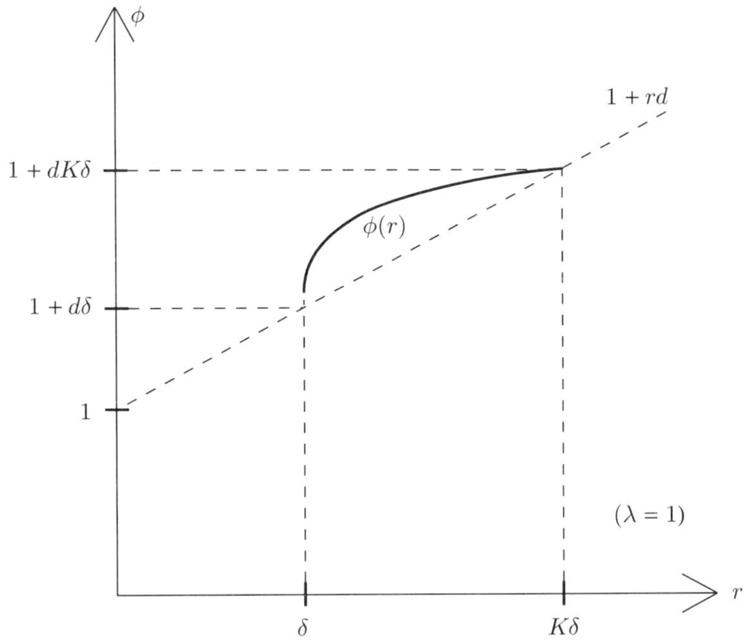

FIGURE 4.5. A solution of Struwe's differential equation.

LEMMA 4.15. *For any* $d, K, \delta, \lambda > 0$ *satisfying* $K \geq e^{4/d^2}$ *and* $4K\delta \leq (\ln K)^{-3/2}$, *there exists a solution* $\phi : [\lambda\delta, K\lambda\delta] \to \mathbb{R}$ *of* (4.8) *with the following properties (see Figure* 4.5*):*

(a) $\phi'(\lambda\delta) = +\infty$ *and* $\lambda + d\lambda\delta \leq \phi(\lambda\delta) < \lambda + dK\lambda\delta$;

(b) $\phi(K\lambda\delta) = \lambda + dK\lambda\delta$ *and* $\phi'(K\lambda\delta) \leq d$;

(c) ϕ *satisfies* (4.9) *and hence is the shape of an i-convex hypersurface cooriented from above.*

PROOF. First note that if ϕ satisfies equation (4.8) and inequality (4.9), then so does the rescaled function $\lambda\phi(r/\lambda)$. Thus it suffices to consider the case $\lambda = 1$. The differential equation (4.8) is equivalent to

$$\left(\frac{1}{\phi'^2}\right)' = -\frac{2\phi''}{\phi'^3} = \frac{1}{r},$$

thus $1/\phi'^2 = \ln(r/\delta)$ for some constant $\delta > 0$, or equivalently, $\phi'(r) = 1/\sqrt{\ln(r/\delta)}$. By integration, this yields a solution ϕ for $r \geq \delta$ which is strictly increasing and concave and satisfies $\phi'(\delta) = +\infty$. Note that $\int_\delta^{K\delta} \phi'(r)dr = \delta K_1$ with

$$K_1 := \int_1^K \frac{du}{\sqrt{\ln u}} < \infty.$$

Fix the remaining free constant in ϕ by setting $\phi(K\delta) := 1 + dK\delta$, thus

$$\phi(\delta) = 1 + dK\delta - K_1\delta.$$

Estimating the logarithm on $[1, K]$ from below by the linear function with the same values at the endpoints,

$$\ln u \geq \frac{\ln K}{K-1}(u-1),$$

we obtain an upper estimate for K_1:

$$K_1 \leq \int_1^K \frac{du}{\sqrt{\frac{\ln K}{K-1}(u-1)}} = \sqrt{\frac{K-1}{\ln K}} \int_0^{K-1} \frac{du}{\sqrt{u}} = \frac{2(K-1)}{\sqrt{\ln K}}. \tag{4.10}$$

By hypothesis we have $\sqrt{\ln K} \geq 2/d$, hence $K_1 \leq d(K-1)$. This implies

$$\phi(\delta) \geq 1 + dK\delta - d(K-1)\delta = 1 + d\delta.$$

Concavity of ϕ implies $\phi(r) \geq 1+dr$ for all $r \in [\delta, K\delta]$, and in particular $\phi'(K\delta) \leq d$. Clearly, $\phi(\delta) < 1 + dK\delta$ because ϕ is increasing. So it only remains to check inequality (4.9). Denoting by $\sim$ equality up to a positive factor, we compute

$$\begin{aligned} \frac{\phi'^3}{2r} - \frac{1}{\phi}(1+\phi'^2) &\geq \frac{\phi'^3}{2r} - \frac{1}{1+dr}(1+\phi'^2) \\ &\sim \frac{\phi'^3}{r}(1+dr) - 2 - 2\phi'^2 \\ &\sim \frac{1}{r} + d - 2\ln(r/\delta)^{3/2} - 2\ln(r/\delta)^{1/2}. \end{aligned} \tag{4.11}$$

The function on the right hand side is decreasing in r. So its minimum is achieved for $r = K\delta$ and has the value

$$\frac{1}{K\delta} - 2(\ln K)^{3/2} - 2(\ln K)^{1/2} + d > \frac{1}{K\delta} - 4(\ln K)^{3/2} \geq 0$$

by hypothesis. Here we have used the inequality $2(\ln K)^{1/2} < d + 2(\ln K)^{3/2}$, which follows for all $d > 0$ from the hypothesis $K \geq e^{4/d^2}$ (arguing separately for the cases $d \leq 2$ and $d > 2$). □

REMARK 4.16. The proof of Lemma 4.15 shows that for given $\delta, K > 0$ and $c \in \mathbb{R}$ the differential equation (4.8) has a unique solution ϕ satisfying $\phi'(\delta) = \infty$ and $\phi(K\delta) = c$, and this solution depends smoothly on δ, K, c. Indeed, the solution ϕ is given by

$$\phi(r) = c + \int_{K\delta}^{r} \frac{1}{\sqrt{\ln(s/\delta)}} ds.$$

To obtain a shape for a hypersurface as in Theorem 4.1, we need to interpolate between the function $\phi(r)$ in Lemma 4.15 (for $\lambda = 1$) near $r = \delta$ and the standard function $S(r) = \sqrt{1 + ar^2}$ near $r = \gamma$, for some given $a > 1$ and $\gamma \in (0, 1)$. This interpolation will occupy the remainder of this section. Since it is rather involved, let us take a moment to explain why it needs to be so complicated.

The straightforward approach would be the following. Given a, γ let us try to find $d, K, \delta > 0$ as in Lemma 4.15 such that $K\delta \leq \gamma$ and the corresponding function ϕ satisfies $\phi(K\delta) \leq S(K\delta)$. Then the graphs of ϕ and S would intersect at a point between δ and $K\delta$ and a smoothing of $\max(\phi, S)$ would yield the desired shape.

Now the condition $\phi(K\delta) = 1 + dK\delta \leq S(K\delta) = \sqrt{1 + a(K\delta)^2}$ implies $aK\delta \geq 2d + d^2K\delta \geq 2d$ and thus $2d/a \leq K\delta \leq \gamma$. This shows that d needs to be small for fixed $a > 1$ and small $\gamma > 0$. On the other hand, the conditions $K \geq e^{4/d^2}$ and $4K\delta \leq (\ln K)^{-3/2}$ in Lemma 4.15 yield

$$\frac{2d}{a} \leq K\delta \leq \frac{1}{4(\ln K)^{3/2}} \leq \frac{1}{4(4/d^2)^{3/2}} = \frac{d^3}{32},$$

and thus $d^2 \geq 64/a$, which contradicts the inequality $2d/a \leq \gamma$ for small γ. Hence this approach fails.

Geometrically, the preceding computation shows that for small γ the graph of ϕ will intersect that of the linear function $L(r) = 1 + dr$ before it meets the graph of S. The next attempt would be to follow ϕ from δ to the intersection with L, then L up to the intersection with S, and then S up to γ. The problem with this is that L is only i-convex for small r and the part of L used in the interpolation fails to be i-convex. This forces us to introduce a further shape, the quadratic function Q defined below. The desired i-convex shape will then be constructed by interpolating from ϕ to L to Q to S.

For numbers $\lambda, a, b, c, d \geq 0$ consider the following functions:

- $S_\lambda(r) = \sqrt{\lambda^2 + ar^2}$ (standard function),
- $Q_\lambda(r) = \lambda + br + cr^2/2\lambda$ (quadratic function),
- $L_\lambda(r) = \lambda + dr$ (linear function).

Let us first determine in which ranges they satisfy the inequalities (4.3) and (4.4).

LEMMA 4.17. *(a) The function $S_\lambda(r)$ is the shape of an i-convex hypersurface for $\lambda \geq 0$, $a > 1$, and $r > 0$.*

(b) The function $Q_\lambda(r)$ is the shape of an i-convex hypersurface for $\lambda > 0$, $b \geq 0$, $c > 1$, and $r > 0$.

(c) The function $Q_\lambda(r)$ is the shape of an i-convex hypersurface for $\lambda > 0$, $b = 4 - c$, $0 \leq c \leq 4$, and $0 < r \leq 2\lambda$.

(d) The function $L_\lambda(r)$ is the shape of an i-convex hypersurface for $\lambda \geq 0$, $d > 1$, and $r > 0$.

(e) The function $L_\lambda(r)$ is the shape of an i-convex hypersurface for $\lambda > 0$, $d > 0$, and $0 < r < \lambda d^3$.

PROOF. First note that by Lemma 4.13(b) we only need to prove the statements for $\lambda = 1$. Set $S := S_1$, $Q := Q_1$, $L := L_1$. We denote by $\sim$ equality up to multiplication by a positive factor.

(a) This holds because $R = S(r)$ describes a level set of the i-convex function $\phi(r, R) = ar^2 - R^2$ for $a > 1$.

(b) Condition (4.3) follows from

$$Q'(r)Q(r) - r = (b + cr)(1 + br + \frac{cr^2}{2}) - r \geq b + cr - r = b + (c-1)r > 0,$$

and condition (4.4) from

$$\begin{aligned}
\mathcal{L}^2(Q) &\geq \frac{c(r^2 - u^2)}{r^2} + \frac{(b+cr)u^2}{r^3} + \frac{(b+cr)^3}{r} - 1 - \frac{(b+cr)^2u^2}{r^2} \\
&\sim cr(r^2 - u^2) + (b+cr)u^2 + r^2(b+cr)^3 - r^3 - r(b+cr)^2u^2 \\
&= (c-1)r^3 + bu^2 + r^2(b+cr)^3 - ru^2(b+cr)^2 \\
&\geq (c-1)r^3 + r^2(b+cr)^3 - r^3(b+cr)^2 \\
&= (c-1)r^3 + r^2(b+cr)^2\big(b + (c-1)r\big) > 0.
\end{aligned}$$

(c) Condition (4.3) follows as in (b) from

$$Q'(r)Q(r) - r \geq b + cr - r = 4 - c(1-r) - r \geq 4 - 4(1-r) = 4r - r > 0.$$

For condition (4.4) it suffices, by (b), to show that

$$\begin{aligned} A &:= (c-1)r + (b+cr)^2\big(b + (c-1)r\big) \\ &= (c-1)r + \big(4 - c(1-r)\big)^2\big(4 - c(1-r) - r\big) > 0. \end{aligned}$$

For $c > 1$ this follows from (b). For $c \leq 1$ we have $4 - c(1-r) \geq 3$ and $4 - c(1-r) - r \geq 3 - r$, hence

$$A \geq -r + 9(3-r) = 27 - 10r > 0$$

for $r \leq 2$.

(d) Condition (4.3) follows from

$$L'(r)L(r) - r = d(1+dr) - r = d + (d^2-1)r > 0,$$

and condition (4.4) from

$$\begin{aligned} \mathcal{L}^2(L) &= \frac{du^2}{r^3} + \frac{d^3}{r} - \frac{1}{1+dr}\Big(1 + \frac{d^2u^2}{r^2}\Big) \\ &\sim (1+dr)du^2 + d^3r^2(1+dr) - r^3 - d^2ru^2 \\ &= du^2 + d^3r^2 + (d^4-1)r^3 > 0. \end{aligned}$$

(e) We will only use the weaker assumption $r(1-d^4) < \lambda d^3$ instead of $r < \lambda d^3$. Condition (4.3) follows from $r(1-d^2) < d^3/(1+d^2)$ via

$$L'(r)L(r) - r = d + (d^2-1)r \geq d - \frac{d^3}{1+d^2} = \frac{d}{1+d^2} > 0,$$

and condition (4.4) as in (d) from

$$\mathcal{L}^2(L) = du^2 + d^3r^2 + (d^4-1)r^3 \geq r^2\big(d^3 + (d^4-1)r\big) > 0. \qquad \square$$

LEMMA 4.18. *(a) For $\lambda, c > 0$ and $d > b > 0$ the functions $Q_\lambda(r)$ and $L_\lambda(r)$ intersect at a unique point $\lambda r_{QL} > 0$, where $r_{QL} = 2(d-b)/c$.*

(b) For $\lambda > 0$ and $a > d^2 > 0$ the functions $L_\lambda(r)$ and $S_\lambda(r)$ intersect at a unique point $\lambda r_{SL} > 0$, where $r_{SL} = 2d/(a-d^2)$.

(c) For $\lambda, b > 0$, $a > c \geq 0$ and $2b^2(a+c)^2 < (a-c)^3$ the functions $S_\lambda(r)$ and $Q_\lambda(r)$ intersect at precisely two points $\lambda r_{SQ}, \lambda r'_{SQ}$ satisfying $0 < r_{SQ} < 4b/(a-c) < r'_{SQ}$. Moreover, the points r_{SQ} and r'_{SQ} depend smoothly on a, b, c.

See Figure 4.6.

PROOF. (a) and (b) are simple computations, so we only prove (c). Again, by rescaling it suffices to consider the case $\lambda = 1$. First observe that for $x > 0$ and $\mu < 1$ we have $\sqrt{1+x} > 1 + \mu x/2$ provided that $1 + x > 1 + \mu x + \mu^2x^2/4$, or equivalently, $x < 4(1-\mu)/\mu^2$. Applying this to $x = ar^2$, we find that $S(r) > 1 + \mu ar^2/2$, provided that

$$r^2 < \frac{4(1-\mu)}{a\mu^2}. \tag{4.12}$$

Hence if

$$1 + \frac{\mu ar^2}{2} = Q(r) = 1 + br + \frac{cr^2}{2}$$

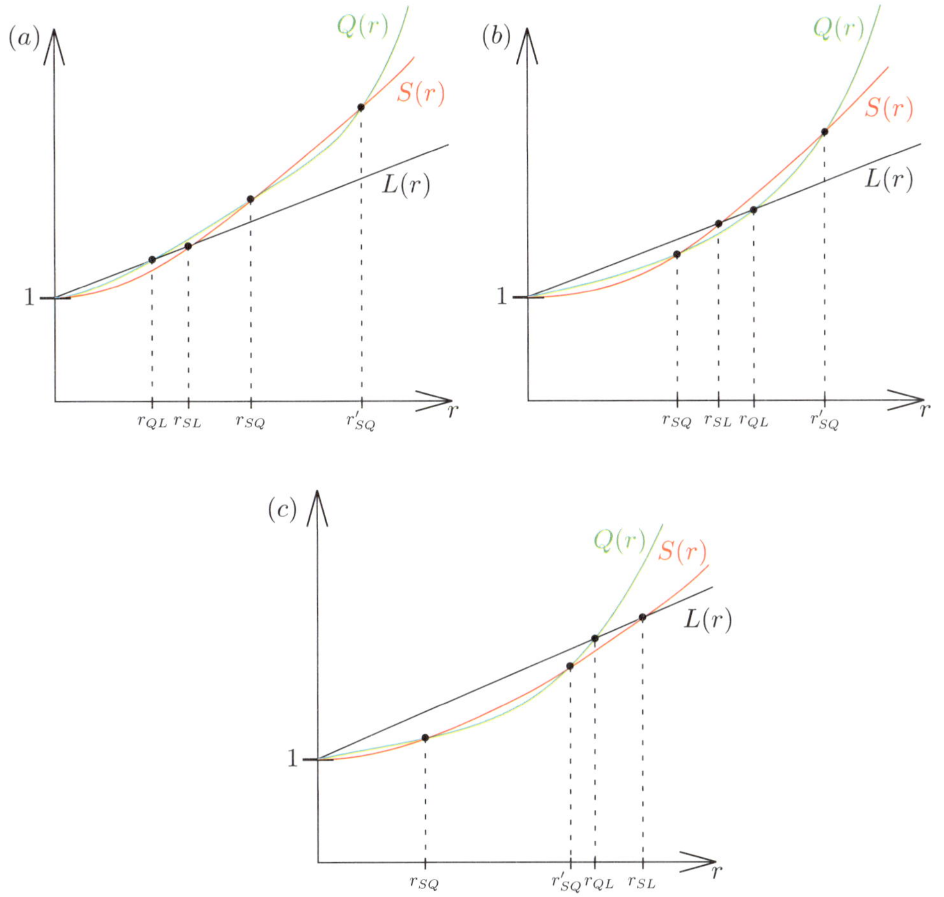

FIGURE 4.6. Intersections of the standard, linear and quadratic shapes.

for some $r > 0$ and $\mu < 1$ satisfying (4.12), then $S(r) > Q(r)$. Assuming $\mu a > c$, we solve the last equation for $r = 2b/(\mu a - c)$. Inequality (4.12) becomes

$$r^2 = \frac{4b^2}{(\mu a - c)^2} < \frac{4(1-\mu)}{a\mu^2},$$

or equivalently,

$$ab^2\mu^2 < (1-\mu)(\mu a - c)^2. \tag{4.13}$$

Now pick $\mu := (a+c)/2a$. The hypothesis $a > c$ implies $\mu < 1$ and $\mu a = (a+c)/2 > c$. With $\mu a - c = (a-c)/2$ and $1 - \mu = (a-c)/2a$, inequality (4.13) becomes

$$ab^2 \left(\frac{a+c}{2a}\right)^2 < \frac{a-c}{2a}\left(\frac{a-c}{2}\right)^2,$$

or equivalently,

$$2b^2(a+c)^2 < (a-c)^3.$$

Assume this inequality holds, so $S(r_+) > Q(r_+)$ at the point

$$r_+ = \frac{2b}{\mu a - c} = \frac{4b}{a - c}.$$

Now $f(r) := Q(r)^2 - S(r)^2$ is a polynomial of degree 4 satisfying $f(0) = 0$ and $f(r) \to +\infty$ as $r \to \pm\infty$. Since $b > 0$, we have $f(r) > 0$ for $r > 0$ close to zero and $f(r) < 0$ for $r < 0$ close to zero, so $f(r_-) = 0$ for some $r_- < 0$. By the preceding discussion we have $f(r_+) > 0$, so f has two more zeroes r_{SQ}, r'_{SQ} with $0 < r_{SQ} < r_+ < r'_{SQ}$. Since the four zeroes of f are distinct, they are all nondegenerate, which implies smooth dependence on the parameters a, b, c. □

Now we combine Lemmas 4.17 and 4.18 to show

LEMMA 4.19. *For every $a > 1$ and $\gamma > 0$ there exists $d \in (0, 1)$ and an increasing i-convex shape $\psi(r)$ which agrees with $S(r) = \sqrt{1 + ar^2}$ for $r \geq \gamma$ and with $L(r) = 1 + dr$ for r close to 0.*

PROOF. Pick any $c \in (1, a)$. Pick $b \in (0, 1)$ such that $2b^2(a+c)^2 < (a-c)^3$ and $4b < \gamma(a-c)$. By Lemma 4.18(c), the i-convex shapes $S(r)$ and $Q(r) = 1+br+cr^2/2$ intersect at a point $0 < r_{SQ} < 4b/(a - c) < \gamma$. Now pick $d \in (b, 1)$ such that $r_{QL} = 2(d - b)/c$ satisfies $r_{QL} < r_{SQ}$ and $r_{QL} < d^3$. By Lemma 4.18(a), the functions $Q(r)$ and $L(r)$ intersect at the point r_{QL}, so we are in the situation of Figure 4.6(a). By Lemma 4.17(e) the function $L(r)$ is i-convex for $r \leq r_{QL}$. Now the desired function is a smoothing of the function which equals $L(r)$ for $r \leq r_{QL}$, $Q(r)$ for $r_{QL} \leq r \leq r_{SQ}$ and $S(r)$ for $r \geq r_{SQ}$. □

REMARK 4.20. The proof of Lemma 4.19 uses only the criteria for i-convexity of the functions S_λ, Q_λ, and L_λ given in Lemma 4.17(a), (b) and (e). Given constants $a > 1$ and $\gamma > 0$, the constants $c := (1+a)/2$ and $b := \gamma(a-1)/16$ satisfy the conditions in the proof of Lemma 4.19 for γ sufficiently small (which we may assume without loss of generality). With this choice, the constant d in Lemma 4.19 satisfies $\gamma(a - 1)/16 < d < 1$.

Now we are ready to prove the main result of this section.

PROPOSITION 4.21. *For every $a > 1$ and $\gamma > 0$ there exists $\delta \in (0, \gamma)$ and an i-convex shape $\chi(r)$ cooriented from above which agrees with $S(r) = \sqrt{1 + ar^2}$ for $r \geq \gamma$ and satisfies $\chi'(\delta) = +\infty$ and $1 < \chi(\delta) < 1 + \gamma$ (see Figure 4.7).*

PROOF. By Lemma 4.19, there exists an increasing i-convex shape $\psi(r)$ which agrees with $S(r) = \sqrt{1 + ar^2}$ for $r \geq \gamma$ and with $L(r) = 1 + dr$ for $r \leq \beta$, for some $d \in (0, 1)$ and $\beta \in (0, \gamma)$. Let $\phi : [\delta, K\delta] \to \mathbb{R}_+$ be the i-convex shape provided by Lemma 4.15, where $K, \delta > 0$ satisfy the conditions in Lemma 4.15 (with $\lambda = 1$ and our given d) and in addition $K\delta < \beta$. Now the desired shape χ is a smoothing of the function which equals ϕ for $r \leq K\delta$ and ψ for $r \geq K\delta$. Note that property (a) in Lemma 4.15 yields

$$1 < \chi(\delta) < 1 + dK\delta < 1 + \gamma.$$ □

REMARK 4.22. The proofs of Lemmas 4.15 and 4.19 show that the i-convex shape χ in Proposition 4.21 can be chosen to depend smoothly on the parameters $a > 1$, $\gamma > 0$ and the sufficiently small $\delta \in (0, \gamma)$. Let us choose a smooth function $(a, \gamma) \mapsto \delta(a, \gamma)$, decreasing in γ, such that $\delta(a, \gamma) \in (0, \gamma)$ is sufficiently small in

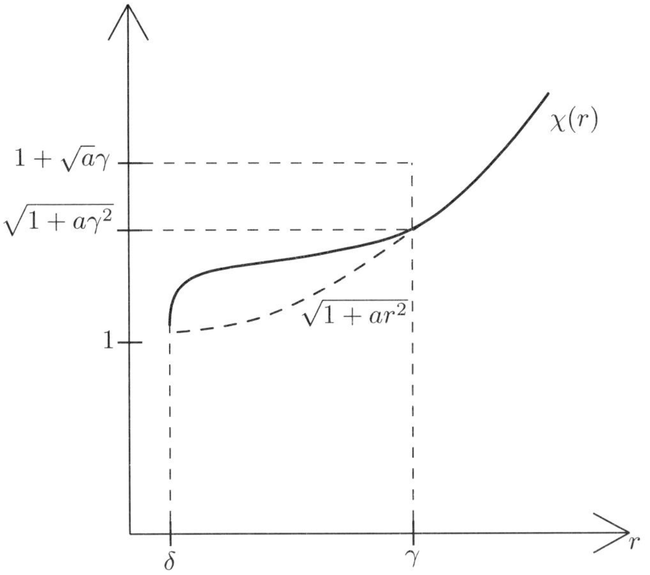

FIGURE 4.7. The i-convex shape χ cooriented from above.

the sense of Proposition 4.21 (in particular $\delta(a,\gamma) \to 0$ as $\gamma \to 0$ for any a). Then we obtain a smooth family of increasing i-convex shapes $\chi_{a,\gamma} : [\delta(a,\gamma), \infty) \to \mathbb{R}$, $a > 1$, $\gamma > 0$ with the properties in Proposition 4.21.

PROOF OF THEOREM 4.1(i)–(iii). With the shape $\chi(r)$ in Proposition 4.21, the desired i-convex hypersurface Σ is given by $\{\chi(r) - R = 0\} \cup \{r = \delta,\ R \leq \chi(\delta)\}$. □

4.6. Families of special shapes

In this section we construct a family of i-convex shapes interpolating between the function in Proposition 4.21 and the standard functions S_λ.

We begin by constructing another family of solutions to Struwe's differential equation (4.8).

LEMMA 4.23. *For any $\delta > 0$ and $d \geq 4$ there exists a solution $\phi : [\delta, 2\delta] \to \mathbb{R}$ of* (4.8) *with the following properties (see Figure 4.8):*

(a) *$\phi'(\delta) = +\infty$ and $\phi(\delta) \geq d\delta$;*
(b) *$\phi(2\delta) = 2d\delta$ and $\phi'(2\delta) \leq d$;*
(c) *ϕ satisfies* (4.9) *and hence is an i-convex shape.*

PROOF. The proof is similar to the proof of Lemma 4.15. By rescaling, it suffices to consider the case $\delta = 1$. Define the solution ϕ by $\phi'(r) := 1/\sqrt{\ln r}$ and $\phi(2) := 2d$, thus

$$\phi(1) = 2d - \int_1^2 \frac{du}{\sqrt{\ln u}}.$$

Estimating the integral as in (4.10) and using $d \geq 4$, we find

$$\phi(1) \geq 2d - \frac{2}{\sqrt{\ln 2}} \geq d + 4 - \frac{2}{\sqrt{\ln 2}} \geq d,$$

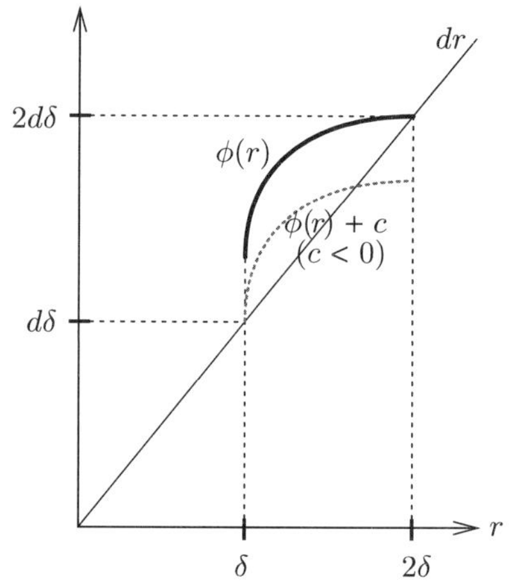

FIGURE 4.8. Another solution of Struwe's differential equation.

since $\sqrt{\ln 2} \geq 1/2$. Concavity of ϕ implies $\phi(r) \geq dr$ for all $r \in [1,2]$, and in particular $\phi'(2) \leq d$. So it only remains to check inequality (4.9). Denoting by $\sim$ equality up to a positive factor, we compute

$$\begin{aligned}\frac{\phi'^3}{2r} - \frac{1}{\phi}(1+\phi'^2) &\geq \frac{\phi'^3}{2r} - \frac{1}{dr}(1+\phi'^2) \\ &\sim d\phi'^3 - 2 - 2\phi'^2 \\ &\sim d - 2(\ln r)^{3/2} - 2(\ln r)^{1/2}.\end{aligned}$$

The function on the right hand side is decreasing in r. So its minimum is achieved for $r = 2$ and has the value

$$d - 2(\ln 2)^{3/2} - 2(\ln 2)^{1/2} > 4 - 2 - 2 = 0,$$

since $d \geq 4$ and $\sqrt{\ln 2} < 1$. □

REMARK 4.24. For ϕ as in Lemma 4.23 and any constant $c \leq 0$, the part of the function $\phi + c$ that lies above the linear function dr is i-convex; see Figure 4.8. Indeed, the last part of the proof applied to $\phi + c$ estimates the quantity in inequality (4.9) by $d - 2(\ln r_1)^{3/2} - 2(\ln r_1)^{1/2}$, where r_1 is the larger intersection point of $\phi + c$ and dr. Since $r_1 \leq 2$, this is positive.

Extend the standard function to $\lambda < 0$ and $a > 1$ by

$$S_\lambda(r) := \sqrt{ar^2 - \lambda^2}, \qquad r \geq |\lambda|/\sqrt{a}.$$

Note that S_λ is the shape of an i-convex hypersurface because its graph is a level set of the i-convex function $\phi(r,R) = ar^2 - R^2$.

We say that a family of i-convex shapes $\phi_\lambda : [\delta,\beta] \to \mathbb{R}_+$ with $\phi'_\lambda(\delta) = \infty$ is *(piecewise) smooth* if their graphs $\{R = \phi_\lambda(r)\}$, extended by the vertical line below $\big(\delta, \phi_\lambda(\delta)\big)$, form a (piecewise) smooth family of smooth curves in the positive quadrant $Q \subset \mathbb{R}^2$.

LEMMA 4.25. *Let $L_\lambda(r) = \lambda + d_\lambda r$, $0 < r \leq \beta$, $0 \leq \lambda \leq 1$, be an increasing smooth family of i-convex shapes, where $\lambda \mapsto d_\lambda$ is decreasing with $d_0 = 8$ and*

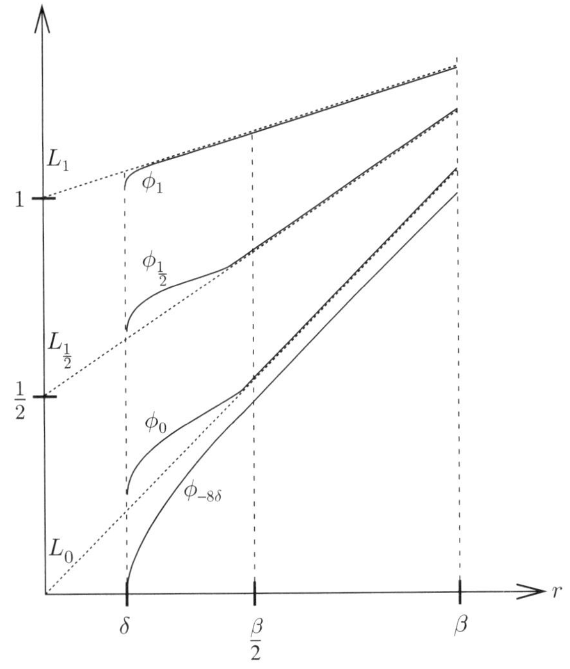

FIGURE 4.9. Bending down linear shapes.

$0 < d_1 \leq 1$. *Then for any sufficiently small* $\delta \in (0, \beta/4)$ *there exists a piecewise smooth family of increasing i-convex shapes* $\phi_\lambda : [\delta, \beta] \to \mathbb{R}$, $-8\delta \leq \lambda \leq 1$, *with the following properties (see Figure 4.9):*

(a) $\phi_{-8\delta}(r) = \sqrt{64r^2 - 64\delta^2}$ *for all* $r \geq \delta$;
(b) $\phi_\lambda(r) = \sqrt{64r^2 - \lambda^2}$ *for* $-8\delta \leq \lambda \leq 0$ *and* $r \geq \beta/2$;
(c) $\phi_\lambda(r) = L_\lambda(r)$ *for* $0 \leq \lambda \leq 1$ *and* $r \geq \beta/2$;
(d) $\phi'_\lambda(\delta) = \infty$ *for all* λ;
(e) $1 < \phi_1(\delta) < 1 + \beta$.

PROOF. **Step 1.** For each $\lambda \in (0, 1]$, set $K_\lambda := e^{4/d_\lambda^2}$. Pick a smooth family of constants $\delta_\lambda > 0$ such that $\lambda\delta_\lambda$ increases with λ and

$$4K_\lambda\delta_\lambda \leq (\ln K_\lambda)^{-3/2}, \qquad K_\lambda\lambda\delta_\lambda < \beta/2.$$

By Lemma 4.15, there exist i-convex solutions $\phi_\lambda : [\lambda\delta_\lambda, K_\lambda\lambda\delta_\lambda] \to \mathbb{R}$ of (4.8) satisfying

- $\phi'_\lambda(\lambda\delta_\lambda) = +\infty$ and $\phi_\lambda(\lambda\delta_\lambda) \geq \lambda + d_\lambda\lambda\delta_\lambda$;
- $\phi_\lambda(K_\lambda\lambda\delta_\lambda) = \lambda + d_\lambda K_\lambda\lambda\delta_\lambda$ and $\phi'_\lambda(K_\lambda\lambda\delta_\lambda) \leq d_\lambda$.

Step 2. From $d_0 = 8$ and $d_1 < 1$, we conclude $K_0 = e^{1/16} < 2$ and $K_1 \geq e^4 > 2$. Hence there exists a $0 < \bar{\lambda} < 1$ with $K_{\bar{\lambda}} = 2$. Set $\bar{\delta} := \bar{\lambda}\delta_{\bar{\lambda}} < \beta/4$. By Lemma 4.23 (with $d = 8$), there exists an i-convex solution $\bar{\phi} : [\bar{\delta}, 2\bar{\delta}] \to \mathbb{R}$ of (4.8) satisfying

- $\bar{\phi}'(\bar{\delta}) = +\infty$ and $\bar{\phi}(\bar{\delta}) \geq 8\bar{\delta}$;
- $\bar{\phi}(2\bar{\delta}) = 16\bar{\delta}$ and $\bar{\phi}'(2\bar{\delta}) \leq 8$.

By Lemma 4.13(a), the functions

$$\bar{\phi}_\lambda := \bar{\phi}(r) + L_\lambda(2\bar{\delta}) - L_0(2\bar{\delta}) \geq \bar{\phi}(r)$$

are i-convex for $0 \leq \lambda \leq \bar{\lambda}$ and $\bar{\delta} \leq r \leq 2\bar{\delta}$. Note that the functions $\phi_{\bar{\lambda}}$ and $\bar{\phi}_{\bar{\lambda}}$ have the same value at $r = 2\bar{\delta}$ and derivative ∞ at $r = \bar{\delta}$. Since they both solve the second order differential equation (4.8), according to Remark 4.16 they coincide on $[\bar{\delta}, 2\bar{\delta}]$. Thus the families constructed above fit together to a continuous family $(\widehat{\phi}_\lambda)_{\lambda \in [0,1]}$ with $\widehat{\phi}_\lambda = \phi_\lambda : [\lambda\delta_\lambda, K_\lambda\lambda\delta_\lambda] \to \mathbb{R}_+$ for $\lambda \geq \bar{\lambda}$, and $\widehat{\phi}_\lambda = \bar{\phi}_\lambda : [\bar{\delta}, 2\bar{\delta}] \to \mathbb{R}_+$ for $\lambda \leq \bar{\lambda}$. Set $\bar{\delta}_\lambda := \lambda\delta_\lambda$ for $\lambda \geq \bar{\lambda}$ and $\bar{\delta}_\lambda := \bar{\delta}$ for $\lambda \leq \bar{\lambda}$ and define $\widetilde{\phi}_\lambda : [\bar{\delta}_\lambda, \beta] \to \mathbb{R}_+$ by

$$\widetilde{\phi}_\lambda(r) := \begin{cases} \widehat{\phi}_\lambda(r) & \text{for } r \leq K_\lambda\delta_\lambda, \\ L_\lambda(r) & \text{for } r \geq K_\lambda\delta_\lambda. \end{cases}$$

After smoothing, the family $\widetilde{\phi}_\lambda$ is i-convex and agrees with L_λ for $r \geq \beta/2$.

Step 3. For $-8\bar{\delta} \leq \tau \leq 0$ consider the functions $\bar{\phi}_\tau := \bar{\phi} + \tau : [\bar{\delta}, 2\bar{\delta}] \to \mathbb{R}_+$. By Remark 4.24, the portion of $\bar{\phi}_\tau$ above the linear function L_0 is i-convex. Thus for $0 < \delta < \bar{\delta}/2$ sufficiently small, the portion of $\bar{\phi}_\tau$ above the function $S_{-8\delta}$ is i-convex. Here $S_\lambda(r) = \sqrt{64r^2 - \lambda^2}$ is the standard function defined above with $a = 64$ and $\lambda \in [-8\delta, 0]$. For $-8\delta \leq \lambda \leq 0$ define $\widetilde{\phi}_\lambda : [\bar{\delta}, \beta] \to \mathbb{R}_+$ by

$$\widetilde{\phi}_\lambda(r) := \begin{cases} \bar{\phi}(r) + S_\lambda(2\bar{\delta}) - S_0(2\bar{\delta}) & \text{for } r \leq 2\bar{\delta}, \\ S_\lambda(r) & \text{for } r \geq 2\bar{\delta}. \end{cases}$$

Since $S_\lambda(r) - S_0(r)$ is increasing in r for $\lambda > 0$, the condition $\bar{\phi}(\bar{\delta}) \geq 8\bar{\delta}$ ensures that $\widetilde{\phi}_\lambda$ lies above S_λ. Thus after smoothing, the family $\widetilde{\phi}_\lambda$ is i-convex for $-8\delta \leq \lambda \leq 1$ and agrees with L_λ if $\lambda \geq 0$ (resp. S_λ if $\lambda \leq 0$) for $r \geq \beta/2$. Now define $\widetilde{\psi}_\lambda : [\delta, \beta] \to \mathbb{R}_+$ by

$$\widetilde{\phi}_\lambda(r) := \begin{cases} S_{-8\delta(r)} & \text{for } r \leq \bar{\delta}_\lambda, \\ \widetilde{\phi}(r) & \text{for } r \geq \bar{\delta}_\lambda. \end{cases}$$

After smoothing, the family $\widetilde{\psi}_\lambda$ is i-convex for $-8\delta \leq \lambda \leq 1$ and satisfies conditions (b)–(d).

Step 4. To arrange condition (a), note that $\widetilde{\psi}_{-8\delta} = \max(S_{-8\delta}, \bar{\phi}_{\bar{\tau}})$ for some $\bar{\tau} < 0$. By the discussion above, the functions $\max(S_{-8\delta}, \bar{\phi}_\tau)$ are i-convex for $-8\bar{\delta} \leq \tau \leq 0$. For δ sufficiently small, we have $\max(S_{-8\delta}, \bar{\phi}_{-8\bar{\delta}}) = S_{-8\delta}$. After rescaling in the parameter λ, this yields a family $\widetilde{\psi}_\lambda$ satisfying conditions (a)–(d).

Step 5. To arrange condition (e), set $\delta_t := (2 - t)\delta_1 + (t - 1)\delta$ for $t \in [1, 2]$ and let $\phi_t : [\delta_t, K_1\delta_t] \to \mathbb{R}$ be the i-convex shape from Lemma 4.15 with $\lambda = 1$ and δ replaced by δ_t. For $\lambda \in [1, 2]$ define $\widetilde{\psi}_\lambda : [\delta, \beta] \to \mathbb{R}_+$ by

$$\widetilde{\phi}_\lambda(r) := \begin{cases} S_{-8\delta(r)} & \text{for } r \leq \delta_\lambda, \\ \phi_\lambda(r) & \text{for } \delta_\lambda \leq r \leq \delta_1, \\ L_1(r) & \text{for } r \geq \delta_1. \end{cases}$$

For $\lambda = 1$ this matches the previous family $\widetilde{\psi}_\lambda$, so rescaling in λ yields the desired family ϕ_λ. $\square$

The following result is a family version of Lemma 4.19.

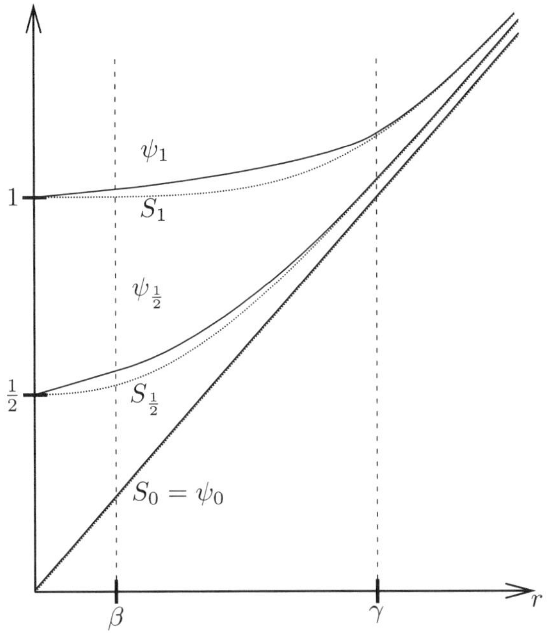

FIGURE 4.10. Interpolation between standard and linear shapes.

LEMMA 4.26. *For any* $\gamma > 0$ *there exists a constant* $0 < \beta < \gamma$ *and a smooth family of increasing i-convex shapes* $\psi_\lambda : \mathbb{R}_+ \to \mathbb{R}_+$, $\lambda \in [0,1]$, *with the following properties (see Figure* 4.10*):*

(a) $\psi_0(r) = 8r$ *for all* r;
(b) $\psi_\lambda(r) = \lambda + d_\lambda r$ *for* $r \le \beta$ *and all* λ, *where* $\lambda \mapsto d_\lambda$ *is decreasing with* $d_0 = 8$ *and* $0 < d_1 \le 1$;
(c) $\psi_\lambda(r) = \sqrt{64r^2 + \lambda^2}$ *for* $r \ge \gamma$ *and all* λ.

PROOF. Set $a := 64$ and $c := 2$. With this choice and $\lambda \in (0,1]$ we consider the functions

$$S_\lambda(r) = \sqrt{\lambda^2 + ar^2}, \qquad Q_{b,\lambda}(r) = \lambda + br + cr^2/2\lambda, \qquad L_{d,\lambda}(r) = \lambda + dr$$

as above. Here the constants b, d will vary in the course of the proof but always satisfy the condition

$$0 < b < d \le b + b^3 < 8. \tag{4.14}$$

Then the numerical condition in Lemma 4.18(c), $2b^2(64+2)^2 < (64-2)^3$, holds because $b < 4$. Hence all the numerical conditions in Lemma 4.18 are satisfied, so the functions $S_\lambda, Q_{b,\lambda}, L_{d,\lambda}$ intersect at points $\lambda r_{QL}(b,d), \lambda r_{SL}(d), \lambda r_{SQ}(b)$ satisfying

$$r_{QL}(b,d) = \frac{2(d-b)}{c}, \qquad r_{SL}(d) = \frac{2d}{a-d^2}, \qquad 0 < r_{SQ}(b) < \frac{4b}{a-c}.$$

By condition (4.14) we have

$$r_{QL}(b,d) \le b^3 < d^3,$$

so the numerical condition in Lemma 4.17(e) is satisfied for $r \le \lambda r_{QL}(b,d)$. It follows that the shape functions $S_\lambda(r)$ and $Q_{b,\lambda}(r)$ are i-convex for all r, and $L_{d,\lambda}(r)$ is i-convex for $r \le \lambda r_{QL}(b,d)$. For each triple (b,d,λ) we consider the function

$$\psi_{b,d,\lambda} := \max\left(S_\lambda, Q_{b,\lambda}, L_{d,\lambda}\right) = \lambda\psi_{b,d,1}(\cdot/\lambda).$$

This function will be i-convex provided that the region where it coincides with $L_{d,\lambda}(r)$ is contained in the interval $[0, \lambda r_{QL}(b,d)]$. We say that $\psi_{b,d,\lambda}$ is of Type

(a) if $r_{QL}(b,d) \le r_{SL}(d) \le r_{SQ}(b)$;
(b) if $r_{SQ}(b) \le r_{SL}(d) \le r_{QL}(b,d)$;
(c) if $r_{SQ}(b) \le r_{QL}(b,d) \le r_{SL}(d)$;

see Figure 4.6 (where we have dropped the parameters b, d, λ). Thus the function $\psi_{b,d,\lambda}$ is i-convex for Types (a) and (b), but not necessarily for Type (c).

After these preparations, we now construct the family ψ_λ in four steps.

Step 1. Consider $\lambda = 1$. Pick a pair (b_1, d_1) satisfying (4.14) and such that

$$r_{QL}(b_1, d_1) = \frac{2(d_1 - b_1)}{c} < r_{SQ}(b_1) < \frac{4b_1}{a-c} < \gamma.$$

Then the shape function $\psi_{b_1,d_1,\lambda}$ is of Type (a) and therefore i-convex for all $\lambda > 0$, and it agrees with S_λ for $r \ge \beta$. Note that in particular we have $r_{SL}(d_1) < \gamma$.

Step 2. Fix a parameter $0 < \lambda^* < \gamma/8$. This condition ensures that for any pair (b,d) satisfying (4.14) we have $\lambda^* r_{QL}(b,d), \lambda^* r_{SQ}(b) < \gamma$. We may assume that b_1 in Step 1 is chosen so small that $b_1^2 < c/(a-b^2)$ for all $b \in [0, b_1]$. Then for any $b \in [0, b_1]$ such that (b, d_1) satisfies (4.14), we have

$$r_{QL}(b, d_1) = \frac{2(d_1 - b)}{c} \le \frac{2b^3}{c} < \frac{2b}{a-b^2} < \frac{2d_1}{a-d_1^2} = r_{SL}(d_1) < \gamma.$$

Let $b_1^* \in (0, b_1]$ be the solution of $b_1^* + (b_1^*)^3 = d_1$. We claim that for all $b \in [b_1^*, b_1]$ the function ψ_{b,d_1,λ^*} is of Type (a) and therefore i-convex. Indeed, by Step 1 this holds for $b = b_1$. Since $r_{SQ}(b)$ depends smoothly on b, if ψ_{b,d_1,λ^*} changes its type there must exist a $b \in [b_1^*, b_1]$ for which $r_{SQ}(b) = r_{SL}(d_1)$. But this implies also $r_{QL}(b, d_1) = r_{SL}(d_1)$, contradicting the preceding inequality.

Step 3. For $b > 0$ consider the function

$$f(b) := \frac{r_{QL}(b,d)}{r_{SL}(d)}\bigg|_{d=b+b^3} = \frac{(d-b)(a-d^2)}{cd}\bigg|_{d=b+b^3} = \frac{b^2\left(a-(b+b^3)^2\right)}{c(1+b^2)}.$$

A short computation shows that $f(0) = 0$, $f(1) > 1$ and $f'(b) > 0$ for all $b \in (0,1)$. Thus there exists a unique $b_2^* \in (0,1)$ with $f(b_2^*) = 1$, i.e., $r_{QL}(b, b+b^3) = r_{SL}(b+b^3)$ precisely for $b = b_2^*$. Since $b_1^* < b_2^*$, the function $\psi_{b,b+b^3,\lambda^*}$ is of Type (a) and therefore i-convex for all $b \in [b_1^*, b_2^*]$. For $b \in [b_2^*, 1]$ we have $r_{QL}(b, b+b^3) \ge r_{SL}(b+b^3)$, so the function $\psi_{b,b+b^3,\lambda^*}$ is of Type (b) and therefore also i-convex. Combining this, we see that the function $\psi_{b,b+b^3,\lambda^*}$ is i-convex for all $b \in [b_1^*, 1]$. Moreover, $\lambda^* r_{SL}(b+b^3) < \gamma$ for all $b \in [b_1^*, 1]$, so $\psi_{b,b+b^3,\lambda^*}(r) = S_\lambda^*(r)$ for $r \ge \gamma$.

Step 4. The previous step leads for $b = 1$ and $d = b + b^3 = 2$ to the function $\psi_{1,2,\lambda^*}$. For $d \in [2, 8]$ define b_d, λ_d by the conditions

$$b_d + b_d^3 = d, \qquad \lambda_d r_{SL}(d) = \gamma,$$

so

$$\lambda_d = \frac{\gamma(a-d^2)}{2d}.$$

Note that $b_2 = 1$, $\lambda_2 > \lambda^*$, and $\psi_{1,\lambda,2}$ is i-convex for all $\lambda \in [\lambda^*, \lambda_2]$ and agrees with S_λ for $r \geq \gamma$. The same holds for the functions ψ_{b_d,d,λ_d} for all $d \in [2,8]$. In the limit $d \to 8$ we find $\lambda_8 = 0$ and thus the linear function

$$\psi_{b_8,8,0}(r) = 8r.$$

Now we combine the homotopies of i-convex functions $\psi_{b,d,\lambda}$ in Steps 1–4: starting from $(b_1, d_1, 1)$, we first decrease λ to (b_1, d_1, λ^*) (Step 1), then decrease b to (b_1^*, d_1, λ^*) (Step 2), next increase (b,d) simultaneously do $(1, 2, \lambda^*)$ (Step 3), and finally increase (b,d) and decrease λ simultaneously to $(b_8, 8, 0)$. By construction, each function $\psi_{b,d,\lambda}$ during this homotopy coincides with the corresponding standard function S_λ for $r \geq \gamma$ and with the linear function L_λ for $r \leq \beta$ for some small $\beta > 0$. Moreover, during the homotopy λ is nonincreasing and d is nondecreasing. Smooth the functions $\psi_{b,d,\lambda}$ and perturb the homotopy such that λ is strictly decreasing from 1 to 0 and d is strictly increasing from $d_1 \leq 1$ to 8. The resulting homotopy, parametrized by $\lambda \in [0,1]$, is the desired family ψ_λ. □

Now we are ready to prove the main result of this section. Recall that

$$S_\lambda(r) = \begin{cases} \sqrt{64r^2 - \lambda^2}: & \lambda < 0, \\ \sqrt{64r^2 + \lambda^2}: & \lambda \geq 0. \end{cases}$$

PROPOSITION 4.27. *For every $\gamma > 0$ and any sufficiently small $\delta \in (0, \gamma)$ there exists a piecewise smooth family of increasing i-convex shapes $\chi_\lambda : [\delta, \infty) \to \mathbb{R}$, $-8\delta \leq \lambda \leq 1$, with the following properties (see Figure 4.11):*

(a) $\chi_{-8\delta}(r) = \sqrt{64r^2 - 64\delta^2}$ for all $r \geq \delta$;
(b) $\chi_\lambda(r) = S_\lambda(r)$ for $r \geq \gamma$ and all λ;
(c) $\chi'_\lambda(\delta) = \infty$ for all λ;
(d) $1 < \chi_1(\delta) < 1 + \gamma$.

PROOF. Let $(\psi_\lambda)_{\lambda \in [0,1]}$ be the family of i-convex shapes from Lemma 4.26 which agree with the standard functions $S_\lambda(r) = \sqrt{64r^2 - \lambda^2}$ for $r \geq \gamma$ and with the linear functions $\lambda + d_\lambda r$ for $r \leq \beta$, for some $\beta \in (0, \gamma)$ and some decreasing family $\lambda \mapsto d_\lambda$ with $d_0 = 8$ and $0 < d_1 \leq 1$. On the other hand, Lemma 4.25 provides us with a family $(\phi_\lambda)_{\lambda \in [-8\delta, 1]}$ which agrees with $\lambda + d_\lambda r$ for $r \geq \beta/2$ and $\lambda \in [0,1]$, and with $S_\lambda(r) = \sqrt{64r^2 - \lambda^2}$ for $r \geq \beta/2$ and $\lambda \in [-8\delta, 0]$, for the given β, d_λ and sufficiently small $\delta \in (0, \beta/4)$. Since $\psi_0(r) = 8r = S_0(r)$, we can define the required family $\chi_\lambda : [\delta, \infty) \to \mathbb{R}$, $-8\delta \leq \lambda \leq 1$, by

$$\chi_\lambda(r) := \begin{cases} \phi_\lambda(r): & r \in [\delta, \beta/2], \lambda \in [-8\delta, 1], \\ \psi_\lambda(r): & r > \beta/2, \lambda \in [0,1], \\ S_\lambda(r): & r > \beta/2, \lambda \in [-8\delta, 0]. \end{cases}$$ □

REMARK 4.28. The proofs of Lemmas 4.25 and 4.26 show that the family χ_λ in Proposition 4.27 can be chosen to depend smoothly on the parameters $\gamma > 0$ and the sufficiently small $\delta \in (0, \gamma)$. Let us choose a smooth decreasing function $\gamma \mapsto \delta(\gamma)$ such that $\delta(\gamma) \in (0, \gamma)$ is sufficiently small in the sense of Proposition 4.27 (in particular $\delta(\gamma) \to 0$ as $\gamma \to 0$). Then we obtain a smooth family of increasing i-convex shapes $\chi_{\lambda,\gamma} : [\delta(\gamma), \infty) \to \mathbb{R}$, $-8\delta(\gamma) \leq \lambda \leq 1$, $\gamma > 0$ with the properties in Proposition 4.27.

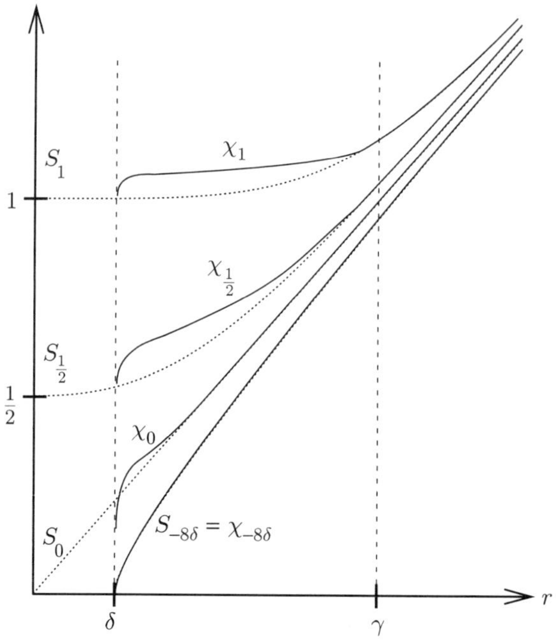

FIGURE 4.11. The family χ_λ of i-convex shapes cooriented from above.

According to Proposition 3.41, the family of i-convex hypersurfaces $\{R = \chi_\lambda(r)\}$ in Proposition 4.27 can be turned into a foliation. The following refinement of Proposition 3.41 shows that this can be done within the class of shapes.

COROLLARY 4.29. *If the hypersurfaces Σ_λ in Proposition* 3.41 *are all given by shapes $R = \phi_\lambda(r)$ in $\mathbb{C}^n$, then so is Σ.*

PROOF. Note that a hypersurface is given by a shape if and only if it is invariant under the group of rotations $G = O(k) \times O(2n-k)$ and transverse to the vector field $R\partial_R$. The latter property is clearly preserved during the proof of Proposition 3.41. For the first property, note that the function ψ constructed in Proposition 3.38 depends only on $M(\Sigma)$ and the Euclidean distance from Σ, both of which are G-invariant if Σ is. Clearly, G-invariance is preserved under taking the maximum. According to Remark 3.14, the smoothing on $\mathbb{C}^n$ can be done by convolution followed by interpolation. Convolution preserves G-invariance if we choose the smoothing kernel G-invariant, and the interpolation can be done in the class of G-invariant functions. □

PROOF OF THEOREM 4.2 (i)–(iii). We first apply Proposition 3.26 to the i-convex functions $\phi(r) = ar^2 - R^2$ and $\psi(r,R) = 64r^2 - R^2$ which coincide to first order along the totally real submanifold $\{r = 0\}$ and have a nondegenerate critical point at the origin. So we find an i-convex function $\vartheta : \mathbb{C}^n \to \mathbb{R}$ with a unique critical point at the origin which coincides with $ar^2 - R^2$ on $\{r \geq \gamma\}$ and with $64r^2 - R^2$ on $\{r \leq \gamma'\}$, for some $\gamma' \in (0,\gamma)$. Thus the level sets $\vartheta = c \geq -1$ coincide along $\{r = \gamma'\}$ with the shapes χ_λ from Proposition 4.27 (extended by S_λ for $\lambda < -8\delta$). By Proposition 3.25 we can modify the hypersurfaces $\Sigma_\lambda = \{R = \chi_\lambda(r)\}$, keeping them fixed near $r = \gamma'$ and near $\lambda = 1$, to a foliation of the region $\{r \le \gamma', R \le \chi_1(r)\}$ by i-convex hypersurfaces $\widetilde{\Sigma}_\lambda$. By Corollary 4.29 we can arrange that the $\widetilde{\Sigma}_\lambda$ are again given by shapes. The function ϑ on $\{\gamma' \le r \le \gamma\}$ extends canonically to a function Ψ on $\{r \le \gamma, R \le \chi_1(r)\}$ with regular i-convex level sets $\widetilde{\Sigma}_\lambda$. □

4.7. Convexity estimates

The i-convex hypersurfaces on $\mathbb{C}^n$ constructed in the previous sections can be transplanted to complex manifolds by holomorphic embeddings, providing J-convex surroundings for *real analytic* totally real submanifolds. In this section we derive quantitative estimates on the normalized modulus of i-convexity of these hypersurfaces which ensure that they remain J-convex under "approximately holomorphic" embeddings. This provides J-convex surroundings for *smooth* totally real submanifolds, and simplifies many subsequent arguments by avoiding real analytic approximations.

We will only consider i-convex shapes $\phi(r)$ for $0 < r \le 1$ and with $0 < \phi \le 2$. By Proposition 4.8 this implies $\phi' > r/\phi \ge r/2$ and

$$M(II_\Sigma) \ge \frac{\phi'}{r\sqrt{1+\phi'^2}} \ge \frac{1/2}{\sqrt{1+r^2/4}} \ge \frac{1/2}{\sqrt{1+1/4}} \ge \frac{1}{3}.$$

Thus the normalized modulus of i-convexity $\mu(\Sigma)$ (see Section 3.7) satisfies

$$3\mu(\Sigma) = \frac{3m(\mathbb{L}_\Sigma)}{\max\{M(II_\Sigma),1\}} \ge \frac{3m(\mathbb{L}_\Sigma)}{\max\{3M(II_\Sigma),1\}} = \frac{m(\mathbb{L}_\Sigma)}{M(II_\Sigma)} =: \bar{\mu}(\Sigma).$$

Therefore, in the following it will suffice to estimate $\bar{\mu}(\Sigma)$ from below. One advantage of $\bar{\mu}$ over μ is that the former is invariant under rescaling $(r, R) \mapsto (\lambda r, \lambda R)$. We will also refer to $\bar{\mu}$ as the normalized modulus of convexity.

Our point of departure is the following consequence of Proposition 4.8.

PROPOSITION 4.30. *Suppose a hypersurface $\Sigma \subset \mathbb{C}^n$ is given by the shape $R = \phi(r)$, cooriented from above, and that $r > 0$, $\phi' > 0$. Then the inequality $\bar{\mu}(\Sigma) > \varepsilon > 0$ for the normalized modulus of i-convexity of Σ is equivalent to the following system of inequalities, stronger than* (4.3) *and* (4.4)*:*

$$\mathcal{L}^{\perp}_{\varepsilon,1}(\phi) := -\varepsilon\frac{|\phi''|}{1+\phi'^2} + \frac{\phi'}{r} - \frac{1}{\phi} > 0, \tag{4.15}$$

$$\mathcal{L}^{\perp}_{\varepsilon,2}(\phi) := (1-\varepsilon)\frac{\phi'}{r} - \frac{1}{\phi} > 0, \tag{4.16}$$

$$\mathcal{L}^{2}_{\varepsilon,1}(\phi) := \frac{s^2\phi''}{r^2} - \varepsilon|\phi''| + \frac{\phi' u^2}{r^3} + \frac{\phi'^3}{r} - \frac{1}{\phi}\Big(1 + \frac{\phi'^2 u^2}{r^2}\Big) > 0, \tag{4.17}$$

$$\mathcal{L}^{2}_{\varepsilon,2}(\phi) := \frac{\phi'' s^2}{r^2} + \frac{\phi' u^2}{r^3} + \frac{\phi'^3}{r} - \frac{1}{\phi}\Big(1 + \frac{\phi'^2 u^2}{r^2}\Big) - \frac{\varepsilon\phi'}{r}(1+\phi'^2) > 0 \tag{4.18}$$

for all (s, u) with $s^2 + u^2 = r^2$. If $\phi' > 0, \phi'' \le 0$ and $0 < \varepsilon < 1$, then the following inequality implies $\bar{\mu}(\Sigma) > \varepsilon$:

$$\phi'' + \frac{(1-\varepsilon)\phi'^3}{r} - \frac{1}{\phi}(1+\phi'^2) - \frac{\varepsilon\phi'}{r} > 0. \tag{4.19}$$

In particular, if $\phi' > \sigma > 0, \phi'' \le 0$ and $0 < \varepsilon \le \min\left(\frac{1}{8}, \frac{\sigma^2}{8}\right)$, then the following inequality is sufficient for $\bar{\mu}(\Sigma) > \varepsilon$:

$$\phi'' + \frac{3\phi'^3}{4r} - \frac{1}{\phi}(1+\phi'^2) > 0. \tag{4.20}$$

PROOF. By Proposition 4.8, the condition $\bar{\mu}(\Sigma) > \varepsilon$ is equivalent to

$$\min\left(\frac{1}{\sqrt{1+\phi'^2}}\left(\frac{\phi'}{r} - \frac{1}{\phi}\right), \mathbb{L}_0\right) > \varepsilon \max\left(\frac{|\phi''|}{(1+\phi'^2)^{\frac{3}{2}}}, \frac{|\phi'|}{r\sqrt{1+\phi'^2}}\right),$$

where $\mathbb{L}_0$ is given by

$$\mathbb{L}_0 = \frac{1}{(1+\phi'^2)^{3/2}}\left(\frac{\phi'' s^2}{r^2} + \frac{\phi' u^2}{r^3} + \frac{\phi'^3}{r} - \frac{1}{\phi}\left(1 + \frac{\phi'^2 u^2}{r^2}\right)\right).$$

Expanding this condition yields the four inequalities (4.15) to (4.18).

If $\phi'' \le 0$ inequality (4.19) clearly implies (4.18) for all ε. To see that it implies (4.17), we divide (4.19) by $1-\varepsilon$ and drop the last term to obtain

$$\frac{\phi''}{1-\varepsilon} + \frac{\phi'^3}{r} - \frac{1}{(1-\varepsilon)\phi}(1+\phi'^2) > 0.$$

Since $1/(1-\varepsilon) > 1+\varepsilon$ this implies (4.17). For (4.16) note that

$$0 < \phi'' + \frac{(1-\varepsilon)\phi'^3}{r} - \frac{1}{\phi}(1+\phi'^2) - \frac{\varepsilon\phi'}{r} < \phi'^2\left(\frac{(1-\varepsilon)\phi'}{r} - \frac{1}{\phi}\right),$$

and hence $\mathcal{L}^{\perp}_{2,\varepsilon} = \frac{(1-\varepsilon)\phi'}{r} - \frac{1}{\phi} > 0$. Furthermore, we have

$$\begin{aligned}
(1+\phi'^2)\mathcal{L}^{\perp}_{1,\varepsilon} &= \varepsilon\phi'' + (1+\phi'^2)\left(\frac{\phi'}{r} - \frac{1}{\phi}\right) > \varepsilon\phi'' + \phi'^2\left(\frac{\phi'}{r} - \frac{1}{\phi}\right) \\
&> \phi'' + \phi'^2\left(\frac{\phi'}{r} - \frac{1}{\phi}\right) \\
&> \phi'' + \frac{(1-\varepsilon)\phi'^3}{r} - \frac{1}{\phi}(1+\phi'^2) - \frac{\varepsilon\phi'}{r} > 0,
\end{aligned}$$

and hence (4.19) implies (4.15).

Finally, the derivative bound $\phi' > \sigma$ implies that $\frac{\varepsilon\phi'}{r} < \frac{\phi'^3}{8r}$ for $\varepsilon \le \frac{\sigma^2}{8}$, and hence (4.20) implies (4.19) for $\varepsilon \le \min\left(\frac{1}{8}, \frac{\sigma^2}{8}\right)$. □

The following is a quantitative version of Lemma 4.15.

LEMMA 4.31. *For any $d, K, \delta, \lambda > 0$ satisfying $K = e^{4/d^2}$ and $8K\delta \le (\ln K)^{-3/2}$ there exists a solution $\phi : [\lambda\delta, K\lambda\delta] \to \mathbb{R}$ of (4.8) which satisfies properties (a) and (b) from Lemma 4.15. In addition, ϕ satisfies (4.20) for $r \in [\lambda\delta, K\lambda\delta)$, and hence the corresponding hypersurface Σ satisfies $\bar{\mu}(\Sigma) > \min\left(\frac{1}{8}, \frac{d^2}{32}\right)$.*

PROOF. By rescaling we need only consider the case $\lambda = 1$. We define ϕ as in the proof of Lemma 4.15, so it satisfies (a) and (b). Moreover, $K = e^{4/d^2}$ implies $\phi'(K\delta) = 1/\sqrt{\ln K} = d/2$. Since ϕ is concave, this shows that $\phi'(r) \ge d/2$ for all $r \in [\delta, K\delta]$. Hence according to Proposition 4.30(iv) with $\sigma = d/2$, inequality (4.20) is sufficient for $\bar{\mu}(\Sigma) > \varepsilon = \min\left(\frac{1}{8}, \frac{d^2}{32}\right)$. Using that ϕ is a solution of (4.8), this reduces to the inequality

$$\frac{\phi'^3}{4r} - \frac{1}{\phi}(1+\phi'^2) > 0.$$

Arguing as in (4.11), we conclude that the bound $8K\delta \le (\ln K)^{-3/2}$ yields this inequality. □

The next lemma is a quantitative version of Lemma 4.17.

LEMMA 4.32. *(a) The function $S_\lambda(r)$ for $\lambda \in \mathbb{R}$, $a > 1$, and $r > \max\{0, -\lambda/\sqrt{a}\}$ satisfies $\bar{\mu} = (a-1)/a$.*

(b) The function $Q_\lambda(r)$ for $\lambda > 0$, $b > 0$, $c > 1$, and $r > 0$ satisfies $\bar{\mu} \ge \min\{1 - 1/c, b^2/(1-b^2)\}$.

(c) The function $Q_\lambda(r)$ for $\lambda > 0$, $b = 4 - c$, $0 \le c \le 1$, and $0 < r \le 2\lambda$ satisfies $\bar{\mu} \ge 1/5$.

(d) The function $L_\lambda(r)$ for $\lambda \ge 0$, $d > 1$, and $r > 0$ satisfies $\bar{\mu} \ge \frac{d^2-1}{d^2(d^2+1)}$.

(e) The function $L_\lambda(r)$ for $\lambda > 0$, $d > 0$, and $0 < r < \lambda d^3$ satisfies $\bar{\mu} \ge \min(d^6, 1)/4$.

PROOF. (a) The Hessian and the complex Hessian of the function $\Psi(r, R) = \frac{1}{2}(ar^2 - R^2)$ defining Σ are

$$\mathrm{Hess}_\Psi = \mathrm{diag}(a, \dots, a, -1, \dots, -1), \qquad H_\Psi = \mathrm{diag}(a-1, \dots, a-1).$$

Thus $m(H_\Psi) = a - 1$, $M(\mathrm{Hess}_\Psi) = 1$, and Lemma 2.24 yields

$$\bar{\mu}(\Sigma) = \frac{m(H_\Psi)}{M(\mathrm{Hess}_\Psi)} = \frac{a-1}{a}.$$

For properties (b)–(e) note that rescaling $(r, R) \mapsto (\lambda r, \lambda R)$ preserves $\bar{\mu}(\Sigma) = m(\Sigma)/M(\Sigma)$, hence we can assume in the following that $\lambda = 1$.

(b) We have $\mathcal{L}^{\perp}_{\varepsilon,1}(Q) > \frac{Q'}{r} - \frac{1}{Q} > 0$ and

$$\mathcal{L}^{\perp}_{\varepsilon,2}(Q) = (1-\varepsilon)\frac{b + cr}{r} - \frac{1}{Q(r)} > (1-\varepsilon)c - 1 \ge 0$$

if $\varepsilon \le 1 - 1/c$. Note that since $Q'' \ge 0$ the inequality $\mathcal{L}^2_{\varepsilon,1}(Q) > 0$ is weaker than $\mathcal{L}^2(Q) > 0$ and hence satisfied according to Lemma 4.17. Arguing as in Lemma 4.17, we get

$$\begin{aligned}
&r^3 \mathcal{L}^2_{\varepsilon,2}(Q) \\
&= (c-1)r^3 + bu^2 + r^2(b+cr)^3 - ru^2(b+cr)^2 - \varepsilon r^2(b+cr)(1+(b+cr)^2) \\
&> (c-1)r^3 + (1-\varepsilon)r^2(b+cr)^3 - r^3(b+cr)^2 - \varepsilon r^2(b+cr) \\
&= (c-1-c\varepsilon)r^3 + r^2(b+cr)^2[(1-\varepsilon)b + (c-1-c\varepsilon)r] - \varepsilon b r^2 =: A.
\end{aligned}$$

The assumption $\varepsilon \le 1 - 1/c$ ensures that $c - 1 - c\varepsilon > 0$, and dropping the corresponding terms we get

$$A \ge r^2 b^2 (1-\varepsilon) b - \varepsilon b r^2 = b r^2[(1-\varepsilon)b^2 - \varepsilon] \ge 0$$

if $\varepsilon \le b^2/(1+b^2)$.

(c) The inequalities $\mathcal{L}^{\perp}_{\varepsilon,1}(Q) > 0$ and $\mathcal{L}^2_{\varepsilon,1} > 0$ follow exactly as in (b). The inequality $\mathcal{L}^{\perp}_{\varepsilon,2}(Q) > 0$ follows from $r \le 2$, $b = 4 - c$, and $c \le 1$ via

$$\begin{aligned} r\mathcal{L}^{\perp}_{\varepsilon,2}(Q) &> (1-\varepsilon)(b+cr) - r = (1-\varepsilon)b + (c - 1 - c\varepsilon)r \\ &\ge (1-\varepsilon)(4-c) - 2(c-1-c\varepsilon) = (1-\varepsilon)(4-c-2c) + 2 \\ &\ge 1 - \varepsilon + 2 > 0. \end{aligned}$$

To show that $\mathcal{L}^2_{\varepsilon,2} > 0$ we first note that $b + cr = 4 - c(1-r) \ge 3$ due to our assumptions. Using this and $r \le 2$, we estimate the term A from (b) by

$$\begin{aligned} A &\ge (c - 1 - c\varepsilon)r^3 + 9r^2[3(1-\varepsilon) - r] - \varepsilon(4-c)r^2 \\ &\ge -r^3 + 27(1-\varepsilon)r^2 - 9r^3 - 4\varepsilon r^2 \sim 27 - 31\varepsilon - 10r \ge 7 - 31\varepsilon > 0 \end{aligned}$$

if $\varepsilon \le 1/5 < 7/31$.

(d) Since $L'' = 0$ inequalities (4.15) and (4.17) are the same as (4.3) and (4.4), and they are verified in Lemma 4.17. For condition (4.16) we compute

$$\begin{aligned} rL(r)\mathcal{L}^{\perp}_{\varepsilon,2}(L) &= (1-\varepsilon)L'(r)L(r) - r = (1-\varepsilon)d(1+dr) - r \\ &= (1-\varepsilon)d + \big[(1-\varepsilon)d^2 - 1)\big]r = (1-\varepsilon)d + Br \end{aligned}$$

with $B := (1-\varepsilon)d^2 - 1$, which is positive if $\varepsilon < 1 - \frac{1}{d^2}$. For condition (4.18) we compute

$$\begin{aligned} \mathcal{L}^2_{\varepsilon,2}(L) &= \frac{du^2}{r^3} + \frac{(1-\varepsilon)d^3 - \varepsilon d}{r} - \frac{1}{1+dr}\Big(1 + \frac{d^2u^2}{r^2}\Big) \\ &\sim (1+dr)du^2 + d^3(1-\varepsilon)r^2(1+dr) - \varepsilon dr^2 - \varepsilon d^2 r^3 - r^3 - d^2ru^2 \\ &= du^2 + [d^3(1-\varepsilon) - \varepsilon d]r^2 + [d^4(1-\varepsilon) - 1 - \varepsilon d^2]r^3 \\ &\ge [d^3(1-\varepsilon) - \varepsilon d]r^2 + [d^4(1-\varepsilon) - 1 - \varepsilon d^2]r^3 \\ &\sim [d^3(1-\varepsilon) - \varepsilon d] + [d^4(1-\varepsilon) - 1 - \varepsilon d^2]r \\ &=: Cr + D. \end{aligned}$$

Now $D > 0$ if and only if $\varepsilon < 1 - \frac{1}{d^2}$, which holds by assumption, and $C > 0$ if and only if $\varepsilon < d^2/(1+d^2)$, which also follows from the assumption on ε because $d^2/(1+d^2) < 1 - 1/d^2$ for all $d > 1$.

(e) Inequalities (4.15) and (4.17) hold as in (d). For condition (4.18) it suffices to show $Cr + D > 0$, where C, D are the expressions defined in (d). We distinguish two cases.

Case 1. $D > 0$. This implies $d > 1$ and $\varepsilon < 1 - 1/d^2$ and thus $C > 0$ as in (d).

Case 2. $D \le 0$. Then using $r < d^3$ we estimate

$$\begin{aligned} Cr + D &\ge [d^3(1-\varepsilon) - \varepsilon d] + [d^4(1-\varepsilon) - 1 - \varepsilon d^2]d^3 \\ &= d^7 - \varepsilon(d + d^3 + d^5 + d^7) > 0 \end{aligned}$$

for $\varepsilon < \min(d^6, 1)/4$.

For condition (4.16) we need to show $(1-\varepsilon)d + Br > 0$, where $B = (1-\varepsilon)d^2 - 1$ is the expression defined in (d). Again we distinguish two cases.

Case 1. $B > 0$. This implies $d > 1$ and $\varepsilon < 1 - 1/d^2 < 1$ and thus $(1-\varepsilon)d > 0$ as well.

Case 2. $B \le 0$. Then using $r < d^3$ and $\varepsilon < 1/2$ we estimate

$$\begin{aligned}(1-\varepsilon)d + Br &\geq (1-\varepsilon)d + \big[(1-\varepsilon)d^2 - 1)\big]d^3 \\ &\sim (1-\varepsilon)(1+d^4) - d^2 > (1+d^4)/2 - d^2 \\ &\sim (1-d^2)^2 \geq 0.\end{aligned}$$

This concludes the proof of Lemma 4.32. □

The following result is a quantitative version of Proposition 4.21.

PROPOSITION 4.33. *For every $a > 1$ and $\gamma > 0$ there exists $\delta \in (0,\gamma)$ and an i-convex shape $\chi(r)$ which agrees with $S(r) = \sqrt{1+ar^2}$ for $r \geq \gamma$ and satisfies $\chi'(\delta) = +\infty$ and $1 < \chi(\delta) < 1+\gamma$ (see Figure 4.7). Moreover, the corresponding hypersurface is J-convex for every complex structure J on $\mathbb{C}^n$ with $\|J - i\|_{C^2} \leq c(a,n)\gamma^{12}$, where $c(a,n)$ is a constant depending only on a and the dimension n.*

PROOF. Let us recall that the required shape $\chi(r)$ in Proposition 4.21 is constructed by smoothing the maximum of four functions: $S(r) = \sqrt{1+ar^2}$, $Q(r) = a + br + cr^2/2$, $L(r) = 1 + dr$ and a solution ϕ of Struwe's equation (4.8). By construction the function ϕ satisfies $\phi' > d$, and according to Remark 4.20 we can choose $c = (a+1)/2$ and $b = \gamma(a-1)/16 < d < 1$. Then Lemma 4.31 ensures that the modulus of convexity of the ϕ-part of the hypersurface satisfies

$$\bar{\mu} > \frac{d^2}{32} > \gamma^2 \frac{(a-1)^2}{2^9}.$$

Now note that the hypotheses in Lemma 4.32(a), (b), and (e) are identical with those in Lemma 4.17(a), (b), and (e). Hence in view of Remark 4.20, the S- , Q-, and L-parts in the construction satisfy the bounds on the moduli of convexity given by Lemma 4.32(a), (b), and (e):

$$\begin{aligned}&\bar{\mu} = (a-1)/a, \\ &\bar{\mu} \geq \min\left\{1 - \frac{1}{c}, \frac{b^2}{1-b^2}\right\} \geq \min\left\{\frac{a-1}{a+1}, \frac{(a-1)^2\gamma^2}{16^2 - (a-1)^2\gamma^2}\right\}, \\ &\bar{\mu} \geq \frac{\min\{d^6, 1\}}{4} \geq \gamma^6 \frac{(a-1)^6}{2^{29}}.\end{aligned}$$

Thus for all parts we have $\mu \geq \bar{\mu}/3 \geq c_a\gamma^6$, where c_a is a constant depending only on a. Now we apply Corollary 4.29 to smooth the maximum of the four functions. By Proposition 3.41, the resulting shape hypersurface is then J-convex for every complex structure J on $\mathbb{C}^n$ with $\|J - i\|_{C^2} \leq c_n(c_a\gamma^6)^2$. □

Similarly, one proves the following quantitative version of Proposition 4.27.

PROPOSITION 4.34. *Let χ_λ, $\lambda \in [-8\delta, 1]$, be the family constructed in Proposition 4.27. Then all the level sets of these functions are J-convex for every complex structure J on $\mathbb{C}^n$ with $\|J - i\|_{C^2} \leq c(a,n)\gamma^{12}$, where $c(a,n)$ is a constant depending only on a and the dimension n.*

PROOF OF THEOREMS 4.1(iv) AND 4.2(iv). This follows from the estimates in Propositions 4.33 and 4.34. □

CHAPTER 5

Some Complex Analysis

In this chapter we collect some definitions and results from the theory of functions of several complex variables. Mostly, we have restricted ourselves to those facts that are directly relevant for this book, so this chapter presents by no means an adequate exposition of the rich and beautiful subject of several complex variables. For such expositions consult one of the many excellent books, such as [**89, 103, 78, 116, 93, 160, 36, 60**].

In particular, we review the notion of holomorphic convexity and its relation with J-convexity (the Levi problem and its generalizations), Grauert's Oka principle and its applications, and the holomorphic filling problem for J-convex CR manifolds. The one subject discussed in greater detail is real analytic approximation because it is important for the purposes of this book.

5.1. Holomorphic convexity

To a subset $K \subset V$ of a complex manifold we associate its *holomorphic hull in* V:

$$\widehat{K}_V := \{x \in V \mid |f(x)| \le \max{}_K |f| \text{ for all holomorphic functions } f : V \to \mathbb{C}\}.$$

Note that this notion depends on the manifold V. If $U \subset V$ is an open subset containing K, then we have $\widehat{K}_U \subset \widehat{K}_V$. In the case $V = \mathbb{C}^n$ we can equivalently replace holomorphic functions by polynomials in the definition and $\widehat{K}_{\mathbb{C}^n}$ is also called the *polynomial hull.* A subset with $\widehat{K}_{\mathbb{C}^n} = K$ is called *polynomially convex.*

EXAMPLE 5.1. (a) Let $K \subset \mathbb{C}^n$ be a compact convex set. Then $\widehat{K}_{\mathbb{C}^n} = K$. Indeed, for any point $z \notin K$ there exists a complex linear function $l : \mathbb{C}^n \to \mathbb{C}$ such that $\mathrm{Re}\,(l(z)) > \max{}_{w\in K}\mathrm{Re}\,(l(w))$. Hence, $|e^{l(z)}| > \max{}_{w\in K} \left|e^{l(w)}\right|$.

(b) Given holomorphic functions $f_1, \dots, f_N : V \to \mathbb{C}$, we call the set $P = P(f_1, \dots, f_N) := \{|f_1| \le 1, \dots, |f_N| \le 1\} \subset V$ an *analytic polyhedron* if it is compact. Clearly, for any analytic polyhedron P we have $\widehat{P}_V = P$.

(c) If $C \subset V$ is a compact complex curve with boundary ∂C, then $C \subset \widehat{\partial C}_V$. Indeed, for any holomorphic function $f : V \to \mathbb{C}$ the maximum principle yields $\max{}_C |f| \le \max{}_{\partial C} |f|$.

A complex manifold V is called *holomorphically convex* if $\widehat{K}_V$ is compact for all compact subsets $K \subset V$.

EXAMPLE 5.2. (a) Any open convex subset $\Omega \subset \mathbb{C}^n$ is holomorphically convex. Indeed, Ω can be exhausted by compact convex sets $K^i \subset \mathrm{Int}\, K^{i+1}$, $i = 1, 2, \dots$ and we have $K^i \subset \widehat{K}^i_\Omega \subset \widehat{K}^i_{\mathbb{C}^n} = K^i$.

(b) The interior of any analytic polyhedron is holomorphically convex. Indeed $\mathrm{Int}\, P(f_1, \dots, f_N)$ can be exhausted by analytic polyhedra

$$P\Big((1+\varepsilon)f_1, \dots, (1+\varepsilon)f_N\Big), \qquad \varepsilon > 0.$$

(c) If V is holomorphically convex and $W \subset V$ is a properly embedded complex submanifold, then W is also holomorphically convex.

Let us call a compact set $K \subset V$ *holomorphically convex* if it can be presented as an intersection of holomorphically convex open domains in V. In other words, K is holomorphically convex if it has arbitrarily small holomorphically convex open neighborhoods.[1]

Note that if a compact set $K \subset V$ satisfies $\widehat{K}_V = K$, then it is holomorphically convex. Indeed, for any neighborhood U of K a simple compactness argument provides an analytic polyhedron $P(f_1, \dots, f_N)$ which contains K and is contained in U. The converse statement is not true; see Corollary 5.14 below.

EXAMPLE 5.3. Suppose that a compact subset $K \subset V$ admits a continuous family of compact complex curves $C_s \subset V$ with $\partial C_s \subset K$ for all $s \in [0,1]$, $C_0 \subset K$, and $C_1 \not\subset K$. Then K is not holomorphically convex. Indeed, otherwise K would have a holomorphically convex open neighborhood $U \subset V$ with $C_1 \not\subset U$. Set $\sigma := \sup\{s \in [0,1] \mid C_s \subset U\} \in (0,1)$. Then by Example 5.2 (c) we have $\bigcup_{s\in[0,\sigma)} C_s \subset \widehat{K}_U$ and hence $\widehat{K}_U$ is not compact.

Polynomially convex sets can be characterized by an approximation property (see e.g. [**160**]):

THEOREM 5.4 (Oka, Weil). *A holomorphically convex compact subset $K \subset \mathbb{C}^n$ is polynomially convex if and only if every holomorphic function on $\mathcal{O}p\, K$ can be approximated uniformly on K by polynomials.*

5.2. Relation to J-convexity

The notion of holomorphic convexity is intimately related to that of J-convexity. *We remind the reader that in this book "J-convexity" without further specification always means "strict J-convexity".*

The following remark will be used repeatedly in the sequel to replace continuous weakly J-convex functions by smooth strict ones.

REMARK 5.5. Suppose V admits a (not necessarily exhausting) J-convex function (this holds, e.g., for open subsets of $\mathbb{C}^n$ or, more generally, of a Stein manifold). Then any exhausting continuous weakly J-convex function $\phi : V \to \mathbb{R}$ can be turned into an exhausting smooth (strictly) J-convex function. Indeed, we can first add a J-convex function to ϕ to make it (strictly) J-convex and then smooth it using Proposition 3.10, keeping it exhausting. On general complex manifolds such smooth approximations need not exist [**59**].

The relation between holomorphic convexity and J-convexity is given in the following theorem.

[1]Warning: There are other definitions of holomorphic convexity for compact sets in the literature; see e.g. [**179**].

THEOREM 5.6. *For an open set $U \subset \mathbb{C}^n$ the following are equivalent:*
(a) U is holomorphically convex;
(b) The continuous function $-\log \mathrm{dist}_{\partial U}$ is weakly i-convex in U;
(c) U admits an exhausting i-convex function.

For the proof of Theorem 5.6, see e.g. [**103**]. Note that the implication $(b) \Longrightarrow (c)$ follows directly from Remark 5.5. The implication $(c) \Longrightarrow (a)$ (and also $(b) \Longrightarrow (a)$) was known as the *Levi problem*. It was solved by Oka in the case of $\mathbb{C}^2$, and independently by Oka, Bremermann, and Norguet for general $\mathbb{C}^n$. Grauert proved in [**77**] a generalization to complex manifolds (see Theorem 5.17 below), which can be stated in slightly stronger form as follows [**103**, Theorem 5.2.10]:

THEOREM 5.7. *Suppose the complex manifold (V, J) admits an exhausting J-convex function $\phi : V \to \mathbb{R}$. Then all sublevel sets of ϕ satisfy*

$$\widehat{\{\phi \le c\}}_V = \{\phi \le c\}.$$

In particular, (V, J) is holomorphically convex.

The following two lemmas allow us to construct exhausting J-convex functions on bounded domains. The first one is elementary:

LEMMA 5.8. *Let V be a complex manifold which possesses a J-convex function, and let $W \subset V$ be a compact domain with smooth J-convex boundary. Then there exists a J-convex function $\phi : W \to \mathbb{R}$ which is constant on ∂W. In particular,* $\mathrm{Int}\, W$ *admits an exhausting J-convex function.*

PROOF. Take a J-convex function $\psi : V \to \mathbb{R}$ and choose a J-convex function ρ defined on a neighborhood $U \supset \partial W$ which is constant on ∂W. Let us assume that $\phi|_{\partial W} = c$ and $U' := \{c - \varepsilon \le \psi \le c\} \subset U$. There exists a function $\sigma : [c - \varepsilon, c] \to \mathbb{R}$ such that $\sigma \circ \rho$ is J-convex on U', $\sigma(c - \varepsilon) < \min \phi|_{\{\psi = c - \varepsilon\}}$ and $\sigma(c) > \max \phi|_{\partial W}$. Then the function $\phi := \mathrm{smooth} \max(\sigma \circ \rho, \psi)$ on W has the required properties. □

For the proof of the following lemma, see [**93**, Theorem 1.5.14].

LEMMA 5.9. *Let Σ be a locally closed weakly i-convex smooth hypersurface in $\mathbb{C}^n$. Let $U \subset \mathbb{C}^n$ be an open tubular neighborhood of Σ such that $U \setminus \Sigma = U_+ \cup U_-$, where $U_- \cup \Sigma$ has Σ as its weakly i-convex boundary. Then, for U sufficiently small, the function $-\log \mathrm{dist}_\Sigma$ is weakly i-convex on U_-.*

COROLLARY 5.10. *Any compact domain $W \subset \mathbb{C}^n$ with smooth weakly i-convex boundary admits an exhausting i-convex function on its interior.*

PROOF. By Lemma 5.9 there exists a collar neighborhood $[-\varepsilon, 0] \times \partial W$ of $0 \times \partial W = \partial W$ such that the function $\phi := -\log \mathrm{dist}_{\partial W}$ is weakly i-convex on $[-\varepsilon, 0) \times \partial W$. Pick any i-convex function $\psi : \mathbb{C}^n \to \mathbb{R}$. Then $\widetilde{\phi} := \phi + \psi$ is i-convex on $[-\varepsilon, 0) \times \partial W$. Pick $c \in \mathbb{R}$ such that $\psi + c > \widetilde{\phi}$ on $\{-\varepsilon\} \times \partial W$. Then $\mathrm{smooth} \max(\widetilde{\phi}, \psi + c)$ defines an exhausting i-convex function on $\mathrm{Int}\, W$. □

REMARK 5.11. Lemma 5.9 also has the following global version (see e.g. [**103**]): Let $W \subset \mathbb{C}^n$ be a compact domain with smooth weakly i-convex boundary ∂W. Then the continuous function $-\log \mathrm{dist}_{\partial W}$ is weakly i-convex on $\mathrm{Int}\, W$.

Plurisubharmonic hull. In analogy with the holomorphic hull, one can define the *plurisubharmonic hull* of a compact set $K \subset V$ in a complex manifold as

$$\widehat{K}_V^{\text{psh}} := \{x \in V \mid \phi(x) \le \max_K \phi \text{ for all continuous weakly } J\text{-convex functions } \phi : V \to \mathbb{R}\}.$$

PROPOSITION 5.12. *Suppose V admits an exhausting J-convex function. Then $\widehat{K}_V^{\text{psh}} = \widehat{K}_V$ for every compact subset $K \subset V$.*

PROOF. The inclusion $\widehat{K}_V^{\text{psh}} \subset \widehat{K}_V$ is trivial because $|f|^2$ is weakly J-convex for every holomorphic function f. To show $\widehat{K}_V \subset \widehat{K}_V^{\text{psh}}$, take $x \notin \widehat{K}_V^{\text{psh}}$. By definition, there exists a continuous weakly J-convex function $\phi : V \to \mathbb{R}$ with $\phi(x) > \max_K \phi$. After adding a small multiple of an exhausting J-convex function and smoothing, we may assume that ϕ is exhausting, smooth, and (strictly) J-convex. Pick a regular value c of ϕ with $\max_K \phi < c < \phi(x)$. By Theorem 5.7, $\widehat{K}_V \subset \widehat{\{\phi \le c\}}_V = \{\phi \le c\}$ does not contain x. □

The following proposition provides weakly J-convex defining functions for sets with $\widehat{K}_V^{\text{psh}} = K$.

PROPOSITION 5.13. *Suppose (V, J) admits an exhausting J-convex function. Then a compact subset $K \subset V$ satisfies $\widehat{K}_V^{\text{psh}} = K$ if and only if there exists an exhausting smooth weakly J-convex function $\psi : V \to \mathbb{R}_{\ge 0}$ such that $\psi^{-1}(0) = K$ and ψ is (strictly) J-convex outside K.*

PROOF. Existence of ψ clearly implies $\widehat{K}_V^{\text{psh}} = K$. Conversely, suppose that $\widehat{K}_V^{\text{psh}} = K$. First note that for every open set U and compact set W with $K \subset U \subset W$, there exists an exhausting J-convex function $\phi : V \to \mathbb{R}$ with $\phi|_K < 0$ and $\phi|_{W \setminus U} > 0$. Indeed, by definition we find for every $x \in W \subset U$ a weakly J-convex function $\phi_x : V \to \mathbb{R}$ with $\phi|_K < 0$ and $\phi(x) > 0$. After adding a small exhausting J-convex function we may assume that ϕ_x is exhausting and (strictly) J-convex. Since $\phi_x > 0$ on a neighborhood N_x of x and finite many such neighborhoods N_{x_i} cover $W \setminus U$, a smoothing of $\max_i \{\phi_{x_i}\}$ gives the desired function.

Using this, we inductively construct a sequence of relatively compact open subsets $V \Supset U_0 \Supset U_1 \Supset U_2 \Supset \cdots$ with $\bigcap_k U_k = K$, and a sequence of exhausting J-convex functions $\phi_k : V \to \mathbb{R}$, $k \ge 1$, satisfying

$$\phi_k|_{\bar{U}_{k+1}} < 0, \qquad \phi_k|_{\bar{U}_{k-1} \setminus U_k} > 0.$$

Pick a decreasing sequence of positive numbers $\varepsilon_k \to 0$ such that the $\psi_k := \varepsilon_k \phi_k$ satisfy $\max_{\partial U_k} \psi_{k+1} < \min_{\partial U_k} \psi_k$; see Figure 5.1.

Pick $0 < \varepsilon < \min_{\bar{U}_0 \setminus U_1} \psi_1$ and set

$$\psi := \begin{cases} 0 & \text{on } K, \\ \max\{\psi_k, \psi_{k+1}\} & \text{on } U_k \setminus U_{k+1}, \\ \max\{\psi_1, \varepsilon\} & \text{on } V \setminus U_1. \end{cases}$$

Note that ψ is smooth along ∂U_k because $\max\{\psi_k, \psi_{k+1}\} = \psi_k = \max\{\psi_k, \psi_{k-1}\}$ there. Moreover, $\psi \ge 0$ and $\psi^{-1}(0) = K$. By choosing the sequence ε_k to decrease sufficiently fast, we can achieve that $\psi(x) \le e^{-1/d(x,K)}$, where d is the distance with respect to some Riemannian metric, which implies smoothness of ψ at K. So a smoothing of the max constructions yields the desired function. □

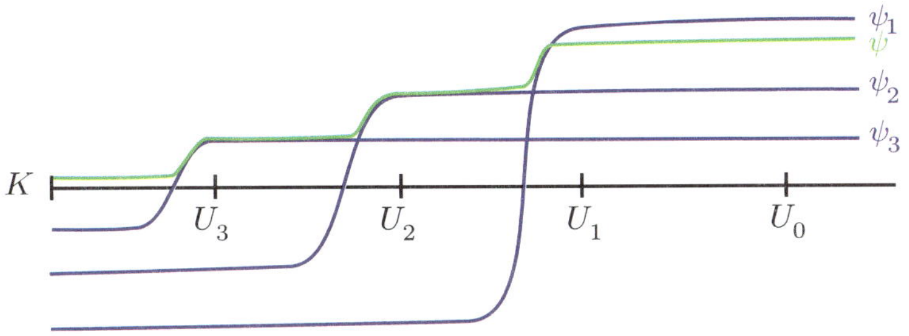

FIGURE 5.1. Construction of the function ψ.

The following corollary illustrates the difference between holomorphic and polynomial convexity.

COROLLARY 5.14. *For a closed totally real submanifold $L \subset \mathbb{C}^n$, the following hold.*

(a) L is holomorphically convex.

(b) If $\dim L = n$ *it is not polynomially convex.*

PROOF. (a) By Proposition 2.15, the squared distance function dist_L^2 is i-convex on some neighborhood U of L, so by Theorem 5.7 $\widehat{L}_U = L$ and L is holomorphically convex.

(b) is a special case of a theorem by Andreotti and Narasimham [**8**], according to which polynomial convexity of L would imply the contradiction $H_n(L;\mathbb{Z}_2) = 0$. Alternatively, this can be proved using symplectic geometry as follows. Suppose L is polynomially convex. Let $\psi : \mathbb{C}^n \to \mathbb{R}_{\geq 0}$ be the exhausting function with $\psi^{-1}(0) = L$ provided by Proposition 5.13. After replacing ψ near L by smooth $\max(\psi, \varepsilon\,\mathrm{dist}_L^2)$ for small $\varepsilon > 0$ we may assume that ψ is (strictly) i-convex. Moreover, we can use Proposition 2.11 to make ψ completely exhausting. Then by Proposition 11.22 below the symplectic form $-dd^{\mathbb{C}}\psi$ is diffeomorphic to the standard form on $\mathbb{C}^n$. Now L is exact Lagrangian for $-dd^{\mathbb{C}}\psi$, i.e., $-d^{\mathbb{C}}\phi|_L \equiv 0$. But this contradicts Gromov's theorem in [**83**] that there are no closed exact Lagrangian submanifolds for the standard symplectic form on $\mathbb{C}^n$. □

5.3. Definitions of Stein manifolds

There exist a number of equivalent definitions of a Stein manifold. We have already encountered two of them.

Affine definition. *A complex manifold V is Stein if it admits a proper holomorphic embedding into some* $\mathbb{C}^N$.

J-convex definition. *A complex manifold V is Stein if it admits an exhausting J-convex function* $f : V \to \mathbb{R}$.

The classical definition rests on the concept of holomorphic convexity.

Classical definition. *A complex manifold V is Stein if it has the following three properties:*

(i) *V is holomorphically convex;*

(ii) *for every* $x \in V$ *there exist holomorphic functions* $f_1, \dots, f_n : V \to \mathbb{C}$ *which form a holomorphic coordinate system at x;*

(iii) *for any* $x \neq y \in V$ *there exists a holomorphic function* $f : V \to \mathbb{C}$ *with* $f(x) \neq f(y)$.[2]

Clearly, the affine definition implies the other two (holomorphic convexity was shown in Example 5.2). The classical definition immediately implies that every compact subset $K \subset V$ can be holomorphically embedded into some $\mathbb{C}^N$. The implication "classical $\Longrightarrow$ affine" is the content of

THEOREM 5.15 (Bishop [**18**], Narasimhan [**144**]). *A Stein manifold* V *in the classical sense of complex dimension* n *admits a proper holomorphic embedding into* $\mathbb{C}^{2n+1}$.

REMARK 5.16. A great deal of research has gone into finding the smallest N such that every n-dimensional Stein manifold embeds into $\mathbb{C}^N$. After intermediate work of Forster, the optimal integer $N = [3n/2] + 1$ was finally established by Eliashberg and Gromov [**50**] and Schürmann [**166**].

The implication "J-convex $\Longrightarrow$ classical" was proved by Grauert in 1958:

THEOREM 5.17 (Grauert [**77**]). *A complex manifold which admits an exhausting* J*-convex function is Stein in the classical sense.*

It is clear from any of the definitions that properly embedded complex submanifolds of Stein manifolds are Stein. We will refer to them as *Stein submanifolds.*

Many results from $\mathbb{C}^n$ generalize to Stein manifolds. For example, the Oka–Weil Theorem 5.4 generalizes to (see [**103**])

THEOREM 5.18. *A holomorphically convex compact subset* K *of a Stein manifold* V *satisfies* $\widehat{K}_V = K$ *if and only if every holomorphic function on* $\mathcal{O}p\, K$ *can be approximated uniformly on* K *by holomorphic functions on* V.

REMARK 5.19. Corollary 5.29 below allows us to generalize Theorem 5.18 to sections of any holomorphic vector bundle over a Stein manifold V.

5.4. Hartogs phenomena

An important new phenomenon in complex dimension $n > 1$ is the existence of open sets $\Omega \subset \mathbb{C}^n$ with the property that all holomorphic functions on Ω extend to some larger set. The first such example was described by Hartogs.

EXAMPLE 5.20 (Hartogs). The domain $\Omega := \operatorname{Int} B^4(1) \setminus B^4(1/2) \subset \mathbb{C}^2$ has the holomorphic hull $\widehat{\Omega} = \operatorname{Int} B^4(1)$ (in particular, Ω is not holomorphically convex). To see this, let $f : \Omega \to \mathbb{C}$ be a holomorphic function. For fixed $z \in \mathbb{C}$, $|z| < 1$, the function $w \mapsto f(z,w)$ on the annulus (or disc) $A_z := \{w \in \mathbb{C} \mid 1/4 - |z|^2 < |w|^2 < 1 - |z|^2\}$ has the Laurent expansion

$$f(z,w) = \sum_{k=-\infty}^{\infty} a_k(z) w^k.$$

The coefficients $a_k(z)$ are given by

$$a_k(z) = \frac{1}{2\pi i} \int_{|\zeta|=r} \frac{f(z,\zeta)}{\zeta^{k+1}} d\zeta$$

[2] In fact, properties (i) and (ii) imply (iii), and (i) and (iii) imply (ii), see [**88**, Section M].

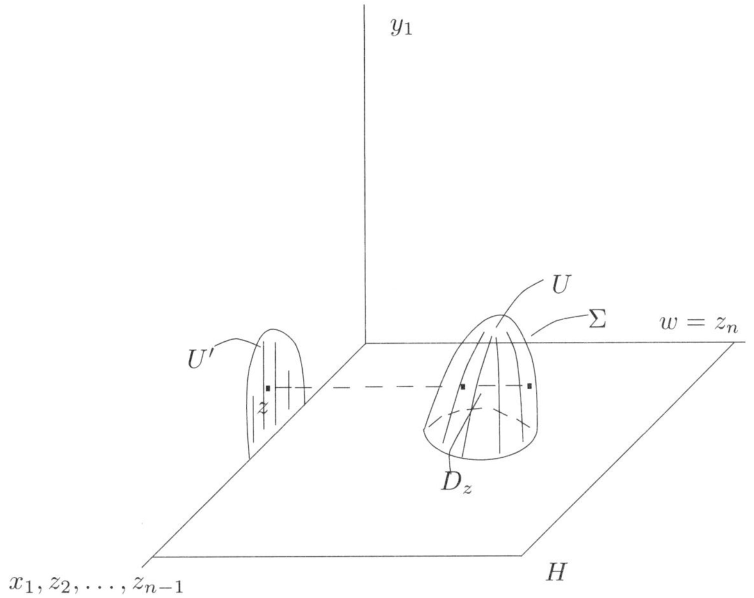

FIGURE 5.2. Holomorphic functions defined near Σ extend holomorphically to the region U between Σ and the hyperplane H.

for any $r > 0$ with $1/4 - |z|^2 < r^2 < 1 - |z|^2$. In particular, $a_k(z)$ depends holomorphically on z with $|z| < 1$. Since A_z is a disc for $|z| > 1/2$, we have $a_k(z) = 0$ for $k < 0$ and $|z| > 1/2$, hence by unique continuation for all z with $|z| < 1$. Thus the Laurent expansion defines a holomorphic extension of f to the ball $\operatorname{Int} B^4(1)$.

Generalizing this example, we have (see [**103**] for a simple proof)

THEOREM 5.21 (Hartogs). *Let Ω be an open subset of $\mathbb{C}^n$, $n > 1$, and let $K \subset \Omega$ be compact with $\Omega \setminus K$ connected. Then every holomorphic function on $\Omega \setminus K$ extends uniquely to a holomorphic function on Ω.*

REMARK 5.22. An open subset $U \subset \mathbb{C}^n$ is called a *domain of holomorphy* if there is no larger unramified domain over $\mathbb{C}^n$ containing U to which all holomorphic functions from U extend holomorphically. It turns out (see e.g. [**103**]) that U is a domain of holomorphy if and only if it is holomorphically convex.

As preparation for a further generalization of Example 5.20, let us consider the following model situation in $\mathbb{C}^n$ with coordinates $z = (z_1, \dots, z_{n-1})$ and $w = z_n$. Let $\Sigma \subset \{y_1 \geq 0\} \subset \mathbb{C}^n$ be a geometrically convex hypersurface with boundary $\partial\Sigma \subset H := \{y_1 = 0\}$; see Figure 5.2.

Denote by U the region between H and Σ and by U' its projection to $\mathbb{C}^{n-1} = \{w = 0\}$. Then for each $z \in U'$ the set $D_z := \{w \in \mathbb{C} \mid (z, w) \in U\}$ is biholomorphic to the unit disc, $\partial D_z \subset \Sigma$, and $D_z \subset \mathcal{O}p\,(\Sigma)$ for z near the point in $\bar{U}'$ where y_1 is maximal. So the same reasoning as in Example 5.20 shows:

Every holomorphic function defined on $\mathcal{O}p\,\Sigma$ extends holomorphically to U.

From this we easily derive the following classical extension result (see e.g. [**89**, Section VII D, Corollary 5]).

THEOREM 5.23. *Let (V, J) be a Stein domain of complex dimension $n \geq 2$ with J-convex boundary ∂V. Then every holomorphic function $\mathcal{O}p\,(\partial V) \to \mathbb{C}$ extends uniquely to a holomorphic function $V \to \mathbb{C}$.*

PROOF. Pick a J-convex Morse function ϕ with regular level set $\partial V = \phi^{-1}(c)$. Let f be a holomorphic function defined on a neighborhood U of ∂V. Consider a level set $\Sigma = \phi^{-1}(b) \subset U$, $b < c$.

By Proposition 2.12, each point $p \in \Sigma$ has a neighborhood $V_p \subset U$ such that $\Sigma \cap V_p$ is biholomorphic to a geometrically convex hypersurface Σ_p in $\mathbb{C}^n$. By the preceding discussion, f can be extended to any region U_p bounded by Σ_p and a hyperplane. Moreover, the proof of Proposition 2.12 shows that the curvature of Σ_p and hence the size of the region U_p can be uniformly bounded below in terms of an upper bound on the second derivatives of ϕ and a lower bound on its gradient and its modulus of J-convexity.

Performing this for all points on $\phi^{-1}(b)$, we holomorphically extend f to $\{\phi \geq a\}$ for some $a < b$, where $b - a$ is bounded below in terms of an upper bound on the second derivatives of ϕ and a lower bound on its gradient and its modulus of J-convexity on $\phi^{-1}(b)$. Continuing this way, we can thus holomorphically extend f to $\{\phi > a\}$, where a is the highest critical value of ϕ. Now we perturb ϕ to a J-convex function $\psi = \phi \circ h$, where h is a small diffeomorphism which equals the identity on $\{\phi \geq b\}$ and maps all critical points of ϕ on level a to $\phi^{-1}(a')$ for some $a' > a$. So all critical points of ϕ on level a lie on the regular level set $\psi^{-1}(a')$. By the preceding argument applied to ψ we holomorphically extend f to the set $\{\psi > a\}$ containing all critical points of ϕ. Then we switch back to ϕ and holomorphically extend f to the next lower critical level of ϕ, and so on until we have extended f to all of V. □

REMARK 5.24. Kohn and Rossi [**114**] prove the following generalization of Theorem 5.23 (under the same assumptions on V): Every function $f : \partial V \to \mathbb{C}$ satisfying the tangential Cauchy–Riemann equations extends to a holomorphic function on V. Moreover, instead of J-convexity they only assume that the Levi form has at least one positive eigenvalue at each point of ∂V. On the other hand, some convexity assumption is clearly necessary: For any closed complex manifold X, $V = \bar{D} \times X$ is a compact complex manifold with Levi-flat boundary and the function $f(z, w) = 1/z$ defined near ∂V has no holomorphic extension to V.

COROLLARY 5.25. *Let V be a Stein domain of complex dimension $n \geq 2$.*

(a) Every holomorphic map $f : \mathcal{O}p\,(\partial V) \to W$ to a Stein manifold W extends uniquely to a holomorphic map $F : V \to W$.

(b) Every biholomorphism $f : \mathcal{O}p\,(\partial V) \to \mathcal{O}p\,(\partial W)$, where W is a Stein domain, extends uniquely to a biholomorphism $F : V \to W$.

PROOF. For (a) pick a proper holomorphic embedding $W \subset \mathbb{C}^N$. By Theorem 5.23, f extends uniquely to a holomorphic map $F : V \to \mathbb{C}^N$. Since $F(\mathcal{O}p\,(\partial V)) \subset W$ and every connected component of V meets ∂V, unique continuation yields $F(V) \subset W$. For (b) simply apply (a) to $f^{-1} : \mathcal{O}p\,(\partial W) \to \mathcal{O}p\,(\partial V)$ to find an inverse of F. □

5.5. Grauert's Oka principle

We discuss in this section some consequences of Grauert's Oka principle:

THEOREM 5.26 (Grauert [**76**]). *Let G be a complex Lie group and $H \subset G$ a closed complex analytic subgroup. Let $P \to V$ be a holomorphic fibration over a Stein manifold V with structure group G and fiber G/H. Then any continuous section $s : V \to P$ is homotopic to a holomorphic one.*

COROLLARY 5.27 (Docquier and Grauert [**38**]). *Let V be a Stein submanifold of a Stein manifold W. Then there exists a neighborhood U of V in W and a holomorphic submersion $U \to V$ fixed on V.*

PROOF. One can view W as a submanifold of $\mathbb{C}^n$. The restriction to W of a submersion defined on $\mathcal{O}p\, V \subset \mathbb{C}^n$ is automatically a submersion on the intersection of this neighborhood with W if the neighborhood is chosen small enough. Hence, it is sufficient to consider the case $W = \mathbb{C}^n$. Consider the holomorphic vector bundles $A = TW|_V = V \times \mathbb{C}^n$, $B = TV$ and $C = A/B$ over V, the holomorphic $GL(n, \mathbb{C})$-principal bundle $E = \text{Iso}(B \oplus C, A)$ and its subbundle $F \to V$ consisting of isomorphisms which restrict to the identity on B. The bundle F is also principal: If $\dim V = k$, then the structure group of F is the subgroup of $GL_n(\mathbb{C})$ preserving $\mathbb{C}^k \subset \mathbb{C}^n$. The bundle F admits a smooth section $s : F \to E$. By Grauert's Theorem 5.26 the section s is homotopic to a holomorphic section, which can be interpreted as a holomorphic fiberwise injective bundle homomorphism $\Phi : C \to A$ transverse to the subbundle $TV \subset A$. This yields a holomorphic map $\overline{\Phi}$ from the total space of the bundle C (which we will still denote by C) to $\mathbb{C}^n$ which sends a vector X in the fiber C_p over a point $p \in V$ to $p + \Phi(X) \in \mathbb{C}^n$. The differential of $\overline{\Phi}$ is an isomorphism along the zero section $V \subset C$, so by the implicit function theorem $\overline{\Phi}$ is a biholomorphism between neighborhoods of the zero section in C and of V in $\mathbb{C}^n$. The bundle projection $\pi : C \to V$ carried by this biholomorphism to $\mathcal{O}p\, V \subset \mathbb{C}^n$ has the required properties. □

REMARK 5.28. Note that the previous argument shows, in particular, that *if $V \subset W$ is a Stein submanifold of a Stein manifold W, then the holomorphic bundle $N = TW|_V/TV$ admits a holomorphic homomorphism $N \to TW|_V$ which is transverse to the subbundle $TV \subset TW|_V$.*

Another useful consequence is

COROLLARY 5.29. *Every holomorphic vector bundle $E \to V$ over a Stein manifold V is holomorphically isomorphic to a subbundle, as well as to a quotient bundle, of the trivial vector bundle $V \times \mathbb{C}^N$, for sufficiently large N.*

PROOF. Consider the holomorphic fibration $P \to V$ of injective complex bundle maps $E \to \mathbb{C}^N$. Its fiber is the complex Stiefel manifold $GL(N; \mathbb{C})/GL(N-k, \mathbb{C})$ of k-frames in $\mathbb{C}^N$, where k is the rank of E, and its structure group is $GL(N, \mathbb{C})$. For large N the bundle P has a continuous section, which by Grauert's Theorem 5.26 is homotopic to a holomorphic section. The resulting injective holomorphic bundle homomorphism $E \to V \times \mathbb{C}^N$ maps E onto a subbundle of $V \times \mathbb{C}^N$. Similarly, Grauert's theorem yields a surjective holomorphic bundle homomorphism $V \times \mathbb{C}^N \to E$, which exhibits E as a quotient bundle of $V \times \mathbb{C}^N$. □

PROPOSITION 5.30. *Let $K \subset V$ be a compact subset with smooth J-convex boundary in a Stein manifold V. Then for every holomorphic vector bundle $\pi : E \to V$ there exists a compact domain $W \subset E$, contained in an arbitrarily small neighborhood of K in E, with smooth J-convex boundary such that $W \cap V = \pi(W) = K$ (where we identify V with the zero section in E).*

PROOF. We use Corollary 5.29 to holomorphically embed E as a subbundle in $V \times \mathbb{C}^N$. Pick a J-convex function $\psi : K \to \mathbb{R}$ with $K = \{\psi \le 0\}$ and regular level set $\partial K = \{\psi = 0\}$. For each $C > 0$ the compact domain $W' := \{(x,z) \in K \times \mathbb{C}^N \mid \psi(x) + C|z|^2 \le 0\} \subset V \times \mathbb{C}^N$ has smooth J-convex boundary and satisfies $W' \cap V = \pi_1(W') = K$, where $\pi_1 : V \times \mathbb{C}^N \to V$ is the projection onto the first factor. Moreover, W' is arbitrarily close to K for large C. Hence $W := E \cap W'$ is the desired domain in E. □

Corollary 5.27 together with Proposition 5.30 implies

COROLLARY 5.31. *For any Stein submanifold $V \subset \mathbb{C}^N$, any compact domain $K \subset V$ with smooth J-convex boundary and any neighborhood U of K in $\mathbb{C}^N$ there exists an arbitrarily small compact domain $W \subset U \subset \mathbb{C}^N$ with smooth J-convex boundary such that $W \cap V = \pi(W) = K$. Here π is a holomorphic submersion from a neighborhood of V in $\mathbb{C}^N$ onto V as constructed in Corollary 5.27.*

In particular, K admits arbitrarily small neighborhoods in $\mathbb{C}^N$ with smooth J-convex boundary.

We also get as a corollary the following analogue of Corollary 5.10 for domains with weakly J-convex boundary in an arbitrary Stein manifold.

COROLLARY 5.32. *Let V be a Stein manifold. Then any compact domain $W \subset V$ with smooth weakly J-convex boundary admits an exhausting J-convex function on its interior.*

PROOF. Let us view V as a submanifold of $\mathbb{C}^N$. Hence TV is a holomorphic subbundle of the trivial bundle $V \times \mathbb{C}^N = T(\mathbb{C}^N)|_V$. Denote by N the quotient bundle $T(\mathbb{C}^N)|_V/TV$. According to Remark 5.28, the bundle N can be realized as a holomorphic subbundle of the trivial bundle $V \times \mathbb{C}^N$ transverse to TV.

Arguing as in the proof of Corollary 5.27, we construct a biholomorphism $\overline{\Phi}$ of a neighborhood Ω of V in N onto a neighborhood $\overline{\Omega}$ of V in $\mathbb{C}^N$. Denote by Σ the total space of the bundle $N|_{\partial W}$. Note that Σ is a weakly pseudo-convex hypersurface in N. Then $\overline{\Sigma} := \overline{\Phi}(\Sigma \cap \Omega)$ is a weakly i-convex hypersurface in $\overline{\Omega}$. Hence, by Lemma 5.9 the function $\phi := -\log \mathrm{dist}_{\overline{\Sigma}}$ is weakly i-convex on the convex side U_- of a sufficiently small tubular neighborhood U of $\overline{\Sigma}$ in $\overline{\Omega}$. Then the restriction $\phi|_{W \cap U_-}$ is weakly J-convex and tends to infinity near ∂W. As in the proof of Corollary 5.10 we now combine this function with an i-convex function $V \to \mathbb{R}$ to obtain an exhausting i-convex function on $\mathrm{Int}\, W$. □

REMARK 5.33. Remark 2.19 provides an alternative proof of Proposition 5.30. It uses some basic facts about curvatures of holomorphic vector bundles; see e.g. [**80**]. Pick an exhausting J-convex function $\phi : V \to \mathbb{R}$. Then $e^{\phi}(\ ,\ )_{\mathrm{st}}$ defines a Hermitian metric on the trivial line bundle $V \times \mathbb{C}$ with negative curvature form $-\partial\bar\partial\phi$. The product metric on the trivial vector bundle $V \times \mathbb{C}^N$ then also has negative curvature, and so does every subbundle of $V \times \mathbb{C}^N$. Thus, by Corollary 5.29, the bundle $\pi : E \to V$ carries a Hermitian metric $|\ |$ of negative curvature. According to Remark 2.19, the function $s(e) = |e|^2$ on E is J-convex outside the zero section. Now let $\psi : K \to \mathbb{R}$ be a J-convex function such that $K = \{\psi \le 0\}$ and $d\psi \ne 0$ along ∂K. Then the function $\Phi := Cs + \psi \circ \pi : \pi^{-1}(K) \to \mathbb{R}$ is J-convex and $W := \{\Phi \le 0\} \subset E$ is the desired domain for large $C > 0$.

Let us remark that Grauert's Oka principle was significantly generalized by Gromov in [**85**] and then further extended by Forstnerič, Lárusson, and others; see [**61**] for a survey of the subject.

5.6. Coherent analytic sheaves on Stein manifolds

Two fundamental results about Stein manifolds are Cartan's Theorems A and B. They are formulated in the language of sheaves; see [**27, 36**] for the relevant definitions and properties. Let V be a complex manifold and $\mathcal{O}$ the sheaf of holomorphic functions on V. For a nonnegative integer p, let $\mathcal{O}^p$ be the sheaf of holomorphic maps to $\mathbb{C}^p$. An *analytic sheaf* is a sheaf of $\mathcal{O}$-modules. A sheaf homomorphism $f : \mathcal{F} \to \mathcal{G}$ between analytic sheaves is called analytic if it is an $\mathcal{O}$-module homomorphism. An analytic sheaf $\mathcal{F}$ is called *coherent* if every $x \in V$ has a neighborhood U such that $\mathcal{F}_U$ equals the cokernel of an analytic sheaf homomorphism $f : \mathcal{O}_U^p \to \mathcal{O}_U^q$, for some nonnegative integers p, q.

Oka's coherence theorem [**154**] states that a subsheaf $\mathcal{F}$ of $\mathcal{O}^p$ is coherent if and only if it is *locally finitely generated*; i.e., for every point $x \in V$ there exists a neighborhood U of x and finitely many sections f_i of $\mathcal{F}_U$ that generate $\mathcal{F}_y$ as an $\mathcal{O}_y$-module for every $y \in U$.

EXAMPLE 5.34. (1) Let $W \subset V$ be a properly embedded complex submanifold of a complex manifold V and $d \geq 0$ an integer. For an open subset $U \subset V$, let $\mathcal{I}_U$ be the ideal of holomorphic functions on U whose d-jet vanishes at all points of $U \cap W$. This defines an analytic sheaf $\mathcal{I}$ on V. We claim that $\mathcal{I}$ is coherent. To see this, let $x \in V$. If $x \notin W$ we find a neighborhood U of x with $U \cap W = \varnothing$ (since $W \subset V$ is closed), hence $\mathcal{I}_U = \mathcal{O}_U$. If $x \in W$ we find a small open polydisc $U \cong \operatorname{Int}\big(B^2(1) \times \cdots \times B^2(1)\big) \subset V$ around x with complex coordinates $(z_1, \dots, z_n)$ in which $W \cap U = \{z_1 = \cdots = z_k = 0\}$. Then the ideal $\mathcal{I}_U$ is generated as an $\mathcal{O}_U$-module by the monomials of degree $(d+1)$ in $z_1, \dots, z_k$, so by Oka's Coherence Theorem [**154**], $\mathcal{I}$ is coherent.

(2) In the situation of (1), fix in addition an integer $e \geq d$ and a properly embedded complex submanifold $Z \subset W$. For $U \subset V$ let $\mathcal{J}_U$ be the ideal of holomorphic functions on U whose d-jet vanishes at all points of $U \cap W$ and whose e-jet vanishes at points of $U \cap Z$. In complex coordinates as above in which $Z = \{z_1 = \cdots = z_\ell = 0\}$, $\ell \geq k$, the ideal $\mathcal{J}_U$ is generated as an $\mathcal{O}_U$-module by the monomials of degree $(e+1)$ in $z_1, \dots, z_\ell$ which have degree at least $(d+1)$ in $z_1, \dots, z_k$. So again, by Oka's theorem, this defines a coherent analytic sheaf $\mathcal{J}$ on V.

REMARK 5.35. The coherence of the sheaves $\mathcal{I}$ and $\mathcal{J}$ in the preceding example can also be proved without Oka's theorem as follows. As above, let $(z_1, \dots, z_n)$ be complex coordinates on a polydisc U in which $W \cap U = \{z_1 = \cdots = z_k = 0\}$. We claim that every $f \in \mathcal{I}_U$ has a unique representation

$$f(z) = \sum_I f_I(z) z^I,$$

where the summation is over all $I = (i_1, \dots, i_k)$ with $i_1 + \cdots + i_k = d + 1$ and $z^I = z_1^{i_1} \cdots z_k^{i_k}$, and the coefficient f_I is a holomorphic function of $z_\ell, \dots, z_n$, where $1 \leq \ell \leq k$ is the largest integer with $i_\ell \neq 0$.

We first prove the claim for $d = 0$ by induction over k. The case $k = 1$ is clear, so let $k > 1$. The function $(z_k, \dots, z_n) \mapsto f(0, \dots, 0, z_k, \dots, z_n)$ vanishes at $z_k = 0$, thus (as in the case $k = 1$) it can be uniquely written as $z_k f_k(z_k, \dots, z_n)$

with a holomorphic function f_k. Since the function $(z_1, \dots, z_n) \mapsto f(z_1, \dots, z_n) - z_k f_k(z_k, \dots, z_n)$ vanishes at $z_1 = \dots = z_{k-1} = 0$, by the induction hypothesis it can be uniquely written as $z_1 f_1(z_1, \dots, z_n) + \dots + z_{k-1} f_{k-1}(z_{k-1}, \dots, z_n)$ with holomorphic functions $f_1, \dots, f_{k-1}$. This proves the case $d = 0$. The general case $d > 0$ follows by induction over d: Using the case $d = 0$, we write $f(z)$ uniquely as $z_1 f_1(z_1, \dots, z_n) + \dots + z_k f_k(z_k, \dots, z_n)$. Now note that the functions $f_1, \dots, f_k$ must vanish to order $d - 1$ at $z_1 = \dots = z_k = 0$ and use the induction hypothesis. This proves the claim.

By the claim, $\mathcal{I}_U$ is the direct sum of copies of the rings $\mathcal{F}_U^\ell$ of holomorphic functions of $z_\ell, \dots, z_n$ for $1 \le \ell \le k$. Since $\mathcal{F}_U^\ell$ is isomorphic to the cokernel of the homomorphism $\mathcal{O}_U^{\ell-1} \to \mathcal{O}_U$, $(f_1, \dots, f_{\ell-1}) \mapsto z_1 f_1 + \dots + z_{\ell-1} f_{\ell-1}$, this proves coherence of $\mathcal{I}$. For coherence of $\mathcal{J}$, note that $f \in \mathcal{J}_U$ has a representation as above with coefficients $f_I(z)$ vanishing to order e along $\{z_1 = \dots = z_\ell = 0\}$ and apply the same argument to represent f_I as a sum over monomials.

Now we can state Cartan's Theorems A and B. Denote by $H^q(V, \mathcal{F})$ the cohomology with coefficients in the sheaf $\mathcal{F}$. In particular, $H^0(V, \mathcal{F})$ is the space of sections in $\mathcal{F}$. Every subsheaf $\mathcal{G} \subset \mathcal{F}$ induces a long exact sequence

$$\dots \to H^q(V, \mathcal{G}) \to H^q(V, \mathcal{F}) \to H^q(V, \mathcal{F}/\mathcal{G}) \to H^{q+1}(V, \mathcal{G}) \to \dots .$$

THEOREM 5.36 (Cartan's Theorems A and B [**27**]). *Let V be a Stein manifold and $\mathcal{F}$ a coherent analytic sheaf on V. Then*

(A) for every $x \in V$, $H^0(V, \mathcal{F})$ generates $\mathcal{F}_x$ as an $\mathcal{O}_x$-module;

(B) $H^q(V, \mathcal{F}) = \{0\}$ for all $q > 0$.

In Section 5.8 we will use the following two consequences of Cartan's Theorem B.

COROLLARY 5.37. *Let $Z \subset W \subset V$ be Stein submanifolds of a Stein manifold V, and let d be a nonnegative integer. Then for every holomorphic function $f : W \cup \mathcal{O}p\,(Z) \to \mathbb{C}$ there exists a holomorphic function $F : V \to \mathbb{C}$ with $F|_W = f$ whose d-jet coincides with that of f at points of Z.*

PROOF. Let $\mathcal{I}$ be the analytic sheaf of holomorphic functions on V that vanish on W and whose d-jet vanishes at points of Z. By Example 5.34, $\mathcal{I}$ is coherent. Thus by Cartan's Theorem B, $H^1(V, \mathcal{I}) = 0$, so by the long exact sequence the homomorphism $H^0(V, \mathcal{O}) \to H^0(V, \mathcal{O}/\mathcal{I})$ is surjective. Now $\mathcal{O}_x/\mathcal{I}_x = \{0\}$ for $x \notin W$, and for $x \in W \setminus Z$ elements of $\mathcal{O}_x/\mathcal{I}_x$ are germs of holomorphic functions on W and for $x \in Z$ elements of $\mathcal{O}_x/\mathcal{I}_x$ are d-jets of germs of holomorphic functions along Z. So f defines a section in $\mathcal{O}/\mathcal{I}$ and we conclude that f is the restriction of a section F in $\mathcal{O}$. □

COROLLARY 5.38. *Every Stein submanifold W of a Stein manifold V is the common zero set of a finite number (at most $(\dim_{\mathbb{C}} V + 1)(\mathrm{codim}_{\mathbb{C}} W + 1)$) of holomorphic functions $f_i : V \to \mathbb{C}$ such that for all $x \in W$ the differentials $d_x f_i : T_x V \to \mathbb{C}$ satisfy $\bigcap_i \ker d_x f_i = T_x W$.*

PROOF. The argument is given in [**28**]. It uses some basic properties of analytic subvarieties; see e.g. [**80, 36**]. An *analytic subvariety* of a complex manifold V is a closed subset $Z \subset V$ that is locally the zero set of finitely many holomorphic functions. Z is a stratified space $Z = Z_0 \cup \dots \cup Z_k$, where Z_i is a (not necessarily closed) complex submanifold of dimension i. Define the (complex) dimension of Z

as the dimension k of the top stratum. If $Z' \subset Z$ are analytic subvarieties of the same dimension, then Z' contains a connected component of the top stratum Z_k of Z.

Now let $W \subset V$ be a Stein submanifold of a Stein manifold V. Pick a set $S_1 \subset V$ containing one point on each connected component of $V \setminus W$. Since S_1 is discrete, $W \cup S_1$ is a Stein submanifold of V. By Corollary 5.37, there exists a holomorphic function $f_1 : V \to \mathbb{C}$ which equals 0 on W and 1 on S_1. The zero set $W_1 := \{f_1 = 0\}$ is an analytic subvariety of V, containing W, such that $W_1 \setminus W$ has dimension $\leq n-1$, where $n = \dim_{\mathbb{C}} V$. Pick a set $S_2 \subset W_1 \setminus W$ containing one point on each connected component of the top stratum of W_1 that is not contained in W. Since each compact set meets only finitely many components of W_1, the set S_2 is discrete, so $W \cup S_2$ is a Stein submanifold of V. By Corollary 5.37, there exists a holomorphic function $f_2 : V \to \mathbb{C}$ which equals 0 on W and 1 on S_2. The zero set $W_2 := \{f_1 = f_2 = 0\}$ is an analytic subvariety of V, containing W, such that $W_2 \setminus W$ has dimension $\leq n-2$. Continuing this way, we find holomorphic functions $f_1, \dots, f_{n+1} : V \to \mathbb{C}$ such that $W \subset W_{n+1} := \{f_1 = \cdots = f_{n+1} = 0\}$ and $W_{n+1} \setminus W$ has dimension ≤ -1. Thus $W_{n+1} \setminus W = \varnothing$ and $W = \{f_1 = \cdots = f_{n+1} = 0\}$.

Finally, we will add more functions to arrange the condition $\bigcap_i \ker d_x f_i = T_x W$ for all $x \in W$. Pick a (discrete) set $S_1 \subset V$ containing one point on each connected component of W. By Corollary 5.37 we find holomorphic functions on V which vanish on W with prescribed complex derivatives at points of S_1. Choosing these derivatives to be linearly independent, we thus find holomorphic functions $g_1, \dots, g_k : V \to \mathbb{C}$ which vanish on W such that $\bigcap_i \ker d_x g_i = T_x W$ for all $x \in S_1$. Now $W_1 := \{x \in W \mid \bigcap_i \ker d_x f_i \neq T_x W\}$ is an analytic subvariety of W, and we continue inductively as above until the dimension becomes negative. □

5.7. Real analytic manifolds

In order to holomorphically attach handles, we need to approximate smooth objects by real analytic ones. In this section we collect the relevant results.

A function $f : U \to \mathbb{R}^m$ on an open domain $U \subset \mathbb{R}^n$ is called *real analytic* if it is locally near each point given by a convergent power series. A *real analytic manifold* is a manifold with an atlas such that all transition functions are real analytic. A submanifold is called real analytic if it is locally the transverse zero set of a real analytic function. Real analytic bundles and sections are defined in the obvious way.

REMARK 5.39. As a special case of the Cauchy–Kovalevskaya theorem (see e.g. [**58**]), the solution of an ordinary differential equation with real analytic coefficients depends real analytically on all parameters.

Complexification. There is a natural functor, called *complexification*, from the real analytic to the holomorphic category. First note that any real analytic function $f : U \to \mathbb{C}^m$, defined on an open domain $U \subset \mathbb{R}^n$, can be uniquely extended to a holomorphic function $f^{\mathbb{C}} : U^{\mathbb{C}} \to \mathbb{C}^m$ on some open domain $U^{\mathbb{C}} \subset \mathbb{C}^n$ with $U^{\mathbb{C}} \cap \mathbb{R}^n = U$. More generally, we have

LEMMA 5.40. *Let V, W be complex manifolds and $M \subset V$ a real analytic totally real submanifold with $\dim_{\mathbb{R}} M = \dim_{\mathbb{C}} V$. Then any real analytic map $f : M \to W$ extends uniquely to a holomorphic map $f^{\mathbb{C}} : \mathcal{O}p\, M \to W$ on a sufficiently small neighborhood of M in V.*

If $\dim_{\mathbb{C}} V = \dim_{\mathbb{C}} W$ *and* f *is a real analytic diffeomorphism of* M *onto a totally real submanifold* $N \subset W$*, then the extension* $f^{\mathbb{C}}$ *is a biholomorphism between* $\mathcal{O}p\, M$ *and* $\mathcal{O}p\, N$.

PROOF. Consider a point $p \in M$. Pick a real analytic coordinate chart $\phi : \mathbb{R}^n \supset U_1 \to M$ and a holomorphic coordinate chart $\psi : \mathbb{C}^n \supset U_2 \to V$, both mapping 0 to p. Complexify $\psi^{-1} \circ \phi$ to a biholomorphic map $\widetilde{\Phi} : \mathbb{C}^n \supset U_1^{\mathbb{C}} \to \mathbb{C}^n$. Then $\Phi = \psi \circ \widetilde{\Phi} : \mathbb{C}^n \supset U \to V$ is a holomorphic coordinate chart mapping $U \cap \mathbb{R}^n$ to M.

Pick a holomorphic coordinate chart $\Psi : \mathbb{C}^m \supset U' \to W$ near $f(p)$ and complexify $\Psi^{-1} \circ f \circ \Phi : U \cap \mathbb{R}^n \to \mathbb{C}^m$ to a holomorphic map $\widetilde{F} : U \to \mathbb{C}^m$. So $F = \Psi \circ \widetilde{F} \circ \Phi^{-1}$ is a holomorphic extension of f to a neighborhood of p in V. By uniqueness of holomorphic extensions, this extension does not depend on the chosen coordinate charts on V and W, and extensions around different points of M fit together to the desired holomorphic extension of f.

The final statement follows from the implicit function theorem and the observation that the complexification of a real isomorphism is a complex isomorphism. □

The following is the fundamental result on complexifications of real analytic manifolds.

THEOREM 5.41 (Bruhat and Whitney [**25**]). *Any real analytic manifold* M *has a* complexification, *i.e., a complex manifold* $M^{\mathbb{C}}$ *with* $\dim_{\mathbb{C}} M^{\mathbb{C}} = \dim_{\mathbb{R}} M$ *which contains* M *as a real analytic totally real submanifold. The germ of a complexification* $M^{\mathbb{C}}$ *is unique in the following sense: if* V, W *are complex manifolds, containing* M *as real analytic and totally real submanifolds, with* $\dim_{\mathbb{C}} V = \dim_{\mathbb{C}} W = \dim_{\mathbb{R}} M$*, then some neighborhoods of* M *in* V *and* W *are biholomorphic.*

Here is a sketch of the proof; see [**25**] for details. Pick a locally finite covering of M by countably many real analytic coordinate charts $\phi_i : \mathbb{R}^n \supset U_i \to M$, so the transition functions

$$\phi_{ij} := \phi_j^{-1} \circ \phi_i : U_{ij} := \phi_i^{-1}\Big(\phi_i(U_i) \cap \phi_j(U_j)\Big) \to U_{ji}$$

are real analytic diffeomorphisms. Now construct open subsets $U_i^{\mathbb{C}} \subset \mathbb{C}^n$ with $U_i^{\mathbb{C}} \cap \mathbb{R}^n = U_i$ and $U_{ij}^{\mathbb{C}} \subset U_i^{\mathbb{C}}$ with $U_{ij}^{\mathbb{C}} \cap \mathbb{R}^n = U_{ij}$ such that the ϕ_{ij} extend to biholomorphic maps $\phi_{ij}^{\mathbb{C}} : U_{ij}^{\mathbb{C}} \to U_{ji}^{\mathbb{C}}$ satisfying the following cocycle conditions:

(i) $\phi_{ji}^{\mathbb{C}} = (\phi_{ij}^{\mathbb{C}})^{-1}$ and $\phi_{ii}^{\mathbb{C}} = \mathrm{Id}$ on $U_{ii}^{\mathbb{C}} = U_i^{\mathbb{C}}$;

(ii) $\phi_{ij}^{\mathbb{C}}$ maps $U_{ijk}^{\mathbb{C}} = U_{ij}^{\mathbb{C}} \cap U_{ik}^{\mathbb{C}}$ biholomorphically onto $U_{jik}^{\mathbb{C}}$ and $\phi_{jk}^{\mathbb{C}} \circ \phi_{ij}^{\mathbb{C}} = \phi_{ik}^{\mathbb{C}} : U_{ijk}^{\mathbb{C}} \to U_{kij}^{\mathbb{C}}$.

Define $M^{\mathbb{C}}$ as the quotient of the disjoint union $\coprod_i U_i^{\mathbb{C}}$ by the equivalence relation $z_i \sim z_j$ if and only if $z_i \in U_{ij}^{\mathbb{C}}$ and $z_j = \phi_{ij}^{\mathbb{C}}(z_i) \in U_{ji}^{\mathbb{C}}$. (This is an equivalence relation because of the cocycle conditions). The inclusions $U_i^{\mathbb{C}} \hookrightarrow \coprod_j U_j^{\mathbb{C}}$ induce coordinate charts $U_i^{\mathbb{C}} \hookrightarrow M^{\mathbb{C}}$ with biholomorphic transition functions. A careful choice of the open sets $U_i^{\mathbb{C}}$ and $U_{ij}^{\mathbb{C}}$ ensures that $M^{\mathbb{C}}$ is Hausdorff. Finally, the uniqueness statement in Theorem 5.41 follows from Lemma 5.40.

Note that as a real manifold, a (sufficiently small) complexification $M^{\mathbb{C}}$ is diffeomorphic to the tangent bundle TM.

Complexification has the obvious functorial properties. For example, if $N \subset M$ is a real analytic submanifold of a real analytic manifold M, then the (sufficiently small) complexification $N^{\mathbb{C}}$ is a complex submanifold of $M^{\mathbb{C}}$.

The crucial observation, due to Grauert [**77**], is that complexifications of real analytic manifolds are in fact Stein.

PROPOSITION 5.42. *Let $M^{\mathbb{C}}$ be a complexification of a real analytic manifold M. Then M possesses arbitrarily small neighborhoods in $M^{\mathbb{C}}$ which are Stein.*

PROOF. By Proposition 2.15, M possesses arbitrary small neighborhoods with exhausting J-convex functions. By Grauert's Theorem 5.17, these neighborhoods are Stein. □

A complexification $M^{\mathbb{C}}$ which is Stein is called a *Grauert tube* of M. Now the basic results about real analytic manifolds follow via complexification from corresponding results about Stein manifolds.

COROLLARY 5.43. *Every real analytic manifold admits a proper real analytic embedding into some $\mathbb{R}^N$.*

PROOF. By Theorem 5.15, a Grauert tube $M^{\mathbb{C}}$ of M embeds properly holomorphically into some $\mathbb{C}^N$. We then restrict this embedding to M. □

COROLLARY 5.44. *Let $P \subset N \subset M$ be properly embedded real analytic submanifolds of a real analytic manifold M, and let d be a nonnegative integer. Then for every real analytic function $f : N \cup \mathcal{O}p\,(P) \to \mathbb{R}$ there exists a real analytic function $F : M \to \mathbb{R}$ with $F|_N = f$ whose d-jet coincides with that of f at points of P.*

PROOF. Let $M^{\mathbb{C}}$ be a Grauert tube of M. After possibly shrinking $N^{\mathbb{C}}$ and $M^{\mathbb{C}}$, we may assume that a complexifications are properly embedded complex submanifold $P^{\mathbb{C}} \subset N^{\mathbb{C}} \subset M^{\mathbb{C}}$, and f complexifies to a holomorphic function $f^{\mathbb{C}}$: $N^{\mathbb{C}} \cup \mathcal{O}p\,(P^{\mathbb{C}}) \to \mathbb{C}$. Corollary 5.37 provides a holomorphic function $G : M^{\mathbb{C}} \to \mathbb{C}$ with $G^{\mathbb{C}}|_{N^{\mathbb{C}}} = f^{\mathbb{C}}$ and whose d-jet agrees with that of $f^{\mathbb{C}}$ at points of $P^{\mathbb{C}}$. Then the restriction of the real part of G to M is the desired function F. □

COROLLARY 5.45. *Every properly embedded real analytic submanifold N of a real analytic manifold M is the common zero set of a finite number (at most $2(\dim_{\mathbb{R}} M + 1)(\mathrm{codim}_{\mathbb{R}} N + 1)$) of real analytic functions $f_i : M \to \mathbb{R}$ such that for all $x \in N$, the differentials $d_x f_i : T_x M \to \mathbb{C}$ satisfy $\bigcap_i \ker d_x f_i = T_x N$.*

PROOF. Complexify N to a properly embedded submanifold $N^{\mathbb{C}} \subset M^{\mathbb{C}}$ of a Grauert tube $M^{\mathbb{C}}$. By Corollary 5.38, $N^{\mathbb{C}}$ is the zero set of at most $(\dim_{\mathbb{R}} M + 1)(\mathrm{codim}_{\mathbb{R}} N + 1)$ holomorphic functions $F_i : M^{\mathbb{C}} \to \mathbb{C}$ satisfying the differential condition. The restrictions of $\mathrm{Re}\, F_i$ and $\mathrm{Im}\, F_i$ to M yield the desired functions f_i. □

REMARK 5.46. H. Cartan [**28**] takes a slightly different route to prove Corollaries 5.44 and 5.45: Define coherent analytic sheaves on real analytic manifolds analogously to the complex analytic case. Cartan proves that for every coherent analytic sheaf $\mathcal{F}$ on M, there exists a coherent analytic sheaf $\mathcal{F}^{\mathbb{C}}$ on a complexification $M^{\mathbb{C}}$ such that $\mathcal{F}^{\mathbb{C}}|_M = \mathcal{F} \otimes \mathbb{C}$. From this he deduces the analogues of theorems A and B in the real analytic category, which imply the corollaries as in the complex analytic case.

We conclude this section with the following extension of Lemma 5.40 to the non-totally real case that will be needed in Chapter 16.

COROLLARY 5.47. *Let U, V, W be complex manifolds and $M \subset V$ a real analytic totally real submanifold with $\dim_{\mathbb{R}} M = \dim_{\mathbb{C}} V$. Then any real analytic map $f : U \times M \to W$ whose restriction to $U \times m$ is holomorphic for all $m \in M$ extends uniquely to a holomorphic map $f^{\mathbb{C}} : U \times V \supset \mathcal{O}p\,(U \times M) \to W$.*

PROOF. Arguing as in the proof of Lemma 5.40, it suffices to consider the case of the open unit balls $U \subset \mathbb{C}^k$, $V \subset \mathbb{C}^\ell$, $W \subset \mathbb{C}^n$ and $M = V \cap \mathbb{R}^\ell$. Then U is foliated by the totally real balls $U_t = U \cap \{\mathrm{Im}\, u = t\}$, $t \in U \cap \mathbb{R}^k$. For each t, the restriction of f to the totally real subspace $U_t \times M \subset (\mathbb{R}^k + it) \times \mathbb{R}^\ell \subset \mathbb{C}^k \times \mathbb{C}^\ell$ extends uniquely to a holomorphic map $f_t^{\mathbb{C}} : Z_t \to W$ on a neighborhood Z_t of $U_t \times M$ in $U \times V$. For $m \in M$ consider the two holomorphic maps $f, f_t^{\mathbb{C}} : Z_t \cap (U \times m) \to W$. Since they agree on the half-dimensional totally real subspace $U_t \times m \subset Z_t \cap (U \times m)$, unique continuation yields $f = f_t^{\mathbb{C}}$ on $Z_t \cap (U \times m)$ for all m, and thus $f = f_t^{\mathbb{C}}$ on Z_t. Again by unique continuation, the extensions $f_t^{\mathbb{C}} : Z_t \to W$ fit together for different t to the desired extension $f^{\mathbb{C}}$. □

5.8. Real analytic approximations

Corollary 5.43 combined with a theorem of Whitney implies that every C^k-function on a real analytic manifold M can be C^k-approximated by real analytic functions. To state the result, equip M with a metric and connection so that we can speak of k-th (covariant) derivatives of functions on M and their norms. We denote by $D^k f$ the vector of derivatives up to order k of a function $f : M \to \mathbb{R}$.

PROPOSITION 5.48. *Let $f : M \to \mathbb{R}$ be a C^k-function on a real analytic manifold. Then for every positive continuous function $h : M \to \mathbb{R}_+$ there exists a real analytic function $g : M \to \mathbb{R}$ such that $|D^k g(x) - D^k f(x)| < h(x)$ for all $x \in M$.*

PROOF. Embed M real analytically into some $\mathbb{R}^N$. Extend f to a C^k-function $F : \mathbb{R}^N \to \mathbb{R}$ and h to a continuous function $H : \mathbb{R}^N \to \mathbb{R}_+$. By a theorem of Whitney [**189**, Lemma 6], there exists a real analytic function $G : \mathbb{R}^N \to \mathbb{R}$ such that $|D^k G(x) - D^k F(x)| < H(x)$ for all $x \in \mathbb{R}^N$. Let g be the restriction of G to M. □

Proposition 5.48 clearly generalizes to sections in real analytic fiber bundles $E \to M$. For this, view the total space of the bundle as a real analytic manifold and note that the image of a map $M \to E$ that is sufficiently C^1-close to a section is again the graph of a section. Thus we have

COROLLARY 5.49. *Let $f : M \to E$ be a C^k-section of a real analytic fiber bundle $E \to M$ over a real analytic manifold M. Then for every positive continuous function $h : M \to \mathbb{R}_+$ there exists a real analytic section $g : M \to E$ such that $|D^k g(x) - D^k f(x)| < h(x)$ for all $x \in M$.*

EXAMPLE 5.50. By Corollary 5.49, every Riemannian metric on a real analytic manifold can be C^k-approximated by a real analytic metric. By Remark 5.39, the exponential map of a real analytic metric is real analytic. Now the standard proof yields real analytic tubular (resp. collar) neighborhoods of compact real analytic submanifolds (resp. boundaries). In particular, this allows us to extend any compact real analytic manifold with boundary to a slightly larger open real analytic manifold.

Corollary 5.49 provides a rather general approximation result. On the other hand, Corollary 5.44 shows that every real analytic function on a properly embedded real analytic submanifold N of a real analytic manifold M can be extended to a real analytic function on M, with prescribed d-jet along a real analytic submanifold P. It is the goal of this section to combine the approximation and extension results.

We begin by introducing some notation. Consider a real vector bundle $E \to N$ and fix an integer $d \geq 0$. The *d-jet bundle of E* is the bundle $J^d E \to N$ whose fiber at $x \in N$ consists of all polynomials of degree d on E_x. Thus a C^k-section of $J^d E$ is a C^k-function $F : E \to \mathbb{R}$ whose restriction to each fiber E_x is polynomial of degree d. Note that $J^0 E = N$ and $J^1 E = E^*$. By taking the Taylor polynomials of degree d on the fibers, every C^{k+d}-function $f : E \to \mathbb{R}$ induces a C^k-section $J^d f$ of $J^d E$ which we call the *fiberwise d-jet of f*. Note that for $d \leq e$ we have a natural projection $J^e E \to J^d E$. Note also that if the bundle E is real analytic then so is $J^d E$.

LEMMA 5.51. *Consider a real analytic vector bundle $E \to N$, integers $k \geq d \geq 0$, and a continuous function $h : E \to \mathbb{R}_+$. Let $f : E \to \mathbb{R}$ be a smooth function whose fiberwise d-jet $J^d f$ is real analytic. Then there exists a smooth function $g : E \to \mathbb{R}$ and arbitrarily small open neighborhoods $U \subset V$ of N in E with the following properties:*

(i) $j^d g = j^d f$ along N;

(ii) $|D^k g(x) - D^k f(x)| < h(x)$ for all $x \in E$;

(iii) g is real analytic on U and $g = f$ outside V.

PROOF. Consider the real analytic bundle $J^{k,d} E \to N$ whose fiber at $x \in N$ consists of all sums of monomials on E_x of degrees between $d+1$ and k. Let $J^{k,d} f = J^k f = J^d f$ be the section of $J^{k,d} E$ defined by f. By Corollary 5.49 there exists a real analytic section F of $J^{k,d} E$ with $|D^k F(x) - D^k J^{k,d} f(x)| < h(x)$ for all $x \in N$. So $G := J^d f + F : E \to \mathbb{R}$ is a real analytic function with $J^d G = J^d f$ and $|D^k G(x) - D^k f(x)| < h(x)$ for all $x \in N$. Since the estimate continues to hold on a neighborhood of N in E, we can interpolate G to f outside a smaller neighborhood to obtain the desired function g. □

Next consider a properly embedded real analytic submanifold N of a real analytic manifold M. Pick a real analytic Riemannian metric on M. Its exponential map yields a real analytic diffeomorphism Φ between a neighborhood of the zero section in the normal bundle $E \to N$ and a neighborhood of N in M. We define the *normal d-jet* $J^d f$ along N of a function $f : M \to \mathbb{R}$ as the fiberwise d-jet of $f \circ \Phi$. Replacing real-valued functions by sections in a bundle, Lemma 5.51 thus yields

COROLLARY 5.52. *Consider a real analytic fiber bundle $E \to M$, a properly embedded real analytic submanifold $N \subset M$, integers $k \geq d \geq 0$, and a continuous function $h : M \to \mathbb{R}_+$. Let $f : M \to E$ be a smooth section whose normal d-jet $J^d f$ along N is real analytic. Then there exists a smooth section $g : M \to E$ and arbitrarily small open neighborhoods $U \subset V$ of N in M with the following properties:*

(i) $j^d g = j^d f$ along N;

(ii) $|D^k g(x) - D^k f(x)| < h(x)$ for all $x \in M$;

(iii) g is real analytic on U and $g = f$ outside V.

THEOREM 5.53. *Consider a real analytic fiber bundle $E \to M$, a properly embedded real analytic submanifold $N \subset M$, integers $k \geq d \geq 0$, and a continuous function $h : M \to \mathbb{R}_+$. Let $f : M \to E$ be a smooth section whose normal d-jet $J^d f$ along N is real analytic. Then there exists a real analytic section $F : M \to E$ with the following properties:*

(i) $j^d F = j^d f$ along N;

(ii) $|D^k F(x) - D^k f(x)| < h(x)$ for all $x \in M$.

PROOF. **Step 1.** As before, it suffices to consider the case of a real valued function $f : M \to \mathbb{R}$. After C^k-approximating f by a smooth function, fixing its normal d-jet along N, we may assume that f is smooth. After applying Corollary 5.52, we may assume that f is real analytic in a neighborhood of N.

Pick any $\ell \geq k \geq d$. By Corollary 5.44 there exists a real analytic function $H : M \to \mathbb{R}$ whose ℓ-jet coincides with that of f at points of N. Then $g := f - H$ vanishes to order ℓ along N. Suppose that we find a real analytic section $G : M \to \mathbb{R}$ that vanishes to order d along N and satisfies $|D^k G(x) - D^k g(x)| < h(x)$ for all $x \in M$. Then the real analytic function $F := G + H : M \to \mathbb{R}$ is the desired approximation f: Its normal d-jet along N satisfies $J^d F = J^d H = J^d f$, and $|D^k F - D^k f| = |D^k G + D^k H - D^k f| = |D^k G - D^k g| < h$ on M.

Step 2. By Step 1 it suffices to prove the theorem under the additional hypothesis that $f : M \to \mathbb{R}$ vanishes to order $\ell := 2d + k + 1$ along N. By Corollary 5.45, there exist real analytic functions $f_1, \dots, f_m : M \to \mathbb{R}$ such that $N = \{\phi_1 = \cdots = \phi_m = 0\}$ and $\bigcap_i \ker d_x f_i = T_x N$ for all $x \in N$. Then $\phi := \phi_1^2 + \cdots + \phi_m^2 : M \to \mathbb{R}$ is real analytic and $N = \phi^{-1}(0)$. Moreover, the Hessian of ϕ at $x \in N$ is positive definite in directions transverse to N, so $\phi \geq \mathrm{dist}_N^2$ for the distance from N with respect to some Riemannian metric on M. Now note that in a neighborhood of each point $p \in N$ we have an estimate $|f(x)| \leq C_p \mathrm{dist}(M, x)^\ell$ and hence $|f(x)|\,|\phi(x)|^{-d} \leq C_p \mathrm{dist}(M, x)^{\ell - 2d} = C_p \mathrm{dist}(M, x)^{k+1}$. This shows that $g := f\phi^{-d}$ defines a C^k-function on M. By Proposition 5.48 there exists a real analytic function $G : M \to \mathbb{R}$ such that $|D^k G - D^k g| < h/(1 + \phi^d)$ on M. Then the real analytic function $F := G\phi^d : M \to \mathbb{R}$ satisfies $|D^k F - D^k f| < h\phi^d/(1 + \phi^d) < h$ on M and vanishes to order d along N, so F is the desired real analytic approximation. □

Theorem 5.53 also has a version with parameters.

COROLLARY 5.54. *Consider a real analytic fiber bundle $E \to M$, a real analytic manifold T with real analytic boundary, and a continuous family of functions $h_t : M \to \mathbb{R}_+$, $t \in T$. Let $f_t : M \to E$, $t \in T$, be a C^k-family of C^k-sections. Suppose that the f_t are real analytic for $t \in \partial T$ and depend real analytically on $t \in \partial T$. Then there exists a family of real analytic sections $F_t : M \to E$, depending real analytically on $t \in T$, with the following properties:*

(i) $F_t = f_t$ for $t \in \partial T$;

(ii) $|D^k F_t(x) - D^k f_t(x)| < h_t(x)$ for all $(t, x) \in T \times M$.

PROOF. By Example 5.50, we can include T in a larger open real analytic manifold $\widetilde{T}$. Extend f_t to a C^k-family $\widetilde{f_t}$ over $\widetilde{T}$ and view $\widetilde{f_t}$ as a C^k-section in the bundle $E \to \widetilde{T} \times M$. Now apply Theorem 5.53 to this section, the function $(t, x) \mapsto h_t(x)$, and the properly embedded real analytic submanifold $\partial T \times M$. □

5.9. Approximately holomorphic extension of maps from totally real submanifolds

Any real homomorphism $\Phi : E_1 \to E_2$ between two complex vector bundles (E_i, J_i) can be canonically presented as a sum $\Phi = \Phi_+ + \Phi_-$, where Φ_+ is complex linear, i.e., $\Phi_+ \circ J_1 = J_2 \circ \Phi_+$, and Φ_- is complex antilinear, i.e., $\Phi_- \circ J_1 = -J_2 \circ \Phi_-$. Indeed, we have $\Phi_+ = \frac{1}{2}(\Phi - J_2 \circ \Phi \circ J_1)$, $\Phi_- = \frac{1}{2}(\Phi + J_2 \circ \Phi \circ J_1)$. Given a smooth map $(V, J) \to (\widetilde{V}, \widetilde{J})$ between two complex manifolds, we set $\overline{\partial} f := (df)_-$. Of course, when the complex manifolds coincide with $\mathbb{C}^n$ and $\mathbb{C}^m$, then $\overline{\partial} f$ is the C^m-valued $(0,1)$-form $\overline{\partial} f = \sum\limits_1^n \frac{\partial f}{\partial \overline{z}_j} d\overline{z}_j$.

The following proposition is a C^k-version of Lemma 5.40.

PROPOSITION 5.55. *Let (V, J) and $(\widetilde{V}, \widetilde{J})$ be two complex manifolds of complex dimensions n and m, respectively. Suppose that $(\widetilde{V}, \widetilde{J})$ is Stein. Let $L \subset V$ be a totally real n-dimensional submanifold (not necessarily real analytic) and let $f : L \to \widetilde{V}$ be a smooth map. Then for any integer $k > 0$ there exists a smooth map $F : \mathcal{O}p\, L \to \widetilde{V}$ such that $F|_L = f$ and $\overline{\partial} F$ vanishes along L together with its k-jet.*

If $m = n$ and f is a diffeomorphism of L onto a totally real submanifold $\widetilde{L} \subset \widetilde{V}$, then F is a diffeomorphism between $\mathcal{O}p\, L$ and $\mathcal{O}p\, \widetilde{L}$ and the complex structures $\widetilde{J}$ and $F_* J$ coincide along $\widetilde{L}$ together with their k-jets.

PROOF. **Step 1.** Let us first consider the case when (V_1, J_1) and (V_2, J_2) coincide with the standard $\mathbb{C}^n$ (resp. $\mathbb{C}^m$) and $L = \mathbb{R}^n \subset \mathbb{C}^n$. We define an extension F of f to $\mathbb{C}^n$ by the formula

$$F(x+iy) := \sum_{|I| \le k+1} \frac{1}{I!} \frac{\partial^{|I|} f}{\partial x^I}(x) i^{|I|} y^I, \tag{5.1}$$

where $I = (i_1, \dots, i_n)$ is a multi-index, $|I| = i_1 + \dots + i_n$, $I! = i_1! \cdots i_n!$, and $y^I = y_1^{i_1} \cdots y_n^{i_n}$. Note that

$$\frac{\partial F}{\partial x_j} = \sum_{|I| \le k+1} \frac{1}{I!} \frac{\partial^{|I|+1} f}{\partial x^I \partial x_j}(x) i^{|I|} y^I,$$

$$\frac{\partial F}{\partial y_j} = \sum_{|I| \le k} \frac{1}{I!} \frac{\partial^{|I|+1} f}{\partial x^I \partial x_j}(x) i^{|I|+1} y^I,$$

$$\frac{\partial F}{\partial x_j} + i \frac{\partial F}{\partial y_j} = \sum_{|I| = k+1} \frac{1}{I!} \frac{\partial^{|I|+1} f}{\partial x^I \partial x_j}(x) i^{|I|} y^I = O(|y|^{k+1}),$$

and hence $\bar{\partial} F$ vanishes along $\mathbb{R}^n$ together with its k-jet.

Step 2. Consider now the case of general complex manifolds (V, J) and $(\widetilde{V}, \widetilde{J})$. Consider on L a real analytic structure compatible with its smooth structure. By Corollary 5.43, there exists a proper real analytic embedding $L \hookrightarrow \mathbb{R}^N$. By Lemma 5.40, this embedding extends to a holomorphic embedding $L^{\mathbb{C}} \hookrightarrow \mathbb{C}^N$ of a complexification of L into $\mathbb{C}^N$. We identify $L, L^{\mathbb{C}}$ with the corresponding subset in $\mathbb{R}^N$ (resp. $\mathbb{C}^N$). Since $\widetilde{V}$ is Stein, by Theorem 5.15 we can view it as a proper complex submanifold of some $\mathbb{C}^M$. Extend f to a smooth map $\widetilde{f} : \mathbb{R}^N \to \widetilde{V} \subset \mathbb{C}^M$ and then extend $\widetilde{f}$ by formula (5.1) to a smooth map $\widetilde{F} : \mathbb{C}^N \to \mathbb{C}^M$. Note that the image of $\widetilde{F}$ need not be contained in $\widetilde{V}$. However, by Corollary 5.27 there exists a neighborhood $\widetilde{U}$ of $\widetilde{V}$ in $\mathbb{C}^M$ which admits a holomorphic projection $\widetilde{\pi} : \widetilde{U} \to \widetilde{V}$. After shrinking $L^{\mathbb{C}}$, we may assume that $\widetilde{F}(L^{\mathbb{C}}) \subset \widetilde{U}$. Then the composition $G := \widetilde{\pi} \circ \widetilde{F}$ is an extension of f to a smooth map $G : L^{\mathbb{C}} \to \widetilde{V}$ such that $\bar{\partial} G$ vanishes along L together with its k-jet.

Step 3. If $L \subset V$ is already real analytic, then $L^{\mathbb{C}}$ is biholomorphic to a neighborhood of L in V and G from Step 2 is the desired extension. For general smooth L, it remains to find an appropriate diffeomorphism $H : L^{\mathbb{C}} \to \mathcal{O}p\, L \subset V$. By Proposition 2.15, the totally real submanifold L has a Stein neighborhood in V and hence we can assume that (V, J) is Stein. Now we apply the construction of Step 2 to the inclusion $h : L \hookrightarrow V$. This yields an extension of h to a smooth diffeomorphism $H : L^{\mathbb{C}} \to \mathcal{O}p\, L \subset V$ such that $\bar{\partial} H$ vanishes along L together with its k-jet. Then $G \circ H^{-1} : \mathcal{O}p\, L \to \widetilde{V}$ is the desired extension of f.

The final statement of the proposition follows from the implicit function theorem. □

5.10. CR structures

We define a *CR structure* on an odd-dimensional manifold M^{2n-1} as a germ of a complex structure J on $\mathcal{O}p\,(0 \times M) \subset \mathbb{R} \times M$. The maximal J-invariant distribution on $0 \times M$ defines a hyperplane distribution ξ on M with complex structure $\bar{J} = J|_\xi$ and the integrability of J implies

$$(5.2) \qquad X, Y \in \xi \Longrightarrow [\bar{J}X, \bar{J}Y] - [X, Y] = \bar{J}([\bar{J}X, Y] + [X, \bar{J}Y]) \in \xi$$

(see e.g. [**108**]). We call the CR structure *J-convex* if $0 \times M$ is a J-convex hypersurface. In this case, ξ is a contact structure, the function $\phi(r, x) = r$ is J-convex on $\mathcal{O}p\,(0 \times M)$, and $\alpha := -d^{\mathbb{C}} \phi|_{0 \times M}$ is a defining contact form for ξ such that $J|_\xi$ is compatible with $d\alpha|_\xi$.

Remark 5.56. (1) Usually (see e.g. [**108**]) a CR structure is defined as a hyperplane distribution ξ on M with a complex structure $\bar{J}$ on ξ satisfying equation (5.2). If (M, ξ, J) are real analytic and satisfy equation (5.2), then $\bar{J}$ extends to an integrable complex structure on $\mathcal{O}p\,(0 \times M) \subset \mathbb{R} \times M$ (see [**108**]). Note that for $\dim M = 3$ the condition (5.2) is vacuous. This implies, in particular, that if M is a real analytic hypersurface in a 4-dimensional almost complex manifold (V, J) with real analytic J, then J can be made integrable in a neighborhood of M without changing the induced CR structure.

(2) In general, a smooth $\bar{J}$ satisfying (5.2) need not extend to a complex structure J on $\mathcal{O}p\,(0 \times M) \subset \mathbb{R} \times M$. For example, for $n = 2$ there exist smooth convex $\bar{J}$ which do not even extend to a neighborhood of a point in $\mathbb{R} \times M$ ([**153, 108**]). For smooth convex $\bar{J}$ and $n \geq 4$, an extension to a neighborhood of a point in $\mathbb{R} \times M$ is always possible ([**118, 5**]), while for smooth convex $\bar{J}$ and $n = 3$ this question remains open. In order to avoid these subtleties, we require extendability to $\mathcal{O}p\,(0 \times M) \subset \mathbb{R} \times M$ in our definition of CR structure.

(3) In the literature "CR structure" often refers to the more general case of a distribution ξ of arbitrary codimension; in this book by "CR structure" we always mean the codimension one case.

Next we discuss the question of fillability of CR structures. A *holomorphic filling* of a closed CR manifold (M, J) is a compact complex manifold $(W, \widetilde{J})$ such that $\partial W = M$ and $\widetilde{J} = J$ on $\mathcal{O}p\,(\partial W)$. It is called a *Stein filling* (resp. *Kähler filling*) if $(W, \widetilde{J})$ is Stein (resp. Kähler).

REMARK 5.57. We define a compact complex manifold (W, J) with boundary as a germ of a slightly larger complex manifold $(\widetilde{W}, \widetilde{J})$ containing W as a smooth submanifold with boundary such that $\widetilde{J}|_W = J$. Such an extension exists whenever (W, J) is an almost complex manifold with J-convex boundary and vanishing Nijenhuis tensor [**29**], but its germ need not be unique [**94**].

First we recall the following result (see [**89**] for the relevant definitions).

THEOREM 5.58 (Rossi). *Given a compact complex manifold (W, J) with J-convex boundary ∂W, there exists a compact Stein space (W', J') with J'-convex boundary and finitely many normal singularities $p_1, \dots, p_k \in \operatorname{Int} W'$ and a holomorphic map $f : W \to W'$ such that $f|_{W \setminus f^{-1}(P)} \to W' \setminus P$ is a biholomorphism. Here we denote by P the set $\{p_1, \dots, p_k\}$ of singular points of W'.*

PROOF. Theorem 4 in Section IX C of [**89**] provides a Stein space (W', J') having all the desired properties except possibly normality. Now (W', J') has a normalization $(\widetilde{W}, \widetilde{J})$ such that the map $f : W \to W'$ factors through a holomorphic map $\widetilde{f} : W \to \widetilde{W}$ [**79**, Chapter 8], and $(\widetilde{W}, \widetilde{J})$ is again Stein by a theorem of Narasimhan [**145**]. □

Combining this result with Hironaka's theorem on resolution of singularities one obtains

THEOREM 5.59. *For any closed convex CR manifold the notions of holomorphic fillability and Kähler fillability coincide.*

PROOF. Clearly, any Kähler filling is a holomorphic filling. Conversely, according to Theorem 5.58, a holomorphic filling of a convex CR manifold M can be turned into a compact Stein space W with J-convex boundary and finitely many normal singularities in its interior. By a theorem of Lempert [**122**], such a Stein space W can be biholomorphically embedded into an affine algebraic variety X. By Hironaka's theorem [**96**], the singularities of X can be resolved. Hence we get a realization of M as a J-convex hypersurface in a smooth projective algebraic variety $\widetilde{X} \to X$ which bounds the preimage $\widetilde{W}$ of W in $\widetilde{X}$, so $\widetilde{W}$ is the desired Kähler filling. □

In complex dimension $n > 2$ we have the following existence theorem for holomorphic fillings.

THEOREM 5.60 (Rossi [**163**]). *For $n > 2$, any closed convex CR manifold (M^{2n-1}, J) is holomorphically (and thus Kähler) fillable.*

On the other hand, in general, for $n > 2$ a (holomorphically fillable) CR structure need not be Stein fillable. Indeed, there are easy homological obstructions to Stein fillability arising, e.g., from the following argument which was explained to the second author by M. Freedman.

LEMMA 5.61. *Let M be a closed manifold of dimension $2n-1$. Suppose that for some coefficient ring R there are cohomology classes $a_i \in H^{d_i}(M; R)$ such that*

$$a_1 \cup \cdots \cup a_k \neq 0, \qquad d_i < n-1, \qquad d_1 + \cdots + d_k > n.$$

Then M is not the boundary of a Stein domain.

PROOF. Suppose $M = \partial W$ for a Stein domain W. Since W has a cell decomposition without cells of index $> n$, it satisfies $H_i(W) = H^i(W) = 0$ for $i > n$ (all (co)homology is with coefficients in R). Now $d_i < n - 1$ implies $H^{d_i+1}(W, \partial W) = H_{2n-d_i-1}(W) = 0$, so by the long exact sequence of the pair $(W, \partial W)$ the pullback map $j^* : H^{d_i}(W) \to H^{d_i}(\partial W)$ is surjective. Thus there exist classes $\alpha_i \in H^{d_i}(W)$ with $j^*\alpha_i = a_i$ and $j^*(\alpha_1 \cup \cdots \cup \alpha_k) = a_1 \cup \cdots \cup a_k \neq 0$. But on the other hand $\alpha_1 \cup \cdots \cup \alpha_k$ vanishes because $H^{d_1+\cdots+d_k}(W) = 0$, so we have a contradiction. □

EXAMPLE 5.62. For $n > 2$ the real projective space $\mathbb{R}P^{2n-1}$ admits no Stein fillable CR structures (although it inherits a CR structure from S^{2n-1}). This follows from Lemma 5.61 because the cup product of $2n-1$ classes of degree 1 (with $\mathbb{Z}_2$ coefficients) is nonzero. Similarly, for $n > 2$ the torus T^{2n-1} does not admit any Stein fillable CR structure. In fact, it would be interesting to know whether T^{2n-1}, $n > 2$, admits a CR structure at all.

The situation for $n = 2$ is drastically different. First of all, a 3-dimensional CR structure need not be holomorphically fillable:

EXAMPLE 5.63 (Rossi [**163**]). Note that for any $\varepsilon \in [0, 1)$ the intersection of the quadric $Q_\varepsilon = \{z_0^2 + z_1^2 + z_2^2 = \varepsilon\} \subset \mathbb{C}^3$ with the boundary ∂B^6 of the unit ball $B^6 \subset \mathbb{C}^3$ is diffeomorphic to $\mathbb{R}P^3$. Let us denote by $\bar{J}_\varepsilon$ the convex CR structure on $\mathbb{R}P^3$ induced by this diffeomorphism.

Hence $(\mathbb{R}P^3, \bar{J}_0)$ is filled by the singular Stein space $W_0 = Q_0 \cap B^6$, while $(\mathbb{R}P^3, \bar{J}_\varepsilon)$ for $\varepsilon \in (0, 1)$ is filled by the smooth Stein domain $W_\varepsilon = Q_\varepsilon \cap B^6$. The pullbacks of $\bar{J}_\varepsilon$ under the quotient map $S^3 \to \mathbb{R}P^3$ yield CR structures J_ε on S^3 depending smoothly on $\varepsilon \in [0, 1)$. We claim that J_ε is not Stein fillable for $\varepsilon > 0$.

To see this, suppose that W is a Stein filling of (S^3, J_ε). The quotient map $S^3 \to \mathbb{R}P^3$ then induces a holomorphic map $\mathcal{O}p\,(\partial W) \to \mathcal{O}p\,(\partial W_\varepsilon)$. By Corollary 5.25, this map extends to a holomorphic map $F : W \to W_\varepsilon$. Since F is a submersion on $\mathcal{O}p\,(\partial W)$, the set $Z \subset W$ where the complex determinant of DF vanishes is a compact codimension one analytic subvariety of W. Since the only compact analytic subvarieties of a Stein manifold are zero dimensional (see e.g. Theorem 5.9 in Chapter II of [**36**]), it follows that Z is empty and hence F is a submersion. As it is a two-to-one covering near ∂W, we see that $F : W \to W_\varepsilon$ is a two-to-one covering. But W_ε is diffeomorphic to the unit disc cotangent bundle of S^2, hence simply connected, so it does not possess any connected two-to-one covering. Since the Stein manifold W is connected, this gives a contradiction.

By contrast, J_0 is isomorphic to the standard CR-structure on the boundary of a ball in $\mathbb{C}^2$ and hence Stein fillable. To see this, consider the holomorphic map $F : \mathbb{C}^2 \to \mathbb{C}^3$ given by the formula

$$F(u_1, u_2) := \Big(u_1^2 + u_2^2, i(u_1^2 - u_2^2), 2u_1u_2\Big), \qquad (u_1, u_2) \in \mathbb{C}^2.$$

Then $F(\mathbb{C}^2) = Q_0$ and $F(B^4) = W_0 = Q_0 \cap B^6$, where $B^4 \subset \mathbb{C}^2$ denotes the ball of radius $1/\sqrt{2}$. The preimage of the origin under F is the origin, while the preimage of any other point in Q_0 is a pair of points $\pm(u_1, u_2)$. Thus F induces a branched holomorphic two-to-one covering $B^4 \to W_0$ and hence a holomorphic filling of (S^3, J_0) by the ball B^4. □

On the other hand, unlike the situation in complex dimension $n > 2$, the following theorem shows that any holomorphically fillable 3-dimensional convex CR structure can be C^∞-perturbed to a Stein fillable one. Recall that a 2-dimensional complex manifold is called *minimal* if it does not contain embedded holomorphic spheres with self-intersection number -1. Any complex manifold can be made minimal by blowing down all such spheres.

THEOREM 5.64 (Bogomolov and de Oliveira [**20**]). *Let (W, J) be a minimal complex manifold of complex dimension 2 with J-convex boundary. Then there exists a deformation J_t of the complex structure $J_0 = J$ such that (W, J_t) is a Stein domain for all sufficiently small $t > 0$.*

Part 2

Existence of Stein Structures

CHAPTER 6

Symplectic and Contact Preliminaries

In this chapter we collect some relevant facts from symplectic and contact geometry. For more details see [**136, 65**].

6.1. Symplectic vector spaces

A *symplectic vector space* (V, ω) is a real vector space V with a nondegenerate skew-symmetric bilinear form ω. Here *nondegenerate* means that $v \mapsto \omega(v, \cdot)$ defines an isomorphism $V \to V^*$. It follows that V has even dimension $2n$. A linear map $\Psi : (V_1, \omega_1) \to (V_2, \omega_2)$ between symplectic vector spaces is called *symplectic* if $\Psi^*\omega_2 \equiv \omega_2(\Psi\cdot, \Psi\cdot) = \omega_1$.

For any vector space U the space $U \oplus U^*$ carries the *standard symplectic structure*

$$\omega_{\rm st}\Big((u, u^*), (v, v^*)\Big) := v^*(u) - u^*(v).$$

In coordinates q_i on U and dual coordinates p_i on U^*, the standard symplectic form is given by

$$\omega_{\rm st} = \sum dq_i \wedge dp_i.$$

Define the *ω-orthogonal complement* of a linear subspace $W \subset V$ by

$$W^\omega := \{v \in V \mid \omega(v, w) = 0 \text{ for all } w \in W\}.$$

Note that $\dim W + \dim W^\omega = 2n$, but $W \cap W^\omega$ need not be $\{0\}$. W is called

- *symplectic* if $W \cap W^\omega = \{0\}$,
- *isotropic* if $W \subset W^\omega$,
- *coisotropic* if $W^\omega \subset W$,
- *Lagrangian* if $W^\omega = W$.

Note that $\dim W$ is even for W symplectic, $\dim W \leq n$ for W isotropic, $\dim W \geq n$ for W coisotropic, and $\dim W = n$ for W Lagrangian. Note also that $(W^\omega)^\omega = W$, and $\big(W/(W \cap W^\omega), \omega\big)$ is a symplectic vector space.

Consider a subspace W of a symplectic vector space (V, ω) and set $N := W \cap W^\omega$. Choose subspaces $V_1 \subset W$, $V_2 \subset W^\omega$ and an isotropic subspace $V_3 \subset (V_1 \oplus V_2)^\omega$ such that

$$W = V_1 \oplus N, \quad W^\omega = N \oplus V_2, \quad (V_1 \oplus V_2)^\omega = N \oplus V_3.$$

Then the decomposition

$$V = V_1 \oplus N \oplus V_2 \oplus V_3$$

induces a symplectic isomorphism

$$\begin{aligned} (V, \omega) &\to (W/N, \omega) \oplus (W^\omega/N, \omega) \oplus (N \oplus N^*, \omega_{\rm st}), \\ v_1 + n + v_2 + v_3 &\mapsto \Big(v_1, v_2, (n, -i_{v_3}\omega)\Big). \end{aligned} \tag{6.1}$$

Every symplectic vector space (V, ω) of dimension $2n$ possesses a *symplectic basis* $e_1, f_1, \dots, e_n, f_n$, i.e., a basis satisfying

$$\omega(e_i, e_j) = \omega(f_i, f_j) = 0, \qquad \omega(e_i, f_j) = \delta_{ij}.$$

Moreover, given a subspace $W \subset V$, the basis can be chosen such that

- $W = \text{span}\{e_1, \dots, e_{k+l}, f_1, \dots, f_k\}$,
- $W^\omega = \text{span}\{e_{k+1}, \dots, e_n, f_{k+l+1}, \dots, f_n\}$,
- $W \cap W^\omega = \text{span}\{e_{k+1}, \dots, e_{k+l}\}$.

In particular, we get the following normal forms:

- $W = \text{span}\{e_1, f_1, \dots, e_k, f_k\}$ if W is symplectic;
- $W = \text{span}\{e_1, \dots, e_k\}$ if W is isotropic;
- $W = \text{span}\{e_1, \dots, e_n, f_1, \dots, f_k\}$ if W is coisotropic;
- $W = \text{span}\{e_1, \dots, e_n\}$ if W is Lagrangian.

This reduces the study of symplectic vector spaces to the *standard symplectic space* $(\mathbb{R}^{2n}, \omega_{\text{st}} = \sum dq_i \wedge dp_i)$.

A pair (ω, J) consisting of a symplectic form ω and a complex structure J on a vector space V is called *compatible* if

$$g_J := \omega(\cdot, J\cdot)$$

is an inner product (i.e., symmetric and positive definite). This is equivalent to saying that

$$H(v, w) := \omega(v, Jw) - i\omega(v, w)$$

defines a Hermitian metric. Therefore, we will also call a compatible pair (ω, J) a *Hermitian structure* and (V, ω, J) a *Hermitian vector space.*

LEMMA 6.1. *(a) The space of symplectic forms compatible with a given complex structure is nonempty and contractible.*

(b) The space of complex structures compatible with a given symplectic form is nonempty and contractible.

PROOF. (a) immediately follows from the fact that the Hermitian metrics for a given complex structure form a convex space.

(b) is a direct consequence of the fact (see [**136**]) that for a symplectic vector space (V, ω) there exists a continuous map from the space of inner products to the space of compatible complex structures which maps each induced inner product g_J to J.

To see this fact, note that an inner product g defines an isomorphism $A : V \to V$ via $\omega(\cdot, \cdot) = g(A\cdot, \cdot)$. Skew-symmetry of ω implies $A^T = -A$. Recall that each positive definite operator P possesses a unique positive definite square root $\sqrt{P}$, and $\sqrt{P}$ commutes with every operator with which P commutes. So we can define

$$J_g := (AA^T)^{-\frac{1}{2}} A.$$

It follows that $J_g^2 = -\text{Id}$ and $\omega(\cdot, J\cdot) = g(\sqrt{AA^T}\cdot, \cdot)$ is an inner product. Continuity of the mapping $g \mapsto J_g$ follows from continuity of the square root. Finally, if $g = g_J$ for some J then $A = J = J_g$. □

Let us call a real subspace $W \subset V$ of a complex vector space (V, J)

- *totally real* if $W \cap JW = \{0\}$,
- *totally coreal* if $W + JW = V$,

- *maximally real* if $W \cap JW = \{0\}$ and $W + JW = V$,
- *complex* if $JW = W$.

Note that $\dim W \leq n$ if W is totally real, $\dim W \geq n$ if W is totally coreal, and $\dim W = n$ if W is maximally real.

For a subspace $W \subset V$ of a Hermitian vector space (V, ω, J) we denote by $W^\perp$ the orthogonal complement with respect to the metric $g_J = \omega(\cdot, J\cdot)$. The following lemma relates the symplectic and complex notions on a Hermitian vector space. It follows easily from the relation $W^\omega = (JW)^\perp = J(W^\perp)$.

LEMMA 6.2. *Let (V, J, ω) be a Hermitian vector space and $W \subset V$ a real subspace. Then*
(a) W isotropic $\Longleftrightarrow JW \subset W^\perp \Longrightarrow W$ totally real,
(b) W coisotropic $\Longleftrightarrow W^\perp \subset JW \Longrightarrow W$ totally coreal,
(a) W Lagrangian $\Longleftrightarrow JW = W^\perp \Longrightarrow W$ maximally real,
(c) W complex $\Longrightarrow W$ symplectic.

6.2. Symplectic vector bundles

The discussion of the previous section immediately carries over to vector bundles. For this, let $E \to M$ be a real vector bundle of rank $2n$ over a manifold. A *symplectic structure* on E is a smooth section ω in the bundle $\Lambda^2 E^* \to M$ such that each $\omega_x \in \Lambda^2 E_x^*$ is a linear symplectic form. A pair (ω, J) of a symplectic and a complex structure on E is called *compatible*, or a *Hermitian structure*, if $\omega(\cdot, J\cdot)$ defines an inner product on E. Lemma 6.1 immediately yields the following facts, where the spaces of sections are equipped with any reasonable topology, e.g., the $C^\infty_{\rm loc}$ topology:

(a) The space of compatible symplectic structures on a complex vector bundle (E, J) is nonempty and contractible.

(b) The space of compatible complex structures on a symplectic vector bundle (E, ω) is nonempty and contractible.

This shows that the homotopy theories of symplectic, complex, and Hermitian vector bundles are the same. In particular, obstructions to trivialization of a symplectic vector bundle (E, ω) are measured by the *Chern classes* $c_k(E, \omega) = c_k(E, J)$ for any compatible complex structure J.

REMARK 6.3. The homotopy equivalence between symplectic, complex, and Hermitian vector bundles can also be seen in terms of their structure groups: The *symplectic group*[1]

$$Sp(2n) := \{\Psi \in GL(2n, \mathbb{R}) \mid \Psi^*\omega = \omega\} = \{\Psi \in GL(2n, \mathbb{R}) \mid \Psi^T J \Psi = J\}$$

and the complex general linear group $GL(n, \mathbb{C})$ both deformation retract onto the unitary group

$$U(n) = Sp(2n) \cap O(2n) = O(2n) \cap GL(n, \mathbb{C}) = GL(n, \mathbb{C}) \cap Sp(2n).$$

We end this section with a normal form for subbundles of symplectic vector bundles.

[1] $Sp(2n)$ is *not* the "symplectic group" $Sp(n)$ considered in Lie group theory; e.g., the latter is compact, while our symplectic group is not.

PROPOSITION 6.4. *Let* (E,ω) *be a rank* $2n$ *symplectic vector bundle and* $W \subset E$ *a rank* $2k+l$ *subbundle such that* $N := W \cap W^\omega$ *has constant rank* l. *Then*

$$(E,\omega) \cong (W/N,\omega) \oplus (W^\omega/N,\omega) \oplus (N \oplus N^*, \omega_{\rm st}).$$

PROOF. Pick a compatible complex structure J on (E,ω). Then

$$V_1 := W \cap JW, \quad V_2 := W^\omega \cap JW^\omega, \quad V_3 := JN$$

are smooth subbundles of E. Now the isomorphism (6.1) of the previous section yields the desired decomposition. □

6.3. Symplectic manifolds

A *symplectic manifold* (V,ω) is a manifold V with a closed nondegenerate 2-form ω. A map $f:(V_1,\omega_1)\to(V_2,\omega_2)$ between symplectic manifolds is called *symplectic* if $f^*\omega_2=\omega_1$, and a symplectic diffeomorphism is called *symplectomorphism.* The following basic result states that every symplectic manifold of dimension $2n$ is locally symplectomorphic to $(\mathbb{R}^{2n},\omega_{\rm st})$. In other words, every symplectic manifold possesses a *symplectic atlas*, i.e., an atlas all of whose transition maps are symplectic.

PROPOSITION 6.5 (Symplectic Darboux theorem). *Let* (V,ω) *be a symplectic manifold of dimension* $2n$. *Then every* $x\in V$ *possesses a coordinate neighborhood* U *and a coordinate map* $\phi: U\to U'\subset\mathbb{R}^{2n}$ *such that* $\phi^*\omega_{st}=\omega$.

The symplectic Darboux theorem is a special case of the symplectic neighborhood theorem which will be proved in the next section. Now let us discuss some examples of symplectic manifolds.

Cotangent bundles. Let $T^*Q \xrightarrow{\pi} Q$ be the cotangent bundle of a manifold Q. The 1-form $\sum p_i dq_i$ is independent of coordinates q_i on Q and dual coordinates p_i on T_q^*Q and thus defines the *Liouville* 1-*form* $\lambda_{\rm st}$ on T^*Q. Intrinsically,

$$(\lambda_{\rm st})_{(q,p)}\cdot v = \langle p, T_{(q,p)}\pi\cdot v\rangle \quad \text{for } v\in T_{(q,p)}T^*Q,$$

where $\langle\ ,\ \rangle$ is the pairing between T_q^*Q and T_qQ. The 2-form $\omega_{\rm st} := d\lambda_{\rm st}$ is clearly closed, and the coordinate expression $\omega_{\rm st}=\sum dq_i\wedge dp_i$ shows that it is also nondegenerate. So $\omega_{\rm st}$ defines the *standard symplectic form* on T^*Q. The standard form on $\mathbb{R}^{2n}$ is a particular case of this construction.

Exact symplectic manifolds. Recall that a Liouville form on an even-dimensional manifold V is a 1-form λ such that $d\lambda$ is symplectic. The pair (V,λ) is then called an *exact symplectic manifold.* For example, the form $p\,dq$ is the standard Liouville form on the cotangent bundle T^*L. An immersion or embedding $\phi: L\to V$ into an exact symplectic manifold (V,λ) is called *exact Lagrangian* if $\phi^*\lambda$ is exact. The vector field X dual to λ with respect to $d\lambda$, i.e., such that $i_Xd\lambda=\lambda$, is called the Liouville field. See Section 11.1 below for detailed discussion of these notions.

Almost complex submanifolds. A pair (ω,J) consisting of a symplectic form and an almost complex structure on a manifold V is called *compatible* if $\omega(\cdot,J\cdot)$ defines a Riemannian metric. It follows that ω induces a symplectic form on every almost complex submanifold $W\subset V$ (which is compatible with $J|_W$).

J-convex functions. If (V,J) is an almost complex manifold and $\phi: V\to\mathbb{R}$ a J-convex function, then the 2-form $\omega_\phi=-dd^{\mathbb{C}}\phi$ is symplectic. Moreover, ω_ϕ is compatible with J if J is integrable (see Section 2.2). In particular, every J-convex function on a Stein manifold induces a symplectic form compatible with J.

Kähler manifolds. A Kähler manifold is a complex manifold (V, J) with a *Kähler metric*, i.e., a Hermitian metric $H = g - i\omega$ on TV such that the 2-form ω is closed. Thus the *Kähler form* ω is a symplectic form compatible with J. Note that every complex submanifold of a Kähler manifold is again Kähler.

The two basic examples of Kähler manifolds are $\mathbb{C}^n$ with the standard complex structure and Hermitian metric, and the complex projective space $\mathbb{C}P^n = (\mathbb{C}^{n+1} \setminus 0)/(\mathbb{C} \setminus 0)$ with the induced complex structure and Hermitian metric (the latter is defined by restricting the Hermitian metric of $\mathbb{C}^{n+1}$ to the unit sphere and dividing out the standard circle action). Passing to complex submanifolds of $\mathbb{C}^n$, we see again that Stein manifolds are Kähler. Passing to complex submanifolds of $\mathbb{C}P^n$, we see that smooth projective varieties are Kähler. This gives us a rich source of examples of closed symplectic manifolds.

REMARK 6.6. While cotangent bundles and Kähler manifolds provide obvious examples of symplectic manifolds, it is not obvious how to go beyond them. The first example of a closed symplectic manifold that is not Kähler was constructed by Thurston [**184**] in 1976. In 1995 Gompf [**69**] proved that every finitely presented group is the fundamental group of a closed symplectic 4-manifold, in stark contrast to the many restrictions on the fundamental groups of closed Kähler surfaces.

REMARK 6.7. A Riemannian metric g on a manifold Q induces a natural almost complex structure J_g on T^*Q, compatible with $\omega_{\rm st}$, which interchanges the horizontal and vertical subspaces defined by the Levi-Civita connection. M. Grueneberg (unpublished) has shown that J_g is integrable if and only if the metric g is flat.

6.4. Moser's trick and symplectic normal forms

An (embedded or immersed) submanifold W of a symplectic manifold (V, ω) is called *symplectic (isotropic, coisotropic, Lagrangian)* if $T_xW \subset T_xV$ is symplectic (isotropic, coisotropic, Lagrangian) for every $x \in W$ in the sense of Section 6.1. In this section we derive normal forms for neighborhoods of such submanifolds.

All the normal forms can be proved by the same technique which we will refer to as *Moser's trick*. It is based on Cartan's formula $L_X\alpha = i_X d\alpha + d i_X\alpha$ for a vector field X and a k-form α. Suppose we are given k-forms α_0, α_1 on a manifold M, and we are looking for a diffeomorphism $\phi : M \to M$ such that $\phi^*\alpha_1 = \alpha_0$. Moser's trick is to construct ϕ as the time-1 map of a time-dependent vector field X_t. For this, let α_t be a smooth family of k-forms connecting α_0 and α_1, and look for a vector field X_t whose flow ϕ_t satisfies

$$\phi_t^*\alpha_t \equiv \alpha_0. \tag{6.2}$$

Then the time-1 map $\phi = \phi_1$ solves our problem. Now equation (6.2) follows by integration (provided the flow of X_t exists, e.g., if X_t has compact support) once its linearized version

$$0 = \frac{d}{dt}\phi_t^*\alpha_t = \phi_t^*(\dot\alpha_t + L_{X_t}\alpha_t)$$

holds for every t. Inserting Cartan's formula, this reduces the problem to the algebraic problem of finding a vector field X_t that satisfies

$$\dot\alpha_t + d i_{X_t}\alpha_t + i_{X_t}d\alpha_t = 0. \tag{6.3}$$

Here is a first application of this method. Here, as well as throughout the book, by *diffeotopy* we denote a smooth family of diffeomorphisms ϕ_t, $t \in [0,1]$, with $\phi_0 = \mathrm{Id}$.

THEOREM 6.8 (Moser's stability theorem). *Let V be a manifold (without boundary but not necessarily compact). Let ω_t, $t \in [0,1]$, be a smooth family of symplectic forms on V which coincide outside a compact set and such that the cohomology class with compact support $[\omega_t - \omega_0] \in H^2_c(V;\mathbb{R})$ is independent of t. Then there exists a diffeotopy ϕ_t with $\phi_t = \mathrm{Id}$ outside a compact set such that $\phi_t^*\omega_t = \omega_0$.*

In particular, this applies if $\omega_t = d\lambda_t$ for a smooth family of 1-forms λ_t which coincide outside a compact set, and in this case there exists a smooth family of functions $f_t : V \to \mathbb{R}$ with compact support such that

$$\phi_t^*\lambda_t - \lambda_0 = df_t.$$

PROOF. For every t the closed 2-form $\dot\omega_t$ is trivial in cohomology with compact support $H^2_c(V;\mathbb{R})$, so there exists a 1-form β_t with compact support such that $d\beta_t = \dot\omega_t$. The forms β_t are not unique, but by an argument of Banyaga [**13**] they can be chosen to depend smoothly on t and to be supported in a fixed compact subset. Now we can solve equation (6.3),

$$0 = \dot\omega_t + d i_{X_t}\omega_t + i_{X_t} d\omega_t = d(\beta_t + i_{X_t}\omega_t)$$

by solving $\beta_t + i_{X_t}\omega_t = 0$, which has a unique solution X_t due to the nondegeneracy of ω_t. Since X_t vanishes outside a compact subset, its flow ϕ_t exists and gives the desired family of diffeomorphisms.

In the case $\omega_t = d\lambda_t$ we pick $\beta_t := \dot\lambda_t$. Then the defining equation for X_t becomes $\dot\lambda_t + i_{X_t} d\lambda_t = 0$, and we find

$$\frac{d}{dt}\phi_t^*\lambda_t = \phi_t^* d(i_{X_t}\lambda_t),$$

which integrates to

$$\phi_t^*\lambda_t - \lambda_0 = d\left(\int_0^t i_{X_s}\lambda_s ds\right). \qquad \square$$

COROLLARY 6.9. *Let W be a compact manifold with (possibly empty) boundary ∂W. Let ω_t, $t \in [0,1]$, be a smooth family of symplectic forms such that the restrictions $\omega_t|_{\partial W}$ and the relative cohomology classes $[\omega_t - \omega_0] \in H^2(W, \partial W; \mathbb{R})$ are independent of t. Then there exists a diffeotopy ϕ_t with $\phi_t|_{\partial W} = \mathrm{Id}$ such that $\phi_t^*\omega_t = \omega_0$.*

PROOF. If $\partial W = \varnothing$ this is a special case of Theorem 6.8. In the case $M = \partial W \neq \varnothing$, we argue as follows. Consider the inclusion $\iota : M \to \mathbb{R}\times M$, $x \mapsto (0,x)$, and the projection $\pi : \mathbb{R}\times M \to M$. Recall from [**87**] that there exists a continuous linear map $P : \Omega^2(\mathbb{R}\times M) \to \Omega^1(\mathbb{R}\times M)$ satisfying the homotopy formula $dP + Pd = \mathrm{Id} - \pi^*\iota^*$. Applying this to the forms $\dot\omega_t$ on a tubular neighborhood $[0,1)\times M$ of M in W, we find a smooth family of 1-forms $\alpha_t := P\dot\omega_t$ vanishing along ∂M (this follows from the explicit formula for P in [**87**]) such that $d\alpha_t = \dot\omega_t$ on $[0,1)\times M$. Pick a cutoff function $f : W \to [0,1]$ which equals 1 near ∂W and 0 outside $[0,1)\times M$ and extend $f\alpha_t$ by 0 over the rest of W. The closed 2-forms $\dot\omega_t = d(f\alpha_t)$ have compact support in $\mathrm{Int}\, W$ and are trivial in cohomology with compact support $H^2_c(\mathrm{Int}\, W;\mathbb{R})$, so by [**13**] there exists a smooth family of 1-forms γ_t with fixed compact support in $\mathrm{Int}\, W$ such that $d\gamma_t = \dot\omega_t - d(f\alpha_t)$. Now we proceed as in the proof of Theorem 6.8 with the 1-forms $\beta_t := \gamma_t + f\alpha_t$. Since the resulting vector field X_t vanishes on ∂W, its flow ϕ_t exists and gives the desired family of diffeomorphisms. □

Our second application of Moser's trick is the following lemma, which is the basis of all the normal form theorems below.

LEMMA 6.10. *Let W be a compact submanifold of a manifold V, and let ω_0, ω_1 be symplectic forms on V which agree at all points of W. Then there exist tubular neighborhoods U_0, U_1 of W and a diffeomorphism $\phi : U_0 \to U_1$ such that $\phi|_W = \mathrm{Id}$ and $\phi^*\omega_1 = \omega_0$.*

PROOF. Set $\omega_t := (1-t)\omega_0 + \omega_1$. Since $\omega_t \equiv \omega_0$ along W, ω_t are symplectic forms on some tubular neighborhood U of W. By a homotopy formula similar to the one used in the previous proof, since $\dot\omega_t = \omega_1 - \omega_0$ is closed and vanishes along W, there exists a 1-form β on U such that $\beta = 0$ along W and $d\beta = \dot\omega_t$ on U. As in the proof of Theorem 6.8, we solve equation (6.3) by setting $\beta + i_{X_t}\omega_t = 0$.

To apply Moser's trick, a little care is needed because U is noncompact, so the flow of X_t may not exist until time 1. However, since $\beta = 0$ along W, X_t vanishes along W. Thus there exists a tubular neighborhood U_0 of W such that the flow $\phi_t(x)$ of X_t exists for all $x \in U_0$ and $t \in [0,1]$, and $\phi_t(U_0) \subset U$ for all $t \in [0,1]$. Now $\phi_1 : U_0 \to U_1 := \phi_1(U_0)$ is the desired diffeomorphism with $\phi_1^*\omega_1 = \omega_0$. □

Now we are ready for the main result of this section.

PROPOSITION 6.11 (Symplectic normal forms). *Let ω_0, ω_1 be symplectic forms on a manifold V and $W \subset V$ a compact submanifold such that $\omega_0|_W = \omega_1|_W$. Suppose that $N := \ker(\omega_0|_W) = \ker(\omega_1|_W)$ has constant rank, and the bundles $(TW^{\omega_0}/N, \omega_0)$ and $(TW^{\omega_1}/N, \omega_1)$ over W are isomorphic as symplectic vector bundles. Then there exist tubular neighborhoods U_0, U_1 of W and a diffeomorphism $\phi : U_0 \to U_1$ such that $\phi|_W = \mathrm{Id}$ and $\phi^*\omega_1 = \omega_0$.*

PROOF. By Proposition 6.4,

$$(TV|_W, \omega_0) \cong (TW/N, \omega_0) \oplus (TW^{\omega_0}/N, \omega_0) \oplus (N \oplus N^*, \omega_{\mathrm{st}}),$$

and similarly for ω_1. By the hypotheses, the terms on the right-hand side are isomorphic for ω_0 and ω_1. More precisely, there exists an isomorphism

$$\Psi : (TV|_W, \omega_0) \to (TV|_W, \omega_1)$$

with $\Psi|_{TW} = \mathrm{Id}$. Extend Ψ to a diffeomorphism $\psi : U_0 \to U_1$ of tubular neighborhoods such that $\psi|_W = \mathrm{Id}$ and $\psi^*\omega_1 = \omega_0$ along W, and apply Lemma 6.10. □

The following normal forms, due to Weinstein, are easy corollaries of this result.

COROLLARY 6.12 (Symplectic neighborhood theorem). *Let ω_0, ω_1 be symplectic forms on a manifold V and $W \subset V$ a compact submanifold such that $\omega_0|_W = \omega_1|_W$ is symplectic, and the symplectic normal bundles $(TW^{\omega_0}, \omega_0)$, $(TW^{\omega_1}, \omega_1)$ over W are isomorphic (as symplectic vector bundles). Then there exist tubular neighborhoods U_0, U_1 of W and a diffeomorphism $\phi : U_0 \to U_1$ such that $\phi|_W = \mathrm{Id}$ and $\phi^*\omega_1 = \omega_0$.*

COROLLARY 6.13 (Isotropic neighborhood theorem). *Let ω_0, ω_1 be symplectic forms on a manifold V and $W \subset V$ a compact submanifold such that $\omega_0|_W = \omega_1|_W = 0$, and the symplectic normal bundles $(TW^{\omega_0}/TW, \omega_0)$, $(TW^{\omega_1}/TW, \omega_1)$*

are isomorphic (as symplectic vector bundles). Then there exist tubular neighborhoods U_0, U_1 *of* W *and a diffeomorphism* $\phi : U_0 \to U_1$ *such that* $\phi|_W = \mathrm{Id}$ *and* $\phi^*\omega_1 = \omega_0$.

COROLLARY 6.14 (Coisotropic neighborhood theorem). *Let* ω_0, ω_1 *be symplectic forms on a manifold* V *and* $W \subset V$ *a compact submanifold such that* $\omega_0|_W = \omega_1|_W$ *and* W *is coisotropic for* ω_0 *and* ω_1. *Then there exist tubular neighborhoods* U_0, U_1 *of* W *and a diffeomorphism* $\phi : U_0 \to U_1$ *such that* $\phi|_W = \mathrm{Id}$ *and* $\phi^*\omega_1 = \omega_0$.

COROLLARY 6.15 (Weinstein's Lagrangian neighborhood theorem [**186**]). *Let* $W \subset (V, \omega)$ *be a compact Lagrangian submanifold of a symplectic manifold. Then there exist tubular neighborhoods* U *of the zero section in* T^*W *and* U' *of* W *in* V *and a diffeomorphism* $\phi : U \to U'$ *such that* $\phi|_W$ *is the inclusion and* $\phi^*\omega = \omega_{\mathrm{st}}$.

PROOF. Since W is Lagrangian, the map $v \mapsto i_v\omega$ defines an isomorphism from the normal bundle $TV/TW|_W$ to T^*W. Extend the inclusion $W \subset V$ to a diffeomorphism $\psi : U \to U'$ of tubular neighborhoods of the zero section in T^*W and of W in V. Now apply the coisotropic neighborhood theorem to the zero section in T^*W and the symplectic forms ω_{st} and $\psi^*\omega$. □

6.5. Contact manifolds and their Legendrian submanifolds

A *contact structure* ξ on a manifold M is a completely nonintegrable tangent hyperplane field. According to the Frobenius condition, this means that for every nonzero local vector field $X \in \xi$, there exists a local vector field $Y \in \xi$ such that their Lie bracket satisfies $[X, Y] \notin \xi$. If α is any 1-form locally defining ξ, i.e., $\xi = \ker \alpha$, this means

$$d\alpha(X, Y) = -\frac{1}{2}\alpha([X, Y]) \neq 0.$$

So the restriction of the 2-form $d\alpha$ to ξ is nondegenerate, i.e., $(\xi, d\alpha|_\xi)$ is a symplectic vector bundle. This implies in particular that $\dim \xi$ is even and $\dim M = 2n+1$ is odd. In terms of a local defining 1-form α, the contact condition can also be expressed as $\alpha \wedge (d\alpha)^n \neq 0$.

A diffeomorphism $f : (M_1, \xi_1) \to (M_2, \xi_2)$ between contact manifolds is called a *contactomorphism* if $f_*\xi_1 = \xi_2$.

REMARK 6.16. If $\dim M = 4k + 3$, the sign of the volume form $\alpha \wedge (d\alpha)^{2k+1}$ is independent of the sign of the defining local 1-form α, so a contact structure defines an orientation of the manifold. In particular, in these dimensions contact structures can exist only on orientable manifolds. On the other hand, a contact structure ξ on a manifold of dimension $4k + 1$ is itself orientable.

Contact structures ξ in this book will always be *cooriented*, i.e., they are globally defined by a 1-form α. In this case the symplectic structure on each of the hyperplanes ξ is defined uniquely up to a positive conformal factor. Moreover, associated to each defining 1-form α is its *Reeb vector field* R_α defined by

$$i_{R_\alpha} d\alpha = 0, \qquad \alpha(R_\alpha) = 1.$$

Given a J-convex hypersurface M (which is by definition cooriented) in an almost complex manifold (V, J), the field ξ of complex tangencies defines a contact structure on M which is cooriented by $J\nu$, where ν is a vector field transverse to M

defining the coorientation. Conversely, any cooriented contact structure ξ arises as a field of complex tangencies on a J-convex hypersurface in an almost complex manifold: Just choose a complex multiplication J on ξ compatible with the symplectic form $d\alpha$ in the sense that $d\alpha(\cdot, J\cdot)$ is a (positive definite) inner product on ξ and extend J arbitrarily to an almost complex structure on $V := M \times (-\epsilon, \epsilon)$.

REMARK 6.17. If $\dim M = 3$, then J can always be chosen integrable. However, in dimensions ≥ 5 this is not always the case; see Remark 6.28 below.

Let $(M, \xi = \ker \alpha)$ be a contact manifold of dimension $2n + 1$. An immersion $\phi : \Lambda \to M$ is called *isotropic* if it is tangent to ξ. Then at each point $x \in \Lambda$ we have $d\phi(T_x\Lambda) \subset \xi_{\phi(x)}$ and $d\alpha|_{d\phi(T_x\Lambda)} = d\big(\alpha|_{\phi(\Lambda)}\big)(x) = 0$. Hence $d\phi(T_xL)$ is an isotropic subspace in the symplectic vector space $(\xi_x, d\alpha)$. In particular,

$$\dim \Lambda \leq \frac{1}{2} \dim \xi = n.$$

Isotropic immersions of the maximal dimension n are called *Legendrian.*

1-jet spaces. Let L be a manifold of dimension n. The space J^1L of 1-jets of functions on L can be canonically identified with $T^*L \times \mathbb{R}$, where T^*L is the cotangent bundle of L. A point in J^1L is a triple (q, p, z) where q is a point in L, p is a linear form on T_qL, and $z \in \mathbb{R}$ is a real number. Let $p\,dq = \sum p_i dq_i$ be the standard Liouville form on T^*L (see Section 6.3). Then the 1-form $dz - p\,dq$ defines the *standard contact structure*

$$\xi_{\rm st} := \ker(dz - p\,dq)$$

on J^1L. A function $f : L \to \mathbb{R}$ defines a section

$$q \mapsto j^1 f(q) := \big(q, df(q), f(q)\big)$$

of the bundle $J^1L \to L$. Since $f^*(dz - p\,dq) = df - df = 0$, this section is a Legendrian embedding in the contact manifold $(J^1L, \xi_{\rm st})$. Consider the following diagram, where all arrows represent the obvious projections.

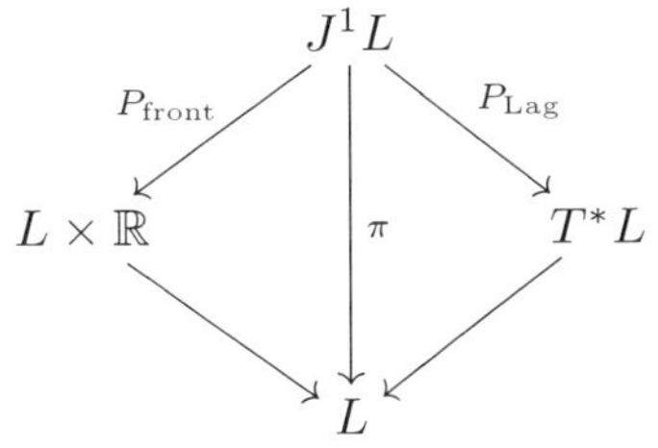

We call $P_{\rm Lag}$ the *Lagrangian projection* and $P_{\rm front}$ the *front projection.* Given a Legendrian submanifold $\Lambda \subset J^1L$, consider its images

$$P_{\rm Lag}(\Lambda) \subset T^*L, \qquad P_{\rm front}(\Lambda) \subset L \times \mathbb{R}.$$

The map $P_{\rm Lag} : \Lambda \to T^*L$ is a Lagrangian immersion with respect to the standard symplectic structure $dp \wedge dq = d(p\,dq)$ on T^*L. Indeed, the contact hyperplanes of $\xi_{\rm can}$ are transverse to the z-direction which is the kernel of the projection $P_{\rm Lag}$. Hence Λ is transverse to the z-direction as well and $P_{\rm Lag}|_\Lambda$ is an immersion. It is Lagrangian because

$$P_{\rm Lag}^*(dp \wedge dq) = d(p\,dq|_\Lambda) = d(dz|_\Lambda) = 0.$$

Conversely, any *exact Lagrangian immersion* $\phi : \Lambda \to T^*L$, i.e., an immersion for which the form $\phi^*(p\,dq)$ is exact, lifts to a Legendrian immersion $\widehat{\phi} : \Lambda \to J^1L$. It is given by the formula $\widehat{\phi} := (\phi, H)$, where H is a primitive of the exact 1-form $\phi^*(p\,dq)$ so that $\widehat{\phi}^*(dz - p\,dq) = dH - \phi^* p\,dq = 0$. The lift $\widehat{\phi}$ is unique up to a translation along the z-axis.

Let us now turn to the front projection. The image $P_{\text{front}}(\Lambda)$ is called the *front* of the Legendrian submanifold $\Lambda \subset J^1L$. If the projection $\pi|_\Lambda : \Lambda \to L$ is nonsingular and injective, then Λ is a graph $\{(q, \alpha(q), f(q)) \mid q \in \pi(\Lambda)\}$ over $\pi(\Lambda) \subset L$. The Legendre condition implies that the 1-form α is given by $\alpha = df$. So

$$\Lambda = \{(q, df(q), f(q)) \mid q \in \pi(\Lambda)\}$$

is the graph of the 1-jet $j^1 f$ of a function $f : \pi(\Lambda) \to \mathbb{R}$. In this case the front $P_{\text{front}}(\Lambda)$ is just the graph of the function f.

In general, the front of a Legendrian submanifold $\Lambda \subset J^1L$ can be viewed as the graph of a multivalued function. Note that since the contact hyperplanes are transverse to the z-direction, the singular points of the projection $\pi|_\Lambda$ coincide with the singular points of the projection $P_{\text{front}}|_\Lambda$. Hence near each of its nonsingular points the front is indeed the graph of a function.

In general, the front can have quite complicated singularities. But when the projection $\pi|_\Lambda : \Lambda \to L$ has only "fold type" singularities, then the front itself has only "cuspidal" singularities along its singular locus as shown in Figure 6.1. Let us discuss this picture in more detail. Consider first the 1-dimensional case when $L = \mathbb{R}$. Then $J^1L = \mathbb{R}^3$ with coordinates (q, p, z) and contact structure $\ker(dz - p\,dq)$. Consider the curve in $\mathbb{R}^3$ given by the equations

$$q = \left(\frac{2p}{3}\right)^2, \qquad z = \left(\frac{2p}{3}\right)^3. \tag{6.4}$$

This curve is Legendrian because $dz = \frac{8}{9}p^2 dp = p\,dq$. Its front is given by (6.4) viewed as parametric equations for a curve in the (q, z)-plane. This is the semicubic parabola $z^2 = q^3$ shown in Figure 6.1.

Generically, any singular point of a Legendrian curve in $\mathbb{R}^3$ looks like this. This means that, after a C^∞-small perturbation of the given curve to another Legendrian curve, there exists a contactomorphism of a neighborhood of the singularity which transforms the curve to the curve described by (6.4) (see [**11**, Chapter 1 §4]). If we want to construct just C^1 Legendrian curves (and any C^1 Legendrian curve can be further C^1-approximated by C^∞ or even real analytic Legendrian curves, see Corollary 6.25), then the following characterization of the front near its cusp points will be convenient. Suppose that the two branches of the front which form the cusp are given locally by the equations $z = f(q)$ and $z = g(q)$, where the functions $f, g : [0, \epsilon) \to \mathbb{R}$ satisfy $f \le g$; see Figure 6.1. Then the front lifts to a C^1 Legendrian curve if and only if

$$f(0) = g(0), \qquad f'(0) = g'(0),$$
$$f''(q) \to -\infty \text{ as } q \to 0, \qquad g''(q) \to +\infty \text{ as } q \to 0.$$

In higher dimensions, suppose that a Legendrian submanifold $\Lambda \subset J^1L$ projects to L with only "fold type" singularities. Then along its singular locus the front consists of the graphs of two functions $f \le g$ defined on an immersed strip $S \times [0, \epsilon)$. Denoting

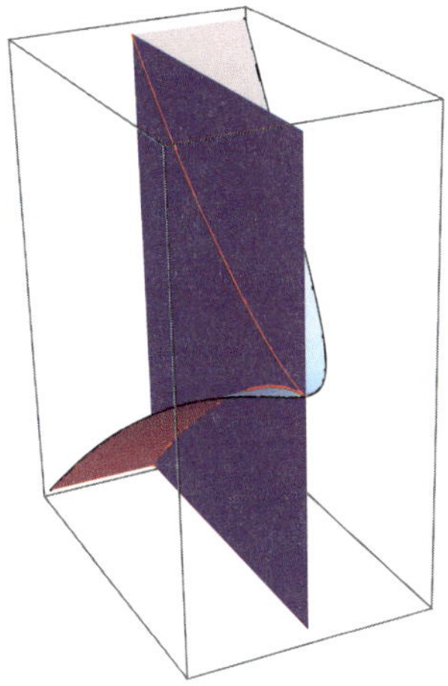

FIGURE 6.1. A Legendrian arc in $\mathbb{R}^3$ whose front projection (onto the shaded region) has a cusp.

coordinates on $S \times [0, \epsilon)$ by (s, t), the front lifts to a C^1 Legendrian submanifold if and only if

$$f(s,0) = g(s,0), \qquad \frac{\partial f}{\partial t}(s,0) = \frac{\partial g}{\partial t}(s,0),$$
$$\frac{\partial^2 f}{\partial t^2}(s,t) \to -\infty \text{ as } t \to 0, \qquad \frac{\partial^2 g}{\partial t^2}(s,t) \to +\infty \text{ as } t \to 0.$$

However, in higher dimensions not all singularities are generically of fold type.

EXAMPLE 6.18. Given a contact manifold $(M, \xi = \ker \alpha)$ and an exact symplectic manifold (V, λ), their product $M \times V$ is a contact manifold with the contact form $\alpha \oplus \lambda$. For example, if $M = J^1N$ and $V = T^*W$ with the canonical contact and Liouville forms, then $M \times V = J^1(N \times W)$ with the canonical contact form. A product $\Lambda \times L$ of a Legendrian submanifold $\Lambda \subset M$ and an exact Lagrangian submanifold $L \subset V$ is a Legendrian submanifold of $M \times V$. In particular, the product of a Legendrian submanifold $\Lambda \subset J^1N$ and an exact Lagrangian submanifold $L \subset T^*W$ is a Legendrian submanifold in $J^1(N \times W)$.

6.6. Contact normal forms

Let $(M^{2n+1}, \xi = \ker \alpha)$ be a contact manifold and $\Lambda^k \subset M$, $0 \le k \le n$, be an isotropic submanifold. The following result is due to Darboux in the case that Λ is a point (see e.g. Appendix 4 of [**10**]); the extension to general Λ is straightforward and left to the reader.

PROPOSITION 6.19 (Contact Darboux theorem). *Near each point on Λ there exist coordinates $(q_1, \dots, q_n, p_1, \dots, p_n, z) \in \mathbb{R}^{2n+1}$ in which $\alpha = dz - \sum p_i dq_i$ and $\Lambda = \mathbb{R}^k \times \{0\}$.*

To formulate a more global result, recall that the form $\omega = d\alpha$ defines a natural (i.e., independent of α) conformal symplectic structure on ξ. Denote the ω-orthogonal on ξ by a superscript ω. Since Λ is isotropic, $T\Lambda \subset (T\Lambda)^\omega$. So the normal bundle of Λ in M is given by

$$TM/T\Lambda = TM/\xi \oplus \xi/(T\Lambda)^\omega \oplus (T\Lambda)^\omega/T\Lambda \cong \mathbb{R} \oplus T^*\Lambda \oplus CSN(\Lambda).$$

Here TM/ξ is trivialized by the Reeb vector field R_α, the bundle $\xi/(T\Lambda)^\omega$ is canonically isomorphic to $T^*\Lambda$ via $v \mapsto i_v\omega$, and $CSN(\Lambda) := (T\Lambda)^\omega/T\Lambda$ denotes the *conformal symplectic normal bundle* which carries a natural conformal symplectic structure induced by ω. Thus $CSN(\Lambda)$ has structure group $Sp(2n-2k)$, which can be reduced to $U(n-k)$ by choosing a compatible complex structure.

Let (M, ξ_M) and (N, ξ_N) be two contact manifolds. A map $f : M \to N$ is called *isocontact* if $f^*\xi_N = \xi_M$, where $f^*\xi_N := \{v \in TM \mid df \cdot v \in \xi_N\}$. Equivalently, f maps any defining 1-form α_N for ξ_N to a defining 1-form $f^*\alpha_M$ for ξ_M. In particular, f must be an immersion and thus $\dim M \leq \dim N$. Moreover, $df : \xi_M \to \xi_N$ is *conformally symplectic*, i.e., symplectic up to a scaling factor. We call a monomorphism $F : TM \to TN$ *isocontact* if $F^*\xi_N = \xi_M$ and $F : \xi_M \to \xi_N$ is conformally symplectic.

PROPOSITION 6.20 (Contact isotropic neighborhood theorem [**187**]). *Let (M, ξ_M), (N, ξ_N) be contact manifolds with $\dim M \leq \dim N$ and $\Lambda \subset M$ an isotropic submanifold. Let $f : \Lambda \to N$ be an isotropic immersion covered by an isocontact monomorphism $F : TM \to TN$. Then there exists an isocontact immersion $g : U \to N$ of a neighborhood $U \subset M$ of Λ with $g|_\Lambda = f$ and $dg = F$ along Λ.*

REMARK 6.21. (a) If f is an embedding, then g is also an embedding on a sufficiently small neighborhood. It follows that a neighborhood of a Legendrian submanifold Λ is contactomorphic to a neighborhood of the zero section in the 1-jet space $J^1\Lambda$ (with its canonical contact structure).

(b) A Legendrian immersion $f : \Lambda \to (M, \xi)$ extends to an isocontact immersion of a neighborhood of the zero section in $J^1\Lambda$.

(c) Suppose that the conformal symplectic normal bundle of an isotropic submanifold Λ is the complexification of a real bundle $W \to \Lambda$ (i.e., the structure group of $CSN(\Lambda)$ reduces from $U(n-k)$ to $O(n-k)$). Then a neighborhood of Λ is contactomorphic to a neighborhood of the zero section in $J^1\Lambda \oplus (W \oplus W^*)$ (with its canonical contact structure, see Example 6.18). In this case (and only in this case) the isotropic submanifold Λ extends to a Legendrian submanifold (the total space of the bundle W).

We will also need the following refinement of the isotropic neighborhood theorem. Following Weinstein [**187**], let us denote by *isotropic setup* a quintuple $(V, \omega, X, \Sigma, \Lambda)$, where

- (V, λ) is a symplectic manifold with Liouville field X and $\omega = d\lambda$,
- $\Sigma \subset V$ is a codimension 1 hypersurface transverse to X,
- $\Lambda \subset \Sigma$ is a closed isotropic submanifold for the contact structure $\ker(\lambda|_\Sigma)$.

Let $(T\Lambda)^\omega/T\Lambda \subset \xi$ be the *symplectic normal bundle* over Λ.

PROPOSITION 6.22 (Weinstein [**187**]). *Let $(V_i, \omega_i, X_i, \Sigma_i, \Lambda_i)$, $i = 0, 1$, be two isotropic setups. Given a diffeomorphism $f : \Lambda_0 \to \Lambda_1$ covered by an isomorphism Φ of symplectic normal bundles, there exists an isomorphism of isotropic setups*

$$F : (U_0, \omega_0, X_0, \Sigma_0 \cap U_0, \Lambda_0) \to (U_1, \omega_1, X_1, \Sigma_1 \cap U_1, \Lambda_1)$$

between neighborhoods U_i of Λ_i in V_i inducing f and Φ.

All the properties discussed in this section also hold for families of isotropic submanifolds. Moreover, any isotropic submanifold with boundary can be extended beyond the boundary to a slightly bigger isotropic submanifold of the same dimension.

A similar homotopy argument proves Gray's stability theorem, which states that on a closed manifold all deformations of a contact structure are diffeomorphic to the original one.

THEOREM 6.23 (Gray's stability theorem [**75**]). *Let $(\xi_t)_{t\in[0,1]}$ be a smooth homotopy of contact structures on a closed manifold M. Then there exists a diffeotopy $\phi_t : M \to M$ with $\phi_0 = \mathrm{Id}$ and $\phi_t^*\xi_t = \xi_0$ for all $t \in [0,1]$.*

More generally, let $(\xi_\lambda)_{\lambda\in D^k}$ be a smooth family of contact structures on a closed manifold M, parametrized by the closed k-dimensional disc D^k. Then there exists a smooth family of diffeomorphisms $\phi_\lambda : M \to M$ with $\phi_0 = \mathrm{Id}$ and $\phi_\lambda^*\xi_\lambda = \xi_0$ for all $\lambda \in D^k$.

Finally, let us mention the following contact version of the isotopy extension theorem (see e.g. [**65**, Theorem 2.6.2]).

PROPOSITION 6.24 (Contact isotopy extension theorem). *Let Λ_t, $t \in [0,1]$, be an isotopy of compact isotropic submanifolds, possibly with boundary, in a contact manifold (M,ξ). Then there exists a smooth family of contactomorphisms $f_t : M \to M$ with $f_0 = \mathrm{Id}$ such that $f_t(\Lambda_0) = \Lambda_t$.*

6.7. Real analytic approximations of isotropic submanifolds

Using the results from Chapter 5, we now derive a result on real analytic approximations of isotropic submanifolds that will be needed later.

COROLLARY 6.25. *Let Λ be a closed isotropic C^k-submanifold ($k \geq 1$) in a real analytic closed contact manifold (M,α) (i.e., the manifold M and the 1-form α are both real analytic). Then there exists a real analytic isotropic submanifold $\Lambda' \subset (M,\alpha)$ arbitrarily C^k-close to Λ.*

Similarly, let $(\Lambda_t)_{t\in[0,1]}$ be a C^k-isotopy of closed isotropic C^k-submanifolds in (M,α) such that Λ_0 and Λ_1 are real analytic. Then there exists a real analytic isotopy of real analytic isotropic submanifolds Λ_t', arbitrarily C^k-close to Λ_t, with $\Lambda_0' = \Lambda_0$ and $\Lambda_1' = \Lambda_1$.

PROOF. Let $\widetilde{\Lambda} \subset M$ be a real analytic submanifold C^k-close to Λ, but not necessarily isotropic. Then $\Lambda = \phi(\widetilde{\Lambda})$ for a C^k-diffeomorphism $\phi : M \to M$ that is C^k-close to the identity. The contact form $\phi^*\alpha$ vanishes on $\widetilde{\Lambda}$ but need not be real analytic. Thus $\phi^*\alpha$ induces a C^k-section in the real analytic vector bundle $T^*M|_{\widetilde{\Lambda}} \to \widetilde{\Lambda}$ which vanishes on the real analytic subbundle $T\widetilde{\Lambda} \subset T^*M|_{\widetilde{\Lambda}}$. Let $\nu \to \widetilde{\Lambda}$ be the normal bundle to $T\widetilde{\Lambda}$ in $T^*M|_{\widetilde{\Lambda}}$ with respect to a real analytic metric and denote by $(\phi^*\alpha)^\nu$ the induced C^k-section in ν. Let β^ν be a real analytic section of ν that is C^k-close to $(\phi^*\alpha)^\nu$ and extend it to a real analytic section β of $T^*M|_{\widetilde{\Lambda}}$ that vanishes on $T\widetilde{\Lambda}$, and hence is C^k-close to $\phi^*\alpha$ along $\widetilde{\Lambda}$. Extend β to a C^k 1-form on M (still denoted by β) that is C^k-close to $\phi^*\alpha$. By construction, β is real analytic along $\widetilde{\Lambda}$ and $\beta|_{\widetilde{\Lambda}} = 0$.

By Theorem 5.53 (with $d = 0$), there exists a real analytic 1-form $\widetilde{\alpha}$ that is C^k-close to β and coincides with β along $\widetilde{\Lambda}$. In particular, $\widetilde{\alpha}|_{\widetilde{\Lambda}} = 0$. By construction,

$\widetilde{\alpha}$ is C^k-close to α. Hence $\alpha_t := (1-t)\widetilde{\alpha} + t\alpha$ is a real analytic homotopy of real analytic contact forms. By Gray's Stability Theorem 6.23, there exists a diffeotopy $\phi_t : M \to M$ and positive functions f_t with $\phi_t^*\alpha = f_t\widetilde{\alpha}$. Now in Moser's proof of Gray's stability theorem (see e.g. [**26**]), the ϕ_t are constructed as solutions of an ODE whose coefficients are real analytic and C^k-small in this case. Hence by Remark 5.39 the ϕ_t are real analytic, C^k-close to the identity, and depend real analytically on t. It follows that $\Lambda' := \phi_1(\widetilde{\Lambda})$ is real analytic, C^k-close to Λ, and $\alpha|_{\Lambda'} = 0$.

The statement about homotopies follows in a similar way using Corollary 5.54. □

REMARK 6.26. (1) Corollary 6.25 remains valid (with essentially the same proof) if the submanifold Λ is not closed, providing a real analytic approximation on a compact subset $K \subset \Lambda$.

(2) If Λ is Legendrian, then Λ' is Legendrian isotopic to Λ: by the Legendrian neighborhood theorem (Proposition 6.20), Λ' is the graph of the 1-jet of a function f in $J^1\Lambda$, and the functions tf provide the isotopy.

6.8. Relations between symplectic and contact manifolds

Symplectic and contact geometries are deeply linked with each other. We describe in this section some basic relations.

Let (M,ξ) be a contact manifold with a cooriented contact structure. Let $N_\xi \subset T^*M$ be the 1-dimensional *conormal* bundle of ξ, i.e., the space of 1-forms annihilating ξ, and N_ξ^+ its $\mathbb{R}^+$-subbundle consisting of forms defining the given coorientation of ξ. The nonintegrability condition for ξ then can be reinterpreted as the condition that the form $\omega_\xi = d(pdq)|_{N_\xi^+}$ is nondegenerate. The symplectic manifold $\mathrm{Symp}(M,\xi) = (N_\xi^+, \omega_\xi)$ is called the *symplectization* of the contact manifold ξ. Note that the form $\lambda_\xi = pdq|_{N_\xi^+}$ is a Liouville form, and the vector field $X_\xi = p\frac{\partial}{\partial p}$ is the corresponding Liouville field. A choice of a contact form α provides a symplectomorphism of $\mathrm{Symp}(M,\xi)$ with $(\mathbb{R}^+ \times M, d(r\alpha))$ and identifies λ_ξ with the 1-form $r\alpha$. Sometimes it is convenient to change the variable $r = e^s, s \in \mathbb{R}$ and thus identify the symplectization with $\big(\mathbb{R} \times M, d(e^s\alpha)\big)$. In this presentation $\lambda_\xi = e^s\alpha$, and the corresponding Liouville field is $X_\xi = \frac{\partial}{\partial s}$. The contact geometry of (M,ξ) can be reinterpreted as the symplectic geometry of $\mathrm{Symp}(M,\xi)$ equivariant with respect to the $\mathbb{R}$-action generated by the flow of the Liouville field X_ξ, or equivalently the geometry of the Liouville manifold (N_ξ^+, λ_ξ). For example, the diffeomorphisms of N_ξ^+ preserving the Liouville form λ_ξ are precisely the lifts of contactomorphisms of (M,ξ).

Conversely, suppose we are given a symplectic manifold (V,ω). A cooriented hypersurface $M \subset V$ is called locally (resp. globally) ω-*convex* (see [**49**]) if there exists a Liouville field X on a neighborhood of M (resp. on all of V) which is positively transverse to M.[2] In both cases the restriction $\alpha = \lambda|_M$ of the Liouville form $\lambda = i_X\omega$ is a contact form on M. If the Liouville field is complete we get a Liouville embedding of the symplectization $(\mathbb{R} \times M, e^s\alpha) \hookrightarrow (V,\lambda)$ by matching the corresponding trajectories of the Liouville fields $\frac{\partial}{\partial s}$ on $\mathbb{R} \times M$ and X on V.

[2]An alternative terminology is *contact type* for locally symplectically convex, and *restricted contact type* for globally symplectically convex; see [**100**].

If $\dim V = 4$, we call M (locally resp. globally) *weakly ω-convex* if it admits a contact form α defining the induced orientation such that $\omega|_\xi = d\alpha|_\xi$, where $\xi = \ker \alpha$. This notion is indeed weaker than ω-convexity; e.g., $\omega|_M$ need not be exact for weakly ω-convex M. It was observed in [**133**, Lemma 2.1] that for $\dim V \geq 6$ the corresponding notion of "weak ω-convexity" would be equivalent to ω-convexity and hence not useful. Massot, Niederkrüger and Wendl [**130**] have recently proposed a different notion of weak ω-convexity in higher dimensions which differs from ω-convexity.

An important special case of the above discussion occurs when $M = \partial V$ and M is cooriented by an outward normal vector field to V. If ∂V is globally (weakly) ω-convex and V is compact the contact structure on M is called *(weakly) symplectically fillable.* In all dimensions there exist many examples of contact manifolds that are not weakly symplectically fillable (see e.g. [**130**]). On the other hand, there exist contact structures on the 3-torus that are weakly but not strongly symplectically fillable [**46**]. With the notion by Massot, Niederkrüger, and Wendl, there also exist contact 5-manifolds that are weakly but not strongly symplectically fillable [**130**]; this question is currently open in dimensions ≥ 7.

A contact manifold (M, ξ) is called *holomorphically fillable* if there exists a compact complex manifold (V, J) with J-convex boundary $M = \partial V$ such that ξ equals the field of complex tangencies on M. In the terminology of Section 5.10, this means that ξ carries a holomorphically fillable CR structure. So Theorem 5.59 implies

COROLLARY 6.27. *Holomorphically fillable contact structures are symplectically fillable.*

REMARK 6.28. Niederkrüger and van Koert [**151**] have shown that every closed contact manifold M carries also a contact structure ξ which is not symplectically fillable. It follows from the preceding corollary and Theorem 5.60 that if $\dim M \geq 5$, then ξ cannot be defined by an (integrable) CR structure.

CHAPTER 7

The h-principles

The *h-principle* for a partial differential equation or inequality asserts, roughly speaking, that a formal solution can be deformed to a genuine solution. This notion first appeared in [**81**] and [**86**]. General references for h-principles are Gromov's book [**84**] and the more recent [**52**].

In this chapter we discuss various h-principles that we need in this book. In the first three sections we collect some relevant h-principles that are available in the literature (mostly [**84, 52**]).

The following three sections are concerned with Legendrian embeddings: In Section 7.5 we show that a formal Legendrian embedding in a contact manifold of dimension ≥ 5 can be deformed to a genuine Legendrian embedding in the same formal class. In Section 7.6 we use the classification of overtwisted contact structures to derive an h-principle for Legendrian knots in such manifolds. In Section 7.7 we describe an h-principle for a remarkable class of "loose" Legendrian knots in dimension ≥ 5 recently found by Murphy.

In the last two sections we combine h-principles for totally real embeddings with those for isotropic contact embeddings to obtain h-principles for totally real discs attached to J-convex boundaries. The resulting Theorems 7.34 and 7.36 are essential ingredients for the existence and deformation of Stein structures discussed in later chapters.

Throughout this chapter, a *knot* denotes a (parametrized) embedding of a connected manifold.

7.1. Immersions and embeddings

We begin by reviewing some facts about smooth immersions and embeddings.

The h-principle for immersions. Let M, N be manifolds. A *monomorphism* $F : TM \to TN$ is a fiberwise injective bundle homomorphism covering a continuous map $M \to N$. Any immersion $f : M \to N$ gives rise to a monomorphism $df : TM \to TN$. We denote by $\mathrm{Mon}(TM, TN)$ the space of monomorphisms and by $\mathrm{Imm}(M, N)$ the space of immersions. Given a (possibly empty) closed subset $A \subset M$ and an immersion $h : \mathcal{O}p\, A \to N$, we denote by $\mathrm{Imm}(M, N; A, h)$ the subspace of $\mathrm{Imm}(M, N)$ which consists of immersions equal to h on $\mathcal{O}p\, A$. Similarly, the notation $\mathrm{Mon}(TM, TN; A, dh)$ stands for the subspace of $\mathrm{Mon}(TM, TN)$ of monomorphisms which coincide with dh on $\mathcal{O}p\, A$. Extending Smale's theory of immersions of spheres [**171, 172**], Hirsch [**97**] proved the following h-principle (see also [**84, 52**]):

THEOREM 7.1 (Smale–Hirsch immersion theorem). *For* $\dim M < \dim N$ *and any immersion* $h : \mathcal{O}p\, A \to N$*, the map* $f \mapsto df$ *defines a homotopy equivalence* *between the spaces* $\mathrm{Imm}(M, N; A, h)$ *and* $\mathrm{Mon}(TM, TN; A, dh)$. *In particular, we have the following special cases:*

(a) Any monomorphism $F \in \mathrm{Mon}(TM, TN; A, dh)$ *is homotopic to the differential* df *of an immersion* $f : M \to N$ *which coincides with* h *on* $\mathcal{O}p\,A$.

(b) Given a homotopy $F_t \in \mathrm{Mon}(TM, TN; A, dh)$, $t \in [0, 1]$, *between the differentials* $F_0 = df_0$ *and* $F_1 = df_1$ *of two immersions* $f_0, f_1 \in \mathrm{Imm}(M, N; A, h)$, *one can find a regular homotopy* $f_t \in \mathrm{Imm}(M, N; A, h)$, $t \in [0, 1]$, *such that the paths* F_t *and* df_t, $t \in [0, 1]$, *are homotopic with fixed ends.*

For example, if a k-dimensional manifold M is parallelizable, i.e., $TM \cong M \times \mathbb{R}^k$, the inclusion $\mathbb{R}^k \hookrightarrow \mathbb{R}^{k+1}$ gives rise to a monomorphism $TM = M \times \mathbb{R}^k \to T(\mathbb{R}^{k+1}) = \mathbb{R}^{k+1} \times \mathbb{R}^{k+1}$, $(x, v) \mapsto (0, v)$. Thus Theorem 7.1 implies that every parallelizable manifold M^k can be immersed into $\mathbb{R}^{k+1}$.

Immersions of half dimension. Next we describe results of Whitney [**191**] on immersions of half dimension. Fix a closed connected manifold M^n of dimension $n \geq 2$ and an oriented manifold N^{2n} of double dimension. Let $f : M \to N$ be an immersion whose only self-intersections are transverse double points. Then if M is orientable and n is even, we assign to every double point $z = f(p) = f(q)$ an integer $I_f(z)$ as follows. Pick an orientation of M. Set $I_f(z) := \pm 1$ according to whether the orientations of $df(T_pM)$ and $df(T_qM)$ together determine the orientation of N or not. Note that this definition depends neither on the order of p and q (because n is even) nor on the orientation of M. Define the *self-intersection index*

$$I_f := \sum_z I_f(z) \in \mathbb{Z}$$

as the sum over all self-intersection points z. If n is odd or M is nonorientable, define $I_f \in \mathbb{Z}_2$ as the number of self-intersection points modulo 2.

THEOREM 7.2 (Whitney [**191**]). *For a closed connected manifold* M^n *and an oriented manifold* N^{2n}, $n \geq 2$, *the following holds.*

(a) The self-intersection index is invariant under regular homotopies.

(b) The self-intersection index of an immersion $f : M \to N$ *can be changed to any given value by a local modification (which is of course not a regular homotopy).*

(c) If $n \geq 3$ *and* N *is simply connected, then any immersion* $f : M \to N$ *is regularly homotopic to an immersion with precisely* $|I_f|$ *transverse double points (where* $|I_f|$ *means* 0 *or* 1 *for* $I_f \in \mathbb{Z}_2$*).*

REMARK 7.3. (i) Whitney states his theorem only for $N = \mathbb{R}^{2n}$, but the proof works without changes for general N (see e.g. [**140**]).

(ii) The theorem continues to hold if M has boundary, provided that for immersions and during regular homotopies no self-intersections occur on the boundary.

(iii) For $n = 1$ Whitney [**191**] defines a self-intersection index $I_f \in \mathbb{Z}$. With this definition, all the preceding results continue to hold for $n = 1$.

Since every immersion of half dimension is regularly homotopic to an immersion with transverse self-intersections ([**190**], see also [**98**]), part (a) allows us to define the self-intersection index for every immersion $f : M \to N$. Since every n-manifold immerses into $\mathbb{R}^{2n}$, parts (b) and (c) imply (the cases $n = 1, 2$ are treated by hand) the following:

COROLLARY 7.4 (Whitney embedding theorem [**191**]). *Every closed* n*-manifold* M^n, $n \geq 1$, *can be embedded in* $\mathbb{R}^{2n}$.

As we will use a similar argument below, let us sketch the proof of Theorem 7.2(c); for details see [**191, 140**]. The claim is well known for $n = 1, 2$, so we will assume that $n \geq 3$. Take an immersion $f : M \to \mathbb{R}^{2n}$ which exists due to general position arguments. Consider two transverse double points $y_i = f(x_i^+) = f(x_i^-)$, $i = 0, 1$. If M is orientable and n is even, we assume that $I_f(y_0) = -I_f(y_1)$. Since $n \geq 3$, we find two disjoint embedded paths $\gamma^\pm$ in M from $x_0^\pm$ to $x_1^\pm$ not meeting any other preimages of double points. Their images $C^\pm = f(\gamma^\pm)$ fit together to an embedded loop $C = C^+ \cup C^-$ in $\mathbb{R}^{2n+1}$ (with corners at y_0 and y_1). Denote by $M^\pm \subset f(M)$ the images under f of tubular neighborhoods of $\gamma^\pm$ in M. Orient $M^\pm$ arbitrarily. We arrange that the intersection numbers of $M^\pm$ at y_0 and y_1 have opposite signs as follows: for M orientable and n even this holds by assumption; for M orientable and n odd it can be achieved by interchanging x_1^+ and x_1^- if necessary; and for M nonorientable we can arrange this by concatenating, if necessary, γ^+ with an orientation reversing loop in M.

Using the simple connectedness of the target, we find an embedded half-disc $\Delta \subset \mathbb{R}^{2n}$ with $\partial_\pm \Delta = C^\pm$. Here the half-disc Δ is diffeomorphic to the lower half-disc $D_- = \{(x_1, x_2) \in \mathbb{R}^2 \mid x_1^2 + x_2^2 \leq 1,\ x_2 \leq 0\}$; see Figure 9.4 below. Using $n \geq 3$, we can arrange that Δ is transverse to $f(M)$ along the boundary and does not meet $f(M)$ in its interior. Such a half-disc Δ is called a *Whitney disc*. The condition that the intersection numbers of $M^\pm$ at y_0 and y_1 have opposite signs allows us to find a diffeomorphism from a neighborhood of Δ into $\mathbb{R}^2 \times \mathbb{R}^{n-1} \times \mathbb{R}^{n-1}$ mapping Δ to $D_- \subset \mathbb{R}^2 \times 0 \times 0$, M^+ to $\partial_+ D_- \times \mathbb{R}^{n-1} \times 0$, and M^- to $\partial_- D_- \times 0 \times \mathbb{R}^{n-1}$. In this model we can now write down an explicit isotopy pulling M^+ away from M^- across Δ, thus obtaining a regular homotopy of f removing the two double points y_0, y_1. Proceeding in this way, we cancel all pairs of double points (with opposite indices if M is orientable and n even) until their number equals $|I_f|$. The elimination procedure just described is sometimes called the *Whitney trick*. Its failure for $n = 2$ is the source of many exotic phenomena in smooth topology that occur in dimension 4 but not in higher dimensions.

Isotopies. Finally, we discuss *isotopies*, i.e., homotopies through embeddings. Consider a closed connected k-manifold M^k and an oriented $(2k+1)$-manifold N^{2k+1}. Let $f_t : M \to N$ be a regular homotopy between embeddings $f_0, f_1 : M \hookrightarrow N$. We define the *self-intersection index* $I_{\{f_t\}}$ of the regular homotopy f_t as the self-intersection index I_F of the immersion $F : M \times I \to N \times I$ given by the formula $(x, t) \mapsto f_t(x)$, $x \in M$, $t \in I = [0, 1]$. The self-intersection index $I_{\{f_t\}}$ is an invariant of f_t in the class of regular homotopies with fixed endpoints f_0, f_1. Recall that $I_{\{f_t\}}$ takes values in $\mathbb{Z}$ if M is orientable and k is odd, and in $\mathbb{Z}_2$ otherwise. In the former case $I_{\{f_t\}}$ remains unchanged when the orientation is switched to the opposite one.

REMARK 7.5. Let us stress the point that in choosing the orientation of $N \times I$, we wrote the interval I as the second factor. Choosing the opposite ordering would result (when k is odd) in switching the sign of $I_{\{f_t\}}$.

The following result is an analogue of Whitney's Theorem 7.2 for isotopies.

THEOREM 7.6. *For $k > 1$ consider a closed connected k-manifold M^k and a simply connected oriented $(2k+1)$-manifold N^{2k+1}. Let $f_t : M \to N$ be a regular homotopy between embeddings $f_0, f_1 : M \hookrightarrow N$. Then f_t can be deformed through regular homotopies with fixed endpoints to an isotopy if and only if $I_{\{f_t\}} = 0$.*

In particular, this implies the following result which was proved by Wu [**194**] and later greatly generalized by Haefliger [**90**].

COROLLARY 7.7. *For $k > 1$ consider a closed connected k-manifold M^k and a simply connected oriented $(2k+1)$-manifold N^{2k+1}. Then any two homotopic embeddings $f_0, f_1 : M \hookrightarrow N$ are isotopic.*

PROOF. Pick a homotopy connecting f_0 and f_1 and deform it to a regular homotopy f_t. By adding new self-intersection points to this homotopy, we can arbitrarily change its self-intersection index (see [**191**]). In particular, we can make $I_{\{f_t\}} = 0$ and then apply Theorem 7.6 to find the desired isotopy. □

The proof of Theorem 7.6 uses the following standard transversality result which is a special case of Thom's multijet transversality theorem; see e.g. [**98**]. The case $\Lambda = [0,1]$ is due to Whitney [**190**].

LEMMA 7.8. *Let M, N, Λ be manifolds and $F : \Lambda \times M \to N$ a smooth map. If $\dim \Lambda + 2\dim M < \dim N$, then F can be C^∞-approximated by a map $\widetilde{F}$ such that $\widetilde{F}(\lambda, \cdot)$ is an embedding for all $\lambda \in \Lambda$. Moreover, if F is already an embedding near a compact subset $K \subset \Lambda \times M$, we can choose $\widetilde{F} = F$ near K.* □

PROOF OF THEOREM 7.6. The argument is an adjustment of the Whitney trick [**191**] explained above. Take two self-intersection points $Y_0 = (y_0, t_0), Y_1 = (y_1, t_1) \in N \times (0,1)$ of the immersion $F : M^k \times [0,1] \to N^{2k+1} \times [0,1]$ defined above. If M is orientable and k is odd, we assume that the intersection indices of these points have opposite signs. Each of the double points y_0, y_1 is the image of two distinct points $x_0^\pm, x_1^\pm \in M$, i.e., we have $f_{t_0}(x_0^\pm) = y_0$ and $f_{t_1}(x_1^\pm) = y_1$. As $k > 1$, we find two embedded paths $\gamma^\pm : [t_0, t_1] \to M$ such that $\gamma^\pm(t_0) = x_0^\pm$, $\gamma^\pm(t_1) = x_1^\pm$, and $\gamma^+(t) \neq \gamma^-(t)$ for all $t \in [t_0, t_1]$. As explained above, we can choose the paths $\gamma^\pm$ such that the (arbitrarily oriented) local branches of $F(M \times [0,1])$ along the images of $\gamma^\pm$ in $N \times [0,1]$ have opposite intersection numbers at Y_0 and Y_1. We claim that there exists a smooth family of paths $\delta_t : [-1,1] \to N$, $t \in [t_0, t_1]$, such that

- $\delta_t(\pm 1) = f_t\big(\gamma^\pm(t)\big)$ for all $t \in [t_0, t_1]$,
- $\delta_{t_0}(s) = y_0$, $\delta_{t_1}(s) = y_1$ for all $s \in [-1,1]$,
- δ_t is an embedding for all $t \in (t_0, t_1)$.

Indeed, a family with the first two properties exists because N is simply connected. Moreover, we can arrange that δ_t is an embedding for $t \neq t_0, t_1$ close to t_0, t_1. Now we can achieve the third property by Lemma 7.8 because $2 \cdot 1 + 1 < 2k + 1$. Define

$$\Delta : [t_0, t_1] \times [-1,1] \to N \times [0,1], \qquad (t,s) \mapsto \big(\delta_t(s), t\big).$$

Then Δ is an embedding on $(t_0, t_1) \times [-1,1]$ and $\Delta(t_0 \times [-1,1]) = Y_0$, $\Delta(t_1 \times [-1,1]) = Y_1$. Thus Δ serves as a Whitney disc for the elimination of the double points Y_0, Y_1 of the immersion F. Due to the special form of Δ, Whitney's elimination construction described above (see [**191, 140**]) can be performed in such a way that the modified immersion $\widetilde{F}$ has the form $\widetilde{F}(x,t) := \big(\widetilde{f}_t(x), t\big)$ for a regular homotopy $\widetilde{f}_t : M \to N$ such that the paths $f_t, \widetilde{f}_t \in \mathrm{Imm}(M,N)$, $t \in [0,1]$, are homotopic. Hence the repeated elimination of pairs of intersection points (of opposite indices if M is orientable and k odd) of the immersion F results in the desired isotopy between f_0 and f_1 if $I_{\{f_t\}} = 0$. □

7.2. The h-principle for isotropic immersions

The following h-principle was proved by Gromov in 1986 ([**84**], see also [**52**]).

Let (M, ξ) be a contact manifold of dimension $2n+1$, Λ a manifold of dimension $k \leq n$, and $A \subset \Lambda$ a closed subset. Let $h : \mathcal{O}p\, A \to M$ be an isotropic immersion. We denote by $\mathrm{Imm}_{\mathrm{isotr}}(\Lambda, M; A, h)$ the space of isotropic immersions $\Lambda \to M$ which coincide with h on $\mathcal{O}p\, A$, and by $\mathrm{Mon}_{\mathrm{isotr}}(T\Lambda, \xi; A, dh)$ the space of *isotropic monomorphisms* $T\Lambda \to \xi$ which coincide with dh on $\mathcal{O}p\, A$. In the case $k = n$ isotropic monomorphisms will also be called *Legendrian monomorphisms* and denoted by $\mathrm{Mon}_{\mathrm{Leg}}(T\Lambda, \xi; A, dh)$.

Note that if we equip ξ with a compatible complex structure, then the space $\mathrm{Mon}_{\mathrm{isotr}}(T\Lambda, \xi; A, dh)$ is a subspace of the space $\mathrm{Mon}_{\mathrm{real}}(T\Lambda, \xi; A, dh)$ of totally real monomorphisms $T\Lambda \to \xi$ which coincide with dh on $\mathcal{O}p\, A$, and the inclusion

$$\mathrm{Mon}_{\mathrm{isotr}}(T\Lambda, \xi; A, dh) \hookrightarrow \mathrm{Mon}_{\mathrm{real}}(T\Lambda, \xi; A, dh)$$

is a homotopy equivalence.

THEOREM 7.9 (Gromov's h-principle for contact isotropic immersions [**84, 52**]). *The map $d : \mathrm{Imm}_{\mathrm{isotr}}(\Lambda, M; A, h) \hookrightarrow \mathrm{Mon}_{\mathrm{isotr}}(T\Lambda, \xi; A, dh)$ is a homotopy equivalence. In particular, we have the following special cases:*

(a) Given $F \in \mathrm{Mon}_{\mathrm{isotr}}(T\Lambda, \xi; A, dh)$ one finds $f \in \mathrm{Imm}_{\mathrm{isotr}}(\Lambda, M; A, h)$ such that df and F are homotopic in $\mathrm{Mon}_{\mathrm{isotr}}(T\Lambda, \xi; A, dh)$. Moreover, f can be chosen C^0-close to the map $\Lambda \to M$ covered by the homomorphism F.

(b) Given two isotropic immersions $f_0, f_1 \in \mathrm{Imm}_{\mathrm{isotr}}(\Lambda, M; A, h)$ and a homotopy $F_t \in \mathrm{Mon}_{\mathrm{isotr}}(T\Lambda, \xi; A, dh)$, $t \in [0,1]$, connecting df_0 and df_1 one finds an isotropic regular homotopy $f_t \in \mathrm{Imm}_{\mathrm{isotr}}(\Lambda, M; A, h)$ connecting df_0 and df_1 such that the paths F_t and df_t, $t \in [0,1]$, are homotopic in $\mathrm{Mon}_{\mathrm{isotr}}(T\Lambda, \xi; A, dh)$ with fixed ends. Moreover, the f_t can be chosen C^0-close to the family of maps $\Lambda \to M$ covered by the homotopy F_t.

Combining the preceding theorem with the Smale–Hirsch Immersion Theorem 7.1 yields

COROLLARY 7.10. *Let Λ, M, A, h be as in Theorem 7.9. Suppose that $f_0 : \Lambda \to M$ is an immersion which coincides with the isotropic immersion h on $\mathcal{O}p\, A$ and F_t is a family of monomorphisms $T\Lambda \to TM$ such that $F_0 = df_0$, $F_t = dh$ on $\mathcal{O}p\, A$ for all $t \in [0,1]$, and $F_1 \in \mathrm{Mon}_{\mathrm{isotr}}(T\Lambda, \xi; A, dh)$. Then there exists a regular homotopy $f_t : \Lambda \to M$ such that*

(i) *$f_1 \in \mathrm{Imm}_{\mathrm{isotr}}(\Lambda, M; A, h)$,*
(ii) *$f_t = h$ on $\mathcal{O}p\, A$ for all $t \in [0,1]$,*
(iii) *there exists a homotopy F_t^s, $s \in [0,1]$, of paths in $\mathrm{Mon}(T\Lambda, TM; A, dh)$ such that $F_t^0 = df_t$ and $F_t^1 = F_t$ for all $t \in [0,1]$, and $F_0^s = df_0$ and $F_1^s \in \mathrm{Mon}_{\mathrm{isotr}}(T\Lambda, \xi; A, dh)$ for all $s \in [0,1]$.*

PROOF. We first use Theorem 7.9 to construct an isotropic immersion $g_2 \in \mathrm{Imm}_{\mathrm{isotr}}(\Lambda, \xi; A, h)$ and a homotopy $F_t \in \mathrm{Mon}_{\mathrm{isotr}}(T\Lambda, TM; A, dh)$, $t \in [1,2]$, such that $F_2 = dg_2$. Next we apply Theorem 7.1 to get a regular homotopy $g_t \in \mathrm{Imm}(\Lambda, M; A, h)$, $t \in [0,2]$, such that $g_0 = f_0$ and the paths dg_t, F_t, $t \in [0,2]$, are homotopic with fixed ends. Let

$$G : [0,2] \times [0,1] \to \mathrm{Mon}(T\Lambda, TM; A, dh), \qquad (t,s) \mapsto G_t^s,$$

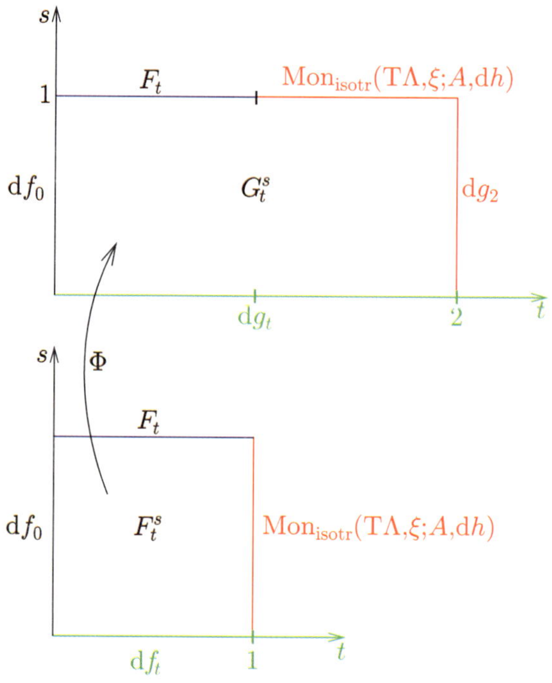

FIGURE 7.1. The families of monomorphisms G and $F = G \circ \Phi$.

be this homotopy, i.e., $G_t^0 = dg_t$, $G_t^1 = F_t$ for all $t \in [0,2]$, and $G_0^s = df_0$, $G_2^s = dg_2$ for all $s \in [0,1]$. The required paths are now defined by $f_t := g_{2t}$, $t \in [0,1]$, and

$$F := G \circ \Phi : [0,1] \times [0,1] \to \mathrm{Mon}(T\Lambda, TM; A, dh), \qquad (t,s) \mapsto F_t^s,$$

where $\Phi : [0,1] \times [0,1] \to [0,2] \times [0,1]$ is any homeomorphism mapping the boundary as follows (see Figure 7.1):

$$[0,1] \times 0 \to [0,2] \times 0, \qquad [0,1] \times 1 \to [0,1] \times 1,$$
$$0 \times [0,1] \to 0 \times [0,1], \qquad 1 \times [0,1] \to (2 \times [0,1]) \cup ([1,2] \times 1).$$ □

7.3. The h-principle for subcritical isotropic embeddings

In this and the next two sections we upgrade, under suitable conditions, the results of the previous section from isotropic immersions to embeddings. We begin with the subcritical case.

Consider a contact manifold (M^{2n+1}, ξ) and a manifold Λ^k of dimension $k \le n$. A *formal isotropic embedding* of Λ into (M, ξ) is a pair (f, F^s), where $f : \Lambda \hookrightarrow M$ is a smooth embedding and $F^s : T\Lambda \to TM$ is a homotopy of monomorphisms over f starting at $F^0 = df$ and ending at an isotropic monomorphism $F^1 : T\Lambda \to \xi$ covering f. In the case $k = n$, we also call this a *formal Legendrian embedding.*

Any genuine isotropic embedding can be viewed as a formal isotropic embedding $(f, F^s \equiv df)$. We will not distinguish between an isotropic embedding and its canonical lift to the space of formal isotropic embeddings, and we will consider formal isotropic isotopies between genuine isotropic embeddings: two isotropic embeddings $f_0, f_1 : \Lambda \hookrightarrow (M, \xi)$ are called *formally isotropically isotopic* if they are isotopic as formal isotropic embeddings.

We will also consider *relative* isotropic embeddings and their isotopies, which are required to coincide with a fixed genuine isotropic embedding on a neighborhood of a closed subset $A \subset \Lambda$. The space of isotropic embeddings which coincide with a given isotropic embedding h on $\mathcal{O}p\, A$ will be denoted by $\mathrm{Emb}_{\mathrm{isotr}}(\Lambda, M; A, h)$, and the corresponding space of formal isotropic embeddings by $\mathrm{Mon}^{\mathrm{emb}}_{\mathrm{isotr}}(T\Lambda, \xi; A, dh)$. With these notations, we have the following h-principle.

THEOREM 7.11 (h-principle for subcritical isotropic embeddings [**84, 52**]). *Consider a contact manifold (M, ξ) of dimension $2n+1$, a manifold Λ of dimension $k < n$, and a closed subset $A \subset \Lambda$. Then the inclusion*

$$\mathrm{Mon}^{\mathrm{emb}}_{\mathrm{isotr}}(\Lambda, M; A, h) \hookrightarrow \mathrm{Emb}_{\mathrm{isotr}}(T\Lambda, \xi; A, dh)$$

is a weak homotopy equivalence. In particular, suppose that two isotropic embeddings $f_0, f_1 \in \mathrm{Emb}_{\mathrm{isotr}}(\Lambda, M; A, h)$ are connected by a formal isotropic isotopy (f_t, F_t^s), $s, t \in [0,1]$ rel $\mathcal{O}p\, A$. Then there exists a genuine isotropic isotopy g_t rel $\mathcal{O}p\, A$ connecting $g_0 = f_0$ and $g_1 = f_1$ which is homotopic to the formal isotropic isotopy (f_t, F_t^s) through formal isotropic isotopies fixed on $\mathcal{O}p\, A$.

7.4. Stabilization of Legendrian submanifolds

The goal of this section is the proof of the following proposition which will play a crucial role in the proof of the Existence Theorem 7.16 for Legendrian embeddings in the next section. Recall from Section 7.1 the definition of the self-intersection index of a smooth immersion.

PROPOSITION 7.12. *For $n \geq 2$ let $\Lambda_0 \subset (M^{2n+1}, \xi = \ker \alpha)$ be a closed orientable Legendrian submanifold and k an integer. Then there exists a Legendrian submanifold $\Lambda_1 \subset M$ and a Legendrian regular homotopy Λ_t, $t \in [0,1]$, such that the self-intersection index of the immersion $L := \bigcup_{t\in[0,1]} \Lambda_t \times \{t\} \subset M \times [0,1]$ equals k* (mod 2 *if n is even).*

REMARK 7.13. By Corollary 7.7, Λ_1 is smoothly isotopic to Λ_0. We will not use this fact; see however Section 7.7 for an elementary proof in the case $k = 0$.

A local construction. The proof of Proposition 7.12 for $n > 1$ is based on a *stabilization* procedure which we will now describe; see Figure 7.2.

Consider the front projection of a (not necessarily closed) orientable Legendrian submanifold $\Lambda_0 \subset \mathbb{R}^{2n+1}$. Suppose that $P_{\mathrm{front}}(\Lambda_0)$ intersects $B^n \times [-1, 2]$ in the two oppositely oriented branches $\{z = 0\}$ and $\{z = 1\}$. Let $f : B^n \to (-1, 2)$ be a function which equals zero near ∂B^n and has no critical points on level 1. Replacing the branch $\{z = 0\}$ over B^n by $\{z = tf(q)\}$, we obtain a family of Legendrian immersions $\Lambda_t \subset \mathbb{R}^{2n+1}$, $t \in [0,1]$. Note that the set $\{q \in B^n \mid f(q) \geq 1\}$ is a smooth n-manifold with boundary. Denote by $\chi(\{f \geq 1\})$ its Euler characteristic.

LEMMA 7.14. *The self-intersection index of the immersion $L := \bigcup_{t\in[0,1]} \Lambda_t \times \{t\} \subset M \times [0,1]$ equals*

$$I_L = (-1)^{n(n-1)/2} \chi(\{f \geq 1\})$$

(mod 2 *if n is even).*

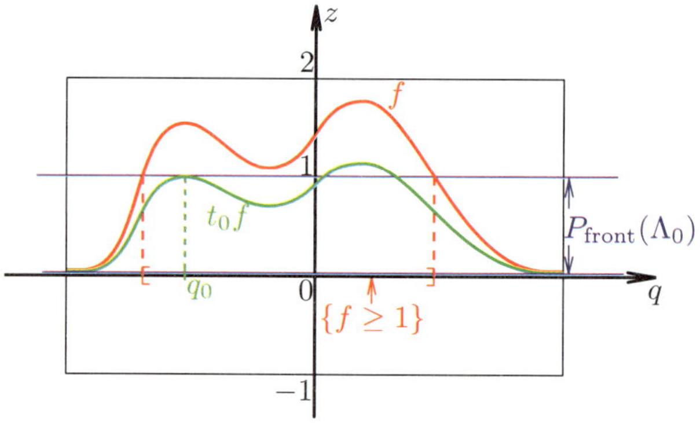

FIGURE 7.2. Stabilization of a Legendrian submanifold.

PROOF. Perturb f such that all critical points above level 1 are nondegenerate and lie on distinct levels. Self-intersections of L occur precisely when $t_0 f$ has a critical point q_0 on level 1 for some $t_0 \in (0,1)$. By the Morse Lemma, we find coordinates near q_0 in which $q_0 = 0$ and f has the form

$$f(q) = a_0 - \frac{1}{2}\sum_{i=1}^{k} q_i^2 + \frac{1}{2}\sum_{i=k+1}^{n} q_i^2,$$

where $a_0 = f(q_0) = 1/t_0$ and k is the Morse index of q_0. The p-coordinates on the branch $\{z = tf(q)\}$ of Λ_t near q_0 are given by

$$p_i = \frac{\partial(tf)}{\partial q_i} = \begin{cases} -tq_i & i \leq k, \\ +tq_i & i \geq k+1. \end{cases}$$

Thus the tangent spaces in $T(\mathbb{R}^{2n+1} \times [0,1]) = \mathbb{R}^{2n+2}$ of the two intersecting branches of L corresponding to $\{z = 1\}$ and $\{z = t_0 f(q)\}$ are given by

$$T_1 = \{p_1 = \cdots = p_n = 0, z = 0\},$$
$$T_2 = \{p_i = -t_0 q_i \text{ for } i \leq k, p_i = +t_0 q_i \text{ for } i \geq k+1, z = a_0 t\}.$$

Without loss of generality (because the self-intersection index does not depend on the orientation of L) suppose that the basis $(\partial_{q_1}, \ldots, \partial_{q_n}, \partial_t)$ represents the orientation of T_1. Since the two branches of Λ_0 are oppositely oriented, the orientation of T_2 is then represented by the basis

$$\big(\partial_{q_1} - t_0\partial_{p_1}, \ldots, \partial_{q_n} + t_0\partial_{p_n}, -(\partial_t + a_0\partial_z)\big).$$

Hence the orientation of (T_1, T_2) is represented by

$$(\partial_{q_1}, \ldots, \partial_{q_n}, \partial_t, -\partial_{p_1}, \ldots, -\partial_{p_k}, \partial_{p_{k+1}}, \ldots, \partial_{p_n}, -\partial_z),$$

which equals $(-1)^{k+n+n(n-1)/2}$ times the complex orientation

$$(\partial_{q_1}, \partial_{p_1}, \ldots, \partial_{q_n}, \partial_{p_n}, \partial_z, \partial_t)$$

of $\mathbb{R}^{2n+2} = \mathbb{C}^{n+1}$. So the local intersection index of L at a critical point q equals

$$I_L(q) = (-1)^{\mathrm{ind}_f(q)+n+n(n-1)/2}$$

(mod 2 if n is even), where $\mathrm{ind}_f(q)$ is the Morse index of q.

On the other hand, for a vector field v on a compact manifold N with boundary which is outward pointing along the boundary and has only nondegenerate zeroes, the *Poincaré–Hopf index theorem* holds: the sum of the indices of v at all its zeroes equals the Euler characteristic of M (see [**87**]). Note that if v is the gradient vector field of a Morse function f, then the index of v at a critical point q of f equals $(-1)^{\mathrm{ind}_f(q)}$. Applying the Poincaré–Hopf index theorem to the gradient of the Morse function $-f$ on the manifold $\{f \geq 1\} = \{-f \leq -1\}$ (which is outward pointing along the boundary because f has no critical point on level 1), we obtain

$$\chi(\{f \geq 1\} = \sum_q \mathrm{ind}_{\nabla(-f)}(q) = \sum_q (-1)^{\mathrm{ind}_{-f}(q)} = \sum_q (-1)^{n-\mathrm{ind}_f(q)}$$
$$= (-1)^{n(n-1)/2} \sum_q I_L(q) = (-1)^{n(n-1)/2} I_L. \qquad \square$$

PROOF OF PROPOSITION 7.12. Since all Legendrian submanifolds are locally isomorphic, a neighborhood in M of a point on Λ_0 is contactomorphic to a neighborhood in $\mathbb{R}^{2n+1}$ of a point on a standard cusp $z^2 = q_1^3$. Thus the front consists of two branches $\{z = \pm q_1^{\frac{3}{2}}\}$ joined along the singular locus $\{z = q_1 = 0\}$. Deform the branches to $\{z = \pm\varepsilon\}$ over a small ball disjoint from the singular locus, thus (after rescaling) creating two parallel branches over a ball as in Lemma 7.14. Now deform Λ_0 to Λ_1 as in Lemma 7.14, for some function $f : B^n \to (-1, 2)$. Hence Proposition 7.12 follows from Lemma 7.14, provided that we can arrange $\chi(\{f \geq 1\}) = k$ for a given integer k if $n > 1$.

So it only remains to find for $n > 1$ an n-dimensional submanifold-with-boundary $N \subset \mathbb{R}^n$ of prescribed Euler characteristic $\chi(N) = k$ (then write $N = \{f \geq 1\}$ for a function $f : N \to [1, 2)$ without critical points on the boundary). Let N_+ be a ball in $\mathbb{R}^n$, thus $\chi(N_+) = +1$. Let N_- be a smooth tubular neighborhood in $\mathbb{R}^n$ of a figure eight in $\mathbb{R}^2$, thus $\chi(N_-) = -1$ (here we use $n \geq 2$!). So we can arrange $\chi(N)$ to be any integer by taking disjoint unions of copies of $N_\pm$. This concludes the proof of Proposition 7.12. $\square$

REMARK 7.15. The preceding proof fails for $n = 1$ because a 1-dimensional manifold with boundary always has Euler characteristic $\chi \geq 0$. Therefore for $n = 1$ the local construction in Lemma 7.14 allows us only to realize *positive* values of the self-intersection index I_L. In Section 7.6 we will see that for *overtwisted* contact structures, one can get around this problem.

7.5. The existence theorem for Legendrian embeddings

The parametric h-principle of the previous section fails for *Legendrian* embeddings: For any $n \geq 1$ there are pairs of Legendrian knots in $\mathbb{R}^{2n+1}$ which are formally but not genuinely Legendrian isotopic [**31, 40**]. However, it turns out that if $n > 1$, then, using the stabilization trick from Section 7.4 and Theorem 7.6, the existence part (i.e., surjectivity on π_0) continues to hold in the Legendrian case $k = n$. For $n = 1$ the analogous claim is false in general, but true in the overtwisted case; see Theorem 7.19 below.

THEOREM 7.16 (Existence theorem for Legendrian embeddings for $n > 1$). *For $n \geq 2$ consider a contact manifold (M, ξ) of dimension $2n + 1$, a simply connected manifold Λ of dimension n, and a closed subset $A \subset \Lambda$. Let (f_0, F_0^s) be a formal Legendrian embedding of Λ into (M, ξ) which is genuine on $\mathcal{O}p\, A$. Then there exists* *a Legendrian embedding $f_1 : \Lambda \hookrightarrow M$ which coincides with f_0 on $\mathcal{O}p\,A$ and can be connected with (f_0, F_0^s) by a formal Legendrian isotopy fixed on $\mathcal{O}p\,A$.*

In the proof we use the following notation. Given two continuous paths $\gamma_1, \gamma_2 : [0,1] \to X$ into a topological space X with $\gamma_1(0) = \gamma_2(1)$, we define their *concatenation* to be the path

$$\gamma_1 \star \gamma_2(t) := \begin{cases} \gamma_1(2t), & t \in [0, \frac{1}{2}], \\ \gamma_2(2t-1), & t \in [\frac{1}{2}, 1]. \end{cases}$$

We will also use the following general position observation.

LEMMA 7.17. *Let (M,ξ) be a contact manifold of dimension $2n+1$ and Λ a manifold of dimension n. Then any Legendrian immersion $f_0 : \Lambda \to (M,\xi)$ can be included into a family of Legendrian immersions $f_t : \Lambda \to (M,\xi)$ C^∞-close to f_0 such that f_1 is an embedding.*

PROOF. Consider a Legendrian immersion $f : \Lambda \to M$. By Proposition 6.20 we can extend f to an isocontact immersion $F : U \to M$ of a neighborhood of the zero section in the 1-jet space $J^1\Lambda$. Then nearby Legendrian immersions correspond to graphs of 1-jets of functions on Λ, hence the claim follows from Thom's jet transversality theorem; see e.g. [**98**]. □

PROOF OF THEOREM 7.16. In what follows we assume that all constructions are done relative to A, and we will not state this explicitly anymore. By applying Corollary 7.10 we can satisfy all the conditions of the theorem, except that f_t will be a regular homotopy rather than an isotopy. By Lemma 7.17, after a C^∞-small isotropic regular homotopy, we may assume that f_1 is a Legendrian embedding. Thus, starting from a formal Legendrian embedding (f_0, F_0^s), we have constructed a regular homotopy $f_t : \Lambda \to M$, $t \in [0,1]$, and a 2-parameter family of monomorphisms $F_t^s : T\Lambda \to TM$ extending the family F_0^s such that $F_t^0 = df_t$ and $F_t^1 \in \mathrm{Mon}_{\mathrm{isotr}}(TM, \xi; A, dh)$ for all t, and $F_1^s = df_1$ for all s.

We will deform the regular homotopy f_t to an isotopy, keeping the end f_0 fixed and changing f_1 via a Legendrian regular homotopy. According to Theorem 7.6, in order to deform the path f_t to an isotopy keeping *both* ends fixed, we need the equality $I_{\{f_t\}} = 0$. (Here the simple connectedness hypothesis in Theorem 7.6 is satisfied because we can perform the whole construction in a tubular neighborhood of $f_0(\Lambda)$ which is simply connected). On the other hand, according to Proposition 7.12, for any Legendrian embedding g_0 there exists a Legendrian regular homotopy g_t to a Legendrian embedding g_1 with any prescribed value of the self-intersection index $I_{\{g_t\}}$. Hence by concatenating f_t, $t \in [0,1]$, with an appropriate Legendrian regular homotopy f_t, $t \in [1,2]$, we obtain a regular homotopy f_t, $t \in [0,2]$, with $I_{\{f_t\}_{t\in[0,2]}} = 0$. We extend F_t^s for $t \in [1,2]$ by df_t and rescale the interval $[0,2]$ back to $[0,1]$. After this, we may hence assume that $I_{\{f_t\}_{t\in[0,1]}} = 0$.

Now Theorem 7.6 provides a 2-parameter family of immersions $g_t^s : \Lambda \to M$, $s,t \in [0,1]$, such that $g_0^s = f_0$ and $g_1^s = f_1$ for all s, $g_t^1 = f_t$, and g_t^0 is an embedding for all $t \in [0,1]$. For each $t \in [0,1]$, let G_t^s, $s \in [0,1]$, be the path of monomorphisms $T\Lambda \to TM$ obtained by concatenating the paths dg_t^s and F_t^s. Then (g_t^0, G_t^s) is a formal Legendrian isotopy connecting (f_0, F_0^s) with the Legendrian knot f_1. □

7.6. Legendrian knots in overtwisted contact manifolds

In dimension 3 there is a dichotomy between tight and overtwisted contact structures, which was introduced in [**41**]. A contact structure ξ on a 3-dimensional manifold M is called *overtwisted* if there exists an embedded disc $D \subset M$ which is tangent to ξ along its boundary ∂D. Equivalently, one can require the existence of an embedded disc with Legendrian boundary ∂D which is transverse to ξ along ∂D. A disc with such properties is called an *overtwisted disc.* Note that any overtwisted disc has a neighborhood foliated by overtwisted discs. Indeed the contact structure in a neighborhood of an overtwisted disc can be given by the normal form $\cos r dz + r \sin r d\phi$, where r, ϕ, z are cylindrical coordinates in $\mathbb{R}^3$ and the overtwisted disc is given in these coordinates as $\{z = 0, r \leq \pi\}$. One then observes that the vector field $\frac{\partial}{\partial z}$ is contact, and hence all parallel discs $\{z = c, r \leq \pi\}$ are overtwisted.

Non-overtwisted contact structures are called *tight.* Bennequin proved in [**16**] that the standard contact structure on $\mathbb{R}^3$ (or S^3) is tight. More generally, (weakly) symplectically fillable contact structures are always tight [**84, 43**]. The sphere S^3 admits a unique tight positive contact structure, the standard one [**44**].

Overtwisted contact structures exhibit remarkable flexibility: their classification up to isotopy coincides with their homotopical classification as plane fields. More precisely, overtwisted contact structures satisfy the following h-principle.

THEOREM 7.18 (Classification of overtwisted contact structures [**41**]).
(a) Any oriented plane field on a closed oriented 3-manifold M is homotopic to a positive contact structure. This contact structure is unique up to isotopy.

(b) Let ξ_0 be an overtwisted contact structure on a closed connected 3-manifold M, and let $D \subset (M, \xi)$ be an overtwisted disc. Let $\mathrm{Cont}_{\mathrm{ot}}(M;\xi_0)$ and $\mathrm{Distr}(M;\xi_0)$ denote the spaces of overtwisted contact structures (resp. tangent plane fields) equal to ξ_0 on $\mathcal{O}p\, D$. Then the inclusion

$$\mathrm{Cont}_{\mathrm{ot}}(M;\xi_0) \hookrightarrow \mathrm{Distr}(M;\xi_0)$$

is a weak homotopy equivalence.

Parts (a) and (b) also hold in relative form for contact structures prescribed near a compact set $A \subset M \setminus D$.

The existence statement in part (a) of the theorem was proved by Lutz [**126**] and Martinet [**128**]. Theorem 7.18 implies the following h-principle for Legendrian knots in overtwisted contact manifolds.

THEOREM 7.19 (Dymara [**39**], Eliashberg and Fraser [**48**]). *Let (M, ξ) be a closed connected overtwisted contact 3-manifold, and $D \subset M$ an overtwisted disc.*

(a) Any formal Legendrian knot (f, F^s) in M is formally Legendrian isotopic to a genuine Legendrian embedding $\widetilde{f} : S^1 \hookrightarrow M \setminus D$.

(b) Let (f_t, F_t^s), $s, t \in [0, 1]$, be a formal Legendrian isotopy in M connecting two genuine Legendrian embeddings $f_0, f_1 : S^1 \hookrightarrow M \setminus D$. Then there exists a Legendrian isotopy $\widetilde{f}_t : S^1 \hookrightarrow M \setminus D$ connecting $\widetilde{f}_0 = f_0$ and $\widetilde{f}_1 = f_1$ which is homotopic to (f_t, F_t^s) through formal Legendrian isotopies with fixed endpoints.

PROOF. (a) After a smooth isotopy we may assume that $L := f(S^1) \subset M \setminus D$. The homotopy F^{1-t}, $t \in [0, 1]$, can be extended to a homotopy ξ_t, $t \in [0, 1]$, of plane fields along L connecting $\xi_0 = \xi|_L$ with a plane field ξ_1 tangent to L. This homotopy can be extended to a homotopy of contact structures on $\mathcal{O}p\, L$. Applying the relative form of Theorem 7.18(a), it can be further extended to a homotopy of contact structures ξ_t on the whole manifold M with $\xi_0 = \xi$ and $\xi_{\mathcal{O}p\, D} = \xi$. Hence, by Gray's Stability Theorem 6.23 there exists a diffeotopy $h_t : M \to M$, $t \in [0,1]$, with $h_0 = \mathrm{Id}$ and $h_t|_{\mathcal{O}p\, D} = \mathrm{Id}$ such that $(h_t)^*\xi = \xi_t$. Then the Legendrian embedding $h_1 \circ f : \Lambda \hookrightarrow (M, \xi)$ is connected to (f, F^s) by the formal Legendrian isotopy $(f_t = h_t \circ f,\ F_t^s = dh_t \circ F^{s(1-t)})$ in $M \setminus D$.

Part (b) can be proven similarly, using Theorem 7.18(b). Again, after a smooth isotopy with fixed endpoints we may assume that $f_t(S^1) \subset M \setminus D$ for all $t \in [0,1]$. Arguing as in (a), we can construct a 2-parameter family of contact structures $\xi_{t,u}$, $t, u \in [0,1]$, such that $\xi_{t,0} = \xi_{0,u} = \xi_{1,u} = \xi$. Gray's Theorem 6.23 yields a 2-parameter family of diffeomorphisms $h_{t,u}$ such that $h_{t,0} = h_{0,u} = h_{1,u} = \mathrm{Id}$, and $h_{t,u}^*\xi = \xi_{t,u}$ for all $t, u \in [0,1]$. Then the Legendrian isotopy $\widetilde{f_t} = h_{t,1} \circ f_t : \Lambda \hookrightarrow (M, \xi)$ connects f_0 and f_1 and is homotopic to the formal Legendrian isotopy (f_t, F_t^s) via the path of formal Legendrian isotopies $(f_{t,u} = h_{t,u} \circ f_t,\ F_{t,u}^s = dh_{t,u} \circ F_t^{s(1-u)})$. □

REMARK 7.20. Let us point out that, in contrast to most other h-principles in this chapter, the Legendrian embeddings in Theorem 7.19 cannot in general be chosen C^0-close to the original smooth embeddings. The reason is that the proof uses an overtwisted disc in an essential way and the original knots may be far from such a disc.

7.7. Murphy's h-principle for loose Legendrian embeddings

While in general the existence of a formal Legendrian isotopy between Legendrian embeddings is far from being sufficient for the existence of a genuine Legendrian isotopy (see e.g. [**31, 40**]), it turns out that there are classes of Legendrian embeddings for which one has an h-principle: the formal condition is sufficient for the existence of a Legendrian isotopy.

We have already encountered this phenomenon in Section 7.6 for Legendrian knots in overtwisted contact 3-manifolds. It turns out that in *any* contact manifold of dimension ≥ 5 there exists a class of Legendrian knots, called *loose*, which satisfy the h-principle: any Legendrian knot Λ can be C^0-approximated by a loose Legendrian knot Λ' in the same formal Legendrian isotopy class, and any two loose Legendrian knots which are formally Legendrian isotopic can be connected by a genuine Legendrian isotopy. This phenomenon was discovered by Emmy Murphy in [**143**].

REMARK 7.21. In [**48**] a Legendrian knot in a 3-dimensional contact manifold is called *loose* if its complement is overtwisted. As we will see below, the higher dimensional loose knots considered in this section exhibit a great deal of similarity with loose knots in dimension 3. However, to avoid any confusion, in this book we will apply the term "loose" only in the sense defined in this section.

In order to define loose Legendrian knots, we need to describe a local model. Throughout this section we assume $n \geq 2$.

Consider first a Legendrian arc λ_0 in the standard contact space $(\mathbb{R}^3, dz - p_1 dq_1)$ with front projection as shown in Figure 7.3, for some $a > 0$. Suppose that the slopes at the self-intersection point are ± 1 and the slope is everywhere in the interval

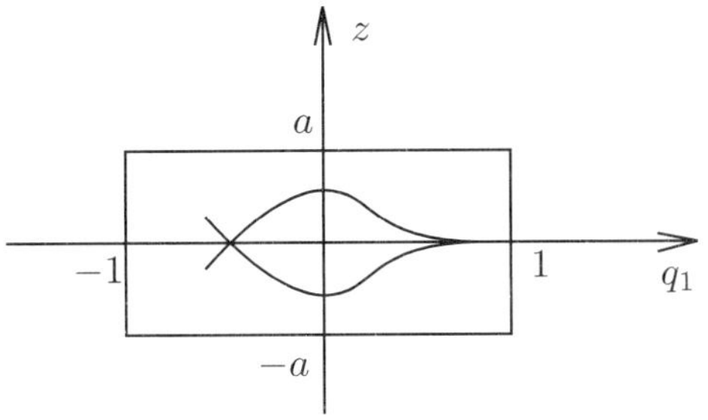

FIGURE 7.3. Front of the Legendrian arc λ_0.

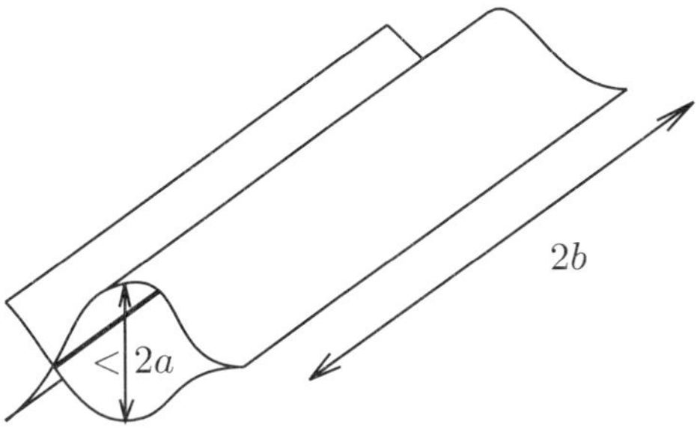

FIGURE 7.4. Front of the Legendrian solid cylinder Λ_0.

$[-1, 1]$, so the Legendrian arc λ_0 is contained in the box

$$Q_a := \{|q_1|, |p_1| \leq 1, |z| \leq a\}$$

and $\partial\lambda_0 \subset \partial Q_a$. Consider now the standard contact space $(\mathbb{R}^{2n+1}, dz - \sum_{i=1}^n p_i dq_i)$, which we view as the product of the contact space $(\mathbb{R}^3, dz - p_1 dq_1)$ and the Liouville space $(\mathbb{R}^{2n-2}, -\sum_{i=2}^n p_i dq_i)$. We set $q' := (q_2, \dots, q_n\}$ and $p' := (p_2, \dots, p_n)$. For $b, c > 0$ we define

$$P_{bc} := \{|q'| \leq b,\ |p'| \leq c\} \subset \mathbb{R}^{2n-2},$$
$$R_{abc} := Q_a \times P_{bc} = \{|q_1|, |p_1| \leq 1,\ |z| \leq a,\ |q'| \leq b,\ |p'| \leq c\}.$$

Let the Legendrian solid cylinder $\Lambda_0 \subset (\mathbb{R}^{2n+1}, dz - \sum_{i=1}^n p_i dq_i)$ be the product of $\lambda_0 \subset \mathbb{R}^3$ with the Lagrangian disc $\{p' = 0, |q'| \leq b\} \subset \mathbb{R}^{2n-2}$. Note that $\Lambda_0 \subset R_{abc}$ and $\partial\Lambda_0 \subset \partial R_{abc}$. The front of Λ_0 is obtained by translating the front of λ_0 in the q'-directions; see Figure 7.4. The pair (R_{abc}, Λ_0) is called a *standard loose Legendrian chart* if

$$a < bc.$$

Given any contact manifold (M^{2n+1}, ξ), a Legendrian submanifold $\Lambda \subset M$ with connected components $\Lambda_1, \dots, \Lambda_k$ is called *loose* if there exist Darboux charts $U_1, \dots, U_k \subset M$ such that $\Lambda_i \cap U_j = \varnothing$ for $i \neq j$ and each pair $(U_i, \Lambda_i \cap U_i)$, $i = 1, \dots, k$, is isomorphic to a standard loose Legendrian chart (R_{abc}, Λ_0).

Given a closed subset $A \subset M$, we say that a Legendrian submanifold $\Lambda \subset M$ is *loose relative to* A if $\Lambda \setminus A$ is loose in $M \setminus A$. A Legendrian embedding $f : \Lambda \hookrightarrow M$ is called *loose* if its image is a loose Legendrian submanifold.

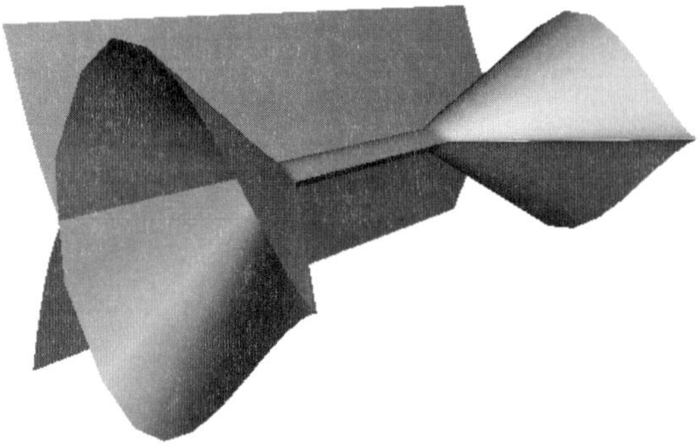

FIGURE 7.5. Shrinking a standard loose Legendrian chart (picture is courtesy of E. Murphy).

REMARK 7.22. (1) Let us stress the point that a link consisting of loose Legendrian knots is not necessarily a loose Legendrian link.

(2) By the contact isotopy extension theorem (Proposition 6.24), looseness is preserved under Legendrian isotopies within a fixed contact manifold. (Note, however, that if $\Lambda_t \subset (M, \xi_t)$, $t \in [0,1]$, is a family of Legendrian knots for *varying* contact structures and M is not closed, then looseness of Λ_0 need not imply looseness of Λ_1.) Since the model Λ_0 above can be extended to a Legendrian disc in standard $\mathbb{R}^{2n-1}$, and any two Legendrian discs are isotopic (shrink the first one to a neighborhood of a point, isotope it to a neighborhood of a point in the second one, and expand again), it follows that *any Legendrian disc is loose*. More precisely, for any closed Legendrian n-disc $D \subset (M^{2n+1}, \xi)$, $n \geq 2$, its interior $D \setminus \partial D$ is loose in $(M \setminus \partial D, \xi)$.

(3) By rescaling q' and p' with inverse factors one can always achieve $c = 1$ in the definition of a standard loose Legendrian chart. However, the inequality $a < bc$ is absolutely crucial in the definition. Indeed, it easily follows from Gromov's isocontact embedding theorem [**83, 52**] that around *any* point in *any* Legendrian submanifold Λ one can find a Darboux neighborhood U such that the pair $(U, \Lambda \cap U)$ is isomorphic to (R_{1b1}, Λ_0) for some sufficiently small $b > 0$.

(4) Figure 7.5 taken from [**143**] shows that the definition of looseness does not depend on the exact choice of the standard loose Legendrian chart (R_{abc}, Λ_0): Given a standard loose Legendrian chart with $c = 1$, the condition $a < b$ allows us to shrink its front in the q'-directions, keeping it fixed near the boundary and with all partial derivatives in $[-1, 1]$ (so the deformation remains in the Darboux chart R_{ab1}), to another standard loose Legendrian chart $(R_{a'b'1}, \Lambda_0')$ with $b' \geq (b-a)/2$ and arbitrarily small $a' > 0$. Moreover, we can arbitrarily prescribe the shape of the cross section λ_0' of Λ_0' in this process. So if a Legendrian submanifold is loose for some model (R_{abc}, Λ_0), then it is also loose for any other model. In particular, fixing b, c we can make a arbitrarily small, and we can create arbitrarily many disjoint standard loose Legendrian charts.

PROPOSITION 7.23 ([**143**]). *The stabilization construction in Proposition* 7.12 *with $k = 0$ makes any Legendrian embedding in dimension ≥ 5 loose without changing its formal Legendrian isotopy class.*

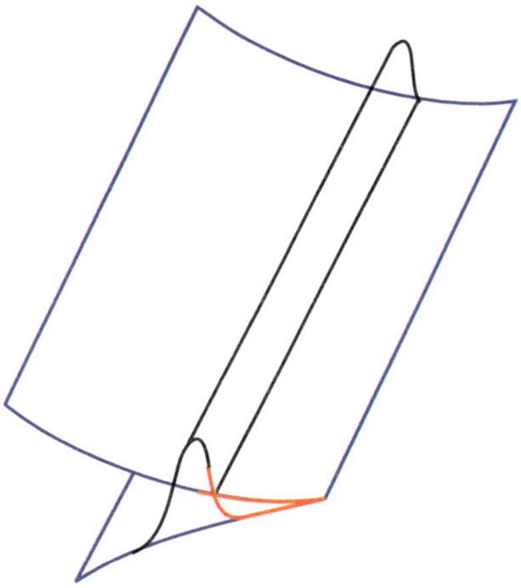

FIGURE 7.6. A standard loose Legendrian chart appears in the stabilization procedure.

PROOF. Let us recall the construction in Proposition 7.12. Given a Legendrian embedding $f_0 : \Lambda \hookrightarrow M$, we choose a Darboux chart with coordinates $(q_1, \dots, q_n, p_1, \dots, p_n, z)$ in which the front of $f_0(\Lambda)$ consists of two branches $\{z = \pm q_1^{3/2}\}$ joined along the singular locus $\{z = q_1 = 0\}$; see Figure 7.6. Deform the lower branch to the graph of a function $\phi(q)$ which is bigger than $q_1^{3/2}$ over a domain $N \subset \mathbb{R}^n$ of Euler characteristic 0 (e.g., diffeomorphic to an annulus $D^{n-1} \times S^1$) disjoint from the singular locus. Performing this construction sufficiently close to the singular locus, we can keep the values and the differential of the function ϕ arbitrarily small. Then the deformation is localized within the chosen Darboux neighborhood, and Figure 7.6 shows that the stabilized Legendrian embedding $f_1 : \Lambda \hookrightarrow M$ is loose.

To show that the stabilized Legendrian embedding f_1 is formally Legendrian isotopic to the original f_0, we reproduce the argument from [**143**]: Since $\chi(N) = 0$, there exists a nowhere vanishing vector field v on N which agrees with $\nabla(\phi - q_1^{\frac{3}{2}})$ near ∂N. Linearly interpolating the p-coordinate of f_1 from $\nabla\phi(q)$ to $\widetilde{v}(q) = v(q) + \nabla q_1^{\frac{3}{2}}$ (keeping the (q, z)-coordinates fixed), then pushing the z-coordinate down to $-q_1^{3/2}$ (keeping (q, p) fixed), and finally linearly interpolating $\widetilde{v}(q)$ to $-\nabla q_1^{3/2}$ (keeping (q, z) fixed) defines a smooth isotopy f_t between f_1 and f_0. On the other hand, the graphs of the functions $t\phi$ define a Legendrian regular homotopy from f_0 to f_1, so their differentials give a path of Legendrian monomorphisms F_t from $F_0 = df_0$ to $F_1 = df_1$. Now note that over the region N all the df_t and F_t project as the identity onto the q-plane, so linearly connecting df_t and F_t yields a path of monomorphisms F_t^s, $s \in [0, 1]$, and hence the desired formal Legendrian isotopy (f_t, F_t^s) from f_0 to f_1. □

REMARK 7.24. Let us stress the point that in dimension 3 any domain $N \subset \mathbb{R}$ has positive Euler characteristic, and hence the above stabilization construction never preserves the formal isotopy class of the Legendrian embedding

Now we can state the main result from [**143**]. Note that part (a) directly follows from Proposition 7.23 and Theorem 7.16.

THEOREM 7.25 (Murphy's h-principle for loose embeddings [**143**]). *Let* (M,ξ) *be a contact manifold of dimension* $2n+1 \geq 5$ *and* Λ *an* n*-dimensional manifold.*

(a) Any formal Legendrian embedding $(f : \Lambda \hookrightarrow M, F^s : T\Lambda \to TM)$ *can be* C^0*-approximated by a loose Legendrian embedding* $\widetilde{f} : \Lambda \hookrightarrow M$ *formally Legendrian isotopic to* (f, F^s).

(b) Any smooth isotopy $f_t : \Lambda \hookrightarrow M$, $t \in [0,1]$, *which begins with a loose Legendrian embedding* f_0 *can be* C^0*-approximated by a Legendrian isotopy starting at* f_0.

(c) Let (f_t, F_t^s), $s,t \in [0,1]$, *be a formal Legendrian isotopy connecting two loose Legendrian knots* f_0 *and* f_1. *Then there exists a Legendrian isotopy* $\widetilde{f}_t$ *connecting* $\widetilde{f}_0 = f_0$ *and* $\widetilde{f}_1 = f_1$ *which is* C^0*-close to* f_t *and is homotopic to the formal isotopy* (f_t, F_t^s) *through formal isotopies with fixed endpoints.*

Theorem 7.25 also holds in relative form. In particular, if in case (c) the formal Legendrian isotopy is genuine on a neighborhood of a closed subset $A \subset \Lambda$ and the Legendrian knots f_0 and f_1 are loose relative to A, then the isotopy $\widetilde{f}_t$ can be chosen equal to f_t over $\mathcal{O}p\, A$.

7.8. Directed immersions and embeddings

In this section we formulate Gromov's h-principle for directed immersions and embeddings and discuss its applications; see [**84**, Section 2.4] and [**52**, Chapter 19]. Given an m-dimensional real vector space V and an integer $k \leq m$, we denote by $G_k(V)$ the Grassmannian of its k-dimensional linear subspaces. A subset $A \subset G_k(V)$ is called *ample* if for any $L \in A$ and any $S \in G_{k-1}(L)$ the convex hull of each component of the set

$$\{v \in V \mid \operatorname{span}(S,v) \in A\}$$

coincides with the whole space V.

More globally, given an m-dimensional manifold M, we denote by $G_k(M)$ the bundle $\bigcup_{x\in M} G_k(T_xM)$ of tangent k-subspaces. A subset $A \subset G_k(M)$ is called *ample* if it is ample fiberwise.

EXAMPLE 7.26. (a) Let $\mathcal{R}^k = \mathcal{R}^k(\mathbb{C}^n) \subset G_k(\mathbb{C}^n)$, $k \leq n$, be the subset of totally real subspaces. Then $\mathcal{R}^k$ is ample.

(b) Let $V = \mathbb{C}^n \times \mathbb{R}$ and $\mathcal{R}^k_{\mathrm{CR}} = \mathcal{R}^k_{\mathrm{CR}}(V) \subset G_k(V)$, $k \leq n$, be the subset which consists of k-dimensional subspaces which project nonsingularly onto totally real subspaces of $\mathbb{C}^n$. Then $\mathcal{R}^k_{\mathrm{CR}}$ is ample.

(c) Let V be a symplectic vector space. Then the set of Lagrangian subspaces is *not* ample, and neither is the (open) set of symplectic subspaces of some given dimension.

One can also consider global versions of the above examples. Namely, if M is an almost complex manifold, then one can consider the subbundle

$$\mathcal{R}^k(M) = \bigcup_{x\in M} \mathcal{R}^k(T_xM) \subset G_k(M).$$

Next consider an *almost CR manifold* (M,ξ,J), i.e., an odd-dimensional manifold equipped with a hyperplane distribution ξ and a complex structure on ξ. Suppose that M is also equipped with a Riemannian metric. Then we obtain an orthogonal splitting $TM = \xi \times \mathbb{R}$ and thus a subbundle

$$\mathcal{R}^k_{\mathrm{CR}}(M) = \bigcup_{x \in M} \mathcal{R}^k_{\mathrm{CR}}(T_xM) \subset G_k(M).$$

Given $A \subset G_k(M)$ and a k-dimensional manifold P, an immersion or embedding $f : P \to M$ is called *A-directed* if for each $p \in P$ we have $df(T_pP) \in A$. A monomorphism $F : TP \to TM$ is called A-directed if for each $p \in P$ we have $F(T_pP) \in A$. For instance, if M is an almost complex manifold, then totally real immersions $P \to M$ of a k-dimensional manifold P are exactly the $\mathcal{R}^k(M)$-directed immersions. Given a Riemannian CR manifold M, we call $\mathcal{R}^k_{\mathrm{CR}}(M)$-directed immersions *CR totally real* (and similarly for embeddings).

Given a subset $A \subset G_k(M) \times [0,1]$, we set $A_t := A \cap \big(G_k(M) \times \{t\}\big)$, $t \in [0,1]$. We call $A \subset G_k(M) \times [0,1]$ *ample* if A_t is ample for each $t \in [0,1]$.

THEOREM 7.27 (h-principle for directed immersions [**52**, Theorem 18.4.1]). *Let $A \subset G_k(M)$ be an open ample set. Then the inclusion*

$$\mathrm{Imm}_{\mathrm{A-dir}}(P,M) \hookrightarrow \mathrm{Mon}_{\mathrm{A-dir}}(TP,TM)$$

of A-directed immersions into A-directed monomorphisms is a weak homotopy equivalence. In particular, we have:

(a) Given any continuous map $f : P \to M$ covered by an A-directed monomorphism $F : TP \to TM$, there exists a C^0-small homotopy $f_t : P \to M$ covered by a homotopy of A-directed monomorphisms $F_t : TP \to TM$ such that $f_0 = f$, $F_0 = F$, f_1 is an A-directed immersion, and $df_1 = F_1$.

(b) If the differentials of two A-directed immersions $f_0, f_1 : P \to M$ are homotopic as A-directed monomorphisms, then there exists an A-directed regular homotopy C^0-close to the given homotopy connecting f_0 and f_1.

The statement also holds in relative form fixed on a neighborhood $\mathcal{O}p\,B$ of a closed subset $B \subset P$.

As a special case, we obtain the following h-principle for totally real immersions. In this book it will only be used, via Theorem 7.38 in the next section, in the proof of Theorem 8.11 which is a special case of the Gromov–Landweber theorem [**82, 120**].

COROLLARY 7.28 (h-principle for totally real immersions [**84, 52**]). *Let (V,J) be an almost complex manifold of dimension $2n$, and L a manifold of dimension $k \le n$. Then the inclusion*

$$\mathrm{Imm}_{\mathrm{real}}(L,V) \hookrightarrow \mathrm{Mon}_{\mathrm{real}}(TL,TV)$$

of totally real immersions into totally real monomorphisms is a weak homotopy equivalence. In particular, any continuous map $f : L \to V$ covered by a totally real monomorphism $F : TL \to TV$ is homotopic to a C^0-close totally real immersion $g : L \to V$ such that dg and F are homotopic through totally real monomorphisms $TL \to TV$. If f is already a totally real immersion on a neighborhood $\mathcal{O}p\,B$ of a closed subset $B \subset L$ and $F = df$ on $TL_{\mathcal{O}p\,B}$, then the homotopy f_t can be chosen fixed on $\mathcal{O}p\,B$.

Directed embeddings. Remarkably, directed *embeddings* in the case of an open ample set A also satisfy an h-principle. To formulate this, we introduce the following terminology analogous to that in Section 7.3. A *formal A-directed* *embedding* of P into M is a pair (f, F^s), where $f : P \hookrightarrow M$ is a smooth embedding and $F^s : TP \to TM$ is a homotopy of monomorphisms over f starting at $F^0 = df$ and ending at an A-directed monomorphism $F^1 : TP \to TM$ covering f. Then every A-directed embedding f gives rise to a formal A-directed embedding $(f, F^s = df)$.

THEOREM 7.29 (h-principle for directed embeddings [**52**, Theorem 19.4.1]). *Let $A \subset G_k(M)$ be an open ample set. Then the inclusion*

$$\mathrm{Emb}_{\mathrm{A-dir}}(P, M) \hookrightarrow \mathrm{Mon}^{\mathrm{emb}}_{\mathrm{A-dir}}(TP, TM)$$

of A-directed embeddings into formal A-directed embeddings is a weak homotopy equivalence. In particular, we have:

(a) Any formal A-directed embedding (f_0, F_0^s) of P into M is connected to a genuine A-directed embedding $f_1 : P \hookrightarrow M$ by a path of formal A-directed embeddings (f_t, F_t^s) such that the f_t are C^0-close to f_0.

(b) Let $A \subset G_k(M) \times [0, 1]$ be an open ample set. Let $f_0, f_1 : P \hookrightarrow M$ be an A_0-directed and an A_1-directed embedding connected by a path of formal A_t-directed embeddings (f_t, F_t^s). Then there exists an isotopy $\widetilde{f}_t$ of A_t-directed embeddings C^0-close to f_t, connecting f_0 and f_1, which is homotopic to (f_t, F_t^s) as paths of formal A_t-directed embeddings with fixed endpoints.

The statement also holds in relative form fixed on a neighborhood $\mathcal{O}p\, B$ of a closed subset $B \subset P$.

In particular, we have the following special cases of this h-principle.

COROLLARY 7.30 (h-principle for totally real embeddings [**84, 52**]). *Let (V, J) be an almost complex manifold of dimension $2n$, and let L be a manifold of dimension $k \leq n$. Then the inclusion*

$$\mathrm{Emb}_{\mathrm{real}}(L, V) \hookrightarrow \mathrm{Mon}^{\mathrm{emb}}_{\mathrm{real}}(TL, TV)$$

of totally real embeddings into formal totally real embeddings is a weak homotopy equivalence. In particular, we have:

(a) Any formal totally real embedding (f_0, F_0^s) of L into V is connected to a genuine totally real embedding $f_1 : L \hookrightarrow V$ by a path of formal totally real embeddings (f_t, F_t^s) such that the f_t are C^0-close to f_0.

(b) Let J_t be a family of almost complex structures on V. Let $f_0, f_1 : L \hookrightarrow V$ be a J_0- (resp. J_1-) totally real embedding connected by a path of formal J_t-totally real embeddings (f_t, F_t^s). Then there exists an isotopy $\widetilde{f}_t$ of J_t-totally real embeddings C^0-close to f_t, connecting f_0 and f_1, which is homotopic to (f_t, F_t^s) as paths of formal J_t-totally real embeddings with fixed endpoints.

The statement also holds in relative form fixed on a neighborhood $\mathcal{O}p\, B$ of a closed subset $B \subset L$.

COROLLARY 7.31. *Let (V, J) be an almost complex manifold and $f : L \hookrightarrow V$ a totally real embedding. Let J_t, $t \in [0, 1]$, be a homotopy of almost complex structures on V with $J_0 = J$. Then there exists an isotopy of embeddings $f_t : L \hookrightarrow V$ such that $f_0 = f$ and f_t is J_t-totally real. If the isotopy f_t is already given on a neighborhood $\mathcal{O}p\, B$ of closed subset $B \subset L$ such that $f_t|_{\mathcal{O}p\, B}$ is J_t-totally real, then f_t can be extended from $\mathcal{O}p\, B$ to L.*

PROOF. There is a family of J_s-totally real monomorphisms $F^s : TL \to TV$ covering f with $F^0 = df$. Hence, we can first apply Corollary 7.30(a) (with the almost complex structure J_1) to find a path of formal J_1-totally real embeddings (f_t, F_t^s) connecting $(f = f_0, F^s)$ to a J_1-totally real embedding (f_1, df_1). After reparametrizing F_t^s in (s,t) we can view this as a path of formal J_t-totally real embeddings connecting the J_0-totally real embedding f_0 to the J_1-totally real embedding f_1. So we can apply Corollary 7.30(b) to find the desired isotopy of J_t-totally real embeddings. □

COROLLARY 7.32 (h-principle for CR totally real embeddings). *Let (M, ξ, J) be a $(2n+1)$-dimensional almost CR manifold, and let Λ be a manifold of dimension $k \leq n$. Then the inclusion*

$$\mathrm{Emb}_{\mathrm{CR-real}}(\Lambda, M) \hookrightarrow \mathrm{Mon}^{\mathrm{emb}}_{\mathrm{CR-real}}(T\Lambda, TM)$$

of CR totally real embeddings into formal CR totally real embeddings is a weak homotopy equivalence. In particular, any formal CR totally real embedding (f_0, F_0^s) of Λ into M is connected to a genuine CR totally real embedding $f_1 : \Lambda \hookrightarrow M$ by a path of formal CR totally real embeddings (f_t, F_t^s) such that the f_t are C^0-close to f_0.

The statement also holds in relative form fixed on a neighborhood $\mathcal{O}p\, B$ of a closed subset $B \subset \Lambda$.

We finish this section with an analogue of Corollary 7.28 for so-called totally real submersions (also called complex submersions of real manifolds, see [**120**]). This result will only be needed for one of the cases of Theorem 8.45.

A linear map $A : L \to V$ between a real vector space L and a complex vector space V is called a *totally real epimorphism* if its complexification $A^{\mathbb{C}} : L \otimes \mathbb{C} \to V$ is surjective. Similarly, a smooth map $f : L \to V$ of a real manifold L to an almost complex manifold (V, J) is called a *totally real submersion* if its differential $TL \to TV$ is a fiberwise totally real epimorphism. A word of caution: a totally real submersion need neither be a submersion, nor does its image have to be totally real.

COROLLARY 7.33 (h-principle for totally real submersions). *Let (V, J) be an almost complex manifold of dimension $2n$, and L a manifold of dimension $m \geq n$. Then the inclusion*

$$\mathrm{Sub}_{\mathrm{real}}(L, V) \hookrightarrow \mathrm{Epi}_{\mathrm{real}}(TL, TV)$$

of totally real submersions into totally real epimorphisms is a weak homotopy equivalence. In particular, any continuous map $f : L \to V$ covered by a totally real epimorphism $F : TL \to TV$ is homotopic to a totally real submersion $g : L \to V$ such that dg and F are homotopic through totally real epimorphisms $TL \to TV$. If f is already a totally real submersion on a neighborhood $\mathcal{O}p\, B$ of a closed subset $B \subset L$ and $F = df$ on $TL|_{\mathcal{O}p\, B}$, then the homotopy f_t can be chosen fixed on $\mathcal{O}p\, B$.

PROOF. It is sufficient to prove an extension statement from a neighborhood of the boundary of a disc to the disc itself. This can be reduced to Corollary 7.28 by suspending the map f to a map $\widehat{f} : D \to V \times \mathbb{C}^{m-n}$ and suspending the totally real epimorphism F to a totally real isomorphism $\widehat{F} : TD \to TV \times \mathbb{C}^{m-n}$, and then projecting the constructed totally real immersion to $V \times \mathbb{C}^{m-n}$ back to V. □

7.9. Discs attached to J-convex boundaries

Theorem 7.34 below, which is a combination of h-principles discussed earlier in this chapter, will play an important role in proving the main results of this book.

Let (V, J) be an almost complex manifold and $W \subset V$ a domain with smooth boundary ∂W. Let L be a (possibly noncompact) manifold with boundary. Let $f : L \hookrightarrow V \setminus \operatorname{Int} W$ be an embedding with $f(\partial L) \supset \partial W$, $\overline{f(L)} \cap \partial W = f(\partial L)$, and which is transverse to ∂W along ∂L. We say in this case that f *transversely* attaches L to W along ∂L. We recall that f attaches L to W *J-orthogonally* if, in addition, $J df(TL|_{\partial L}) \subset T(\partial W)$. Note that this implies that $df(\partial L)$ is tangent to the distribution $\xi = T(\partial W) \cap JT(\partial W)$. In particular, if ∂W is J-convex, then $f(\partial L)$ is an isotropic submanifold for the contact structure ξ.

THEOREM 7.34. *Suppose that (V, J) is an almost complex manifold of real dimension $2n$, and $W \subset V$ is a domain with smooth J-convex boundary. Suppose that an embedding $f : D^k \hookrightarrow V$, $k \le n$, transversely attaches D^k to W along ∂D^k. If $k = n = 2$, we assume, in addition, that the induced contact structure on ∂W is overtwisted. Then there exists an isotopy $f_t : D^k \hookrightarrow V$, $t \in [0,1]$, through embeddings transversely attaching D^k to W, such that $f_0 = f$ and f_1 is totally real and J-orthogonal to ∂W. Moreover, in the case $k = n > 2$ we can arrange that the Legendrian embedding $f_1|_{\partial D^k} : \partial D^k \hookrightarrow \partial W$ is loose, while for $n = 2$ we can arrange that the complement $\partial W \setminus f_t(\partial D^2)$ is overtwisted for all $t \in [0,1]$.*

The proof uses the following homotopical lemma.

LEMMA 7.35. *Consider $f : D^k \hookrightarrow V$ as in Theorem 7.34. Then there exists a homotopy of monomorphisms $F_t : TD^k \to TV$, $t \in [0,1]$, covering f such that $F_0 = df$, F_1 is totally real and, in addition,*

(a) *$F_1(T\partial D^k) \subset \xi$,*
(b) *$F_t(T\partial D^k) \subset T\partial W$ for all $t \in [0,1]$.*

PROOF. We write $D = D^k$. Let us fix an outward normal vector field $\mathbf{r}$ to ∂D in D and an inward pointing vector field $\mathbf{n}$ along ∂W such that $J\mathbf{n} \in T\partial W$. After an isotopy of f we may assume that $df(\mathbf{r}) = \mathbf{n}$ along ∂D. Consider the bundle $\operatorname{Mon}(TD, f^*TV) \to D$ and its subbundle $\operatorname{Mon}_{\mathrm{real}}(TD, f^*TV)$ of totally real monomorphisms. Similarly, over the boundary ∂D we have the bundles

$$\operatorname{Mon}_{\mathrm{real}}(T\partial D, f^*\xi) \subset \operatorname{Mon}(T\partial D, f^*T\partial W).$$

Sending $\mathbf{r}$ to $\mathbf{n}$ defines natural inclusions

$$\operatorname{Mon}(T\partial D, f^*T\partial W) \subset \operatorname{Mon}(TD, TV)|_{\partial D} \quad \text{and}$$
$$\operatorname{Mon}_{\mathrm{real}}(T\partial D, f^*\xi) \subset \operatorname{Mon}_{\mathrm{real}}(TD, TV)|_{\partial D}.$$

Note that df defines a section in $\operatorname{Mon}(TD, TV)$ which restricts to a section in $\operatorname{Mon}(T\partial D, f^*T\partial W)$ over ∂D.

After picking a metric and orthogonalization, we may assume that the fiber of the bundle $\operatorname{Mon}(TD, TV)$ is the Stiefel manifold $V_{2n,k}$ of orthogonal k-frames in $\mathbb{R}^{2n}$ (see Appendix A.2) and its structure group is $O(2n)$. Similarly, we may assume that the fiber of the bundle $\operatorname{Mon}_{\mathrm{real}}(TD, TV)$ is the Stiefel manifold $V^{\mathbb{C}}_{n,k}$ of unitary k-frames in $\mathbb{C}^n$ and its structure group is $U(n)$, and similarly for the bundles over ∂D.

The $U(n-1)$-bundle $\mathrm{Mon}_{\mathrm{real}}(T\partial D, f^*\xi) \to S^{k-1}$ is obtained by gluing two trivial bundles via a map $g : S^{k-2} \to U(n-1)$. Since the bundle extends over D as the $U(n)$-bundle $\mathrm{Mon}_{\mathrm{real}}(TD, TV)$, g lies in the kernel of the map $\pi_{k-2}U(n-1) \to \pi_{k-2}U(n)$, which is trivial by Corollary A.10. Thus we obtain compatible trivializations of all the bundles

$$\mathrm{Mon}_{\mathrm{real}}(TD, f^*TV) \cong D \times V^{\mathbb{C}}_{n,k} \subset \mathrm{Mon}(TD, f^*TV) \cong D \times V_{2n,k},$$
$$\mathrm{Mon}_{\mathrm{real}}(T\partial D, f^*\xi) \cong \partial D \times V^{\mathbb{C}}_{n-1,k-1} \subset \mathrm{Mon}(T\partial D, f^*T\partial W) \cong \partial D \times V_{2n-1,k-1}.$$

With these trivializations, df defines a map $(D^k, \partial D^k) \to (V_{2n,k}, V_{2n-1,k-1})$ which we want to deform to a map $(D^k, \partial D^k) \to (V^{\mathbb{C}}_{n,k}, V^{\mathbb{C}}_{n-1,k-1})$. This is possible by Corollary A.8(a) which asserts $\pi_k(V_{2n,k}, V_{2n-1,k-1}) = 0$ for $k \le 2n-2$, i.e., in particular for $k \le n$ and $n \ge 2$. □

PROOF OF THEOREM 7.34. We write $D = D^k$. Let F_t be the homotopy from Lemma 7.35. The isotopy f_t is constructed in two steps.

Step 1. The restriction $F_t|_{T(\partial D)}$ gives us a homotopy of monomorphisms $\widetilde{F}_t : T(\partial D) \to T(\partial W)$ covering $f|_{\partial D}$ such that $\widetilde{F}_0 = df|_{T\partial D}$ and $\widetilde{F}_1 : T(\partial D) \to \xi$ is totally real. Hence, we can apply Theorem 7.16 if $n > 2$, and Theorem 7.19 if $n = 2$, to find an isotopy $g_t : \partial D \hookrightarrow \partial W$ such that

(α) $g_0 = f|_{\partial D}$,
(β) g_1 is isotropic, and
(γ) the path of monomorphisms $dg_t : T(\partial D) \to T(\partial W)$, $t \in [0,1]$, is homotopic to $\widetilde{F}_t$ in the class of paths of monomorphisms beginning at dg_0 and ending at a totally real monomorphism $T(\partial D) \to \xi$.

When $n > 2$ Theorem 7.16 allows us to make the isotopy g_t C^0-small and using Theorem 7.25 we can arrange $g_1(\partial D)$ to be loose, while when $n = 2$ the complement $\partial W \setminus g_t(\partial D)$ can be made overtwisted by Theorem 7.19.

We extend the isotopy g_t to an isotopy $f_t : D \hookrightarrow V \setminus \mathrm{Int}\, W$ of smooth embeddings transversely attached to W such that $f_0 = f$. Note that any subspace of T_pV, $p \in \partial W$, which is transverse to ∂W and intersects $\xi_p \subset T_p\partial W$ along a totally real subspace is totally real itself. Hence, we can further deform the disc $f_1(D)$ near $f_1(\partial D)$ through totally real discs, keeping the boundary fixed, to make it J-orthogonally attached to ∂W.

Step 2. We claim that there exists a homotopy of monomorphisms $G_t : TD \to TV$, $t \in [0,1]$ such that

a) $G_0 = df_1 : TD \to TV$,
b) G_1 is totally real, and
c) $G_t = df_1$ on $TD|_{\partial D}$ for all $t \in [0,1]$.

Indeed, consider first the homotopy $\widetilde{G}_t : TD \to TV$,

$$\widetilde{G}_t := \begin{cases} df_{1-2t}, & t \in [0, \frac{1}{2}]; \\ F_{2t-1}, & t \in (\frac{1}{2}, 1]. \end{cases}$$

The homotopy $\widetilde{G}_t$ satisfies the above conditions a) and b), but not c). However, in view of property (γ) the path $\widetilde{G}_t|_{T\partial D}$ is homotopic through paths with fixed ends to a path of totally real monomorphisms, and hence the homotopy $\widetilde{G}_t$ can be modified to a homotopy G_t satisfying condition c) as well. More explicitly, (γ)

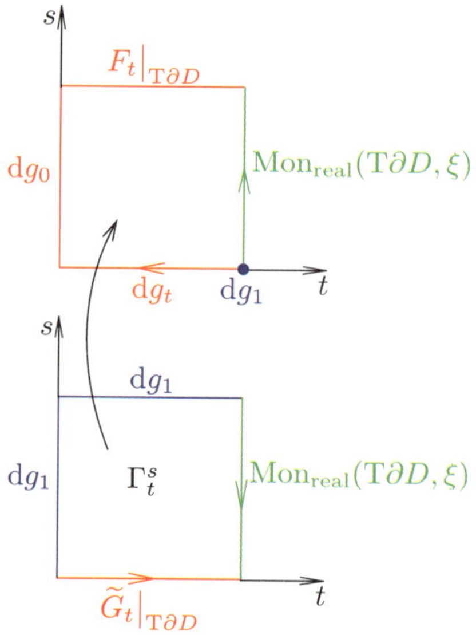

FIGURE 7.7. Construction of the family of monomorphisms Γ_t^s.

allows us to pick a continuous family of monomorphisms $\Gamma_t^s : T(\partial D) \to T(\partial W)$, $s, t \in [0,1]$, such that $\Gamma_t^0 = \widetilde{G}_t|_{T\partial D}$, $\Gamma_t^1 = \Gamma_0^s = df_1|_{T\partial D}$, and $\Gamma_1^s : T(\partial D) \to \xi$ is totally real for all $s \in [0,1]$; see Figure 7.7.

We extend Γ_t^s from $T(\partial D)$ to $TD_{\partial D}$ sending the outward normal $\mathbf{r}$ along ∂D to the inward normal $\mathbf{n}$ along ∂W, so that it satisfies $\Gamma_t^0 = \widetilde{G}_t$, $\Gamma_t^1 = \Gamma_0^s = df_1$, and Γ_1^s is J-orthogonal and totally real along ∂D. After rescaling in the unit disc D, we may assume that $\widetilde{G}_t(x)$ is independent of the radius for $x \in D$ with $|x| \geq 1/2$. Then the desired homotopy $G_t : TD \to TV$ can be defined by

$$G_t(x) := \begin{cases} \widetilde{G}_t(2x), & |x| \in [0, \frac{1}{2}]; \\ \Gamma_t^{2|x|-1}(x), & |x| \in (\frac{1}{2}, 1]. \end{cases}$$

It remains to apply Gromov's h-principle for totally real embeddings, Corollary 7.30(a). It provides an isotopy of embeddings $f_t : D \to V \setminus \mathrm{Int}\, W$, $t \in [1,2]$, fixed along ∂D together with its differential, such that $f_2 : D \to V \setminus \mathrm{Int}\, W$ is totally real and J-orthogonal to ∂W. Finally, note that the isotopy provided by Corollary 7.30(a) can also be chosen C^0-small. This concludes the proof of Theorem 7.34. □

In Section 14.3 we will need the following 1-parametric version of Theorem 7.34 in the *flexible* situation.

THEOREM 7.36. *Let J_t, $t \in [0,1]$, be a family of almost complex structures on a $2n$-dimensional manifold V. Let $W \subset V$ be a domain with smooth boundary which is J_t-convex for all $t \in [0,1]$. Suppose that $k \leq n$. Let $f_t : D^k \hookrightarrow V \setminus \mathrm{Int}\, W$, $t \in [0,1]$, be an isotopy of embeddings transversely attached to ∂W along ∂D^k. Suppose that for $i = 0,1$ the embedding f_i is J_i-totally real and J_i-orthogonally attached to ∂W along ∂D^k. Suppose that either $k < n$ or $k = n > 2$ and the*

Legendrian embeddings $f_i|_{\partial D}$, $i = 0, 1$ *are* loose. *Then there exists a* 2*-parameter family of embeddings* $f_t^s : D^k \hookrightarrow V \setminus W$ *with the following properties:*

- f_t^s *is transversely attached to* W *along* ∂D^k *and* C^0*-close to* f_t *for all* $t, s \in [0,1]$;
- $f_t^0 = f_t$ *for all* $t \in [0,1]$ *and* $f_0^s = f_0$, $f_1^s = f_1$ *for all* $s \in [0,1]$;
- f_t^1 *is* J_t*-totally real and* J_t*-orthogonally attached to* ∂W *along* ∂D^k *for all* $t \in [0,1]$.

The proof uses the following homotopical lemma.

LEMMA 7.37. *Consider* $f_t : D^k \hookrightarrow V \setminus \mathrm{Int}\, W$ *as in Theorem* 7.36*, where we allow the critical case* $k \le n$*. Then there exists a* 2*-parameter family of monomorphisms* $F_t^s : TD \to TV$, $t, s \in [0,1]$*, covering* f_t *such that* $F_t^0 = df_t$, $F_0^s = df_0$, $F_1^s = df_1$, F_t^1 *is totally real and, in addition,*

(a) $F_t^1(T\partial D) \subset \xi$, *and*
(b) $F_t^s(T\partial D) \subset T\partial W$.

PROOF. As shown in the proof of Lemma 7.35, we can trivialize the relevant bundles of monomorphisms over $D := D^k$ and ∂D. So $F_t := df_t$ defines a homotopy of maps

$$F_t : (D, \partial D) \to (V_{2n,k}, V_{2n-1,k-1}), \qquad t \in [0,1],$$

with endpoints

$$F_0, F_1 : (D, \partial D) \to (V^{\mathbb{C}}_{n,k}, V^{\mathbb{C}}_{n-1,k-1})$$

which we want to deform into $(V^{\mathbb{C}}_{n,k}, V^{\mathbb{C}}_{n-1,k-1})$ with fixed ends at $t = 0, 1$. By Corollary A.8(a) we have $\pi_k(V^{\mathbb{C}}_{n,k}, V^{\mathbb{C}}_{n-1,k-1}) = 0$ for $k \le n$ and $n \ge 2$. After contracting F_0, F_1 it thus suffices to consider the case that $F_0 \equiv F_1 \equiv v \in V_{n-1,k-1}$ is constant. After a further deformation, we may assume that $F_t(0) \equiv v$ for the origin $0 \in D$ and all $t \in [0,1]$, and collapsing $\{0,1\} \times D \cup [0,1] \times \{0\} \subset [0,1] \times D$ to a point p, we obtain a map

$$\bar{F} : (D^{k+1}, \partial D^{k+1}, p) \to (V_{2n,k}, V_{2n-1,k-1}, v)$$

which we need to contract to the point v. This is possible by Corollary A.8(a) because $\pi_{k+1}(V^{\mathbb{C}}_{n,k}, V^{\mathbb{C}}_{n-1,k-1}) = 0$ provided that $k + 1 \le 2n - 2$. This condition is satisfied if $k \le n$ and $n > 2$, or if $k < n$ and $n = 2$.

It remains to treat the case $k = n = 2$. Note that in the first step of the preceding argument we have chosen a contraction of F_0 in $\pi_2(V^{\mathbb{C}}_{2,2}, V^{\mathbb{C}}_{1,1}) = 0$. Two such contractions differ by an element in $\pi_3(V^{\mathbb{C}}_{2,2}, V^{\mathbb{C}}_{1,1})$. So by varying the contractions, we can change the resulting class $[\bar{F}] \in \pi_3(V_{4,2}, V_{3,1})$ by adding classes in $\pi_3(V^{\mathbb{C}}_{2,2}, V^{\mathbb{C}}_{1,1})$. Since by Corollary A.8(b) the map $\pi_3(V^{\mathbb{C}}_{2,2}, V^{\mathbb{C}}_{1,1}) \to \pi_3(V_{4,2}, V_{3,1})$ is an isomorphism, we can arrange $[\bar{F}] = 0$ and thus conclude the proof. □

PROOF OF THEOREM 7.36. With Lemma 7.37, the rest of the proof of Theorem 7.36 is parallel to the proof of Theorem 7.34: First we apply either the h-principle for subcritical isotropic embeddings (Theorem 7.11) if $k < n$, or Murphy's h-principle for loose Legendrian embeddings (Theorem 7.25) if $k = n > 2$ in order to make $f_t|_{\partial D}$ an isotropic isotopy. Then we deform f_t, as in the proof of Theorem 7.34, to make it J_t-orthogonal to ∂W along ∂D. Finally, we apply the h-principle for totally real embeddings, Corollary 7.30(b), to make f_t totally real. □

Finally, in Section 8.3 we will need the following version of Theorem 7.34 in which we remove the condition of J-convexity of the boundary, but obtain instead of J-orthogonality the following weaker notion of "J-transversality". Let us assume that W is endowed with a Riemannian metric for which J acts as an orthogonal operator and denote by $\mathbf{n}$ the unit outward normal vector to the boundary of W. Then we say that f attaches L to W *J-transversely* if $\mathbf{n} \in df(TL|_{\partial L})$ and $df(TL|_{\partial L})$ is totally real. Note that this implies that $f(\partial L)$ is a CR totally real submanifold of ∂W in the sense of Section 7.8 above.

THEOREM 7.38. *Suppose that (V, J) is an almost complex manifold of dimension $2n$ and $W \subset V$ is a domain with smooth boundary (not necessarily J-convex). Suppose that an embedding $f : D^k \to V$, $k \leq n$, transversely attaches D^k to W along ∂D^k in V. Then there exists an isotopy $f_t : D^k \hookrightarrow V$, $t \in [0,1]$, through embeddings C^0-close to f and transversely attaching D^k to W, such that $f_0 = f$ and f_1 is totally real and J-transverse to ∂W.*

PROOF. The proof repeats the proof of Theorem 7.34, using Corollary 7.32 instead of Theorem 7.16. □

CHAPTER 8

The Existence Theorem

In this chapter we prove Theorem 1.5 from the Introduction on the existence of Stein structures in complex dimension $\neq 2$. The proof combines the techniques developed in earlier chapters: in Section 8.2 we use the i-convex model functions from Chapter 4 to extend J-convex functions over discs, and in Section 8.3 we use the h-principles from Chapter 7 to extend complex structures over discs. The Existence Theorem 1.5, as well as an ambient version due to Gompf, is then proved in Section 8.4.

Sections 8.5 to 8.7 contain various refinements of results in the previous sections that will not be used in the remainder of the book. In Sections 8.2 and 8.6 we refine the J-convex surroundings from Section 8.2 and discuss some applications to holomorphic convexity. In Section 8.7 we derive some holomorphic approximation results due to Forstnerič and Slapar. Finally, in Section 8.8 we prove a variant of Kallin's lemma which will be needed in Section 16.2.

8.1. Some notions from Morse theory

In this chapter we use the following notions from Morse theory; for more details see Chapter 9.

Recall that a function $\phi : V \to \mathbb{R}$ is called *Morse* if all its critical points are nondegenerate, and the *(Morse) index* of a critical point of ϕ is the maximal dimension of a subspace on which the Hessian is negative definite. A vector field X is called *gradient-like* for ϕ if it satisfies

$$X \cdot \phi \geq \delta(|X|^2 + |d\phi|^2)$$

for some $\delta > 0$, where $|X|$ is the norm with respect to some Riemannian metric on V and $|d\phi|$ is the dual norm. The *stable manifold* W_p^- (with respect to X) of a critical point p of ϕ is the set of all points converging to p under the forward flow of X. The *skeleton* of a Morse function (with respect to X) is the union of all stable manifolds.

By the Morse lemma (see [**139**]), near a nondegenerate critical point p of ϕ of index k there exist coordinates u_i in which ϕ has the form

$$\phi(u) = \phi(p) - u_1^2 - \cdots - u_k^2 + u_{k+1}^2 \cdots + u_m^2.$$

We will use the following easy consequence, as well as a refinement given in Lemma 9.29 below.

COROLLARY 8.1. *Let $\phi : V \to \mathbb{R}$, $\phi' : V' \to \mathbb{R}$ be Morse functions with gradient-like vector fields X, X' and critical points p, p' of the same index and value with stable manifolds $W_p^-, W_{p'}^-$. Then there exists a diffeomorphism $f : \mathcal{O}p\, W_p^- \to \mathcal{O}p\, W_{p'}^-$ such that $\phi' = \phi \circ f$.*

PROOF. By the Morse lemma there exists a diffeomorphism $f : \mathcal{O}p\, p \to \mathcal{O}p\, p'$ such that $\phi' = \phi \circ f$. It extends uniquely to a diffeomorphism $\mathcal{O}p\, W_p^- \to \mathcal{O}p\, W_{p'}^-$ with $\phi' = \phi \circ f$ and mapping trajectories of X to trajectories of X'. □

A *cobordism* is a compact oriented manifold W with oriented boundary $\partial W = \partial_+ W \amalg \partial_- W$, where the orientation agrees with the boundary orientation for $\partial_+ W$ and is opposite to it for $\partial_- W$. We allow one or both of $\partial_\pm W$ to be empty. A *Morse cobordism* (W, ϕ) is a cobordism W with a Morse function $\phi : W \to \mathbb{R}$ having $\partial_\pm W = \phi^{-1}(c_\pm)$ as regular level sets. We call a Morse cobordism (W, ϕ) *elementary* if ϕ admits a gradient-like vector field X such that no two critical points of ϕ are connected by an X-trajectory. In that case each stable manifold W_p^- is an embedded disc which we will refer to as the *stable disc* of p.

8.2. Surrounding stable discs

In this section we prove our two main results about J-convex surroundings. The first one, Theorem 8.4, states that a J-orthogonally attached totally real disc can be surrounded by J-convex hypersurfaces. It is a crucial ingredient in the proof of the existence of Stein structures in Section 8.4. The second one, Theorem 8.5, states that a stable disc of a J-convex Morse function can be surrounded by deforming the level sets of the given function. It is the basis for the holomorphic approximations in Section 8.7 and for the deformations of J-convex functions studies in Chapter 10.

Let (V, J) be a complex manifold, possibly with boundary.

DEFINITION 8.2. Given a subset $A \subset V$ and a neighborhood $U \subset V$ of A, we say that a J-convex hypersurface $\Sigma \subset U$ *surrounds A in U* if it is the J-convex boundary of a domain in U containing A. We say that *A can be surrounded by J-convex hypersurfaces* if J-convex surrounding hypersurfaces exist in arbitrarily small neighborhoods of A.

EXAMPLE 8.3. In Section 2.7 we saw that the following sets can be surrounded by J-convex hypersurfaces:

(i) totally real submanifolds;
(ii) the zero section of a negative holomorphic line bundle, and hence any properly embedded complex codimension 1 submanifold with negative normal bundle.

Theorem 4.1 provides a solution of the surrounding problem for a more subtle case: the set $A = \{ar^2 - R^2 \le -1\} \cup \{r = 0\} \subset \mathbb{C}^n$ can be surrounded by i-convex hypersurfaces, where $a > 1$ and

$$r := \sqrt{x_1^2 + \cdots + x_n^2 + y_{k+1}^2 + \cdots + y_n^2}, \quad R := \sqrt{y_1^2 + \cdots + y_k^2}$$

for some fixed $k \le n$ and complex coordinates $x_1 + iy_1, \ldots, x_n + iy_n$ in $\mathbb{C}^n$.

The following two main results of this section generalize this model case to a solution of the surrounding problem for totally real discs suitably attached to J-convex domains.

THEOREM 8.4. *Let (W, J) be a complex manifold with compact J-concave boundary $\partial_- W$. Let $\Delta \subset W$ be a totally real disc J-orthogonally attached to $\partial_- W$. Then $\partial_- W \cup \Delta$ can be surrounded by J-convex hypersurfaces. Moreover, if we are given*

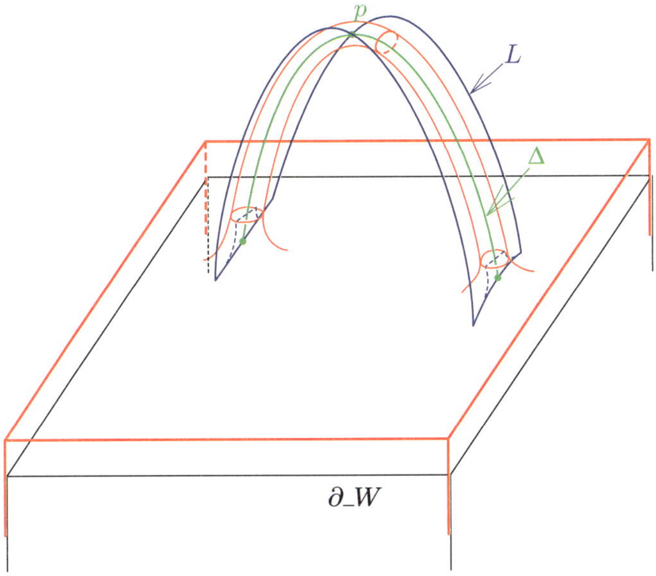

FIGURE 8.1. Surrounding a J-orthogonally attached totally real disc.

a totally real submanifold $L \supset \Delta$ *of dimension* $\dim L > \dim \Delta$ *which is also* J-*orthogonally attached to* $\partial_- W$, *then the surrounding hypersurface can be chosen* J-*orthogonal to* L; *see Figure* 8.1.

THEOREM 8.5. *Let* (W, J) *be a complex manifold with compact* J-*concave boundary* $\partial_- W$. *Let* $\phi : W \to \mathbb{R}$ *be a* J-*convex Morse function which is constant on* $\partial_- W$ *and has no critical points on* $\partial_- W$. *Let* p *be a critical point of* ϕ *such that all trajectories of* $\nabla_\phi \phi$ *reach* $\partial_- W$ *in backward time and denote by* $\Delta \subset W$ *the stable disc of* p. *Then for any neighborhood* $U \supset \partial_- W \cup \Delta$ *there exists a* J-*lc function* $\psi : W \to \mathbb{R}$ *with the following properties (see Figure* 8.2*):*

(a) ψ *is equal to* ϕ *near* $\partial_- W$ *and outside a neighborhood* $N \subset U$ *of* Δ, *and target equivalent to* ϕ *near* Δ;
(b) $\psi|_N$ *has the unique critical point* p *with stable disc* Δ;
(c) *some level set* $\{\psi = c\}$ *surrounds* $\partial_- W \cup \Delta$ *in* U.

Moreover, there exists a homotopy ψ_t, $t \in [0,1]$, *of* J-*convex functions with properties (a) and (b) connecting* $\psi_0 = \phi$ *and* $\psi_1 = \psi$. *If we are given a totally real manifold* $L \supset \Delta$ *of dimension* $\dim L > \dim \Delta$ *which is* J-*orthogonal to all regular level sets of* ϕ, *then the same property can be arranged for the functions* ψ_t.

REMARK 8.6. We will see in Section 10.2 that the J-orthogonality condition in Theorem 8.4 can be replaced by the weaker condition that $\partial\Delta \subset \partial_- W$ is isotropic (i.e., tangent to the field of complex tangencies). The same remark applies to Theorem 8.23 and Corollary 8.26 below.

Theorem 8.5 will be proved below using Corollary 4.4. Theorem 8.4 can be proved following the same scheme using the (easier) Theorem 4.1. We have chosen a different approach, formally deducing Theorem 8.4 from Theorem 8.5. For this, we need to construct a J-convex function ϕ which has the given disc Δ as stable

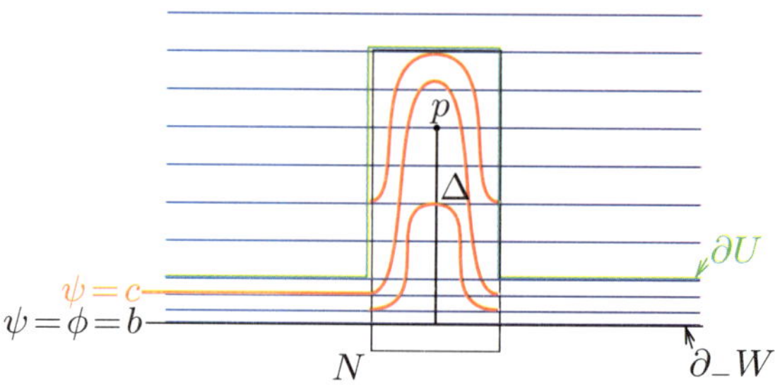

FIGURE 8.2. Surrounding a stable disc of a J-convex function.

disc of some critical point. In particular, the gradient of ϕ needs to be tangent to Δ and its regular level sets J-orthogonal to Δ (see Remark 2.23). The construction is based on the following lemma which is also of independent interest.

LEMMA 8.7. *Let (V, J) be a complex manifold, $L \subset V$ a compact totally real submanifold (possibly with boundary), $\phi : V \to \mathbb{R}$ a smooth function, and X a nowhere vanishing vector field which is tangent to L and gradient-like for ϕ. Let $K \subset L$ be a compact subset such that on $\mathcal{O}p\, K \subset V$ the function ϕ is J-convex, $\nabla_\phi \phi = X$, and $L \cap \mathcal{O}p\, K$ is J-orthogonal to the level sets of ϕ. Then there exists a J-convex function $\widetilde{\phi}$ on $\mathcal{O}p\, L$ which agrees with ϕ on $L \cup \mathcal{O}p\, K$ such that L is J-orthogonal to the level sets of $\widetilde{\phi}$, and $\nabla_{\widetilde{\phi}} \widetilde{\phi} = X$ along L.*

PROOF. After extending L, we may assume without loss of generality that $\dim_{\mathbb{R}} L = \dim_{\mathbb{C}} V =: n$. After pulling back by an approximately holomorphic map along L provided by Proposition 5.55, it suffices to consider the case $L = i\mathbb{R}^n \subset \mathbb{C}^n$ and $J = i$. By hypothesis, L is i-orthogonal to the level sets of ϕ on $\mathcal{O}p\, K$. We can deform ϕ, keeping it fixed on $L \cup \mathcal{O}p\, K$, to make all its level sets i-orthogonal to L, i.e., tangent to $\mathbb{R}^n$. Then the Taylor expansion of ϕ at $(0, y) \in L$ has the form

$$\phi(x, y) = \vartheta(y) + Q_y(x) + o(|x|^2),$$

where $\vartheta(y) = \phi(0, y)$ and $Q_y(x)$ is a quadratic form in x with coefficients depending on y. Let $H_y\vartheta$ denote the Hessian quadratic form of ϑ at a point y. Then the Hessian $H_{(0,y)}\phi$ and the complex Hessian $H^{\mathbb{C}}_{(0,y)}\phi$ are given by

$$H_{(0,y)}\phi(x, y') = Q_y(x) + H_y\vartheta(y'),$$
$$H^{\mathbb{C}}_{(0,y)}\phi(x, y') = Q_y(x) + H_y\vartheta(y') + Q_y(y') + H_y\vartheta(x) = S_y(x) + S_y(y'),$$

where $S_y(y') = Q_y(y') + H_y\vartheta(y')$ is the restriction of $H^{\mathbb{C}}_{(0,y)}\phi$ to $i\mathbb{R}^n$. By assumption, S_y is positive definite on a neighborhood of K in L and X is the gradient of $\phi|_L$ with respect to the metric S_y on this neighborhood. Elsewhere on L, the vector field X is gradient-like for ϕ. Let us extend the metric S_y from the neighborhood of K to a metric $\widetilde{S}_y$ on L in such a way that X is the gradient of $\phi|_L$ with respect to the metric $\widetilde{S}_y$. We view this metric again as a family of quadratic forms $\widetilde{S}_y$ on $\mathbb{R}^n$ and define

$$\widetilde{\phi}(x, y) := \phi(x, y) + \widetilde{S}_y(x) - S_y(x), \ (x, y) \in \mathbb{C}^n.$$

This function coincides with ϕ on $L \cup \mathcal{O}p\, K$. Since it agrees to first order with ϕ along L, its level sets are still i-orthogonal to L. Its complex Hessian at $(0, y) \in L$

is given by

$$H^{\mathbb{C}}_{(0,y)}\widetilde{\phi}(x,y') = S_y(x) + S_y(y') + \widetilde{S}_y(x) + \widetilde{S}_y(y') - S_y(x) - S_y(y') = \widetilde{S}_y(x) + \widetilde{S}_y(y').$$

Hence $\widetilde{\phi}$ is i-convex near L and its gradient with respect to the metric $g_{i,\widetilde{\phi}}$ coincides with X along L. □

REMARK 8.8. A similar argument can be used to prove a parametric version of Lemma 8.7.

COROLLARY 8.9. *Let (V, J) be a complex manifold, $L \subset V$ a compact totally real submanifold (possibly with boundary), and $\phi : V \to \mathbb{R}$ a Morse function whose restriction to L is Morse with the same indices. Let $K \subset L$ be a compact subset such that on $\mathcal{O}p\, K \subset V$ the function ϕ is J-convex, $\nabla_\phi \phi$ is tangent to $L \cap \mathcal{O}p\,(K)$, and $L \cap \mathcal{O}p\, K$ is J-orthogonal to all regular level sets of ϕ. Then there exists a J-convex Morse function $\widetilde{\phi}$ on $\mathcal{O}p\, L$ which agrees with ϕ on $L \cup \mathcal{O}p\, K$ such that L is J-orthogonal to all regular level sets of $\widetilde{\phi}$ and $\nabla_{\widetilde{\phi}}\widetilde{\phi}$ is tangent to L.*

PROOF. Consider a critical point $p \in L \setminus K$ of index $k \le \ell = \dim L$. Pick an embedding $f : \mathbb{R}^\ell \supset \mathcal{O}p\, 0 \hookrightarrow L$ such that

$$f^*\phi(x_1, \dots, x_\ell) = -x_1^2 - \cdots - x_k^2 + \frac{1}{2}(x_{k+1}^2 + \cdots + x_\ell^2).$$

Use Proposition 5.55 to extend f to an embedding $F : \mathbb{C}^n \supset \mathcal{O}p\, 0 \hookrightarrow V$ such that F^*J agrees with i to second order along $\mathbb{R}^\ell$. Note that the gradient vector field $\nabla_{i,\psi}\psi$ of the function

$$\psi(z_1, \dots, z_n) = -x_1^2 - \cdots - x_k^2 + 2(y_1^2 + \cdots + y_k^2) + \frac{1}{2}(|z_{k+1}|^2 + \cdots + |z_n|^2)$$

with respect to the complex structure i is tangent to $\mathbb{R}^k$, and hence the same holds for the gradient $\nabla_{F^*J,\psi}\psi$ with respect to the complex structure F^*J. Moreover, $\mathbb{R}^k$ is i-orthogonal, and hence F^*J-orthogonal, to all regular level sets of ϕ. Thus $F_*\psi$ provides an extension of $\phi|_L$ to a J-convex function on $\mathcal{O}p\, p$ whose gradient (with respect to J) is tangent to L, and such that L is J-orthogonal to its regular level sets. We choose such extensions near all critical points in $L \setminus K$, extend the gradient of $F_*\psi$ to any gradient-like vector field X for ϕ tangent to L, and then apply Lemma 8.7. □

PROOF OF THEOREM 8.4 (ASSUMING THEOREM 8.5). Pick a J-convex function ϕ without critical points in a tubular neighborhood of $\partial_- W$ having $\partial_- W$ as a level set such that Δ is J-orthogonal to its level sets and tangent to $\nabla_\phi\phi$. Using Corollary 8.9 we can extend ϕ to a J-convex Morse function on a tubular neighborhood W' of $\partial_- W \cup \Delta$ with a unique critical point $p \in \Delta$ of index $\dim \Delta$ and such that $\nabla_\phi\phi$ is tangent to Δ. It follows that Δ is the stable manifold of p, so we are in the position to apply Theorem 8.5 to the manifold W'. □

So it remains to prove Theorem 8.5. For this, let us introduce some notation that will be used throughout the rest of this chapter. For $t > 0$ we denote by $D_t := \{R \le t,\ r = 0\} \subset \mathbb{R}^k$ the k-disc of radius t, and we abbreviate $D = D_1$. For $\varepsilon > 0$ we also introduce the k-handle

$$H_\varepsilon := \{R \le 1 + \varepsilon, r \le \varepsilon\} \subset \mathbb{C}^n$$

and its imaginary part $H^y_\varepsilon := H_\varepsilon \cap i\mathbb{R}^n$. Note that for $k = n$, we have $H^y_\varepsilon = D_{1+\varepsilon}$.

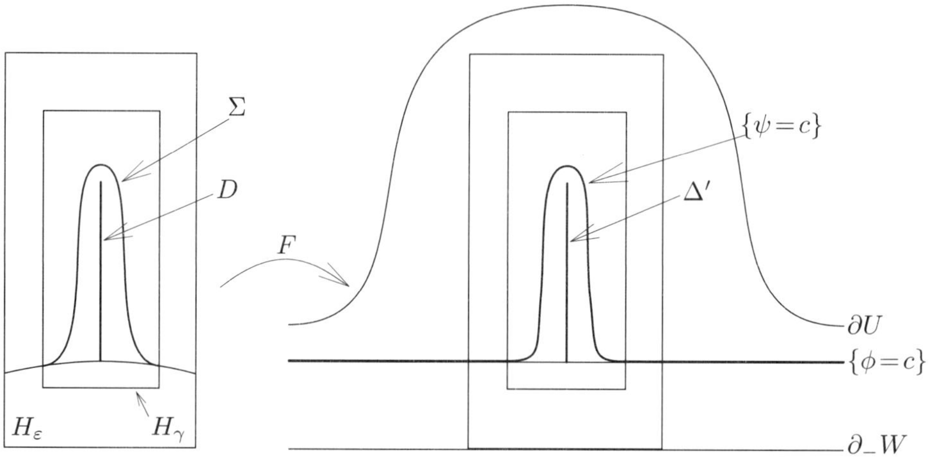

FIGURE 8.3. Implanting a model function near a stable disc.

PROOF OF THEOREM 8.5. The strategy of the proof is to deform the J-convex function ϕ to a standard quadratic function near the disc Δ and then implant the family of J-convex functions from Corollary 4.4; see Figure 8.3. We split the construction into five steps. Step 5 is the main step, while Steps 1–4 are preparatory.

Step 1. Recall from Lemma 2.21 that the stable disc Δ is ω_ϕ-isotropic. If $k < n$ we extend Δ to an n-dimensional ω_ϕ-Lagrangian submanifold $\widehat{\Delta} \subset W$. If we were given a totally real extension L of Δ which is J-orthogonal to the regular level sets of ϕ, and hence ω_ϕ-isotropic, then we can choose $\widehat{\Delta}$ such that it contains a neighborhood of Δ in L. Note that tangency of $\nabla_\phi \phi$ to Δ and the Lagrangian condition on $\widehat{\Delta}$ imply $-d\phi(Jv) = \omega_\phi(\nabla_\phi\phi, v) = 0$ for all $v \in T_x\widehat{\Delta}$, $x \in \Delta$, so $\widehat{\Delta}$ is J-orthogonal to the level sets of ϕ along Δ.

Step 2. Take a slightly bigger regular value $c > b := \phi|_{\partial_- W}$ such that there are no critical values in $[b, c]$ and $\widetilde{W} := \{\phi \leq c\} \subset U$. Set $\Delta' := \Delta \cap \{\phi \geq c\}$. Without loss of generality we may assume that $\phi(p) = 0$ and $c = -1$. Fix any $a > 1$ and consider the i-convex quadratic function $Q(r, R) = ar^2 - R^2$ on the k-handle $H_\varepsilon \subset \mathbb{C}^n$.

The Morse function $\phi|_{\widehat{\Delta}}$ has a unique critical point of index k and value 0 and equals $b < -1$ on $\widehat{\Delta} \cap \partial_- W$. Hence by Corollary 8.1, for a sufficiently small $\varepsilon > 0$ there exists an embedding $f : H_\varepsilon^y \hookrightarrow \widehat{\Delta}$ such that $f(D) = \Delta'$ and $\phi \circ f = Q$.

Step 3. Using Proposition 5.55, we can extend the embedding f to an embedding $F : H_\varepsilon \hookrightarrow U$ such that the 14-jet of the pull-back complex structure $\widetilde{J} = F^*J$ coincides with the standard complex structure i along H_ε^y. (The choice of the order 14 will become clear in Step 5 below.) Hence, we have an estimate

$$\|\widetilde{J} - i\|_{C^2} \leq Cr^{13}. \tag{8.1}$$

This estimate implies that, after possibly shrinking ε, we may assume that both functions $\phi \circ F$ and Q are i-convex as well as $\widetilde{J}$-convex on H_ε.

Now the crucial observation is that the 1-jets of the functions $F^*\phi$ and Q coincide along $D_{1+\varepsilon}$. To see this, consider a tangent vector $v \in T_z H_\varepsilon^y = i\mathbb{R}^n$ at

$z \in D_{1+\varepsilon}$. Since $F^*\phi = Q$ on H_ε^y we have $d_z(F^*\phi)(v) = d_zQ(v)$. On the other hand,

$$d_z(F^*\phi)(iv) = d\phi \circ d_zF(iv) = d\phi\Big(J\, d_zF(v)\Big) = 0,$$

where the second equality holds because dF is complex linear along $D_{1+\varepsilon}$, and the last equality follows from $d_zF(v) \in T\widehat{\Delta}$ and J-orthogonality of $\widehat{\Delta}$ to the level sets of ϕ along Δ.

Step 4. In view of the last observation, we can apply Proposition 3.26 to the functions $\phi \circ F$ and Q with the complex structure $\widetilde{J}$ and the totally real manifold H_ε^y. It provides a $\widetilde{J}$-convex function $\widetilde{\psi}$ on H_ε which coincides with $\phi \circ F$ near ∂H_ε and with Q on H_γ for some $\gamma < \varepsilon$. Moreover, $d\widetilde{\psi} = d(\phi \circ F)$ along H_ε^y, which ensures $\widetilde{J}$-orthogonality of H_ε^y to the levels sets of $\widetilde{\psi}$.

Step 5. Consider now the family of functions $\Psi_t : H_\gamma \to \mathbb{R}$ constructed in Corollary 4.4. For γ sufficiently small, estimate (8.1) implies inequality (4.1) for the complex structure $\widetilde{J}$, so the functions Ψ_t are $\widetilde{J}$-lc. Since $\Psi_t = Q = \phi \circ F$ near ∂H_γ, we can extend $\Psi_t \circ F^{-1}$ by ϕ outside $F(H_\gamma)$ to a family of J-lc functions $\psi_t : W \to \mathbb{R}$.

Since the level sets of the shape functions Ψ_t are i-orthogonal, and hence $\widetilde{J}$-orthogonal, to H_ε^y, the level sets of ψ_t are J-orthogonal to $\widetilde{\Delta}$, and hence to L. By construction, the functions ψ_t agree to first order along Δ with target reparametrizations of ϕ. So we can modify the ψ_t once more using Proposition 3.26 to make them target equivalent to ϕ near on a neighborhood of Δ. Then the family ψ_t and the function $\psi = \psi_1$ have the desired properties in Theorem 8.5. $\square$

REMARK 8.10. The same proofs also yield parametric versions of Theorems 8.4 and 8.5.

8.3. Existence of complex structures

In this section we prove the following result on approximation of almost complex structure by integrable ones, which is a special case of a theorem of Gromov and Landweber.

THEOREM 8.11 (Gromov [**82**], Landweber [**120**]).

(a) Let (W, J) be a $2n$-dimensional almost complex manifold which admits an exhausting Morse function ϕ without critical points of index $> n$. Let L be the skeleton of ϕ (with respect to some gradient-like vector field). Then J can be C^0-approximated by an the almost complex structure which coincides with J outside a neighborhood of L and is integrable on $\mathcal{O}p\, L$. In particular, J is homotopic to an integrable complex structure.

(b) Let (W, J, ϕ) be a $2n$-dimensional almost complex Morse cobordism, where the function ϕ has no critical points of index $> n$. Suppose that J is integrable near $\partial_- W$. Let L be the skeleton of ϕ (with respect to some gradient-like vector field). Then J can be C^0-approximated by an almost complex structure which coincides with J on $\mathcal{O}p\,(\partial_- W)$ and outside a neighborhood of L and is integrable on $\mathcal{O}p\,(L \cup \partial_- W)$. In particular, J is homotopic to an integrable complex structure via a homotopy fixed on $\mathcal{O}p\, \partial_- W$.

The proof is based on the following proposition.

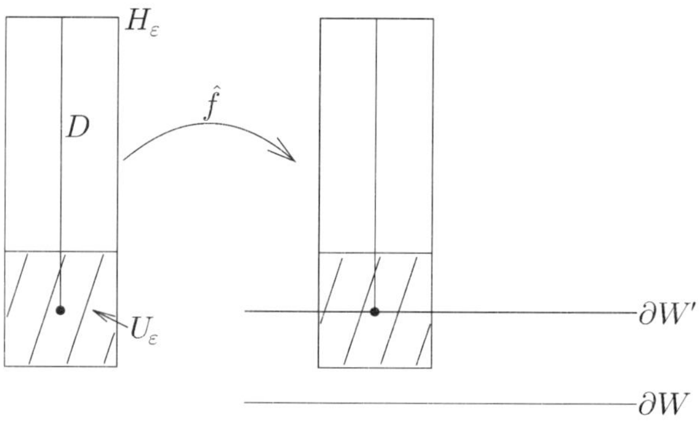

FIGURE 8.4. Holomorphically attaching a standard handle.

PROPOSITION 8.12. *Let (W, J) be an almost complex manifold of dimension $2n$ with compact boundary $\partial_- W$ near which J is integrable. Let $\Delta \subset W$ be an embedded totally real k-disc, $k \leq n$, transversely attached to $\partial_- W$ along $\partial\Delta \subset \partial_- W$. Then there exists a C^0-small perturbation $\widetilde{J}$ of J which is integrable on $\mathcal{O}p\,(\partial_- W \cup \Delta)$, coincides with J on $\mathcal{O}p\,(\partial_- W)$ and outside a neighborhood of Δ, and for which Δ is totally real.*

PROOF. We use the notation introduced in Section 8.2. Namely, D stands for the unit k-disc $\{R \leq 1, r = 0\} \subset \mathbb{C}^n$ and we denote by $D_{1+\varepsilon}$ the disc $\{R \leq 1+\varepsilon, r = 0\} \supset D$. We denote by H_ε the k-handle $H_\varepsilon = \{R \leq 1+\varepsilon, r \leq \varepsilon\} \subset \mathbb{C}^n$, and by H_ε^y its imaginary part $H_\varepsilon \cap i\mathbb{R}^n$. When $k = n$, we have $H_\varepsilon^y = D_{1+\varepsilon}$.

Pick a tubular neighborhood $W' \subset W$ of $\partial_- W$ with real analytic interior boundary $\partial_+ W'$ such that J is integrable near W' and the smaller disc $\Delta' := \Delta \setminus \operatorname{Int} W'$ is transversely attached to W'. Pick a diffeomorphism $f : D_{1+\varepsilon} \to \Delta$ such that $f(D) = \Delta'$. Extend f to a totally real embedding $H_\varepsilon^y \to W$. Using Theorem 5.53, we can find an embedding $\widetilde{f} : H_\varepsilon^y \hookrightarrow W$ which is C^∞-close to f, maps ∂D to $\partial_+ W'$, and is real analytic on $\mathcal{O}p\,(\partial D) \subset H_\varepsilon^y$. In particular, $\widetilde{f}$ still transversely attaches D to W'.

By Lemma 5.40 (after shrinking ε) we can extend $\widetilde{f}$ to an embedding $\widehat{f} : H_\varepsilon \hookrightarrow W$ which is biholomorphic on $U_\varepsilon := \{r \leq \varepsilon, 1-\varepsilon \leq R \leq 1+\varepsilon\}$ and whose differential is complex linear along H_ε^y; see Figure 8.4.

Thus, for ε small enough, the complex structure $\widehat{f}_* i$ is C^0-close to J on $\widehat{f}(H_\varepsilon)$ and coincides with J on $\widehat{f}(U_\varepsilon)$, where i denotes the standard complex structure on $H_\varepsilon \subset \mathbb{C}^n$. So we can define an integrable complex structure on $W' \cup \widehat{f}(H_\varepsilon)$ by J on W' and by $\widehat{f}_* i$ on $\widehat{f}(H_\varepsilon)$. We extend this complex structure from $W' \cup \widehat{f}(H_\varepsilon)$ to an almost complex structure $\widehat{J}$ on the whole manifold W which is C^0-close to J and coincides with J outside a neighborhood of $W' \cup \widehat{f}(H_\varepsilon)$. By construction, $\widehat{J}$ is integrable on a neighborhood of $W' \cup \widetilde{f}(D)$ and agrees with J on W'. Finally, pick a diffeomorphism $g : W \to W$ which is C^∞-close to the identity, maps $\widetilde{f}(D)$ to $f(D) = \Delta'$, and equals the identity near $\partial_- W$ and outside a neighborhood of $W' \cup \widetilde{f}(D)$. Then $\widetilde{J} := g_* \widehat{J}$ is the desired almost complex structure. □

REMARK 8.13. A similar proof yields a parametric version of Proposition 8.12.

Combining Proposition 8.12 and Theorem 7.38, we obtain

COROLLARY 8.14. *Let (V, J) be an almost complex manifold of dimension $2n$ with compact boundary $\partial_- W$ near which J is integrable. Let $\Delta \subset V$ be an embedded k-disc, $k \leq n$, transversely attached to $\partial_- W$ along $\partial\Delta \subset \partial_- W$. Then J can be C^0-approximated by an almost complex structure $\widetilde{J}$ which is integrable on $\mathcal{O}p\,(W \cup \Delta)$, coincides with J on $\mathcal{O}p\,(\partial_- W)$ and outside a neighborhood of Δ, and for which Δ is totally real.*

PROOF. Fix a neighborhood U of Δ. We first use Theorem 7.38 to find a C^0-small isotopy of Δ in U through embedded discs transversely attached to $\partial_- W$ to a totally real disc Δ'. Then we apply Proposition 8.12 to find a C^0-small perturbation J' of J which is integrable on $\mathcal{O}p\,(W' \cup \Delta')$ and coincides with J on $\mathcal{O}p\, W'$ and outside U, for some slightly larger domain $W' \subset W$. Finally, we obtain $\widetilde{J}$ by pushing forward J' under a diffeomorphism which is isotopic to the identity, equals the identity near $\partial_- W$ and outside U, and maps $\Delta' \setminus W'$ onto $\Delta \setminus W'$. □

PROOF OF THEOREM 8.11. (a) After a C^∞-small perturbation of ϕ (keeping the gradient-like vector field and thus the skeleton fixed), we may assume that no two critical points have the same value. We order the critical points $p_0, p_1, \dots$ by increasing value and set $L_j := \bigcup_{i \leq j} W^-_{p_i}$. Then each L_j is compact, $L_j \setminus L_{j-1} = W^-_{p_j}$ is the stable manifold of p_j, and the skeleton is $L = \bigcup_j L_j$. We deform J to make it integrable near the minimum $L_0 = \{p_0\}$. Proceeding inductively, suppose that for some $j \geq 1$ we already made J integrable on $\mathcal{O}p\, L_{j-1}$. Choose a compact domain Ω with smooth boundary such that $L_{j-1} \subset \operatorname{Int}\Omega$ and J is already integrable on Ω, and such that $L_j \setminus \operatorname{Int}\Omega$ is a disc transversely attached to Ω. (Such a domain can be obtained by picking a regular level c_j between $\phi(p_{j-1})$ and $\phi(p_j)$ and moving the set $\{\phi \leq c_j\}$ under the backward flow of the gradient-like vector field for a sufficiently long time.) Hence we can apply Corollary 8.14 to make J integrable on $\mathcal{O}p\, L_j$.

For part (b) we move down the value $\phi|_{\partial_- W}$ (without changing the skeleton) until it is the minimum. Then we set $L_0 := \partial_- W$ and $L_j := L_0 \cup \bigcup_{i=1}^{j} W^-_{p_i}$ for the critical points $p_1, p_2, \dots$ and proceed inductively as in part (a), starting with the hypothesis that J is already integrable near $\partial_- W$. □

8.4. Existence of Stein structures in complex dimension $\neq 2$

In this section we prove two of the main theorems in this book. The first one is equivalent to the Existence Theorem 1.5 from the Introduction and was proved in [**42**].

THEOREM 8.15. *Let V^{2n} be an open smooth manifold of dimension $2n \neq 4$ with an almost complex structure J and an exhausting Morse function ϕ without critical points of index $> n$. Then V admits a Stein structure. More precisely, J is homotopic through almost complex structures to an integrable complex structure $\widetilde{J}$ such that ϕ is $\widetilde{J}$-lc.*

The second theorem concerns the realization of Stein manifolds of given topology within a given ambient complex manifold.

THEOREM 8.16 (Gompf [**72, 73**]). *Let V^{2n} be an open smooth manifold of dimension $2n \neq 4$ with an (integrable) complex structure J and an exhausting*

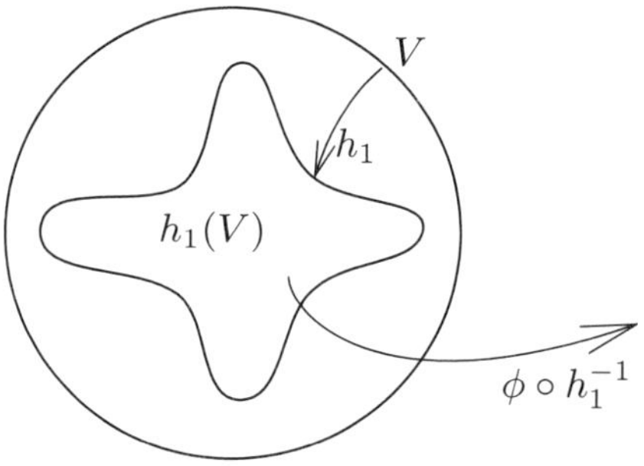

FIGURE 8.5. Constructing a Stein manifold within a given ambient complex manifold.

Morse function ϕ without critical points of index $> n$. Then J is homotopic through (integrable) complex structures to a complex structure $\widetilde{J}$ which is Stein.

More precisely, there exists an isotopy $h_t : V \hookrightarrow V$ with $h_0 = \mathrm{Id}$ such that $\phi \circ h_1^{-1}$ is J-lc on $h_1(V)$. In particular, $h_1(V) \subset V$ is Stein with the induced complex structure J and $J_t = h_t^* J$ is a homotopy of complex structures on V such that $J_0 = J$ and ϕ is J_1-lc; see Figure 8.5.

The statements of both theorems are false for $n = 2$; see Chapter 16 for further discussion. We will prove later in this chapter (Theorem 8.46) a stronger version of Theorem 8.16 which allows us to prescribe the *Stein* homotopy class of the complex manifold $h_1(V)$ within the given almost complex homotopy class.

Theorems 8.15 and 8.16 have the following versions for cobordisms. Here a *Stein cobordism* (W, J, ϕ) is a Morse cobordism (W, ϕ) with a complex structure J for which ϕ is J-convex. We stress the point that in the cobordism versions we allow also the case $n = 2$. However, the price we have to pay here is the assumption that the induced contact structure on $\partial_- W$ is overtwisted.

THEOREM 8.17. *Let (W^{2n}, ϕ) be a Morse cobordism of dimension $2n$ with an almost complex structure J. Suppose that ϕ has no critical points of index $> n$, and near $\partial_- W$ the structure J is integrable and ϕ is J-lc. If $n = 2$ suppose, in addition, that the contact structure induced by J on $\partial_- W$ is overtwisted. Then W admits a Stein cobordism structure. More precisely, J is homotopic through almost complex structures which agree with J near $\partial_- W$ to an integrable complex structure $\widetilde{J}$ such that ϕ is $\widetilde{J}$-lc.*

In the case $n = 2$ the above theorem implies the following version without any assumption about $\partial_- W$.

COROLLARY 8.18. *Let (W, ϕ) be a 4-dimensional Morse cobordism such that the function ϕ has no critical points of index > 2. Let J be an almost complex structure on W. Suppose that $\partial_- W \neq \varnothing$. Then J is homotopic to an integrable complex structure $\widetilde{J}$ for which the function ϕ is $\widetilde{J}$-convex.*

PROOF. There exists an overtwisted contact structure ξ on $\partial_- W$ in the same homotopy class of plane fields as the field of complex tangencies induced by the almost complex structure J; see Section 7.6. According to Remark 5.56, we can deform J to make it integrable near $\partial_- W$ and to induce the contact structure ξ on $\partial_- W$. Then we apply Theorem 8.17. $\square$

Finally, we state the ambient version of Theorem 8.17.

THEOREM 8.19. *Let (W^{2n}, ϕ) be a Morse cobordism of dimension $2n$ with an (integrable) complex structure J and such that ϕ has no critical points of index $> n$ and ϕ is J-lc near $\partial_- W$. If $n = 2$ suppose, in addition, that the contact structure induced by J on $\partial_- W$ is overtwisted. Then J is homotopic through (integrable) complex structures fixed near $\partial_- W$ to a complex structure $\widetilde{J}$ which is Stein.*

More precisely, there exists an isotopy $h_t : W \hookrightarrow W$ fixed near $\partial_- W$ with $h_0 = \mathrm{Id}$ such that $\phi \circ h_1^{-1}$ is J-lc on $h_1(W)$. In particular, $h_1(W) \subset W$ is Stein with the induced complex structure J and $J_t = h_t^* J$ is a homotopy of complex structures on W such that $J_0 = J$ and ϕ is J_1-lc.

We now turn to the proofs of these theorems. First note that Theorem 8.11 reduces Theorems 8.15 and 8.17 to Theorems 8.16 and 8.19, respectively. Hence, we only need to prove the latter two theorems.

The following lemma will serve as the main inductive step for the proofs of Theorems 8.16 and 8.19. Recall that a Morse cobordism (W, ϕ) is called *elementary* if ϕ admits a gradient-like vector field X such that no two critical points of ϕ are connected by an X-trajectory.

LEMMA 8.20. *Let (W, ϕ) be an* elementary *Morse cobordism of dimension $2n$ without critical points of index $> n$. If $n = 2$ we suppose, in addition, that the contact structure induced by J on $\partial_- W$ is overtwisted. Let J be an integrable complex structure on W such that ϕ is J-lc near $\partial_- W$. Then there exists an isotopy $h_t : W \hookrightarrow W$ with $h_0 = \mathrm{Id}$ and $h_t = \mathrm{Id}$ near $\partial_- W$ for all $t \in [0,1]$ such that the function ϕ is $h_1^* J$-lc. If $n = 2$ then one can additionally arrange that the contact structure induced on $\partial_+ W$ by the complex structure $h_1^* J$ is overtwisted.*

PROOF. Let X be a gradient-like vector field for ϕ such that no two critical points of ϕ are connected by an X-trajectory. Then the stable disc Δ of each critical point p meets no other critical points, and therefore meets the J-convex hypersurface $\partial_- W$ transversely along a sphere $\partial\Delta$. By assumption we have $\dim \Delta \leq n$. The hypothesis that $\partial_- W$ is overtwisted in the case $n = 2$ allows us to apply Theorem 7.34 to construct an isotopy of embedded discs Δ_t transversely attached to $\partial_- W$ with $\Delta_0 = \Delta$ and such that Δ_1 is totally real and J-orthogonal to $\partial_- W$. If $n = 2$ one can arrange that the contact structure on $\partial_- W$ is overtwisted in the complement of $\partial\Delta_1$. Since the stable discs of different critical points do not intersect, we can do the above modification independently for all stable discs. To simplify the notation, we will assume that the cobordism contains just one critical point p. The proof in the general case follows exactly the same scheme.

We find a diffeotopy $h_t : W \to W$ with $h_0 = \mathrm{Id}$ and $h_t|_{\mathcal{O}p\,\partial_- W)} = \mathrm{Id}$ and a family of gradient-like vector fields X_t for $\phi \circ \phi_t^{-1}$ starting with $X_0 = X$ for which Δ_t is the stable disc of the critical point $p_t = \phi_t(p)$. After renaming ϕ_1, p_1, Δ_1 back to ϕ, p, Δ we may hence assume that the stable disc Δ of p is totally real and J-orthogonal to $\partial_- W$. After a further modification of Δ near $\partial_- W$ we may assume in addition that Δ is tangent to $\nabla_\phi \phi$ and J-orthogonal to the level sets of ϕ near $\partial_- W$.

According to Corollary 8.9, the function $\phi|_{\mathcal{O}p\,(\partial_- W)}$ can be extended to a J-convex function $\widetilde{\phi}$ on a neighborhood $\widetilde{W}$ of $\partial_- W \cup \Delta$ having p as its unique critical

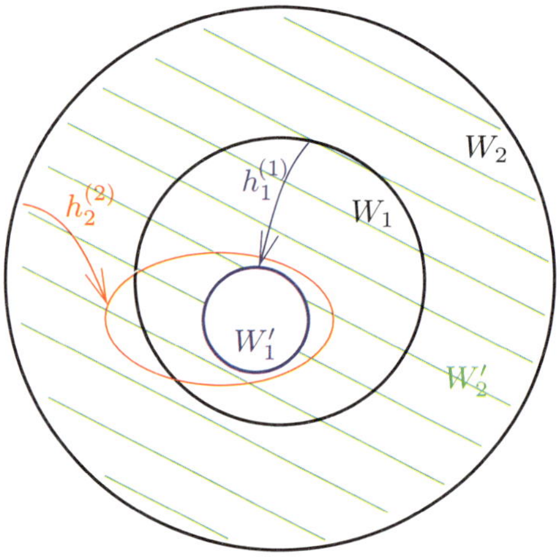

FIGURE 8.6. Inductive construction of the shrinking isotopy h_t.

point with stable disc Δ (with respect to $\nabla_{\widetilde{\phi}}\widetilde{\phi}$). Then we apply Theorem 8.5 to deform $\widetilde{\phi}$ to a J-lc function ϕ' on a neighborhood W' of $\partial_- \cup \Delta$ such that

- $\phi' = \phi$ on $\mathcal{O}p\,(\partial_- W)$,
- ϕ' has p as its unique critical point with stable disc Δ (with respect to $\nabla_{\phi'}\phi'$, cf. Remark 3.33),
- $\phi'|_{\partial_+ W'}$ is constant of value $\phi'(\partial_+ W') = \phi(\partial_+ W)$.

(In fact, this conclusion does not really require Theorem 8.5. It can also be derived from the easier Theorem 8.4 by applying the maximum construction to the function $\widetilde{\phi}$ and a suitable J-convex function having as level sets the surrounding hypersurfaces provided by Theorem 8.4.)

According to Lemma 9.29 below, there exists an isotopy $h_t : W \hookrightarrow W$ with $h_0 = \mathrm{Id}$, $h_t = \mathrm{Id}$ near $\partial_- W$, $h_1(W) = W'$, and $\phi' \circ h_1 = \phi$. Let us also note that in the case $n = 2$ the induced contact structure on $\partial_+ W'$ is overtwisted. Indeed (see Lemma 11.4 below), the gradient flow of the function ϕ' defines a contactomorphism between $\partial_- W \setminus \partial\Delta$ and $\partial_+ W' \setminus \partial\Delta'$, where Δ' is the unstable disc of p. By construction $\partial_- W \setminus \partial\Delta$ is overtwisted, and hence so is $\partial_+ W'$. $\square$

PROOF OF THEOREM 8.16. According to Lemma 9.28 below, there exists an increasing sequence of regular values $c_k \to \infty$ of ϕ with $c_0 < \inf \phi$ such that

$$\left(W_k := \phi^{-1}([c_{k-1}, c_k]), \phi|_{W_k}\right)$$

is an elementary Morse cobordism for all $k = 1, \dots$. We will inductively extend the required isotopy h_t over the elementary cobordisms W_k, $k = 1, \dots$.

First we apply Lemma 8.20 to construct an isotopy $h_t : W_1 \to W_1$, $t \in [0,1]$, such that the function $\phi \circ h_1^{-1}$ is J-lc on $W_1' = h_1(W_1)$. We extend the isotopy h_t (keeping the same notation) to all of V such that $h_t|_{V \setminus W_2}$ is the identity. Set

$$h_t^{(1)} := h_t|_{W_2} : W_2 \to W_2, \quad \phi_1 := \phi \circ (h_1^{(1)})^{-1} : W_2 \to \mathbb{R}, \quad W_2' := W_2 \cup (W_1 \setminus W_1');$$

see Figure 8.6.

Next we apply Lemma 8.20 to the elementary Morse cobordism (W_2', ϕ_1) and find an isotopy $h_t' : W_2' \to W_2'$, fixed on $\partial_- W_2' = \partial_+ W_1'$, such that the function $\phi_2 := \phi_1 \circ (h_1')^{-1}$ is J-lc. We extend h_t' (keeping the same notation) by the identity over W_1'. Then the isotopy

$$h_t^{(2)} := h_t^{(1)} \circ (h_1^{(1)})^{-1} \circ h_t' \circ h_1^{(1)} : W_2 \to W_2,$$

$t \in [0,1]$, has the following properties:

- $h_0^{(2)} = \mathrm{Id}$;
- $h_t^{(2)} = h_t^{(1)}$ on W_1;
- $h_1^{(2)} = h_1' \circ h_1^{(1)}$ and hence $\phi_1 \circ (h_1^{(2)})^{-1} = \phi_2$.

We extend the isotopy $h_t^{(2)}$ (keeping the same notation) to all of V such that $h_t^{(2)}|_{V \setminus W_3}$ is the identity. Note that $\phi_2 = \phi \circ (h_1^{(2)})^{-1}$ is J-lc on $h_1^{(2)}(W_2)$.

Continuing this process, we inductively construct isotopies $h_t^{(k)} : V \to V$, $t \in [0,1]$, $k = 1, \dots$ with the following properties:

- $h_0^{(k)} = \mathrm{Id}$ and $h_t^{(k)} = \mathrm{Id}$ on $V \setminus W_{k+1}$;
- $h_t^{(k+1)} = h_t^{(k)}$ on W_k;
- $\phi \circ (h_1^{(k)})^{-1}$ is J-lc on $h_1^{(k)}(W_k)$.

In view of the second property, the sequence $h_t^{(k)}$ stabilizes and hence converges as $k \to \infty$ to an isotopy $h_t : V \to V$. By the other two properties, $h_0 = \mathrm{Id}$ and the function $\phi \circ h_1^{-1} : V \to \mathbb{R}$ is J-convex. □

PROOF OF THEOREM 8.19. The proof is analogous to the preceding one but simpler, decomposing (W, ϕ) into finitely many elementary Morse cobordisms. □

8.5. J-convex surrounding functions

In this and the following section we put the results of Section 8.2 into a more global context and discuss some applications to holomorphic convexity. These two sections also serve as preparation for the holomorphic approximation results in Section 8.7.

DEFINITION 8.21. Let $A \subset V$ be a compact subset. A weakly J-convex exhausting function $\phi : V \to [0, \infty)$ is called a *J-convex surrounding function for A in V* if

- $\phi|_A = 0$,
- ϕ is (strictly) J-convex in $V \setminus A$,
- ϕ has no critical points in $U \setminus A$ for some neighborhood $U \subset V$ of A.

We say that A *admits a local J-convex surrounding function* if it admits a J-convex surrounding function on a neighborhood $U \subset V$ of A.

The following result, which follows directly from Theorem 5.7, relates these notions to notions of holomorphic convexity in Chapter 5.

PROPOSITION 8.22. *If a compact set $A \subset V$ admits a J-convex surrounding function, then its holomorphic hull in V satisfies $\widehat{A}_V = A$ (so A is polynomially convex in the case $V = \mathbb{C}^n$). In particular, if A admits a local J-convex surrounding function, then it is holomorphically convex. If V is Stein and A admits a local J-convex surrounding function, then A admits a fundamental system of Stein neighborhoods.*

In Section 2.7 we saw that the sets in Example 8.3 admit local J-convex surrounding functions whenever they are compact.

We have the following improvement of Theorems 8.4 and 8.5.

THEOREM 8.23. *(a) Under the assumptions of Theorem 8.4, the set $\partial_-W \cup \Delta$ admits a local J-convex surrounding function.*

(b) Under the assumptions of Theorem 8.5, the set $\partial_-W \cup \Delta$ admits a (global) J-convex surrounding function.

PROOF. (a) Let $A := \partial_-W \cup \Delta$. By Corollary 8.9 there exists a J-convex function ψ on a neighborhood U of A which has a unique critical point $p \in \Delta$ with stable disc Δ. Moreover, we can arrange $\psi \geq 0$ and $\psi|_{\partial_-W} \equiv 0$. Pick any neighborhood $U_1 \Subset U$ of A. By Theorem 8.5 we find a J-lc function $\psi_1 : U \to \mathbb{R}$ which equals ψ near ∂_-W and outside U_1, has unique critical point p, and such that $A \subset \{\psi_1 \leq c_1\} \subset U_1$ for some $c_1 > 0$. Pick a smaller neighborhood $U_2 \Subset \{\psi_1 < c_1\}$ of A. Again by Theorem 8.5, we find a J-lc function $\psi_2 : U \to \mathbb{R}$ which equals ψ_1 near ∂_-W and outside U_2, has unique critical point p, and such that $A \subset \{\psi_2 \leq c_2\} \subset U_2$ for some $0 < c_2 < c_1$.

We continue this process for a sequence of neighborhoods $U_1 \Supset U_2 \Supset \cdots$ of A with $\bigcap_{i\in\mathbb{N}} U_i = A$ and a sequence of values $c_1 > c_2 > \cdots$ converging to 0. In the limit we obtain a smooth J-lc function $U \setminus A \to [0, \infty)$ without critical points which extends to a continuous function $\phi : U \to [0, \infty)$ with $\phi|_A \equiv 0$. According to Proposition 8.29 below, we can make the function ϕ smooth on U and J-convex on $U \setminus A$ by a suitable target reparametrization.

(b) This follows from the same construction, starting with the given J-convex function $\phi : V \to \mathbb{R}$. □

The proof of Theorem 8.23(b) also shows

COROLLARY 8.24. *Let (W, J) be a complex manifold with compact J-concave boundary ∂_-W. Let $\phi : W \to [0, \infty)$ be an exhausting J-convex Morse function with regular level set $\phi^{-1}(0) = \partial_-W$ and finitely many critical points $p_1, \dots, p_k$. Suppose that no critical points are connected by a gradient trajectory, and denote by $\Delta_1, \dots, \Delta_k$ their stable discs. Then $\partial_-W \cup \Delta_1 \cup \cdots \cup \Delta_k$ admits a J-convex surrounding function $\psi : W \to [0, \infty)$ without critical points outside $\partial_-W \cup \Delta_1 \cup \cdots \cup \Delta_k$ which agrees with ϕ outside a neighborhood of $\Delta_1 \cup \cdots \cup \Delta_k$.*

Next, we generalize Theorem 8.23(a) to totally real submanifolds other than discs.

COROLLARY 8.25. *Let (W, J) be a complex manifold with compact J-concave boundary ∂_-W, and $L \subset W$ be a compact totally real submanifold attached J-orthogonally to ∂_-W along ∂L. Then $\partial_-W \cup L$ admits a local J-convex surrounding function.*

PROOF. Pick a Morse function $\phi : L \to \mathbb{R}$ with regular level set $\partial L = \phi^{-1}(0)$ and critical points p_i of values $0 < \phi(p_1) < \cdots < \phi(p_m)$ and Morse indices k_i. According to Corollary 8.9, the function ϕ can be extended to a J-convex Morse function ψ on a neighborhood $U \supset \partial_-W \cup L$ with $\psi|_{\partial_-W} \equiv 0$ and such that L is the union of the stable manifolds of the critical points p_i of ψ for the gradient vector field $\nabla_\psi \psi$.

Pick any neighborhood $U_1 \Subset U$ of $\partial_- W \cup L$. Inductively applying Theorem 8.5 to the pair (U, ψ) and the stable discs of the critical points p_i we construct a J-convex function $\psi_1 : U \to \mathbb{R}$ which equals ψ near $\partial_- W$ and outside U_1, has the same critical points as ψ, and such that one of its level sets surrounds $\partial_- W \cup L$ in U_1. The important fact which allows us to proceed inductively is that in each application of Theorem 8.5 the manifold L remains J-orthogonal to the level sets of the new function. Now the construction of the local J-convex surrounding function can be completed as in the proof of Theorem 8.23. □

The preceding corollary extends to totally real immersions. We say that two totally real submanifolds L_1, L_2 of the same dimension in an almost complex manifold (V, J) *intersect J-orthogonally* at p if $JT_pL_1 = T_pL_2$.

COROLLARY 8.26. *Let (W, J) be a complex manifold with compact J-concave boundary $\partial_- W$. Let $f : L \to W$ be a totally real immersion of a compact manifold L, with finitely many J-orthogonal interior self-intersection points and J-orthogonal to $\partial_- W$ along ∂L. Then $\partial_- W \cup f(L)$ admits a local J-convex surrounding function.*

PROOF. Pick any open neighborhood $U \subset W$ of $\partial_- W \cup f(L)$. Let L_1, L_2 be the two local branches of $f(L)$ at a self-intersection point p. By J-orthogonality of the intersection, there exists a local holomorphic coordinate map $g : B \to U$ from the unit ball B in $\mathbb{C}^n$ mapping 0 to p, $\mathbb{R}^{k_1}$ to T_pL_1, and $i\mathbb{R}^{k_2}$ to T_pL_2, where k_i is the dimension of L_i near p. After precomposing g with the map $z \mapsto \delta z$ for sufficiently small δ, we may assume that the preimages of the T_pL_i are C^2-close to $\mathbb{R}^{k_1}$ resp $i\mathbb{R}^{k_2}$. Since $\mathbb{R}^{k_1}$ and $i\mathbb{R}^{k_2}$ are i-orthogonal to ∂B, we can find a domain $B' \subset \mathbb{C}^n$ whose boundary is C^2-close to ∂B, hence i-convex, and intersects each $g^{-1}(L_i)$ i-orthogonally. Its image $B(p) := g(B')$ is contained in U, and the boundary $\partial B(p)$ is J-convex and intersects L_1 and L_2 J-orthogonally. Construct such balls around all self-intersection points $p_1, \dots, p_m$, disjoint from each other and from $\partial_- W$. Then $W' := W \setminus \big(B(p_1) \cup \dots \cup B(p_m)\big)$ has compact J-concave boundary $\partial_- W'$ to which the totally real submanifold $f(L) \cap W'$ is attached J-orthogonally. Hence Corollary 8.25 provides a local J-convex surrounding function for $\partial_- W \cup f(L) \cup \bigcup_i B(p_i)$ in U. Now we proceed inductively as in the proof of Theorem 8.23, making the neighborhood U and the balls $B(p_i)$ smaller at each step, to find the desired local J-convex surrounding function for $\partial_- W \cup f(L)$. □

In particular, for $\partial_- W = \varnothing$ we obtain

COROLLARY 8.27. *Let (V, J) be a complex manifold and $f : L \to V$ a totally real immersion of a closed manifold L with finitely many J-orthogonal self-intersections. Then $f(L)$ admits a local J-convex surrounding function.*

REMARK 8.28. In the case that L is real analytic near its double points, an alternative proof of the last corollary can be given by combining the surroundings of totally real embeddings in Proposition 2.15 with the surroundings near the double points provided by Lemma 4.12.

It remains to prove the following technical result that was used in the proof of Theorem 8.23.

PROPOSITION 8.29. *Let (V, J) be a complex manifold and $\phi : V \to \mathbb{R}_{\geq 0}$ a nonconstant continuous function such that $K = \phi^{-1}(0)$ is compact and $\phi|_{V \setminus K}$ is* *smooth with compact regular J-convex level sets. Then there exists a smooth function $f : \mathbb{R} \to \mathbb{R}_{\geq 0}$ such that $f \equiv 0$ on $\mathbb{R}_{\leq 0}$, $f' > 0$ on $\mathbb{R}_+$, $\psi = f \circ \phi$ is smooth (with zero set K), and $\psi|_{V \setminus K}$ is (strictly) J-convex.*

The proof is based on two lemmas about real-valued functions.

LEMMA 8.30. *Let $g : [0, 1] \to \mathbb{R}_{\geq 0}$ be a continuous function with $g^{-1}(0) = \{0\}$. Then there exists a smooth function $f : \mathbb{R} \to \mathbb{R}_{\geq 0}$ satisfying $f \equiv 0$ on $\mathbb{R}_{\leq 0}$, $f' > 0$ on $\mathbb{R}_+$, and $f \leq g$ on $[0, 1]$.*

PROOF. For $n \in \mathbb{N}$ set $a_n := \min_{[1/n,1]} g$ and define a piecewise constant function $h : \mathbb{R}_+ \to \mathbb{R}_+$ by

$$h(t) := \min \{a_n, e^{-n}\}, \qquad t \in \left[\frac{1}{n}, \frac{1}{n-1}\right), \quad n \in \mathbb{N}.$$

Smooth h to a smooth function $f : \mathbb{R}_+ \to \mathbb{R}_+$ satisfying $f' > 0$ and $f \leq h \leq g$, and extend f by 0 over $\mathbb{R}_{\leq 0}$. Since $f(t) \leq h(t) \leq e^{-1/t}$ for $t > 0$, the function f is smooth at $t = 0$. □

LEMMA 8.31. *Let V be a manifold and $\phi : V \to \mathbb{R}_{\geq 0}$ a nonconstant continuous function such that $K = \phi^{-1}(0)$ is compact and $\phi|_{V \setminus K}$ is smooth with compact level sets. Then there exists a smooth function $f : \mathbb{R} \to \mathbb{R}_{\geq 0}$ such that $f \equiv 0$ on $\mathbb{R}_{\leq 0}$, $f' > 0$ on $\mathbb{R}_+$, and $f \circ \phi$ is smooth (with zero set K).*

PROOF. After rescaling, we may assume that $[0, 1] \subset \phi(V)$. Pick a Riemannian metric on V and denote by $d(x, y)$ the corresponding distance function. Define a continuous function $d : [0, 1] \to \mathbb{R}_{\geq 0}$ by

$$d(t) := \min \{d(x, y) \mid x \in K, y \in \phi^{-1}(t)\}.$$

This function satisfies $d^{-1}(0) = \{0\}$. Define $g : [0, 1] \to \mathbb{R}_{\geq 0}$ by $g(0) := 0$ and $g(t) := e^{-1/d(t)}$ for $t > 0$. Then g is continuous with $g^{-1}(0) = \{0\}$, so by Lemma 8.30 there exists a smooth function $f : \mathbb{R} \to \mathbb{R}_{\geq 0}$ satisfying $f \equiv 0$ on $\mathbb{R}_{\leq 0}$, $f' > 0$ on $\mathbb{R}_+$, and $f(t) \leq g(t) = e^{-1/d(t)}$ for $t \in [0, 1]$. It remains to show smoothness of the function $\psi := f \circ \phi$ at points of K. So let $x \in K$ and $y \in V$ with $\phi(y) = t \in [0, 1]$. Then $d(t) \leq d(x, y)$ and thus

$$\psi(y) - \psi(x) = f(t) \leq e^{-1/d(t)} \leq e^{-1/d(x,y)},$$

which implies smoothness of ψ at x. □

PROOF OF PROPOSITION 8.29. After applying Lemma 8.31, we may assume that ϕ is smooth. Moreover, after rescaling, we may assume that $[0, 1] \subset \phi(V)$. A short computation as in the proof of Lemma 2.7 shows that $\psi = f \circ \phi$ is J-convex on $\phi^{-1}((0, 1])$ provided that

$$f''(t)\|d\phi(x)\|^2 - f'(t)\|dd^{\mathbb{C}}\phi(x)\| > 0$$

for all $x \in V$ with $\phi(x) = t \in (0, 1]$. Pick smooth functions $a, b : (0, 1] \to \mathbb{R}_+$ with

$$a(t) < \min_{\phi^{-1}(t)} \|d\phi\|^2, \qquad b(t) > \max_{\phi^{-1}(t)} \|dd^{\mathbb{C}}\phi\|.$$

Then $\psi = f \circ \phi$ is J-convex on $\phi^{-1}((0, 1])$ if f solves the differential equation $a(t)f''(t) = b(t)f'(t)$, i.e.,

$$\frac{d}{dt} \log f'(t) = \frac{b(t)}{a(t)} =: c(t), \qquad t \in (0, 1].$$

The solution with $f'(1) = 1$ satisfies

$$f'(t) = e^{-\int_t^1 c(s)ds} := d(t) > 0.$$

By choosing the function a sufficiently small, we can ensure that $c(t) \to \infty$ as $t \to 0$ so fast that $d(t) \le e^{-1/t}$, so d extends to a smooth function on $(-\infty, 1]$ with $d \equiv 0$ on $\mathbb{R}_{\le 0}$. Then $f(t) := \int_0^t d(t)dt$ is the desired function on $(-\infty, 1]$. Finally, we extend f over $[1, \infty)$ by Lemma 2.7. □

8.6. J-convex retracts

Consider a compact set $A \subset V$ which admits a J-convex surrounding function $\phi : V \to [0, \infty)$ without critical points in $V \setminus A$. Then pushing down along gradient trajectories of ϕ yields an isotopy $h_t : V \hookrightarrow V$, $t \in [0, \infty)$, such that

- $h_0 = \mathrm{Id}$ and $h_t|_A = \mathrm{Id}$ for all $t \in [0, \infty)$,
- $\bigcap_{t \in [0,\infty)} h_t(V) = A$,
- the isotopy h_t maps level sets of ϕ to level sets, so in particular the function $\phi \circ h_t^{-1}$ is J-lc for all $t \in [0, \infty)$.

More generally, consider a closed (not necessarily compact) set $A \subset V$ and an exhausting weakly J-convex function $\phi : V \to \mathbb{R}$ without critical points in $V \setminus A$. We say that $A \subset V$ is a *J-convex retract adapted to ϕ* if there exists an isotopy $h_t : V \hookrightarrow V$, $t \in [0, \infty)$, with the following properties:

- $h_0 = \mathrm{Id}$ and $h_t|_A = \mathrm{Id}$ for all $t \in [0, \infty)$;
- $\bigcap_{t \in [0,\infty)} h_t(V) = A$;
- the function $\phi \circ h_t^{-1}$ is J-lc for all $t \in [0, \infty)$.

Note that if A is noncompact, the exhausting function ϕ has to be unbounded (in particular nonconstant) on A. The example to keep in mind is the skeleton of an exhausting J-convex Morse function. The J-convex retract A in the following theorem may not be exactly the skeleton, but it shares many of its properties.

We say that a closed subset $A \subset V$ admits a *totally real stratification by affine strata* if A can be presented as a countable union $A = \bigcup_{i \in \mathbb{N}} A_i$ such that each A_i is the image of a totally real injective immersion $\mathbb{R}^{k_i} \hookrightarrow V$.

THEOREM 8.32. *Let (V, J, ϕ) be a Stein manifold with exhausting J-convex Morse function $\phi : V \to \mathbb{R}$.*

(a) If ϕ has finitely many critical points, then there exists a compact subset $A \subset V$ which admits a finite totally real stratification by affine strata and a J-convex surrounding function $\psi : V \to [0, \infty)$ without critical points in $V \setminus A$.

(b) In general, there exists a closed J-convex retract $A \subset V$ adapted to ϕ which admits a totally real stratification by affine strata.

PROOF. After adding a constant, we may assume that $\min \phi = 0$. Pick an increasing sequence $c_0 < c_1 < \cdots$ of regular values of ϕ such that $c_0 < 0$, $c_j \to \infty$, and each (c_j, c_{j+1}) contains at most one critical value. For simplicity we will assume that each cobordism $W_j := \{c_j \le \phi \le c_{j+1}\}$ contains at most one critical point p_j of ϕ; the general case differs only in the notation. We will also assume that ϕ has a unique local minimum p_0. Set $V_i := \bigcup_{j=0}^{i} W_j$; see Figure 8.7.

(a) Let us first consider the case when ϕ has finitely many critical points, so the domain $\{\phi \ge c_{k+1}\}$ contains no critical points for some k. Choose a neighborhood $U_0 \subset V_0$ of p_0, pick $\widetilde{c}_1 < c_1$ such that $p_0 \in \widetilde{V}_0 := \{\phi \le \widetilde{c}_1\} \subset U_0$, and set

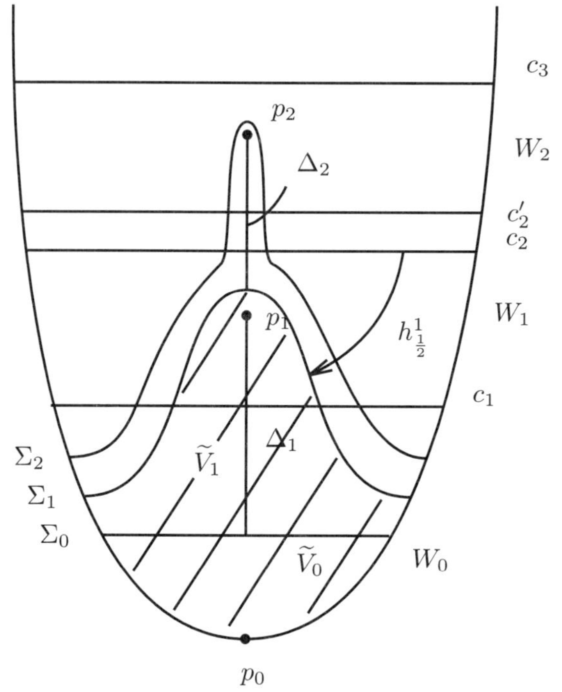

FIGURE 8.7. Constructing a J-convex retract.

$\widetilde{W}_1 := V_1 \setminus \operatorname{Int} \widetilde{V}_0$ and $\Sigma_0 := \{\phi = \widetilde{c}_1\}$. Let Δ_1 be the stable disc of the critical point p_1 in $\widetilde{W}_1$ for the function ϕ. Choose a neighborhood $U_1 \subset V_1$ of $\widetilde{V}_0 \cup \Delta_1$ and apply Theorem 8.5 to $\phi|_{\widetilde{W}_1}$ to construct a J-lc function $\phi_1 : V \to \mathbb{R}$ with the following properties:

- ϕ_1 equals ϕ outside U_1 and on $\widetilde{V}_0$;
- $\phi_1|_{U_1 \setminus \widetilde{V}_0}$ has the unique critical point p_1;
- some level set Σ_1 of ϕ_1 surrounds $\widetilde{V}_0 \cup \Delta_1$ in U_1.

Denote by $\widetilde{V}_1$ the domain bounded by Σ_1 in V_1. Set $\widetilde{W}_2 := V_2 \setminus \operatorname{Int} \widetilde{V}_1$ and consider the stable disc Δ_2 of the critical point p_2 in $\widetilde{W}_2$ for the function ϕ_1. Note that $\Delta_2 \cap W_2$ coincides with the stable disc of p_2 in W_2 for the function ϕ.

We continue this process inductively. Choose a neighborhood $U_2 \subset V_2$ of $\widetilde{V}_1 \cup \Delta_2$ and use Theorem 8.5 to further modify ϕ_1 to a J-lc function ϕ_2 one of whose level sets Σ_2 surrounds $\widetilde{V}_1 \cup \Delta_2$ in U_2, etc. This process terminates at the k-th step to give a J-lc function $\phi^{(1)} := \phi_k$ one of whose level sets Σ_k bounds a domain $V^{(1)} := \widetilde{V}_k$ which contains all the critical points p_j of ϕ_k.

Next we repeat the whole process for the domain $V^{(1)}$ with the function $\phi^{(1)}$, choosing smaller neighborhoods of the stable discs. It is important to observe that the stable disc of the critical point p_j for the function $\phi^{(1}$ in $W_j^{(1)}$ contains the stable disc of the same critical point p_j for the function ϕ_j in W_j. As a result of the second cycle we produce a J-lc function $\phi^{(2)} := \phi_k^{(1)}$, one of whose level sets $\Sigma_k^{(1)}$ bounds a domain $V^{(2)} := \widetilde{V}_k^{(1)}$ which contains all the critical points p_j of $\phi^{(2)} := \phi_k^{(1)}$ (the critical points remain the same for all functions in the construction).

We continue this process inductively, each time surrounding the stable discs by smaller neighborhoods so that their widths tends to 0. As a result, we construct a sequence of J-lc functions $\phi^{(m)} : V \to \mathbb{R}$, $m \in \mathbb{N}$, and domains $V \supset V^{(1)} \supset V^{(2)} \cdots$ such that

(i) for each $n > m$ the function $\phi^{(n)}$ is equal to $\phi^{(m)}$ on $V \setminus V^{(m)}$;
(ii) all the functions $\phi^{(m)}$ have the same critical points p_j as the function ϕ;
(iii) each $V^{(m)}$ can be presented as the union $V^{(m)} = \bigcup_{i=1}^k V_i^{(m)}$ such that $\partial V_i^{(m)}$ is a regular level set of $\phi^{(m)}$ and each cobordism $W_i^{(m)} := V_i^{(m)} \setminus \operatorname{Int} V_{i-1}^{(m)}$ contains the unique critical point p_i;
(iv) the stable disc $\Delta_i^{(m)}$ of the critical point p_i for the function $\phi^{(m)}$ in $\widetilde{W}_i^{(m)}$ is contained in the stable disc $\Delta_i^{(m+1)}$ of p_i for the function $\phi^{(m+1)}$ in $W_i^{(m+1)}$;
(v) the set $A := \bigcap_{m=1}^\infty V^{(m)}$ is compact and admits a finite totally real stratification $A = \bigcup_{j=0}^k A_j$ by the affine strata $A_j := \bigcup_{m=1}^\infty \Delta_j^{(m)}$ for $j > 0$ and $A_0 := \{p_0\}$.

The desired J-convex surrounding function is obtained by a target reparametrization (using Proposition 8.29) of the function which coincides with $\phi^{(m)}$ on $V^{(m-1)} \setminus V^{(m)}$, $m = 1, \dots$, where we set $V^{(0)} := V$. This concludes the proof in the case when the function ϕ has finitely many critical points.

(b) In the case of infinitely many critical points, our first inductive process works for constructing the domain $V^{(1)} := \bigcup_{k=1}^\infty \widetilde{V}_k$ and for constructing the function $\phi^{(1)}$ on $V^{(1)}$, but not on $V \setminus V^{(1)}$. Instead, at this step we will construct an isotopy $h_t^{(1)} : V \hookrightarrow V$ with the following properties:

- $h_1^{(1)}(V) = V^{(1)}$;
- $\phi \circ (h_1^{(1)})^{-1} = \phi^{(1)}$;
- the function $\phi \circ (h_t^{(1)})^{-1}$ is J-lc for all $t \in [0, 1]$.

The isotopy is constructed as follows. Take regular values $c_k' > c_k$ of the original function ϕ such that there are no critical values of ϕ in $[c_k, c_k']$ and set $V_k' := \{\phi \le c_k'\}$. Using the notation of the above construction, consider the cobordism $\widetilde{W}_2 = V_2 \setminus \operatorname{Int} \widetilde{V}_1$ and the J-lc function ϕ_1 which is constant on the boundary components of $\widetilde{W}_2$. There exists a diffeotopy $h_t^1 : V \to V$, $t \in [0, \frac{1}{2}]$, which maps level sets of the function ϕ_1 to level sets and such that $h_0^1 = \mathrm{Id}$, $h_{\frac{1}{2}}^1(V_2) = \widetilde{V}_1$, and $h_t^1 = \mathrm{Id}$ on $\widetilde{V}_0$ and outside V_2' for all $t \in [0, \frac{1}{2}]$.

For $k \in \mathbb{N}$ set $d_k := \sum_{i=1}^k \frac{1}{2^i}$, $k = 0, 1, \dots$. As in the construction of h_t^1 above, we construct for each $k \ge 2$ diffeotopies $h_t^k : V_{k+1}' \to V_{k+1}'$, $t \in [d_{k-1}, d_k]$, with the following properties:

- h_t^k maps level sets of the function ϕ_k to level sets;
- $h_0^k = \mathrm{Id}$ and $h_{d_{k+1}}^k(V_{k+1}) = \widetilde{V}_k$;
- $h_t^k = \mathrm{Id}$ on $\widetilde{V}_{k-1}$ and outside V_{k+1}' for all $t \in [d_k, d_{k+1}]$.

Define the diffeotopy $h_t^{(1)} : V \to V$, $t \in [0,1)$, by the formula $h_t^{(1)} = h_t^1$ for $t \in [0, \frac{1}{2})$ and

$$h_t^{(1)} = h_t^{k+1} \circ h_{d_{k+1}}^k \circ h_{d_k}^{k-1} \circ \cdots \circ h_{\frac{1}{2}}^1, \quad \text{for } t \in [d_k, d_{k+1}],\ k \geq 1.$$

Note that there exists a limit $h_1^{(1)} = \lim_{t \to 1} h_t^{(1)}$ because the diffeotopy $h_t^{(1)}$ stabilizes on compact sets. However, the limit map $h_1^{(1)}$ is not onto but maps V diffeomorphically to $\operatorname{Int} V^{(1)}$. In other words, $h_t^{(1)}$ can be defined for all $t \in [0,1]$ as an *isotopy* rather than a diffeotopy. Now it is clear that we can inductively continue this process for the functions $\phi^{(k)}$, $k = 1, \dots$, and find isotopies $h_t^{(j)} : \operatorname{Int} V^{(j-1)} \hookrightarrow \operatorname{Int} V^{(j-1)}$, $j = 2, \dots$, parametrized by $t \in [j-1, j]$ and such that $h_{j-1}^{(j)} = \mathrm{Id}$ and $h_j^{(j)}(\operatorname{Int} V^{(j-1)}) = \operatorname{Int} V^{(j)}$. Finally, we define the desired isotopy $h_t : V \hookrightarrow V$ inductively by $h_0 = \mathrm{Id}$ and $h_t = h_t^{(j)} \circ h_{j-1}$ for $t \in [j-1, j]$ and $j \in \mathbb{N}$. □

8.7. Approximating continuous maps by holomorphic ones

In this section we apply our previous results to problems of approximating continuous maps by holomorphic ones. For example, we will obtain the following holomorphic approximation theorem, proven by Forstnerič and Slapar in [**63**] (see also [**62, 60**]), as a consequence of results of Hörmander and Wermer and Theorem 8.4.

THEOREM 8.33. *Let (V, J) be a Stein manifold, $W \subset V$ a compact domain with smooth J-convex boundary, and $L \subset V \setminus \operatorname{Int} W$ a totally real submanifold J-orthogonally attached to W. Then any C^k-function $f : (\mathcal{O}p\, W) \cup L \to \mathbb{C}$ which is holomorphic on $\mathcal{O}p\, W$ can be C^k-approximated uniformly on $W \cup L$ by holomorphic functions on $\mathcal{O}p\,(W \cup L)$.*

REMARK 8.34. (1) Let us emphasize that Theorem 8.33 provides only approximations of the derivatives of f in directions tangent to L and not in the normal directions.

(2) Corollary 5.29 allows us to generalize Theorem 8.33 to sections of any holomorphic vector bundle over a Stein manifold V.

COROLLARY 8.35. *Let (V, J) be a Stein manifold with exhausting J-convex Morse function $\phi : V \to \mathbb{R}$. Let c be a regular value of ϕ, $W = \{\phi \leq c\}$, and $(\Delta, \partial\Delta) \subset (V \setminus \operatorname{Int} W, \partial W)$ the stable disc of a critical point of ϕ in $V \setminus \operatorname{Int} W$. Then any continuous function $f : \mathcal{O}p\,(W \cup \Delta) \to \mathbb{C}$ which is holomorphic on $\mathcal{O}p\, W$ can be C^0-approximated uniformly on $W \cup \Delta$ by holomorphic functions on V.*

PROOF. According to Theorem 8.23 and Proposition 8.22, the set $A := W \cup \Delta$ satisfies $\widehat{A}_V = A$. Hence the generalized Oka–Weil Theorem 5.18 allows us to approximate a holomorphic function on $\mathcal{O}p\, A$ by a holomorphic function on V. □

The proof of Theorem 8.33 is based on the following uniform estimate for solutions of the $\overline{\partial}$-equation, which is a combination of results by Hörmander [**102**] and Hörmander and Wermer [**104**].

THEOREM 8.36 (Hörmander and Wermer). *Let $\Omega \subset \mathbb{C}^n$ be a bounded* open *domain with smooth J-convex boundary. Then given a smooth closed $(0,1)$-form g*

on $\mathcal{O}p\,\overline{\Omega}$ there exists a smooth solution $f : \Omega \to \mathbb{C}$ of the equation $\overline{\partial} f = g$ which satisfies for each integer $k \geq 0$ an estimate

$$|D^k f(z)| \leq \frac{C}{\operatorname{dist}(z, \partial\Omega)^{n+k}} ||g||_{C^k(\Omega)}, \tag{8.2}$$

for any $z \in \Omega$. Here the left-hand side in this inequality is the pointwise norm of the k-jet of the function f at z, and the constant C depends only on k and the diameter of the domain Ω.

PROOF. Theorem 2.2.3 in [**102**] (with $\varphi = 0$) provides an L^2-bound

$$||f||_{L^2(\Omega)} \leq C||g||_{L^2(\Omega)},$$

where the constant C depends only on the diameter of Ω. On the other hand, Lemma 4.4 in [**104**] gives a pointwise bound

$$|f(z)| \leq C \left(\frac{1}{\operatorname{dist}(z, \partial\Omega)^n} ||f||_{L^2(\Omega)} + \operatorname{dist}(z, \partial\Omega) ||g||_{C^0(\Omega)} \right).$$

These two bounds together with the obvious bound $||g||_{L^2(\Omega)} \leq C||g||_{C^0(\Omega)}$ imply the estimate (8.2) for $k = 0$. The estimate for higher k follows from this via Lemma 8.37 below: If we denote by Ω_ε the set of points in Ω of distance $\geq \varepsilon$ from $\partial\Omega$, then Lemma 8.37 yields for any $z \in \Omega_{2\varepsilon}$ an estimate

$$|D^k f(z)| \leq C_k \left(\frac{||f||_{C^0(\Omega_\varepsilon)}}{\varepsilon^k} + \frac{||g||_{C^k(\Omega)}}{\varepsilon^{k-1}} \right),$$

which combined with (8.2) for $k = 0$ yields (8.2) for any k. □

It remains to prove the lemma used in the proof of Theorem 8.36. Consider the polydisc $P^n_\varepsilon := \{z \in \mathbb{C}^n \mid |z_1|, \dots, |z_n| \leq \varepsilon\}$ of radius $\varepsilon > 0$ and the torus $T^n_\varepsilon := \{z \in \mathbb{C}^n \mid |z_1| = \dots = |z_n| = \varepsilon\}$.

LEMMA 8.37. *For each integer $k \geq 0$ there exists a constant C_k depending only on k such that every smooth function $f : \mathcal{O}p\, P^n_{2\varepsilon} \to \mathbb{C}$, $0 < \varepsilon < 1$, satisfies the estimate*

$$|D^k f(0)| \leq C_k \left(\frac{||f||_{C^0(T^n_\varepsilon)}}{\varepsilon^k} + \frac{||\overline{\partial} f||_{C^k(P^n_{2\varepsilon})}}{\varepsilon^{k-1}} \right). \tag{8.3}$$

PROOF. We first consider the case $n = 1$. By the inhomogeneous Cauchy integral formula (see e.g. [**103**]), we have

$$f(z) = \frac{1}{2\pi i} \int_{|\zeta|=\varepsilon} \frac{f(\zeta) d\zeta}{\zeta - z} + \frac{1}{2\pi i} \int_{|\zeta| \leq \varepsilon} \frac{g(\zeta) d\zeta \wedge d\bar{\zeta}}{\zeta - z} =: I_1(z) + I_2(z)$$

for $|z| < \varepsilon$, where we have set $g := \frac{\partial f}{\partial \bar{z}}$. Let $D = \frac{\partial^{i+j}}{\partial z^i \partial \bar{z}^j}$ be any partial derivative of order $i + j = k$. Applying D to both sides of the Cauchy integral formula yields $|Df(0)| \leq |DI_1(0)| + |DI_2(0)|$. The standard estimate for the Cauchy integral I_1 gives us

$$|DI_1(0)| \leq \frac{1}{2\pi} \int_{|\zeta|=\varepsilon} \frac{k!|f(\zeta) d\zeta|}{|\zeta - z|^{k+1}} \leq \frac{k!||f||_{C^0(T^1_\varepsilon)}}{\varepsilon^k}. \tag{8.4}$$

To estimate the second term, we pick a smooth cutoff function $\alpha : [0,\infty) \to [0,1]$ which equals 1 on $[0,1]$ and 0 outside $[0,2)$ and consider the integral

$$I_3(z) := \int_{|\zeta|\le 2\varepsilon} \frac{\alpha(\frac{|\zeta|}{\varepsilon})g(\zeta)d\zeta\wedge d\bar\zeta}{\zeta - z} = \int_{\mathbb{C}} \frac{\alpha(\frac{|\zeta|}{\varepsilon})g(\zeta)d\zeta\wedge d\bar\zeta}{\zeta - z}$$
$$= \int_{\mathbb{C}} \frac{\alpha(\frac{|z+u|}{\varepsilon})g(z+u)du\wedge d\bar u}{u}.$$

Differentiating the last integral, we get an estimate

$$|DI_3(0)| \le \frac{C}{\varepsilon^k}\|g\|_{C^k(P^1_{2\varepsilon})}\int_{|u|\le 2\varepsilon}\frac{|du\wedge d\bar u|}{|u|} \le \frac{C}{\varepsilon^{k-1}}\|g\|_{C^k(P^1_{2\varepsilon})}. \tag{8.5}$$

Differentiating the difference

$$I_4(z) := I_3(z) - I_2(z) = \int_{\varepsilon\le|\zeta|\le 2\varepsilon} \frac{\alpha(\frac{|\zeta|}{\varepsilon})g(\zeta)d\zeta\wedge d\bar\zeta}{\zeta - z},$$

we get a similar estimate

$$|DI_4(0)| \le \int_{\varepsilon\le|\zeta|\le 2\varepsilon}\frac{C\|g\|_{C^k(P^1_{2\varepsilon})}|d\zeta\wedge d\bar\zeta|}{|\zeta - z|^{k+1}} \le \frac{C}{\varepsilon^{k-1}}\|g\|_{C^k(P^1_{2\varepsilon})}. \tag{8.6}$$

Combining estimates (8.4), (8.5) and (8.6) yields (8.3) in the case $n = 1$.

The general case follows by induction on n. For $n \ge 2$ consider any partial derivative D in $z_1, \bar z_1, \dots, z_n, \bar z_n$ of order $k \ge 1$. After reordering the coordinates, if necessary, we can write $D = D_1D_2$, where D_1 is a partial derivative of order k_1 in $z_1, \bar z_1$ and D_2 is a partial derivative of order k_2 in the remaining variables such that $k_1 + k_2 = k$. Applying the induction hypothesis for fixed $z_1 \in T^1_\varepsilon$ to the function $f(z_1,\cdot) : P^{n-1}_{2\varepsilon} \to \mathbb{C}$, we obtain the estimate

$$|D_2 f(z_1, 0, \dots, 0)| \le C_{k_2}\left(\frac{\|f\|_{C^0(T^n_\varepsilon)}}{\varepsilon^{k_2}} + \frac{\|g\|_{C^{k_2}(P^n_{2\varepsilon})}}{\varepsilon^{k_2-1}}\right), \tag{8.7}$$

where we have set $g := \bar\partial f = g_1 d\bar z_1 + \cdots + g_n d\bar z_n$. Differentiating $\frac{\partial f}{\partial \bar z_1} = g_1$ we obtain $\frac{\partial}{\partial \bar z_1} D_2 f(z_1, 0, \dots, 0) = D_2 g_1(z_1, 0, \dots, 0)$. Applying the case $n = 1$ with the operator D_1 to this equation we get

$$|D_1D_2 f(0)| \le C_{k_1}\left(\frac{\|D_2 f(\cdot, 0, \dots, 0)\|_{C^0(T^1_\varepsilon)}}{\varepsilon^{k_1}} + \frac{\|D_2 g_1(\cdot, 0, \dots, 0)\|_{C^{k_1}(P^1_{2\varepsilon})}}{\varepsilon^{k_1-1}}\right),$$

which together with (8.7) yields the desired estimate. □

REMARK 8.38. Theorem 8.36 can be extended to domains in an arbitrary Stein manifold V in the following way: Embed V into some $\mathbb{C}^N$ and measure distances and diameters in $\mathbb{C}^N$. *Then for any bounded open domain $\Omega \subset V$ with smooth J-convex boundary there exists a solution f of the equation $\bar\partial f = g$ which satisfies estimate* (8.2) *with a constant C which depends only on the diameter of Ω.* Indeed, according to Corollary 5.27 there exists a neighborhood U of V in $\mathbb{C}^N$ which admits a holomorphic retraction $\pi : U \to V$. Then the $(0,1)$-form $g' := \pi^* g$ on U is $\bar\partial$-closed. Let $\Omega \subset V$ be a bounded open domain with smooth J-convex boundary. By Corollary 5.31, there exists a bounded open domain $\Omega' \subset U$ with smooth J-convex boundary such that $\pi(\Omega') = \Omega$ and $\operatorname{diam}(\Omega') \le 2\operatorname{diam}(\Omega)$. Thus we can apply

Theorem 8.36 to the form g' on Ω', and then restrict the solution of the $\overline{\partial}$-equation back to Ω.

PROOF OF THEOREM 8.33. First, we observe that it is sufficient to consider the case $V = \mathbb{C}^n$. Indeed, we can embed V in some $\mathbb{C}^n$, extend the function f to a neighborhood of V in $\mathbb{C}^n$, and replace W by a neighborhood of W with J-convex boundary in $\mathbb{C}^n$. Furthermore, using induction over a handlebody decomposition of L as in the proof of Corollary 8.25, we need only consider the case when $L = \Delta$ is a disc.

It is sufficient to consider the case when f is a C^∞-function. Using Proposition 5.55 we can find a function $\widetilde{f} : \mathcal{O}p\,(W \cup \Delta) \to \mathbb{C}$ which coincides with f on $(\mathcal{O}p\,W) \cup \Delta$ and such that $\overline{\partial} f$ vanishes at points of Δ together with its $(n+2k)$-jet. Suppose that $\widetilde{f}$ is defined on a neighborhood $U \supset W \cup \Delta$ and holomorphic on an open set $U_1 \supset W$ with $\overline{U}_1 \subset U$. Let us pick a slightly larger compact domain $W_1 \subset U_1$ with smooth J-convex boundary ∂W_1 to which $\Delta_1 := \Delta \setminus W_1$ is J-orthogonally attached along $\partial\Delta_1$.

According to Theorem 8.23, the set $W_1 \cup \Delta_1$ admits a local J-convex surrounding function. In particular, there exists a family Ω_ε, $\varepsilon \in (0, \varepsilon_0]$, of bounded open domains with smooth J-convex boundary such that

- $\Omega_\varepsilon \subset \Omega_{\varepsilon'}$ if $\varepsilon < \varepsilon'$,
- $\bigcap_{\varepsilon>0} \Omega_\varepsilon = W_1 \cup \Delta_1$,
- $\partial\Omega_\varepsilon \setminus U_1 = \{z \in U \mid \mathrm{dist}_\Delta(z) = \varepsilon\} \setminus U_1$.

(The last property can be extracted from the proof of Theorem 8.23, in which the hypersurfaces $\partial\Omega_\varepsilon$ are defined near Δ by shapes as shown in Figure 4.1).

Set $g = \overline{\partial}\widetilde{f}$. This is a closed $(0,1)$-form on U which vanishes on U_1. It also vanishes along Δ_1 together with its $(n+2k)$-jet, so we have

$$||g||_{C^k(\Omega_\varepsilon)} = o(\varepsilon^{n+k}). \tag{8.8}$$

By construction of Ω_ε, for ε sufficiently small we have $\mathrm{dist}(z, \partial\Omega_\varepsilon) \geq \varepsilon$ for all $z \in W \cup \Delta$ (with equality if $z \in \Delta_1$). Hence, according to Theorem 8.36, the equation $\overline{\partial} h_\varepsilon = g$ on Ω_ε has a solution h_ε which satisfies the estimate

$$||h_\varepsilon||_{C^k(W\cup\Delta)} \leq \frac{C}{\varepsilon^{n+k}} ||g||_{C^k(\Omega_\varepsilon)}, \tag{8.9}$$

where the constant C is independent of ε. Then (8.8) and (8.9) imply that

$$||h_\varepsilon||_{C^k(W\cup\Delta)} \underset{\varepsilon\to 0}{\to} 0.$$

Thus the function $f_\varepsilon := \widetilde{f} - h_\varepsilon$ is holomorphic on Ω_ε and satisfies

$$||\widetilde{f} - f_\varepsilon||_{C^k(W\cup\Delta)} = ||h_\varepsilon||_{C^k(W\cup\Delta)} \underset{\varepsilon\to 0}{\to} 0.$$

This concludes the proof of Theorem 8.33. □

In the remainder of this section we discuss applications of Theorem 8.33. The first one is the following approximation result. Recall that a Stein manifold (V, J) is said to be of *finite type* if it admits an exhausting J-convex function with only finitely many critical points.

COROLLARY 8.39. *Let (V, J) be a Stein manifold and $f : V \to \mathbb{C}$ a continuous function.*

(a) Suppose V is of finite type. Then for every $\varepsilon > 0$ there exists a sublevel set $W = \{\phi < c\}$ of an exhausting J-convex function $\phi : V \to \mathbb{R}$ without critical points in $V \setminus W$, and a globally defined holomorphic function $g : V \to \mathbb{C}$ satisfying

$$||g - f||_{C^0(W)} < \varepsilon.$$

(b) For general V, any positive function $\varepsilon : V \to \mathbb{R}$, and any exhausting J-convex function $\phi : V \to \mathbb{R}$ there exist an isotopy $h_t : V \hookrightarrow V$, $t \in [0,1]$, such that $h_0 = \mathrm{Id}$ and $\phi \circ h_t^{-1}$ is J-lc for all $t \in [0,1]$, and a holomorphic function $g : W = h_1(V) \to \mathbb{C}$ such that

$$|g(x) - f(x)| < \varepsilon(x), \quad x \in W.$$

PROOF. Take any exhausting J-convex Morse function $\phi : V \to \mathbb{R}$. Pick an increasing sequence $c_0 < c_1 < \cdots$ of regular values of ϕ such that $c_0 < \min \phi$ and each cobordism $W_i := \{c_{i-1} \le \phi \le c_i\}$ is elementary. We will assume that each W_i contains exactly one critical point p_i of ϕ. The general case differs only in the notation. We will also assume that ϕ has a unique local minimum p_0. Let $V_j = \bigcup_{i=1}^{j} W_i$.

(a) Let us first consider the case when ϕ has finitely many critical points, so that the domain $\{\phi \ge c_{k+1}\}$ contains no critical points for some k. Fix some $\varepsilon > 0$. Choose a holomorphic C^0-approximation g_0 of f near p_0. By choosing the regular value c_1 sufficiently close to the minimum, we can assume that g_0 is defined on $\mathcal{O}p\, W_0 = \mathcal{O}p\, V_0$ and satisfies $||f - g_0||_{C^0(V_0)} < \frac{\varepsilon}{2}$. We extend g_0 elsewhere on V as a continuous function $\frac{\varepsilon}{2}$-close to f. Let Δ_1 denote the stable disc of p_1 in W_1. According to Theorem 8.33 there exists a neighborhood $U_1 \supset W_0 \cup \Delta_1$ and a holomorphic function $g_1 : U_1 \to \mathbb{C}$ such that $||g_1 - g_0||_{\mathbb{C}^0(U_1)} < \frac{\varepsilon}{4}$. We extend the function g_1 (after shrinking U_1 is necessary) to a continuous function on the whole manifold V satisfying the estimate $||g_1 - g_0||_{\mathbb{C}^0(V)} < \frac{\varepsilon}{4}$.

Next, we apply Theorem 8.5 to construct a J-convex function $\phi_1 : V \to \mathbb{R}$ which is target equivalent to ϕ on a smaller neighborhood $U_1' \Subset U_1$, $U_1' \supset W_0 \cup \Delta_1$ and outside U_1, with no critical points in $U_1 \setminus U_1'$ and such that one of its level sets Σ_1 surrounds $W_0 \cup \Delta_1$ in U_1. Denote by V_1' the domain bounded in V by Σ_1 and set $W_2' := W_2 \setminus \mathrm{Int}\, V_1'$. Denote by Δ_2 the stable disc of the critical point p_2 for the function ϕ_1 in W_2'. We again apply Theorem 8.33 to construct a holomorphic approximation g_2 of g_1 on a neighborhood $U_2 \supset V_1' \cup \Delta_2$ such that $||g_2 - g_1||_{C^0(U_2)} < \frac{\varepsilon}{8}$. Applying Theorem 8.5 again, we construct a J-convex function $\phi_2 : V \to \mathbb{R}$ which is target equivalent to ϕ_1 on a smaller neighborhood $U_2' \Subset U_2$, $U_2' \supset V_1' \cup \Delta_2$ and outside U_2, with no critical points in $U_2 \setminus U_2'$ and such that one of its level sets Σ_2 surrounds $V_1' \cup \Delta_2$ in U_2. Now we denote by V_2' the domain bounded in V_2 by Σ_2, set $W_3' := W_3 \setminus \mathrm{Int}\, V_2'$, denote by Δ_3 the stable disc of the critical point p_3 in W_3' for ϕ_2, and continue the process inductively.

If ϕ has finitely many critical points, the process terminates at the k-th step. The holomorphic function g_k defined on $\mathcal{O}p\, V_k'$ satisfies the estimate $||g_k - f||_{C^0(V_k')} < \varepsilon$. The set V_k' is a sublevel set of the exhausting J-convex function function $\phi_k : V \to \mathbb{R}$ which has no critical points in the complement of $W = \mathrm{Int}\, V_k'$. By Theorem 5.7 the holomorphic hull of V_k' in V equals V_k', hence Theorem 5.18 provides a globally defined holomorphic function $g : V \to \mathbb{C}$ satisfying $||g - f||_{C^0(V_k')} < \varepsilon$.

(b) If the number of critical points of ϕ is infinite, one needs to make the following modification to the process. Instead of a constant $\varepsilon > 0$, we fix a positive function $\varepsilon : V \to \mathbb{R}_+$ and then at each step we choose the required holomorphic approximation g_k to satisfy the estimate

$$||g_k - g_{k-1}||_{C^0(V'_k)} < \frac{1}{2^{k+1}} \min_{V'_k} \varepsilon(x).$$

The required holomorphic approximation $g := \lim_{k\to\infty} g_k$ is now defined on the open set $W := \bigcup_{k=1}^{\infty} \operatorname{Int} V'_k \subset V$. The existence of an isotopy $h_t : V \hookrightarrow V$, $t \in [0,1]$, such that $h_1(V) = W$ and the function $\phi \circ h_t^{-1}$ is J-lc for all $t \in [0,1]$ can be shown as in the proof of Theorem 8.32(b). □

Corollary 8.39 can be generalized to maps to arbitrary complex manifolds. We will need for this the following

LEMMA 8.40. *Let (X, J) be any complex manifold. Then for a sufficiently large integer N there exists a C^∞-small isotopy $h_t : X \hookrightarrow X \times \mathbb{C}^N$, $t \in [0,1]$, of the inclusion $h_0 : X = X \times 0 \hookrightarrow X \times \mathbb{C}^N$ such that $h_1(X)$ is totally real. In particular, $h_1(X)$ has arbitrarily small Stein neighborhoods in $X \times \mathbb{C}^N$.*

PROOF. In the space $\operatorname{Hom}_{\mathbb{R}}(\mathbb{C}^n, \mathbb{C}^N)$ the set of linear maps which are complex linear on at least one complex line, i.e., whose graph contains a complex line, is a stratified subset of codimension $N - 2n + 2$. Hence, if $\dim_{\mathbb{C}} X = n$, then Thom's transversality theorem ensures that when $N > 4n - 2$ the graph of a generic map $X \to \mathbb{C}^N$ is totally real. Hence the lemma follows from Proposition 2.15. □

COROLLARY 8.41. *Let (V, J) be a Stein manifold, (Y, I) any complex manifold, and $f : V \to Y$ a continuous map.*

(a) Suppose V is of finite type. Then for every $\varepsilon > 0$ there exists a sublevel set $W = \{\phi < c\}$ of an exhausting J-convex function $\phi : V \to \mathbb{R}$ without critical points in $V \setminus W$, and a holomorphic function $g : W \to Y$ satisfying

$$||g - f||_{C^0(W)} < \varepsilon.$$

(b) For general V, any positive function $\varepsilon : V \to \mathbb{R}$ and any exhausting J-convex function $\phi : V \to \mathbb{R}$ there exists an isotopy $h_t : V \hookrightarrow V$, $t \in [0,1]$, such that $h_0 = \mathrm{Id}$ and $\phi \circ h_t^{-1}$ is J-lc for all $t \in [0,1]$, and a holomorphic function $g : W = h_1(V) \to \mathbb{C}$ such that

$$|g(x) - f(x)| < \varepsilon(x), \quad x \in W.$$

REMARK 8.42. Note that, in contrast to Corollary 8.39(a), the holomorphic map g in Corollary 8.41 is not defined globally but only on the set W.

PROOF. Suppose first that (Y, I) is Stein. Take a proper holomorphic embedding $h : Y \hookrightarrow \mathbb{C}^N$ and apply Corollary 8.39 to the composition $h \circ f : V \to \mathbb{C}^N$. According to Corollary 5.27, there exists a neighborhood U of $\widetilde{Y} := h(Y)$ in $\mathbb{C}^N$ which admits a holomorphic retraction $\pi : U \to \widetilde{Y}$. If the holomorphic approximation $g : W \to \mathbb{C}^N$ of $h \circ f$ provided by Corollary 8.39 is good enough, the image $g(W)$ is contained in U and hence can be projected back to $\widetilde{Y}$, so the desired approximation is $h^{-1} \circ \pi \circ g$.

For the case of a general Y we first use Lemma 8.40 to find a smooth embedding $h : Y \hookrightarrow Y \times \mathbb{C}^N$, which is C^∞-close to the inclusion $Y = Y \times 0 \hookrightarrow Y \times \mathbb{C}^N$ and such that the image $h(Y)$ has a Stein neighborhood U. Then we construct a holomorphic approximation $g : W \to U$ of $h \circ f$ and project it back to Y under the projection $Y \times \mathbb{C}^N \to Y$. □

Corollary 8.41 implies the following result, which is a stronger form of Forstnerič and Slapar's Theorem 1.1 in [**63**].

THEOREM 8.43. *Let (V, J) be a Stein manifold and (Y, I) any other complex manifold. Let $\phi : V \to \mathbb{R}$ be an exhausting J-convex function. Then given a continuous map $f : V \to Y$ there exists an isotopy $h_t : V \hookrightarrow V$, $t \in [0,1]$, with $h_0 = \mathrm{Id}$, and a holomorphic map $g : h_1(V) \to Y$ such that*

- *the function $\phi \circ h_t^{-1}$ is J-lc for all $t \in [0,1]$,*
- *the map g is homotopic to $f|_{h_1(V)}$.*

In particular, there exists a homotopy of Stein structures J_t on V with $J_0 = J$ for which ϕ is J_t-lc, and a homotopy $f_t : V \to Y$ with $f_0 = f$ such that f_1 is J_1-holomorphic.

PROOF. Let $h_t : V \hookrightarrow V$ be the isotopy and $g : h_1(V) \to Y$ the holomorphic approximation provided by Corollary 8.41. Then the function ϕ is J_t-lc for each of the complex structures $J_t := h_t^* J$ on V. If g is sufficiently close to $f|_{h_1(V)}$, they are connected by a homotopy $g_t : h_1(V) \to Y$ with $g_0 = f|_{h_1(V)}$ and $g_1 = g$. So we can homotope $f : V \to Y$ via $f \circ h_t$ to $g \circ h_1$ and then via $g_t \circ h_1$ to the map $f_1 := g \circ h_1 : V \to Y$ which is J_1-holomorphic. □

REMARK 8.44. Our proof of Theorem 8.43 is essentially the same as the one given by Forstnerič and Slapar in [**63**], with one major difference: we use as the main technical tool Theorem 8.5, while they use a result analogous to Theorem 8.4. A result which is essentially equivalent to Theorem 8.4 was proven in [**42**]. Theorem 8.5 was first announced in [**47**], but its proof has never been published before. Using this stronger technical tool, we can remove the unnecessary constraint $n \neq 2$ in Theorem 1.1(i) in [**63**] and also upgrade a homotopy of complex structures to a Stein homotopy.

By using Theorem 8.5 one can similarly strengthen other results from [**63**]. In particular, in combination with the h-principles for totally real immersions (Corollary 7.28), submersions (Corollary 7.33), and embeddings (Corollary 7.30), one can prove the following result similar to Theorem 1.4 in [**63**].

THEOREM 8.45. *Let (V, J, ϕ) and (Y, I) be as in Theorem 8.43.*

(a) Let $f : V \to Y$ be a continuous map covered by a complex homomorphism $F : TV \to TY$ of maximal rank. Then the holomorphic map $g : (h_1(V), J) \to (Y, I)$ constructed in Theorem 8.43 can be chosen to be a holomorphic immersion or submersion with dg homotopic to $F|_{h_1(V)}$ in the class of complex homomorphisms of maximal rank.

(b) If f is an embedding and $F : TV \to TY$ is a complex injective homomorphism covering f which is homotopic to df through real injective homomorphisms, then g can be made a holomorphic embedding isotopic to the embedding $f|_{h_1(V)}$.

PROOF. The proof follows the lines of the proof of Corollary 8.39 with the following modification: In each induction step, before applying the Approximation Theorem 8.33, we use one of the appropriate h-principles for totally real immersions (Corollary 7.28), submersions (Corollary 7.33), or embeddings (Corollary 7.30) to find a C^0-small homotopy (resp. isotopy) fixed near $\partial\Delta_k$ of the map $g_{k-1}|_{\Delta_k}$ to a totally real immersion/submersion (resp. embedding). Then we use the C^1-approximation provided by Theorem 8.33 to approximate g_{k-1} by a holomorphic map $g_k : \mathcal{O}p\,(V'_{k-1} \cup \Delta_k) \to Y$. Since $g_k|_{\Delta_k}$ is a totally real immersion/submersion/embedding (by C^1-closeness), the map g_k is a holomorphic immersion/submersion/embedding. $\square$

We will introduce later on in Section 11.6 the notion of a *Stein homotopy.* In particular, a family of Stein structures (V, J_t) which share the same exhausting J-lc function ϕ is a Stein homotopy. Then Theorem 8.45 implies the following strengthened version of Theorem 8.16.

THEOREM 8.46. *Let (V, J) be a complex manifold. Then given any Stein structure $\widetilde{J}$ homotopic to J as an almost complex structure, there exists an isotopy $h_t : V \hookrightarrow V$, $t \in [0, 1]$, with $h_0 = \mathrm{Id}$, such that $h_1^* J$ is a Stein structure on V in the same Stein homotopy class as $\widetilde{J}$.*

Indeed, to prove this one simply applies Theorem 8.45 to the identity map $(V, \widetilde{J}) \to (V, J)$.

8.8. Variations on a theme of E. Kallin

In this section we prove a lemma closely related to Kallin's lemma in [**109**] (see also [**37**]). It will only be used in Section 16.2 below.

For $C > 0$, consider the quadratic function $Q_C : \mathbb{C}^n \to \mathbb{R}$ given by the formula

$$Q_C(x_1, \dots, x_n, y_1, \dots, y_n) = \sum_1^n x_i^2 + \sum_1^{n-1} y_j^2 - C y_n^2.$$

For $C < 1$ the function Q_C is i-convex, while for $C > 1$ it is not. However, each level set $\{Q_C = -a\}$, $a > 0$, is the union of the two convex, and hence i-convex, hypersurfaces

$$y_n = \pm \sqrt{\frac{a + x_1^2 + \cdots + y_{n-1}^2}{C}}$$

for *any* $C > 0$ (being cooriented by ∇Q_C). Denote by X the vector field

$$X = \sum_1^n x_i \frac{\partial}{\partial x_i} + \sum_1^{n-1} y_j \frac{\partial}{\partial y_j} - y_n \frac{\partial}{\partial y_n}.$$

LEMMA 8.47. *For any $a, C > 0$ there exists an i-convex Morse function $\psi : \mathbb{C}^n \to \mathbb{R}$ with the following properties:*

(i) *ψ has a unique critical point at the origin, of index 1, with stable manifold (with respect to $\nabla_\psi \psi$) $\{x_1 = \cdots = x_n = y_1 = \cdots = y_{n-1} = 0\}$ and unstable manifold $\{y_n = 0\}$;*
(ii) *ψ has the hypersurface $\{Q_C = -a\}$ as one of its level sets and is convex in the region $\{Q_C \le -a\}$;*
(iii) *$d\psi(X) > 0$ outside the origin;*
(iv) *$\psi(z_1, \dots, z_{n-1}, z_n) = \psi(z_1, \dots, z_{n-1}, \bar{z}_n)$.*

PROOF. For $n = 1$, let us take any smooth function $\phi : \mathbb{C} \to \mathbb{R}$ with properties (i)–(iv) which is equal to $x^2 - \frac{1}{2} y^2$ near the origin. Then after an appropriate

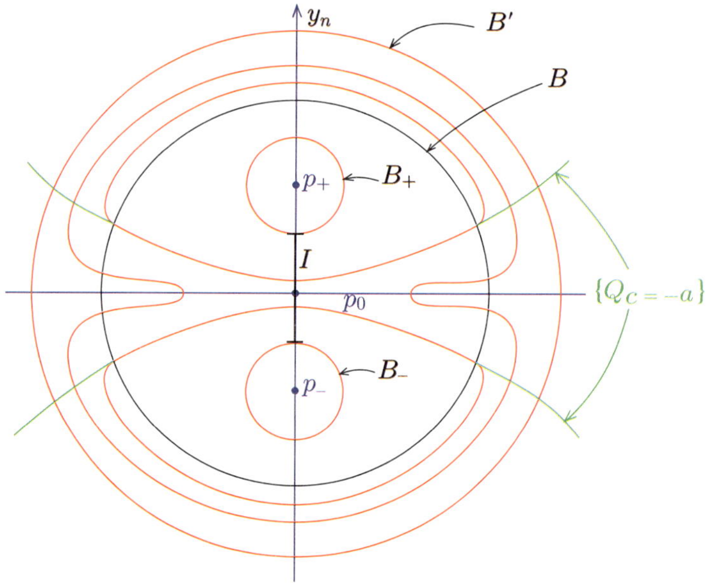

FIGURE 8.8. The i-convex function ϕ.

target reparametrization, the function ϕ becomes i-convex (see Lemma 2.7). In the general case we define the required function by the formula

$$\psi(z) := \sum_1^{n-1} |z_j|^2 + \phi(z_n). \qquad \square$$

REMARK 8.48. The hypersurface $\{Q_C = -a\}$ in the above lemma can be replaced by any hypersurface of the form $\{|y_n| = H(x_1, \dots, x_n, y_1, \dots, y_{n-1})\}$ for a convex function H.

COROLLARY 8.49. *Consider two disjoint balls $B_\pm \subset \mathbb{C}^n$. Let $I \subset \mathbb{C}^n$ be the unique straight line segment connecting ∂B_+ and ∂B_- and perpendicular to both boundaries. Then there exists an exhausting i-convex function $\phi : \mathbb{C}^n \to \mathbb{R}$ with the following properties (see Figure 8.8):*

- *$\phi(z) = f(|z|^2)$ outside a large ball containing $B_+ \cup B_-$;*
- *$\phi|_{B_+ \cup B_-}$ is convex and has $\partial B_+ \cup \partial B_-$ as one of its level sets;*
- *ϕ has exactly three critical points: two local minima $p_\pm \in B_\pm$, and an index 1 critical point $p_0 \in I$ with stable disc I (with respect to $\nabla_\phi \phi$).*

PROOF. After a unitary rotation we may assume that $I = \{x_1 = \dots = x_n = y_1 = \dots = y_{n-1} = 0, |y_n| \le b\}$ for some $b > 0$. Pick a ball B around the origin containing $B_+ \cup B_-$ in its interior, and a larger ball $B' \supset B$. Pick $C > 1$ sufficiently large and $a > 0$ sufficiently small so that $B_+ \cup B_- \subset \{Q_C < -a\}$, where Q_C is the quadratic function defined above. Moreover, we pick a so small that the vector field X above satisfies $X \cdot |z|^2 > 0$ on the region $\{Q_C \ge -a\} \cap (B' \setminus B)$. Let $\psi : \mathbb{C}^n \to \mathbb{R}$ be the i-convex function provided by Lemma 8.47.

Take a convex increasing function $f : \mathbb{R} \to \mathbb{R}$ such that $F(z) := f(|z|^2) < \psi$ on B and $F(z) > \psi$ outside B', and define $G := \text{smooth}\max(\psi, F)$. As both functions

ψ and F are invariant with respect to the involution $\sigma : \mathbb{C}^n \to \mathbb{C}^n$ defined by $\sigma(z_1, \dots, z_n) = (z_1, \dots, z_{n-1}, \bar{z}_n)$, the function G can also be taken to be invariant with respect to σ. Since $X \cdot \psi > 0$ and $X \cdot F > 0$ on $\{Q_C \geq -a\} \cap (B' \setminus B)$, the function $G|_{\{Q_C \geq -a\}}$ has a unique critical point at the origin, of index 1 and with stable manifold contained in I. On the region $\{Q_C \leq -a\}$ both functions ψ and F are convex, so according to Remark 3.24 the function $G|_{\{Q_C \leq -a\}}$ is also convex and has exactly two critical points, the local minima in the two components of the domain $\{Q_C \leq -a\} \cap B'$. In particular, one of the level sets of G bounds two convex components $\Omega_\pm$ containing the balls $B_\pm$. Hence there exists a convex function $H : \Omega_- \cup \Omega_+ \to \mathbb{R}$ which has $\partial B_+ \cup \partial B_-$ and $\partial\Omega_- \cup \partial\Omega_+$ as level sets. By reparametrizing G, if necessary, we can arrange that $G > H$ on $\partial\Omega_- \cup \partial\Omega_+$ and $G < H$ on $B_- \cup B_+$. Then the function $\phi := \text{smooth}\max(G, H)$ has all the required properties provided that I is contained in the stable manifold of the origin, which we can ensure, e.g., by making the whole construction invariant under the symmetry $(x_1, \dots, y_n) \mapsto (-x_1, \dots, -y_{n-1}, y_n)$. □

Corollary 8.49 implies the following special case of a lemma of E. Kallin [**109**]:

COROLLARY 8.50. *The union $B_+ \cup B_-$ of two disjoint balls in $\mathbb{C}^n$, and hence the union of any two disjoint compact convex sets with smooth boundary in $\mathbb{C}^n$, is polynomially convex. Moreover, the union $B_+ \cup B_- \cup I$ (with I as in Corollary 8.49) is polynomially convex.*

PROOF. By Corollary 8.49, $B_+ \cup B_-$ is a sublevel set of an exhausting i-convex function $\phi : \mathbb{C}^n \to \mathbb{R}$ and hence polynomially convex by Theorem 5.7. Given two disjoint compact convex sets with smooth boundary, we pick two disjoint balls $B_\pm \supset K_\pm$ and modify the exhausting i-convex function $\phi : \mathbb{C}^n \to \mathbb{R}$ provided by Corollary 8.49 as a convex function inside $B_+ \cup B_-$ such that $\partial K_+ \cup \partial K_-$ becomes a level set. For the last statement we use Theorem 8.23 to deform the function $\phi : \mathbb{C}^n \to \mathbb{R}$ from Corollary 8.49 to an i-convex surrounding function and apply Proposition 8.22. □

Part 3

Morse–Smale Theory for J-Convex Functions

CHAPTER 9

Recollections from Morse Theory

In this chapter we recollect some results about smooth functions and vector fields that will be needed in the second half of the book. We discuss local normal forms of functions near critical points and of vector fields near zeroes. A crucial notion is that of a Lyapunov pair consisting of a function and a gradient-like vector field. We discuss deformations of Lyapunov pairs near critical points and prove a smooth version of the J-convex surroundings in Chapter 8.2.

In Sections 9.6 and 9.7 we introduce the notions about cobordisms that will play a central role in the discussion of Stein and Weinstein cobordisms in Part 4 of the book: Smale cobordisms and homotopies, elementary cobordisms, profiles, and holonomy. In Section 9.8 we sketch a proof of Smale's h-cobordism theorem, based on four geometric lemmas for which we will later prove Stein and Weinstein analogues in Chapters 10 and 12. Finally, we discuss the two-index theorem of Hatcher and Wagoner and pseudo-isotopies to which we will return in Chapter 14.

Throughout this chapter, V denotes a smooth manifold and W a cobordism, both of dimension m.

9.1. Critical points of functions

Let $\phi : V \to \mathbb{R}$ be a smooth function, and let $p \in V$ be a critical point of ϕ, i.e., $d_p\phi = 0$. The Hessian $\mathrm{Hess}_p\phi$ defines a symmetric bilinear form on T_pV. The *nullity* of ϕ at p is the dimension of $\ker \mathrm{Hess}_p\phi := \{v \in T_pV \mid \mathrm{Hess}_p\phi(v,w) = 0$ for all $w \in T_pV\}$. The *(Morse) index* at p is the maximal dimension of a subspace on which the quadratic form $v \mapsto \mathrm{Hess}_p(v,v)$ is negative definite. The critical point p is called *nondegenerate* if its nullity is zero. It is well-known (see e.g. [**140**]) that a generic function is *Morse*, i.e., has only nondegenerate critical points.

LEMMA 9.1 (Morse Lemma [**139**]). *Near a nondegenerate critical point p of ϕ of index k there exist smooth coordinates $u \in \mathbb{R}^m$ mapping p to 0 in which ϕ has the form*

$$\phi(u) = \phi(p) - u_1^2 - \cdots - u_k^2 + u_{k+1}^2 \cdots + u_m^2. \tag{9.1}$$

More precisely, this means that for a function ϕ on a neighborhood of $0 \in \mathbb{R}^m$ there exists a diffeomorphism g between neighborhoods of 0 such that $g^*\phi$ has the form (9.1).

REMARK 9.2. (1) If the function ϕ on a neighborhood of $0 \in \mathbb{R}^n$ already satisfies $\phi(x_1,\dots,x_k,0,\dots,0) = \phi(p) - x_1^2 - \cdots - x_k^2$, then we can choose the diffeomorphism g to satisfy $g(x_1,\dots,x_k,0,\dots,0) = (x_1,\dots,x_k,0,\dots,0)$. To see this, apply the proof of the Morse lemma in [**139**] to find new coordinates $u_1,\dots,u_m$ near 0 in which $\phi(u) = \phi(p) - u_1^2 - \cdots - u_k^2 + u_{k+1}^2 \cdots + u_m^2$. Inspection of the proof shows that $u_i = x_i$ on $\mathbb{R}^k \times \{0\}$.

(2) The Morse lemma also holds with parameters as follows: For a compact manifold (possibly with boundary) K, let $\phi_z : V \to \mathbb{R}$, $z \in K$ be a smooth family of functions with a nondegenerate critical of index k at p for all z. Then there exists a smooth family of diffeomorphisms $g_z : (U, 0) \to (V_z, p)$ from a neighborhood $U \subset \mathbb{R}^m$ of 0 onto neighborhoods $V_z \subset V$ of p such that for all $z \in K$,

$$\phi_z \circ g_z(u) = \phi_z(p) - u_1^2 - \cdots - u_k^2 + u_{k+1}^2 \cdots + u_m^2.$$

The next lemma shows that near a degenerate critical point one can always split off the nondegenerate directions.

LEMMA 9.3. *Near a critical point p of ϕ index k and nullity ℓ there exist smooth coordinates $(x_1, \dots, x_{m-k-\ell}, y_1, \dots, y_k, z_1, \dots, z_\ell) \in \mathbb{R}^m$ in which ϕ has the form*

$$\phi(x, y, z) = x_1^2 + \cdots + x_{m-k-\ell}^2 - y_1^2 \cdots - y_k^2 + \psi(z)$$

with a smooth function $\psi(z)$.

PROOF. Set $B := \mathrm{Hess}_p \phi$ and $n := m - \ell$. Identify a neighborhood of p in V with a neighborhood of 0 in $\mathbb{R}^m = \mathbb{R}^n \oplus \mathbb{R}^\ell$ such that $\mathbb{R}^\ell = \ker B$. Define a function F on a neighborhood of 0 in $\mathbb{R}^m$ by

$$F(w, z) := \frac{\partial \phi}{\partial w}(w, z).$$

Since $\frac{\partial F}{\partial w}(0, 0) = \frac{\partial^2 \phi}{\partial w^2}(0, 0)$ is invertible, the zero set $F^{-1}(0)$ is a graph $w = w(z)$ over $\mathbb{R}^\ell$. After applying a diffeomorphism near $0 \in \mathbb{R}^m$, we may assume $F^{-1}(0) = \mathbb{R}^\ell$. Consider the smooth family of functions $\phi_z = \phi(\cdot, z) : \mathbb{R}^n \to \mathbb{R}$, $z \in \mathbb{R}^\ell$ near 0. By construction, each ϕ_z has a nondegenerate critical point of index k at $w = 0$. Now Lemma 9.3 follows from the parametrized Morse Lemma in Remark 9.2. □

Let us now describe the critical points that occur in a generic 1-parameter family of functions $\phi_t : V \to \mathbb{R}$, $t \in \mathbb{R}$. A critical point p of a function $\phi : V \to \mathbb{R}$ is called *embryonic* if $\ker \mathrm{Hess}_p \phi$ is 1-dimensional and the third derivative of f in the direction of $\ker \mathrm{Hess}_p \phi$ is nonzero. We say that a 1-parameter family of functions $\phi_t : V \to \mathbb{R}$, $t \in \mathbb{R}$, has a *birth-death type* critical point $p \in V$ at $t = 0$ if p is an embryonic critical point of ϕ_0 and $(0, p)$ is a nondegenerate critical point of the function $(t, x) \mapsto \phi_t(x)$.

With a family of functions $\phi_t : V \to \mathbb{R}$, $t \in \mathbb{R}$, one can associate its *profile* (or Cerf diagram). This is the subset $C(\{\phi_t\}) \subset \mathbb{R} \times \mathbb{R}$ such that $C(\{\phi_t\}) \cap (t \times \mathbb{R})$ is the set of critical values of the function ϕ_t. If ϕ_t is a family of Morse functions, then $C(\{\phi_t\})$ is a collection of graphs of smooth functions. Part (b) of the following theorem shows that birth-death points correspond to cusps of the profile.

THEOREM 9.4 (Whitney). *(a) Near an embryonic critical point p of ϕ of index $k-1$ there exist coordinates $(x, y, z) \in \mathbb{R}^{m-k} \oplus \mathbb{R}^{k-1} \oplus \mathbb{R}$ in which ϕ has the form*

$$\phi(x, y, z) = \phi(p) + |x|^2 - |y|^2 + z^3.$$

(b) Suppose that p is a birth-death type critical point of index $k-1$ for the family of functions $\phi_t : V \to \mathbb{R}$, $t \in \mathbb{R}$, at $t = 0$. Then there exist families of local diffeomorphisms $f_t : \mathcal{O}p\, p \to \mathcal{O}p\, 0 \subset \mathbb{R}^m$ and $g_t : \mathcal{O}p\, \phi_0(p) \to \mathcal{O}p\, 0 \subset \mathbb{R}$, $t \in \mathcal{O}p\, 0$, such that the family of functions $\psi_t = g_t \circ \phi_t \circ f_t^{-1}$ has the form

$$\psi_t(x, y, z) = |x|^2 - |y|^2 + z^3 \pm tz \tag{9.2}$$

for $(x, y, z) \in \mathbb{R}^{m-k} \oplus \mathbb{R}^{k-1} \oplus \mathbb{R}$.

(c) Let $\phi_t, \widetilde{\phi}_t : V \to \mathbb{R}$ be two families of functions with birth-death type critical points $p, \widetilde{p}$ at $t = 0$ of the same index and with the same profile. Then there exist a family of local diffeomorphisms $h_t : \mathcal{O}p\, p \to \mathcal{O}p\, \widetilde{p}$, $t \in \mathcal{O}p\, 0$, such that $\widetilde{\phi}_t \circ h_t = \phi_t$.

(d) A generic 1-parameter family of functions $\phi_t : V \to \mathbb{R}$ has only nondegenerate and birth-death type critical points.

In particular, part (a) shows that embryonic critical points are isolated. We say that a birth-death type critical point p is of *birth type* if the sign in front of t in formula (9.2) is minus, and of *death type* otherwise. Note that near a birth type critical point a pair of nondegenerate critical points of indices k and $k - 1$ appears at $t = 0$, and near a death type critical point such a pair disappears.

PROOF. Part (d) follows from a standard transversality argument. Using Lemma 9.3, we can reduce parts (a)–(c) to the case $m = 1$ of functions $\mathbb{R} \to \mathbb{R}$. For $m = 1$ this result is essentially proved in [**192**]. In the present formulation it can be proved as follows.

For (a), it is easy to see that every function $\mathbb{R} \to \mathbb{R}$ with an embryonic point equals z^3 in a suitable coordinate z; see e.g. [**129**, Proposition III 1.2].

For (b), consider a family of functions $\phi_t : \mathbb{R} \to \mathbb{R}$ with a birth-death type critical point p at $t = 0$. It follows, e.g., from [**129**, Theorem IV 6.1] that there exists a family of local diffeomorphisms $f_t : \mathcal{O}p\, p \to \mathcal{O}p\, 0 \subset \mathbb{R}$ such that

$$\phi_t \circ f_t^{-1}(z) = z^3 \pm a(t)z + b(t)$$

with smooth functions $a, b : \mathbb{R} \to \mathbb{R}$ such that $a(0) = 0$ and $sa'(0) > 0$. With $g_t(y) := y - b(t)$, we then have $\widetilde{\phi}(z) := g_t \circ \phi_t \circ f_t^{-1}(z) = z^3 \pm a(t)z$. Finally, the diffeomorphisms $\widetilde{f}_t(z) = \left(\frac{t}{a(t)}\right)^{1/2} z$ and $\widetilde{g}_t(y) = \left(\frac{t}{a(t)}\right)^{3/2} y$ transform $\widetilde{\phi}_t$ into $\widetilde{g}_t \circ \widetilde{\phi}_t \circ \widetilde{f}_t^{-1}(z) = z^3 \pm tz$.

For (c), consider two families of functions $\phi_t, \widetilde{\phi}_t : \mathbb{R} \to \mathbb{R}$ with death type critical points at $t = 0$ and equal profiles (the birth case is similar). By the discussion in (b), after composing $\phi_t, \widetilde{\phi}_t$ with suitable families of diffeomorphisms $f_t, \widetilde{f}_t : \mathbb{R} \to \mathbb{R}$, we may assume $\phi_t(z) = z^3 + a(t)z + b(t)$ and $\widetilde{\phi}_t(z) = z^3 + \widetilde{a}(t)z + \widetilde{b}(t)$. Equality of the profiles implies $a(t) = \widetilde{a}(t)$ and $b(t) = \widetilde{b}(t)$ for all $t \leq 0$, hence $\phi_t = \widetilde{\phi}_t$ for $t \geq 0$. We look for the desired family of local diffeomorphisms in the form $h_t(z) = z + g_t(z)$, $t \in \mathbb{R}$, where $g_t \equiv 0$ for $t \leq 0$. Then $\widetilde{\phi}_t \circ h_t = \phi_t$ is equivalent to

$$g_t(z)^3 + 3zg_t(z)^2 + \big(3z^2 + \widetilde{a}(t)\big)g_t(z) = \big(a(t) - \widetilde{a}(t)\big)z + b(t) - \widetilde{b}(t).$$

Recall that $\widetilde{a}(0) = 0$ and $\widetilde{a}'(0) > 0$, so the coefficient $3z^2 + \widetilde{a}(t)$ is positive for all $t > 0$. Looking at the discriminant, one sees that this third order equation has a unique real solution $g_t(z)$ for all $t \geq 0$ which depends smoothly on $t \geq 0$ and z. Since the right hand side vanishes to infinite order at $t = 0$, the solution $g_t(z)$ also vanishes to infinite order at $t = 0$ and hence extends smoothly by zero to $t < 0$. □

9.2. Zeroes of vector fields

Let X be a smooth vector field on V and $p \in V$ be a zero of X. The differential $D_pX : T_pV \to T_pV$ induces a splitting into invariant subspaces

$$T_pV = E_p^+ \oplus E_p^- \oplus E_p^0,$$

where E_p^+ (resp. E_p^-, E_p^0) is spanned by the generalized eigenvectors corresponding to eigenvalues with positive (resp. negative, vanishing) real part. The dimension of E_p^- is called the *(Morse) index*[1] of X at p. Denote by $X^s : V \to V$, $s \in \mathbb{R}$, the flow of X.

THEOREM 9.5 (Center manifold theorem [**4**]). *Let $p \in V$ be a zero of a C^{r+1}-vector field X, $r \in \mathbb{N}$. Then there exist the following local X^s-invariant manifolds through p:*

- $W_p^{0\pm}$ *tangent to* $E_p^0 \oplus E_p^\pm$ *of class* C^{r+1};
- $W_p^\pm \subset W_p^{0\pm}$ *tangent to* $E_p^\pm$ *of class* C^r;
- $W_p^0 = W_p^{0+} \cap W_p^{0-}$ *tangent to* E_p^0 *of class* C^{r+1}.

The $W_p^\pm$ are unique, and they are smooth (resp. real analytic) if X is.

W_p^- (resp. W_p^+, W_p^0, W_p^{0-}, W_p^{0+}) is called the *local stable (resp. unstable, center, center-stable, center-unstable) manifold* at p. The center, center-stable, and center-unstable manifolds are in general not unique, and they need not be smooth even if X is. By the center manifold theorem we can choose C^r-coordinates $Z = (x, y, z) \in E_p^+ \oplus E_p^- \oplus E_p^0$ in which $W_p^\pm$ and $W_p^{0\pm}$ correspond to $E_p^\pm$ (resp. $E_p^0 \oplus E_p^{0\pm}$). In these coordinates X is of the form

$$X(x,y,z) = (A^+x + O(|x|\,|Z|), A^-y + O(|y|\,|Z|), A^0z + O(|z|\,|Z| + |x|\,|y|) \tag{9.3}$$

with linear maps A^+ (resp. A^-, A^0) all of whose eigenvalues have positive (resp. negative, zero) real part. (The specific form of the higher order terms follows from tangency of X to $W_p^\pm$ and $W_p^{0\pm}$.)

A zero p of a vector field X is called *nondegenerate* if all its eigenvalues are nonzero. It is called *hyperbolic* if $E_p^0 = \{0\}$, i.e., all eigenvalues of D_pX have nonzero real part. In this case we have *global* stable and unstable manifolds characterized by

$$W_p^\pm = \{x \in V \mid \lim_{s \to \mp\infty} X^s(x) = p\}. \tag{9.4}$$

They are injectively immersed (but not necessarily embedded) in V. For a hyperbolic zero the local representation (9.3) simplifies to

$$X(x,y) = (A^+x + O(|x|\,|Z|), A^-y + O(|y|\,|Z|)). \tag{9.5}$$

Let us call a zero p *embryonic* if E_p^0 is 1-dimensional and the restriction of X to a center manifold W_p^0 has nonvanishing second derivative at p (for some local coordinate on $W_p^0 \cong \mathbb{R}$; the definition depends neither on this local coordinate nor on the choice of W_p^0). It follows that in suitable coordinates $Z = (x, y, z) \in \mathbb{R}^{m-k} \otimes \mathbb{R}^{k-1} \otimes \mathbb{R}$ near p the vector field is of the form

$$\begin{aligned} X(x,y,z) = \Big(&A^+x + O(|x|\,|Z|), A^-y + O(|y|\,|Z|),\\ &z^2 + O\big(|z|\,(|x| + |y| + |z|^2) + |x|\,|y|\big)\Big) \end{aligned} \tag{9.6}$$

with linear maps A^+, A^- all of whose eigenvalues have positive (resp. negative) real part.

[1]Not to be confused with the topological index of a vector field at an isolated zero.

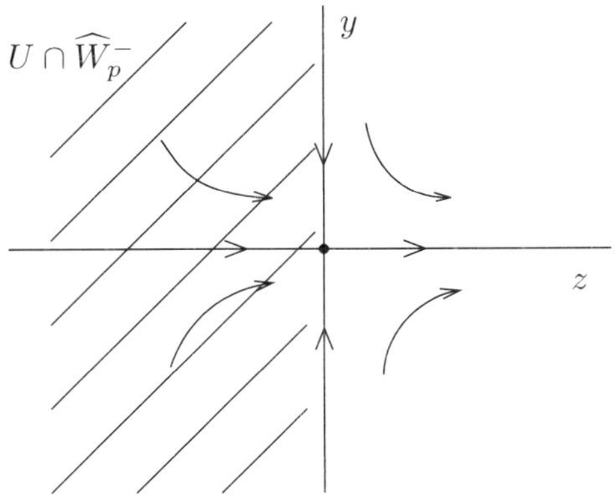

FIGURE 9.1. The flow near an embryonic zero.

LEMMA 9.6. *Let p be an embryonic zero of a smooth vector field X. Then*

$$\widehat{W}_p^{\pm} := \{x \in V \mid \lim_{s \to \mp\infty} X^s(x) = p\}$$

is an injectively immersed smooth manifold with boundary $W_p^{\pm}$.

PROOF (cf. [**175**]). Pick coordinates $Z = (x, y, z)$ on a neighborhood U of p in which X is of the form (9.6). We claim that

$$U \cap \widehat{W}_p^- = \{(x, y, z) \in U \mid x = 0,\ z \leq 0\};$$

see Figure 9.1. Since this is a smooth submanifold of U with boundary $U \cap W_p^- = \{(x, y, z) \in U \mid x = z = 0\}$, the claim implies the statement for $\widehat{W}_p^-$ by invariance under the flow of X, and the statement for $\widehat{W}_p^+$ is proved analogously.

To prove the claim, consider a flow line $\big(x(t), y(t), z(t)\big)$ starting at $t = 0$ at $(x_0, y_0, z_0) \in U$. It follows from (9.6) that $x(t) \to 0$ as $t \to \infty$ if and only if $x(t) \equiv 0$. Moreover, the second component decays exponentially, $|y(t)| \leq e^{-\lambda t}$, for some $\lambda > 0$. Inserting this in the equation for z yields the estimate (for possibly smaller $\lambda > 0$)

$$\dot{z} \geq z^2/2 - e^{-\lambda t}|z|. \tag{9.7}$$

Moreover, the equation for z in (9.6) shows that $z(t)$ cannot change its sign.

If $z_0 > 0$, inequality (9.7) yields $\frac{d}{dt} \ln z \geq z/2 - e^{-\lambda t} \geq -e^{-\lambda t}$, which integrates to $\ln z(t) - \ln z_0 \geq (e^{-\lambda t} - 1)/\lambda \geq -1/\lambda$ and hence $z(t) \geq z_0 e^{-1/\lambda} > 0$. Thus $z(t)$ does not tend to 0 as $t \to \infty$, and hence $(0, y_0, z_0) \notin \widehat{W}_p^-$.

If $z_0 < 0$, we have for every $t_1 \geq 0$ the following dichotomy: either the right hand side of (9.7) is $\geq z^2/4$, in which case z grows like $z(t) \geq \big(1/z(t_1) - (t - t_1)/4\big)^{-1}$ for $t \geq t_1$, or the right hand side of (9.7) is $\leq z^2/4$, which means that $z(t_1) \geq -4e^{-\lambda t_1}$. This shows that $z(t) \to 0$ as $t \to \infty$, and hence $(0, y_0, z_0) \in \widehat{W}_p^-$. □

We say that a 1-parameter family X_t, $t \in (-\varepsilon, \varepsilon)$ of vector fields near $p \in V$ is of *birth-death type* if p is an embryonic zero of X_0 and the section $(t, Z) \mapsto X_t(Z)$ is transverse to the zero section of the bundle $TM \to \mathbb{R} \times M$ at $(0, p)$. It follows that in suitable coordinates $Z = (x, y, z) \in \mathbb{R}^{m-k} \otimes \mathbb{R}^{k-1} \otimes \mathbb{R}$ near p the family is of the form

$$X_t(x,y,z) = \Big(A_t^+x + O(|x|\,|Z|),\ A_t^-y + O(|y|\,|Z|),\\ z^2 \pm t + O\big((|z|+|t|)\,(|x|+|y|+|z^2\pm t|) + |x|\,|y|\big)\Big) \tag{9.8}$$

with smooth families of linear maps $A_t^\pm$ all of whose eigenvalues have positive (resp. negative) real part. (The specific form of the higher order terms follows from tangency of the vector field $\widehat{X}(t,Z) = \big(0, X_t(Z)\big)$ on $\mathbb{R}\times M$ to $\{0\}\times W_p^\pm$ and $\mathbb{R}\times W_p^0$, plus the fact that in suitable coordinates the zero set of $\widehat{X}$ is the curve $\{x=y=z^2\pm t=0\}$; see [**175**].)

We say that the family is of *birth type* if the sign in $z^2\pm t$ in (9.8) is minus, and of *death type* otherwise. Note that in a birth type family a pair of hyperbolic zeroes of indices k and $k-1$ appears at $t=0$ and in a death type family such a pair disappears.

LEMMA 9.7. *(a) A generic vector field has only hyperbolic zeroes.*

(b) In a generic 1*-parameter family of vector fields without nonconstant periodic orbits, only birth-death type degeneracies appear.*

PROOF. (a) This follows from general transversality arguments.

(b) In a generic 1-parameter family of vector fields only two types of degeneracies appear (see [**11**, §§32 – 33]): the first type corresponds to birth-death type; the second type corresponds to a Hopf bifurcation in which a nonconstant periodic orbit appears or disappears at $t=0$, which is excluded by the hypothesis of (b). □

9.3. Gradient-like vector fields

In the previous two subsections we have studied functions and vector fields independently. Now we will look at them jointly. We call a smooth function $\phi : V\to\mathbb{R}$ a *Lyapunov function* for X, and X *gradient-like* for ϕ, if

$$X\cdot\phi \ge \delta(|X|^2 + |d\phi|^2) \tag{9.9}$$

for some positive function $\delta : V\to\mathbb{R}_+$, where $|X|$ is the norm with respect to some Riemannian metric on V and $|d\phi|$ is the dual norm. We call ϕ a *weak Lyapunov function* for a vector field X, and X *weakly gradient-like* for ϕ, if zeroes of X coincide with critical points of ϕ and $X\cdot\phi > 0$ outside the zeroes of X. A pair (X,ϕ) consisting of a vector field and a (weak) Lyapunov function will be called a *(weak) Lyapunov pair.*

By the Cauchy–Schwarz inequality, condition (9.9) implies

$$\delta|X| \le |d\phi| \le \frac{1}{\delta}|X|. \tag{9.10}$$

In particular, zeroes of X coincide with critical points of ϕ, so every Lyapunov pair is also a weak Lyapunov pair.

LEMMA 9.8. *(a) If X_0, X_1 are (weakly) gradient-like vector fields for ϕ, then so is $f_0X_0 + f_1X_1$ for any nonnegative functions f_0, f_1 with $0<\varepsilon\le f_0+f_1\le 1/\varepsilon$.*

(b) If ϕ_0,ϕ_1 are (weak) Lyapunov functions for X, then so is $\lambda_0\phi_0+\lambda_1\phi_1$ for any nonnegative constants λ_0,λ_1 with $\lambda_0+\lambda_1>0$.

In particular, the following spaces are convex cones and hence contractible:

- *the space of (weak) Lyapunov functions for a given vector field X;*
- *the space of (weakly) gradient-like vector fields for a given function ϕ.*

PROOF. The condition on a weak Lyapunov pair is obviously preserved under positive combinations of the functions or vector fields. To see that condition (9.9) is preserved under positive combinations (with changing δ) of vector fields, consider two vector fields X_0, X_1 satisfying $X_i \cdot \phi \geq \delta_i(|X_i|^2 + |d\phi|^2)$ and nonnegative functions f_0, f_1 with $f_0 + f_1 \geq \varepsilon > 0$. Then the vector field $X = f_0X_0 + f_1X_1$ satisfies (9.9) with $\delta := \min\left\{\frac{\delta_0}{2f_0}, \frac{\delta_1}{2f_1}, f_0\delta_0 + f_1\delta_1\right\}$:

$$\begin{aligned} X \cdot \phi &\geq f_0\delta_0|X_0|^2 + f_1\delta_1|X_1|^2 + (f_0\delta_0 + f_1\delta_1)|d\phi|^2 \\ &\geq 2\delta(|f_0X_0|^2 + |f_1X_1|^2) + \delta|d\phi|^2 \\ &\geq \delta(|X|^2 + |d\phi|^2). \end{aligned}$$

Positive combinations of functions are treated analogously. □

Lyapunov pairs near critical points. Consider a Lyapunov pair (X, ϕ) and a (possibly degenerate) zero p of X. Then p is also a critical point of ϕ, so in coordinates Z near $p = \{Z = 0\}$, we have

$$X(Z) = AZ + O(|Z|^2), \qquad \phi(Z) = \phi(p) + \frac{1}{2}B(Z, Z) + O(|Z|^3)$$

with the linear map $A := D_pX$ and the symmetric bilinear form $B := \mathrm{Hess}_p\phi$. Gradient-likeness

$$X \cdot \phi(Z) = B(Z, AZ) + O(|Z|^3) \geq \delta\Big(|AZ|^2 + |B(Z, \cdot)|^2\Big) + O(|Z|^3)$$

yields

$$B(v, Av) \geq \delta\Big(|Av|^2 + |B(v, \cdot)|^2\Big). \tag{9.11}$$

LEMMA 9.9. *Suppose a linear map $A : V \to V$ and a symmetric bilinear form $B : V \times V \to \mathbb{R}$ satisfy* (9.11). *Then:*

(a) all nonzero eigenvalues of A have nonzero real part;

(b) there exists an A-invariant splitting $V = E^+ \oplus E^- \oplus E^0$, where

$$E^0 = \ker A, \qquad E^{\pm} = \{v \mid \lim_{t \to \mp\infty} e^{tA}v = 0\};$$

(c) B is positive definite on E^+ and negative definite on E^-;

(d) A is nondegenerate if and only if B is nondegenerate, moreover, in this case condition (9.11) *is equivalent to an inequality*

$$B(v, Av) \geq \beta|v|^2, \qquad \beta > 0.$$

PROOF. (a) Extend A $\mathbb{C}$-linearly to the complexified space $V \otimes \mathbb{C}$, and extend B to $V \otimes \mathbb{C}$ by

$$B(x + iy, x' + iy') := \Big(B(x, x') + B(y, y')\Big) + i\Big(B(y, x') - B(x, y')\Big).$$

Thus B is $\mathbb{C}$-linear in the first and $\mathbb{C}$-antilinear in the second argument, $B(v, w) = \overline{B(w, v)}$, and $\mathrm{Re}\, B(v, Av) \geq \delta|Av|^2$. Let $0 \neq v \in V \otimes \mathbb{C}$ be an eigenvector of A to an eigenvalue $\lambda \in \mathbb{C}$, i.e., $Av = \lambda v$. Then

$$\lambda B(v, Av) = B(Av, Av) = \overline{B(Av, Av)} = \bar{\lambda}\overline{B(v, Av)}.$$

Suppose now that $\lambda \neq 0$ is purely imaginary. Then it follows that $B(v, Av) = -\overline{B(v, Av)}$, so with $v = x + iy$, $x, y \in V$, we find

$$0 = \mathrm{Re}\, B(v, Av) = B(x, Ax) + B(y, Ay) \geq \delta(|Ax|^2 + |Ay|^2).$$

But then $Ax = Ay = 0$, which implies $0 = Av = \lambda v$ and hence (since $\lambda \neq 0$) $v = 0$, in contradiction to the assumption $v \neq 0$.

(b) This follows from Sections 22.2 and 22.3 in [**9**].

(c) The flow e^{tA} preserves $E^\pm$ and satisfies $\frac{d}{dt}\phi(e^{tA}Z) \geq \delta|Ae^{tA}Z|^2 > 0$ for $0 \neq Z \in E^\pm$. For $0 \neq Z \in E^+$ it follows that

$$\phi(Z) - 0 = \int_{-\infty}^0 \frac{d}{dt}\phi(e^{tA}Z) > 0,$$

and similarly $\phi(Z) < 0$ for $0 \neq Z \in E^-$.

(d) Nondegeneracy of A or B gives an estimate $B(v, Av) \geq \beta|v|^2$, $\beta > 0$, which in turn implies nondegeneracy of A and B. □

REMARK 9.10. Suppose that X is the gradient of ϕ with respect to a positive definite but not necessarily symmetric $(2,0)$ tensor field g, i.e., $d\phi(v) = g(X, v)$ for all $v \in TV$ and $g(v, v) > 0$ for all $v \neq 0$. Then X is gradient-like for ϕ. At a zero p of X, we have $\mathrm{Hess}_p\phi(v, w) = g_p(D_pX \cdot v, w)$.

If g is symmetric (i.e., a Riemannian metric), then so is the bilinear form $\mathrm{Hess}_p\phi(\cdot, D_pX\cdot) = g_p(D_pX\cdot, D_pX\cdot)$ and all eigenvalues of D_pX are real.

REMARK 9.11. By Lemma 9.9(a), for a Lyapunov pair (X, ϕ) each nondegenerate zero of X is hyperbolic. Lemma 9.9(d) can be rephrased as follows: For a Lyapunov pair (X, ϕ), a zero p of X is nondegenerate if and only if it is a nondegenerate critical point of ϕ. Moreover, in this case gradient-likeness is equivalent to an inequality

$$X \cdot \phi(Z) \geq \beta|Z|^2, \qquad \beta > 0,$$

in coordinates Z near $p = \{Z = 0\}$.

We use this criterion to derive a technical result about perturbations of Lyapunov functions.

LEMMA 9.12. *(a) Let p be a hyperbolic critical point for a Lyapunov pair (X, ϕ). If (Y, ψ) is another Lyapunov pair with p as hyperbolic critical point and ψ sufficiently C^2-close to ϕ, then ψ is also a Lyapunov function for X near p.*

(b) Let p be an embryonic critical point for a Lyapunov pair (X, ϕ). If (Y, ψ) is another Lyapunov pair with p as embryonic critical point with the same null direction and ψ sufficiently C^3-close to ϕ, then ψ is also a Lyapunov function for X near p.

PROOF. (a) Write

$$X(Z) = AZ + O(|Z|^2)$$

in coordinates near p, where all eigenvalues of A have nonzero real part. Then ϕ is a Lyapunov function for X near p if and only if

$$\phi(Z) = \frac{1}{2}\langle Z, BZ\rangle + O(|Z|^3), \qquad BA > 0.$$

This implies the assertion of the lemma because $BA > 0$ is a C^2-open condition on ϕ.

(b) Pick coordinates $Z = (w, z)$ near p in which X has the form (see (9.6) with $w = (x, y)$)

$$X(w, z) = \Big(Aw + O(|w|\,|Z|), z^2 + O\big(z(|w| + z^2) + |w|^2\big)\Big).$$

Claim. ϕ is a Lyapunov function for X near p if and only if it has the form

$$\phi(w,z) = \frac{1}{2}\langle w, Bw\rangle + \frac{1}{3}cz^3 + 2z^2\langle d, w\rangle + O(|w|^2|Z|) + O(|Z|^4),$$

where the coefficients B, c, d satisfy

$$\begin{pmatrix} BA & A^t d \\ d^t A & c \end{pmatrix} > 0.$$

From this claim the assertion of the lemma follows because ψ has a Taylor expansion of the same form and the positivity condition on the coefficients is C^3-open. To prove the claim, let us write the general Taylor expansion up to the second order of a function ϕ with critical point at p:

$$\phi(w,z) = \frac{1}{2}\langle w, Bw\rangle + az^2 + z\langle b, w\rangle + O(|Z|^3).$$

The condition $X \cdot \phi > 0$ restricted to $\{z = 0\}$ and $\{w = 0\}$ yields $BA > 0$ and $a = 0$. If b were nonzero, we could pick (w, z) with $z\langle b, Aw\rangle < 0$ and obtain the contradiction

$$X \cdot \phi(t^2 w, tz) = t^3 z\langle b, Aw\rangle + O(t^4) < 0$$

for $t > 0$ sufficiently small. Thus $b = 0$ and the Taylor expansion of ϕ up to third order has the form in the claim. Then

$$X \cdot \phi(w,z) = \langle w, BAw\rangle + cz^4 + 2z^2\langle d, Aw\rangle + O(|w|^2|Z|) + O(|w|z^3) + O(|Z|^5).$$

This is a quadratic form in the variables (w, z^2) plus terms of order $5/2$ and higher, so it is positive if and only if the quadratic form is positive, which is precisely the positivity condition in the claim. □

Next we show that a sufficiently nondegenerate Lyapunov pair can be put into standard form near critical points without changing the stable and unstable manifolds.

PROPOSITION 9.13. *Let (X, ϕ) be a Lyapunov pair with nondegenerate or embryonic critical points. Then there exists a homotopy of Lyapunov pairs (X_t, ϕ_t) with the following properties:*

- *(X_t, ϕ_t) agrees with (X, ϕ) for $t = 0$ and outside a neighborhood of the critical points;*
- *for all t, X_t has the same nondegenerate (resp. embryonic) critical points and the same stable, unstable, and center manifolds as X_0;*
- *near each nondegenerate critical point of index k there exist coordinates $x_1, \dots, x_n$ in which*

$$X_1 = -\sum_{i=1}^{k} x_i \partial_{x_i} + \sum_{j=k+1}^{n} x_j \partial_{x_j}, \qquad \phi_1 = -\frac{1}{2}\sum_{i=1}^{k} x_i^2 + \frac{1}{2}\sum_{j=k+1}^{n} x_j^2;$$

- *near each embryonic critical point of index $k-1$ there exist coordinates $x_1, \dots, x_n$ in which*

$$X_1 = x_1^2 \partial_{x_1} - \sum_{i=2}^{k} x_i \partial_{x_i} + \sum_{j=k+1}^{n} x_j \partial_{x_j}, \qquad \phi_1 = \frac{x_1^3}{3} - \frac{1}{2}\sum_{i=2}^{k} x_i^2 + \frac{1}{2}\sum_{j=k+1}^{n} x_j^2.$$

Similarly, each birth-death family (X_t, ϕ_t) *can be modified through birth-death families to one which near the birth-death point at* $t = 0$ *looks like*

$$X_t = (x_1^2 \pm t)\partial_{x_1} - \sum_{i=2}^{k} x_i \partial_{x_i} + \sum_{j=k+1}^{n} x_j \partial_{x_j}, \quad \phi_t = \frac{x_1^3}{3} \pm t x_1 - \frac{1}{2}\sum_{i=2}^{k} x_i^2 + \frac{1}{2}\sum_{j=k+1}^{n} x_j^2.$$

PROOF. Near a nondegenerate zero p of index k pick local coordinates $Z = (x, y) \in \mathbb{R}^\ell \oplus \mathbb{R}^k$, $\ell = m - k$, in which X is given by (9.5). Note that in these coordinates $W_p^+ = \mathbb{R}^\ell \oplus 0$ and $W_p^- = 0 \oplus \mathbb{R}^k$. The linear vector field $AZ = (A^+x, A^-y)$ has the same stable and unstable manifolds and is also gradient-like for ϕ in a sufficiently small neighborhood of p, so by Lemma 9.8 the same holds for the vector fields $X_t(Z) := \big(1 - t\rho(|Z|)\big)X(Z) + t\rho(|Z|)AZ$, where $t \in [0,1]$ and $\rho : [0,\infty) \to [0,1]$ equals 0 near 0 and 1 on $[\varepsilon, \infty)$ for sufficiently small $\varepsilon > 0$. After renaming X_1 back to X, we may hence assume that $X(Z) = AZ$ near p. A similar argument allows us to replace ϕ by its quadratic part $\frac{1}{2}B(v,v)$ (here we need to choose the cutoff function ρ more carefully such that $r\rho'(r) \leq 1$). By Lemma 9.9 the symmetric bilinear form B satisfies $B(Z, AZ) \geq \beta|Z|^2$ for some $\beta > 0$, and its restrictions $B^\pm$ to $W_p^\pm$ are positive (resp. negative) definite.

Consider the split quadratic form $B_1(Z,Z) := B^+(x,x) + B^-(y,y)$ and the family $B_t(Z) := (1-t)B(Z,Z) + tB_1(Z,Z)$, $t \in [0,1]$. Since $B(Z, AZ)$ and $B_1(Z, AZ) = B^+(x, A^+x) + B^-(y, A^-y)$ are both positive, so is $B_t(Z, AZ)$ for all $t \in [0,1]$. This allows us (again via cutoff away from p) to replace B by B_1. Now consider the family of linear maps $A_t(Z) := (1-t)AZ + t(x, -y)$, which satisfies $B(Z, A_tZ) > 0$ for all $t \in [0,1]$. So we can replace the linear vector field A by the standard vector field $A_1(Z) = (x, -y)$. Finally, we linearly interpolate for this vector field from B_1 to the standard quadratic form $B_2(Z) = |x|^2 - |y|^2$. Renaming $y = (x_1, \dots, x_k)$ and $(x = x_{k+1}, \dots, x_m)$, the pair (A_1, B_2) has the desired standard form.

The proofs in the embryonic and birth-death case are similar and will be omitted. □

The following corollary shows that a Lyapunov pair can be arbitrarily altered near a hyperbolic or embryonic critical point. In Section 12.4 we will prove a version of this result for Weinstein structures.

COROLLARY 9.14. *Let* $p \in V$ *be a hyperbolic (resp. embryonic) critical point of a Lyapunov pair* (X, ϕ). *Let* $(X_{\mathrm{loc}}, \phi_{\mathrm{loc}})$ *be a Lyapunov pair on a neighborhood* V_{loc} *of* p *such that* p *is a hyperbolic (resp. embryonic) critical point of* ϕ_{loc} *of value* $\phi_{\mathrm{loc}}(p) = \phi(p)$ *and Morse index* $\mathrm{ind}_p(\phi_{\mathrm{loc}}) = \mathrm{ind}_p(\phi)$. *Then there exists a homotopy of Lyapunov pairs* (X_t, ϕ_t) *on* V *with the following properties:*

(i) $(X_0, \phi_0) = (X, \phi)$ *and* $(X_t, \phi_t) = (X, \phi)$ *outside* V_{loc};

(ii) X_t *has a unique hyperbolic (resp. embryonic) zero at* p *in* V_{loc} *for all* t;

(iii) $(X_1, \phi_1) = (X_{\mathrm{loc}}, \phi_{\mathrm{loc}})$ *near* p;

(iv) *if* $W_p^-(X_{\mathrm{loc}}) = W_p^-(X)$ *(resp.* $\widehat{W}_p^-(X_{\mathrm{loc}}) = \widehat{W}_p^-(X)$*), then* $W_p^-(X_t) = W_p^-(X)$ *(resp.* $\widehat{W}_p^-(X_t) = \widehat{W}_p^-(X)$*) for all* t.

PROOF. After moving (X, ϕ) by a diffeotopy, we may assume that p has the same stable, unstable, and center manifolds with respect to X and X_{loc}. By Proposition 9.13, there exists a homotopy (X_t, ϕ_t), $t \in [0, 1/2]$, with properties (i), (ii) and (iv) such that $(X_{1/2}, \phi_{1/2})$ has standard form near p. Reversing the argument in Proposition 9.13, there exists a homotopy (X_t, ϕ_t), $t \in [1/2, 1]$, with properties (i), (ii) and (iv) such that $(X_1, \phi_1) = (X_{\rm loc}, \phi_{\rm loc})$ on a smaller neighborhood of p. □

From gradient-like to gradient vector fields. By Remark 9.10, imaginary eigenvalues of D_pX provide an obstruction to C^1-approximating X by a gradient vector field. But we have the following C^0-approximation result.

LEMMA 9.15. *Let (X, ϕ) be a Lyapunov pair on V, and let Z be the zero set of X. Then for every neighborhood U of Z there exists a Riemannian metric g on V such that $\nabla_g\phi$ agrees with X outside U and is arbitrarily C^0-close to X on U. This construction also works smoothly for families and relative to a subset where X is already a gradient.*

PROOF. Pick a reference metric g_0 for which condition (9.9) holds. It implies on $V \setminus Z$ the uniform estimates $\delta|X| \leq |\nabla_{g_0}\phi| \leq |X|/\delta$ for the lengths and $\cos\theta \geq \delta^2$ for the angle θ between X and $\nabla_{g_0}\phi$. Thus we can pick a metric g_1 on $V \setminus Z$ of uniformly bounded distance from g_0 for which $\nabla_{g_1}\phi = X$. Modify g_1 inside U to a metric g which smoothly extends over Z and still has bounded distance from g_0. Then $\nabla_g\phi$ agrees with X outside U and in U it satisfies an estimate $|X - \nabla_g\phi| \leq |X| + C|d\phi|$, which can be made arbitrarily small by choosing U small. □

COROLLARY 9.16. *Let $(X_\lambda, \phi_\lambda)_{\lambda\in\Lambda}$ be a smooth family of Lyapunov pairs on V, and ψ_λ any family of functions C^k-close to ϕ_λ, $k \geq 1$. Then there exists a family of metrics g_λ such that the family $(\nabla_{g_\lambda}\psi_\lambda, \psi_\lambda)$ is connected to $(X_\lambda, \phi_\lambda)$ by a homotopy of families of Lyapunov pairs that is C^0-small in the vector fields and C^k-small in the functions.*

PROOF. First linearly homotope (with fixed functions) from $(X_\lambda, \phi_\lambda)$ to $(\nabla_{g_\lambda}\phi_\lambda, \phi_\lambda)$, with the metrics g_λ provided by Lemma 9.15, and then through gradient pairs (with fixed metrics) from $(\nabla_{g_\lambda}\phi_\lambda, \phi_\lambda)$ to $(\nabla_{g_\lambda}\psi_\lambda, \psi_\lambda)$. □

Existence of Lyapunov functions. The question of existence of a (weak) Lyapunov function for a vector field X separates into two issues: local existence near the zero set of X and global existence. Assuming local existence near the zero set, Sullivan [**181**] gives a necessary and sufficient criterion for the existence of a global (weak) Lyapunov function in terms of foliation cycles. The simplest obstruction to a weak Lyapunov function is a nonconstant periodic orbit of X.

The following lemma settles the local existence question near a hyperbolic or birth-death type zero.

LEMMA 9.17. *(a) Near each hyperbolic zero a vector field admits a Lyapunov function.*

(b) For a birth or death type family X_t near p there exists a neighborhood U of p and a smooth family of Lyapunov functions $\phi_t : U \to \mathbb{R}$ for X_t.

PROOF. (a) Consider coordinates in which X has the form (9.5). By [**9**, Theorem 22.3] there exist quadratic forms $Q^\pm$ on $E_p^\pm$ which are Lyapunov for the linear maps $A^\pm$. Then $\phi(x, y) := Q^+(x) + Q^-(y)$ is a Lyapunov function for X.

(b) Consider coordinates in which X_t has the form (9.8). Let $Q_t^\pm$ be a smooth family of quadratic forms on $E_p^\pm$ as in (a) that are Lyapunov for $A_t^\pm$. Then

$$\phi_t(x,y,z) := Q_t^+(x) + Q_t^-(y) + \frac{1}{3}z^3 - tz$$

is a smooth family of Lyapunov functions for X_t. □

9.4. Smooth surroundings

In this section we discuss a smooth version of the J-convex surroundings in Chapter 8.2.

By a *flow box* (W,X) we will mean a compact manifold with corners whose boundary is a union $\partial W = \partial_+ W \cup \partial_- W \cup \partial_v W$ of three codimension 1 manifolds, called the *positive* (resp. *negative, vertical*) *boundary*, together with a vector field X which is inward (resp. outward) pointing along $\partial_\pm W$, and tangent to $\partial_v W$ without zeroes on $\partial_v W$. Note that the case $\partial_v W = \varnothing$ corresponds to a cobordism. We denote by X^t the flow of X and define the *skeleton*

$$\mathrm{Skel}(W,X) := \bigcap_{t\geq 0} X^{-t}(W) := \{x \in W \mid X^t(x) \in W \text{ for all } t \geq 0\}.$$

For $x \in W$ define the *ω-limit set*

$$\omega(x) := \bigcap_{t\geq 0} \overline{X^{\geq t}(x)} = \{y \in W \mid X^{t_k}(x) \to y \text{ for some sequence } t_k \to \infty\}.$$

For each zero p of X define

$$\widehat{W}_p^- := \{x \in W \mid p \in \omega(x)\} = \{x \in W \mid X^{t_k}(x) \to p \text{ for some sequence } t_k \to \infty\}.$$

LEMMA 9.18. *If X admits a weak Lyapunov function ϕ, then $\omega(x) \subset \mathrm{Zero}(X)$ for each $x \in W$. If in addition X has finitely many zeroes, then*

$$\mathrm{Skel}(W,X) = \bigcup_{p\in \mathrm{Zero}(X)} \widehat{W}_p^-.$$

The set $\widehat{W}_p^-$ agrees with the stable manifold W_p^- if p is hyperbolic, and with the manifold $\widehat{W}_p^-$ in Lemma 9.6 if p is embryonic.

PROOF. For $y \notin \mathrm{Zero}(X)$, we have $X \cdot \phi \geq \varepsilon > 0$ on some neighborhood U of y. This implies that a trajectory $X^t(x)$ passing close to y will have $\phi(X^{t_0}(x)) > \phi(y)$ for some $t_0 > 0$ and hence cannot get close to y for $t > t_0$, so $y \notin \omega(x)$. This proves $\omega(x) \subset \mathrm{Zero}(X)$. The second statement follows from the equivalence of $x \in \mathrm{Skel}(W,X)$ and $\omega(x) \neq \varnothing$, and the last statement follows from equation (9.4) and Lemma 9.6. □

In the following, (W,X) is a flow box with skeleton $\Delta := \mathrm{Skel}(W,X)$, and $\phi : W \to [a_-,a_+]$ is a function satisfying $X \cdot \phi > 0$ outside Δ with constant values $\phi|_{\partial_\pm W} \equiv a_\pm$.

The following is a (much simpler) smooth version of Theorem 8.5; see Figure 8.2.

PROPOSITION 9.19. *Let (W,X,ϕ,Δ) be as above. Fix an open neighborhood U of $\partial_- W \cup \Delta$ and a regular value $c \in (a_-,a_+)$ of the function ϕ such that there are no critical values of ϕ in $[c,a_+]$. Then there exists a diffeotopy $h_t : W \to W$ with the following properties:*

- $h_0 = \mathrm{Id}$ *and* $h_t = \mathrm{Id}$ *on* $\mathcal{O}p\,(\partial W \cup \Delta)$;

- *h_t preserves trajectories of X;*
- *$h_1(\{\phi \le c\}) \subset U$.*

In particular, $\phi_t := \phi \circ h_t^{-1}$, $t \in [0,1]$, is a family of functions satisfying $X \cdot \phi_t > 0$ outside Δ such that the level set $\{\phi_1 = c\}$ surrounds $\partial_- W \cup \Delta$ in U.

PROOF. Pick a smooth function $\rho : W \to [0,1]$ which equals 0 on $\mathcal{O}p\,(\partial W \cup \Delta)$ and 1 on $\{\phi \le c\} \setminus U$. Then the vector field $Y := \rho X$ is complete, i.e., its flow Y^t is defined for all $t \in \mathbb{R}$, and $Y^t = \mathrm{Id}$ on $\mathcal{O}p\,(\partial W \cup \Delta)$. Moreover, by definition of the skeleton there exists $T > 0$ such that $Y^{-T}(\{\phi \le c\}) \subset U$. Hence the isotopy $h_t := Y^{-Tt}$, $t \in [0,1]$, has the required properties. □

In the next section we will need a more precise version of smooth surroundings. We call a subset $A \subset W$ *backward invariant* if $X^t(A) \subset A$ for all $t \le 0$. For a compact backward invariant subset $A \subset W$, we define its *exit set*

$$\partial_+ A := \overline{\{x \in A \mid \inf\{t > 0 \mid X^t(x) \notin A\} = 0\}}.$$

Thus every forward orbit that exits A exits through $\partial_+ A$. Note that $\phi(x) > \min_{\partial_+ A} \phi$ for every $x \notin A$ whose backward orbit meets A.

LEMMA 9.20. *Let (W, X, ϕ, Δ) be as above. Fix a compact backward invariant neighborhood A of Δ and set $c := \min_{\partial_+ A} \phi$. Let $g : [a_-, a_+] \to [a_-, a_+]$ be a diffeomorphism which equals the identity near $\partial[a_-, a_+]$ and satisfies $g(x) \le x$ for $x \ge c$. Then for every compact neighborhood $A' \subset \mathrm{Int}\, A$ of Δ there exists a function $\psi : W \to \mathbb{R}$ satisfying $X \cdot \psi > 0$ outside Δ with the following properties:*

- *$\psi = \phi$ on $\mathcal{O}p\,\partial W$ and outside A;*
- *$\psi = g \circ \phi$ on A'.*

PROOF. Let us rescale X such that $X \cdot \phi \equiv 1$ outside A. For $t \in [0,1]$ define diffeomorphisms $g_t(x) := (1-t)x + tg(x)$ on $[a_-, a_+]$ and functions $f_t : [a_-, a_+] \to \mathbb{R}$,

$$f_t\big(g_t(x)\big) := \dot g_t(x) = g(x) - x.$$

Note that $f_t = 0$ near $\partial[a_-, a_+]$ and $f_t(x) \le 0$ for $x \ge c$. Pick a smooth function $\rho : W \to [0,1]$ which equals 1 outside A and 0 on A'. Define a family of diffeomorphisms $h_t : W \to W$ as the solution of

$$\dot h_t = (f_t \circ \phi\, \rho X)(h_t).$$

Note that h_t moves backward along trajectories of X in the region $\{\phi \ge c\}$. Hence, by definition of c, $h_1(x) \notin A$ implies that $h_t(x) \notin A$ for all $t \in [0,1]$. For such x we have $\rho \circ h_t(x) = 1$ and thus

$$\frac{d}{dt}(\phi \circ h_t(x) = d\phi \cdot (f_t \circ \phi X)\big(h_t(x)\big) = f_t \circ \phi \circ h_t(x).$$

Since $\frac{d}{dt}(g_t \circ \phi)(x) = \dot g_t \circ \phi(x) = f_t \circ g_t \circ \phi(x)$, the paths $\phi \circ h_t(x)$ and $g_t \circ \phi(x)$ satisfy the same differential equation with the same initial condition $\phi(x)$ and hence coincide, so in particular $\phi \circ h_1(x) = g \circ \phi(x)$ whenever $h_1(x) \notin A$. This shows that the function $\psi := g \circ \phi \circ h_1^{-1}$ agrees with ϕ outside A. On A' we have $\rho = 0$, hence $h_1 = \mathrm{Id}$ and $\psi = g \circ \phi$. □

Replacing a given neighborhood by a smaller backward invariant one and choosing $g(\max_\Delta \phi) < b$, this implies the following improved version of Proposition 9.19.

COROLLARY 9.21. *Let (W, X, ϕ, Δ) be as above. Then for every neighborhood $U \subset W$ of Δ and every $b \in (a_-, a_+)$ there exists a function $\psi : W \to \mathbb{R}$ satisfying $X \cdot \psi > 0$ outside Δ with the following properties:*

- *$\psi = \phi$ on $\mathcal{O}p\, \partial W$ and outside U;*
- *ψ is target equivalent to ϕ near Δ;*
- *$\psi|_\Delta < b$.* □

REMARK 9.22. The last three results continue to hold if $\Delta \subset W$ is any compact backward invariant subset containing $\mathrm{Skel}(W, X)$. To see this, pick a smooth function $f : W \to [0, \infty)$ which vanishes exactly on Δ. Then Δ is the skeleton of the vector field fX and $fX \cdot \phi > 0$ outside Δ, so we can apply the results to the quadruple (W, fX, ϕ, Δ).

9.5. Changing Lyapunov functions near critical points

In this section we show that a weak Lyapunov function can be put into any prescribed form near a hyperbolic or birth-death type zero. The following proposition will be used repeatedly in the manipulations of Weinstein structures in Chapter 12.

PROPOSITION 9.23. *(a) Let X be a vector field on V with a hyperbolic or embryonic zero p. Let $\phi : V \to \mathbb{R}$ be a weak Lyapunov function for X and $\phi^{\mathrm{loc}} : U \to \mathbb{R}$ a weak Lyapunov function on a neighborhood U of p with $\phi(p) = \phi^{\mathrm{loc}}(p)$. Then there exists a weak Lyapunov function $\psi : V \to \mathbb{R}$ which agrees with ϕ outside U and with ϕ^{loc} near p.*

(b) Let X_t, $t \in [-\varepsilon, \varepsilon]$ be a smooth family of vector fields on V with a birth or death type zero p. Let $\phi_t : V \to \mathbb{R}$ be a smooth family of weak Lyapunov functions for X_t and $\phi_t^{\mathrm{loc}} : U \to \mathbb{R}$ a smooth family of weak Lyapunov functions on a neighborhood U of p with $\phi_t(p) = \phi_t^{\mathrm{loc}}(p)$ for all t. Then there exists a smooth family of weak Lyapunov functions $\psi_t : V \to \mathbb{R}$, $t \in [-\varepsilon, \varepsilon]$ which agrees with ϕ_t outside U and with ϕ_t^{loc} near p.

REMARK 9.24. (1) In case (a), $\phi_u := (1 - u)\phi + u\psi$, $u \in [0, 1]$ is a smooth family of weak Lyapunov functions with $\phi_0 = \phi$, $\phi_u = \phi$ outside U, and $\phi_1 = \phi^{\mathrm{loc}}$ near p.

(2) By Lemma 9.17, in case (a) we can choose ϕ^{loc} to be Lyapunov, so ψ is Lyapunov near p. Hence, in case (a) any weak Lyapunov function can be made Lyapunov by a local deformation near the zeroes of X.

Analogous remarks apply to case (b).

The proof of Proposition 9.23 requires some preparation. Consider $(X, \phi, \phi^{\mathrm{loc}})$ as in the proposition with hyperbolic or embryonic zero p of value $\phi(p) = b$. Pick a regular value $a < b$ such that $\Delta := W_p^- \cap \{\phi \geq a\}$ is a smoothly embedded disc in the hyperbolic case, and $\Delta := \widehat{W}_p^- \cap \{\phi \geq a\}$ is a smoothly embedded half-disc in the embryonic case, where $\widehat{W}_p^\pm$ is defined as in Lemma 9.6. We choose a so close to b that $\Delta \subset U$, so ϕ^{loc} is defined near Δ.

We first show that we can interpolate between ϕ and ϕ^{loc} near Δ.

LEMMA 9.25. *In the notation above, there exists a weak Lyapunov function $\chi : \mathcal{N} \to [a, \infty)$ on a neighborhood $\mathcal{N}$ of Δ which agrees with ϕ near $\mathcal{N} \cap \phi^{-1}(a)$ and with ϕ^{loc} near p.*

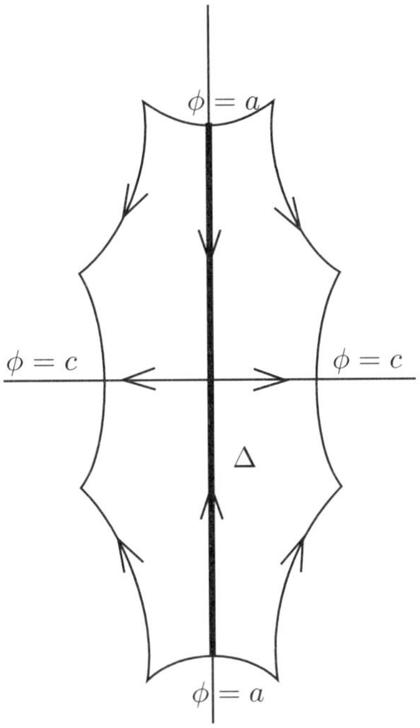

FIGURE 9.2. The flow box W near a hyperbolic zero.

PROOF. Pick a sufficiently small $\delta > 0$. If p is hyperbolic, X has no critical points on the set $\Delta \cap \{\phi \geq a+\delta\} \cap \{\phi^{\mathrm{loc}} \leq b-\delta\}$ and is transverse to its boundary. If p is embryonic, X has no critical points on the set $\Delta \cap \{\phi \geq a+\delta\} \cap \{\phi^{\mathrm{loc}} \leq b-\delta\}$, is transverse to the boundary components $\Delta \cap \{\phi = a+\delta\}$ and $\Delta \cap \{\phi^{\mathrm{loc}} = b-\delta\}$, and is tangent to the boundary component $W_p^- \cap \{\phi \geq a+\delta\} \cap \{\phi^{\mathrm{loc}} \leq b-\delta\}$. Hence in either case we can use the flow of X to construct a Lyapunov function χ on Δ which agrees with ϕ for $\phi \leq a+\delta$ and with ϕ^{loc} for $\phi^{\mathrm{loc}} \geq b-\delta$. Applying the same argument to a small neighborhood of Δ yields the desired function χ. □

PROOF OF PROPOSITION 9.23. (a) We use the notation above. After applying Lemma 9.25 and shrinking the neighborhood U of Δ, we may assume that $\phi = \phi^{\mathrm{loc}}$ near $U \cap \phi^{-1}(a)$.

We construct a flow box $W \subset U$ as in Section 9.4 as follows; see Figure 9.2 for the hyperbolic case. Pick a tubular neighborhood T_- with smooth boundary $S_- = \partial T_-$ of $\Delta \cap \phi^{-1}(a)$ in the level set $\phi^{-1}(a)$. Note that $S_- \cong S^{k-1} \times S^{m-k-1}$ if p is hyperbolic of index k, and $S_- \cong S^{m-2}$ if p is embryonic. Let $\Sigma \cong [a, c] \times S_-$ be the hypersurface in V obtained by flowing S_- under X up to some regular level $\phi = c > b$. Then $S_+ := \Sigma \cap \phi^{-1}(c)$ is the boundary of a tubular neighborhood T_+ of $\Delta \cap \phi^{-1}(c)$ in the level set $\phi^{-1}(c)$. The union $T_- \cup T_+ \cup \Sigma$ bounds a compact subset $W \subset V$ containing Δ. By choosing T_- small and c close to b, we can arrange that $W \subset U$. Note that W is a flow box as in Section 9.4 with positive/negative boundary $\partial_\pm W = T_\pm$ and vertical boundary Σ (to which X is tangent).

In a similar way we construct smaller backward invariant compact neighborhoods $A'' \subset A' \subset A \subset W$ of Δ. Pick $\varepsilon > 0$ such that $a + 2\varepsilon < b < c - 2\varepsilon$ and

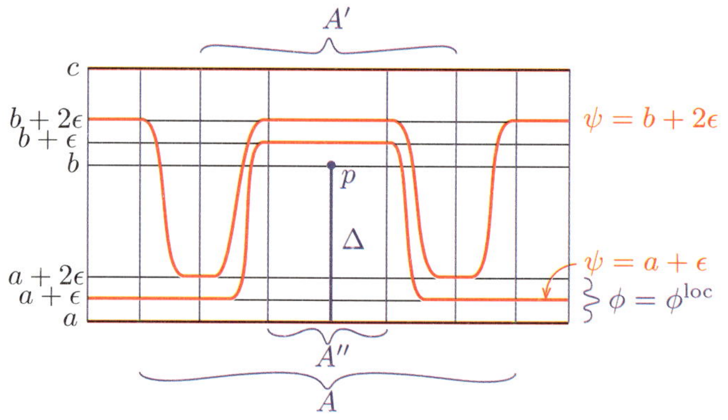

FIGURE 9.3. Smooth surroundings of a stable disc.

$\phi^{\rm loc} = \phi$ on $W \cap \{a \le \phi \le a + 2\varepsilon\}$. Now we apply Lemma 9.20 twice to construct the following smooth surroundings (see Figure 9.3):

- a weak Lyapunov function $\psi_1 : W \to \mathbb{R}$ which agrees with ϕ outside A and with $g_1 \circ \phi$ on A';
- a weak Lyapunov function $\psi_2 : W \to \mathbb{R}$ which agrees with ψ_1 outside A' and with $g_2 \circ \psi_1 = g \circ \phi$ on A''.

Here $g_1, g_2, g = g_2 \circ g_1 : [a, c] \to [a, c]$ are diffeomorphisms which equal the identity near $\partial[a, c]$ and have the following properties:

- $g_1(a + \varepsilon) = a + \varepsilon$, $g_1(a + 2\varepsilon) = b + 2\varepsilon$ and $g_1(x) \ge x$ for all x;
- $g_2(b + \varepsilon) = a + \varepsilon$ and $g_2(b + 2\varepsilon) = a + 2\varepsilon$.
- $g(b + \varepsilon) = a + \varepsilon$, $g(b + 2\varepsilon) = b + 2\varepsilon$ and $g(x) \le x$ for all x.

(We choose g and g_1 with these properties and define $g_2 := g \circ g_1^{-1}$.)

After target reparametrization above the level b, we may assume $\max_W \phi^{\rm loc} \le b + \varepsilon$. Then near $\partial A'$ we have $\psi_2(x) = g_1 \circ \phi(x) \ge \phi(x) = \phi^{\rm loc}(x)$ if $\phi(x) \le a + 2\varepsilon$, and $\psi_2(x) \ge g_1(a + 2\varepsilon) = b + 2\varepsilon > \phi^{\rm loc}(x)$ if $\phi(x) \ge a + 2\varepsilon$, hence $\psi_2 \ge \phi^{\rm loc}$ near $\partial A'$. On A'' we have $\psi_2(x) = g \circ \phi(x) \le \phi(x) = \phi^{\rm loc}(x)$ if $\phi(x) \le a + 2\varepsilon$, and $\psi_2(x) \le g(b + \varepsilon) = a + \varepsilon < \phi^{\rm loc}(x)$ if $a + \varepsilon \le \phi(x) \le b + \varepsilon$, hence $\psi_2 \le \phi^{\rm loc}$ on $A'' \cap \{\phi \le b + \varepsilon\}$. This shows that the function $\max(\psi_2, \phi^{\rm loc})$ agrees with $\phi^{\rm loc}$ near Δ and with ϕ near $\partial A'$, so we can extend it by ϕ to a function on V. According to Remark 3.21, this function can be smoothed to a weak Lyapunov function $\psi : V \to \mathbb{R}$ for X with the desired properties. □

9.6. Smale cobordisms

Recall from Section 8.1 that a *cobordism* W is an oriented compact smooth manifold with cooriented boundary ∂W. Its boundary splits as a disjoint union $\partial W = \partial_- W \cup \partial_+ W$ where the coorientation is provided, respectively, by the inward or outward normal vector field.

DEFINITION 9.26. A *Lyapunov cobordism* is a triple (W, ϕ, X), where W is a cobordism, $\phi : W \to \mathbb{R}$ is a smooth function constant on $\partial_\pm W$, and X is a gradient-like vector field for ϕ which points inward along $\partial_- W$ and outward along $\partial_+ W$. In particular, ϕ has no critical points on ∂W.

A *Smale cobordism* is a Lyapunov cobordism (W, ϕ, X) for which the function ϕ is Morse, so (W, ϕ) is a *Morse cobordism* in the sense of Section 8.1.

A Lyapunov cobordism (W, ϕ, X) is called *elementary* if there are no X-trajectories between different critical points of ϕ.

Note that if (W, ϕ, X) is an elementary Smale cobordism, then the stable manifold of each critical point p is a disc D_p^- which intersects $\partial_- W$ along a sphere $S_p^- = \partial D_p^-$. We call D_p^- and S_p^- the *stable disc (resp. sphere)* of p. Similarly, the unstable manifolds and their intersections with $\partial_+ W$ are called *unstable discs and spheres.* At an embryonic critical point, the manifolds $\widehat{W}_p^\pm$ in Lemma 9.6 give rise to *(un-)stable half-discs* $\widehat{D}_p^\pm$ *and hemispheres* $\widehat{S}_p^\pm$.

DEFINITION 9.27. An *admissible partition* of a Lyapunov cobordism (W, ϕ, X) is a finite sequence $m = c_0 < c_1 < \cdots < c_N = M$ of regular values of ϕ, where we denote $\phi|_{\partial_- W} = m$ and $\phi|_{\partial_+ W} = M$, such that each subcobordism $W_k = \{c_{k-1} \leq \phi \leq c_k\}$, $k = 1, \dots, N$, is elementary.

LEMMA 9.28. *Any Lyapunov cobordism with only Morse or embryonic critical points admits an admissible partition into elementary cobordisms.*

Similarly, for any exhausting function ϕ with only Morse or embryonic critical points and gradient-like vector field X on a noncompact manifold V, one can find regular values $c_0 < \min \phi < c_1 < \cdots \to \infty$ such that the cobordisms $W_k = \{c_{k-1} \leq \phi \leq c_k\}$, $k = 1, \dots,$ are elementary. If ϕ has finitely many critical points, then all but finitely many of these cobordisms have no critical points.

PROOF. We prove the second statement, the first one being analogous but simpler. Critical points of ϕ are Morse or embryonic, hence isolated. Since ϕ is also exhausting, its set of critical values is discrete and bounded below. So we can order the critical values as a sequence $\inf \phi = d_1 < d_2 < \cdots$ which is either finite or tends to infinity. Pick regular values c_k such that $c_0 < d_1 < c_1 < d_2 < c_2 \cdots$. Then all critical points in the cobordism $W_k = \{c_{k-1} \leq \phi \leq c_k\}$ have value d_k, so there are no X-trajectories between critical points and the cobordism is elementary. □

Equivalence of elementary Smale cobordisms.

LEMMA 9.29. *Let (W, X, ϕ) be an elementary Smale cobordism with critical points $p_1, \dots, p_k$ and skeleton (= union of the stable discs) $\Delta = \bigcup_{i=1}^k D_{p_i}^-$. Let*

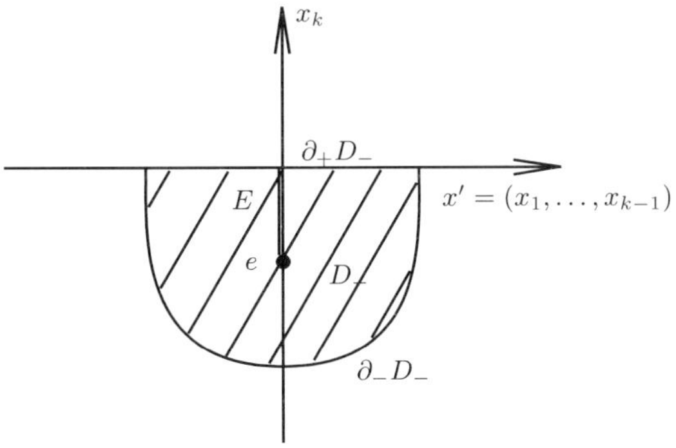

FIGURE 9.4. The lower half-disc.

(W', X', ϕ') be another Smale cobordism with the following properties:

- *$W' \subset W$ and $\partial_- W = \partial_- W'$;*
- *(X', ϕ') has the same critical points and stable discs as (X, ϕ);*
- *$\phi' = \phi$ on $\mathcal{O}p\,(\partial_- W)$, $\phi'(\partial_+ W') = \phi(\partial_+ W)$, and $\phi'(p_i) = \phi(p_i)$ for all $i = 1, \dots, k$.*

Then there exists an isotopy $h_t : W \hookrightarrow W$, $t \in [0,1]$, with $h_0 = \mathrm{Id}$, $h_t(\Delta) = \Delta$ and $h_t = \mathrm{Id}$ on $\mathcal{O}p\,(\partial_- W)$, such that $h_1(W) = W'$ and $\phi = \phi' \circ h_1$. Moreover, the construction can be done smoothly in families.

PROOF. **Step 1.** Applying the Morse Lemma 9.1 and Remark 9.2 near each critical point and extending the diffeomorphisms to all of W, we find a diffeotopy $h_t : W \to W$ preserving Δ, fixed on $\partial_- W$, and such that $\phi' \circ h_1 = \phi$ on $\bigcup_{i=1}^k \mathcal{O}p\, p_i$. After renaming $\phi' \circ h_1$ back to ϕ' and modifying the gradient-like vector field X', we may hence assume that $(X', \phi') = (X, \phi)$ on $\bigcup_{i=1}^k \mathcal{O}p\, p_i$.

Step 2. The identity on $\bigcup_{i=1}^k \mathcal{O}p\, p_i$ extends to a unique diffeomorphism $h_1 : \mathcal{O}p\,\Delta \to \mathcal{O}p\,\Delta$ mapping trajectories of X to trajectories of X' and such that $\phi' \circ h_1 = \phi$ on $\mathcal{O}p\,\Delta$. Following the trajectories for shorter times allows us to connect h_1 to the identity by an isotopy $h_t : \mathcal{O}p\,\Delta \to \mathcal{O}p\,\Delta$ fixed on $\bigcup_{i=1}^k \mathcal{O}p\, p_i$. Then we can adjust h_t near $\partial_- W$ and extend it to a diffeotopy $h_t : W \to W$, fixed on $\bigcup_{i=1}^k \mathcal{O}p\, p_i \cup \mathcal{O}p\,(\partial_- W)$ and preserving Δ, such that $\phi' \circ h_1 = \phi$ on $\mathcal{O}p\,\Delta$. After

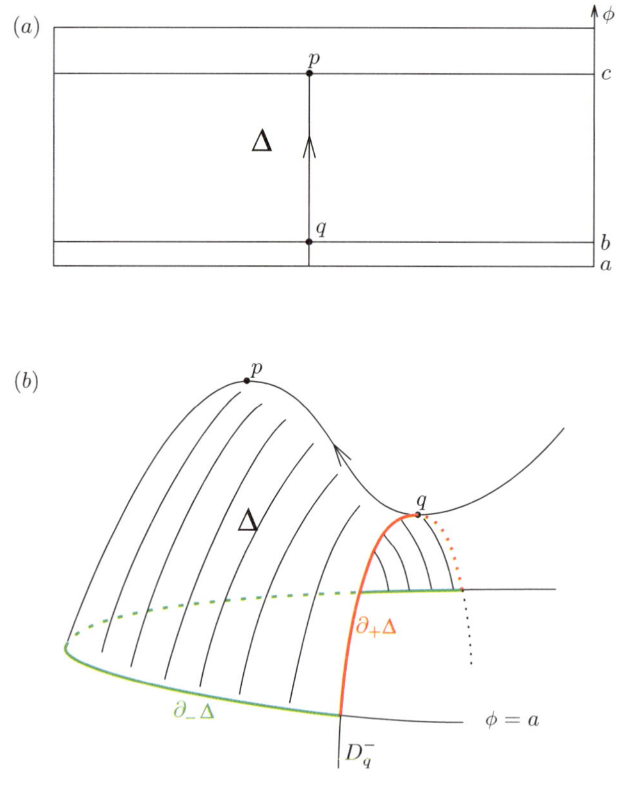

FIGURE 9.5. A cancellation pair of critical points.

renaming $\phi' \circ h_1$ back to ϕ' and modifying the gradient-like vector field X', we may hence assume that $(X', \phi') = (X, \phi)$ on a neighborhood U of $\partial_- W \cup \Delta$.

Step 3. The identity on $\mathcal{O}p\,(\partial_- W \cup \Delta)$ extends to a unique diffeomorphism $h_1 : W \to W'$ mapping trajectories of X to trajectories of X' and such that $\phi' \circ h_1 = \phi$. By Proposition 9.19, there exists an isotopy $g_t : W \hookrightarrow W$, fixed on $\mathcal{O}p\,(\partial_- W \cup \Delta)$ and preserving trajectories of X, such that $g_0 = \mathrm{Id}$ and $g_1(W) \subset U$. Similarly, there exists an isotopy $g_t' : W' \hookrightarrow W'$, fixed on $\mathcal{O}p\,(\partial_- W \cup \Delta)$ and preserving trajectories of X', such that $g_0' = \mathrm{Id}$ and $g_1'(W') \subset U$. As the embeddings $g_1, g_1' \circ h_1 : W \hookrightarrow W$ are both fixed on $\mathcal{O}p\,(\partial_- W \cup \Delta)$ and preserve trajectories of X, they can be connected by an isotopy f_t with the same properties by sliding along trajectories of X. The composition of the isotopies g_t, f_t and the inverse of $g_t' \circ h_t$ gives the desired isotopy $h_t : W \hookrightarrow W$ from $h_0 = \mathrm{Id}$ to h_1. □

If $W' = W$ in Lemma 9.29 the map $h_1 : W \to W$ is a diffeomorphism, but the h_t cannot in general be chosen to be diffeomorphisms. For example, this happens if ϕ and ϕ' have no critical points but define different pseudo-isotopy classes; see Section 9.10.

Cancellation pairs. We conclude this section by describing the setup for cancellation of a pair of critical points. We will return to this setup in Chapters 10 and 13.

Let $\mathbb{R}^k = \mathbb{R}^{k-1} \times \mathbb{R}$ be the space with coordinates $(x_1, \dots, x_k)$, and let $D \subset \mathbb{R}^k$ be the unit disc. We denote by D_- the *lower half-disc* $D \cap \{x_k \le 0\}$, and set $\partial_+ D_- = D_- \cap \{x_k = 0\}$ and $\partial_- D_- = \partial D \cap D_-$, so that we have $\partial D_- = \partial_- D_- \cup \partial_+ D_-$; see Figure 9.4. Further let $e := (0, \dots, 0, -1/2) \in D_-$ and $E := \{(0, \dots, 0, x_k) \in D_- \mid -1/2 < x_k < 0\}$.

Consider now a Smale cobordism (W, X, ϕ) with precisely two critical points p, q of index k and $k-1$, respectively, that are connected by a unique X-trajectory along which W_q^+ intersect transversely. Recall that the stable manifold of q is an embedded disc $W_q^- = D_q^-$. Let Δ be the closure of the stable manifold W_p^- of p in W. See Figure 9.5(a) for a schematic and (b) for a more realistic picture with $a = \phi|_{\partial_- W}$, $b = \phi(q)$, and $c = \phi(p)$. Note that Δ is just the skeleton of (W, X).

LEMMA 9.30. *Suppose that near p and q the pair (X, ϕ) has the standard form from Proposition 9.13. Then Δ is a smoothly embedded half-disc with upper boundary $\partial_+ \Delta = D_q^-$ and lower boundary $\partial_- \Delta = \Delta \cap \partial_- W$. More precisely, there exists a smooth embedding $\alpha : D_- \hookrightarrow W$ such that*

- $\alpha(D_- \setminus \partial_+ D_-) = W_p^-$, $\alpha(\partial_+ D_-) = W_q^-$, *and* $\alpha(\partial_- D_-) \subset \partial_- W$;
- $\alpha(0) = q$, $\alpha(e) = p$, *and* $\alpha(E) = W_p^- \cap W_q^+$.

PROOF. By hypothesis, in suitable coordinates $(x_1, \dots, x_m)$ near q the vector field X is given by

$$X = -\sum_{i=1}^{k-1} x_i \partial_{x_i} + x_k^2 \partial_{x_k} + \sum_{j=k+1}^{m} x_j \partial_{x_j}.$$

In the following discussion, the indices i, j always range over $i = 1, \dots, k-1$ and $j = k+1, \dots, m$. In these coordinates, the stable and unstable manifolds are given by $W_q^- = \{x_k = x_j = 0\}$ and $W_q^+ = \{x_i = 0\}$. Moreover, every trajectory converging to q as $t \to -\infty$ is a ray emanating from the origin in W_q^+. After a rotation in W_q^+, we may assume that the trajectory from q to p corresponds to

$\{x_i = x_j = 0, x_k > 0\}$ in these coordinates. By the transversality assumption, $W_p^- \cap \{x_k = 1\}$ can be locally written as the graph $\{x_j = g_j(x_i), x_k = 1\}$ of smooth functions $g_j : \mathbb{R}^{k-1} \to \mathbb{R}$ with $g_j(0) = 0$. Then W_p^- is the image of $W_p^- \cap \{x_k = 1\}$ under the flow $(x_i, x_k, x_j) \mapsto (e^{-t}x_i, e^t x_k, e^t x_i)$,

$$\begin{aligned} W_p^- &= \{(e^{-t}x_i, e^t, e^t x_j) \mid x_j = g_j(x_i),\ t \in \mathbb{R}\} \\ &= \{(x_i, x_k, x_j) \mid x_j = x_k g_j(x_k x_i) \mid x_k > 0\}. \end{aligned}$$

Thus W_p^- is the graph $\{x_j = G_j(x_i, x_k)\}$ of the functions $G_j(x_i, x_k) = x_k g_j(x_k x_i)$ over the open half-space $\{x_k > 0\}$, which obviously extends smoothly to the closed half-space $\{x_k \geq 0\}$.

The preceding discussion shows that Δ is a smooth submanifold with boundary W_q^- near q. Applying the backward flow of X, this shows smoothness of Δ near all points of W_q^-. On the other hand, $\operatorname{Int}\Delta = W_p^-$ is smooth by Theorem 9.5. This proves that Δ is a smoothly embedded half-disc with upper boundary $\partial_+\Delta = D_q^-$ and lower boundary $\partial_-\Delta = \Delta \cap \partial_- W$, from which the existence of a parametrization α with the desired properties easily follows. □

EXAMPLE 9.31. If (X, ϕ) is not standard near q, the set Δ need not even be C^1-embedded. For example, suppose that $\dim W = 3$ and in coordinates $Z = (x, y, z)$ near q the vector field is given by

$$X = -ax\partial_x + by\partial_y + z\partial_z, \qquad a, b > 0.$$

Moreover, suppose that the trajectory from q to p corresponds to $\{x = y = 0, z > 0\}$ in these coordinates. By the transversality assumption, $W_p^- \cap \{z = 1\}$ can be locally written as the graph $\{y = g(x), z = 1\}$ of a function $g : \mathbb{R} \to \mathbb{R}$ with $g(0) = 0$. Then W_p^- is the image of $W_p^- \cap \{z = 1\}$ under the flow $(x, y, z) \mapsto (e^{-at}x, e^{bt}y, e^t z)$,

$$W_p^- = \{(e^{-at}x, e^{bt}y, e^t) \mid y = g(x),\ t \in \mathbb{R}\} = \{(x, y, z) \mid y = z^b g(z^a x) \mid z > 0\}.$$

For the function $G(x, z) := z^b g(z^a x)$, we compute

$$\frac{\partial G}{\partial z} = bz^{b-1}g(z^a x) + axz^{a+b-1}g'(z^a x) = xz^{a+b-1}\left(b\frac{g(z^a x)}{z^a x} + ag'(z^a x)\right).$$

As $z \to 0$ the term in brackets tends to $(b + a)g'(0)$, so $\frac{\partial G}{\partial z}$ does not extend continuously to $z = 0$ if $g'(0) \neq 0$ and $a + b < 1$.

9.7. Morse and Smale homotopies

DEFINITION 9.32. A smooth family (W, ϕ_t, X_t), $t \in [0, 1]$, of Lyapunov cobordism structures is called *Smale homotopy* if there is a finite set $A \subset (0, 1)$ with the following properties:

- for each $t \in A$ the function ϕ_t has a unique birth-death type critical point e_t such that $\phi_t(e_t) \neq \phi_t(q)$ for all other critical points q of ϕ_t;
- for each $t \notin A$ the function ϕ_t is Morse.

In this case we call the underlying (W, ϕ_t) a *Morse homotopy.*

REMARK 9.33. (a) Note the slight abuse of language because (ϕ_t, X_t) is *not* a Smale cobordism structure for $t \in A$.

(b) By Theorem 9.4 and Corollary 9.16, any family (W, ϕ_t, X_t) of Lyapunov cobordism structures such that (ϕ_0, X_0) and (ϕ_1, X_1) are Smale can be turned into a Smale homotopy by a perturbation fixed near $t = 0, 1$ (C^0-small in the vector fields and C^∞-small in the functions).

(c) It will sometimes be convenient to allow the domain W_t to vary by an isotopy of submanifolds in an ambient equidimensional manifold. We can consider this a homotopy with fixed domain W_0 by pulling back the structures under a family of diffeomorphisms $W_0 \to W_t$.

DEFINITION 9.34. A Smale homotopy $\mathfrak{S}_t = (W, X_t, \phi_t)$, $t \in [0, 1]$ is called an *elementary Smale homotopy* of Type I, IIb, IId, respectively, if the following holds:

- Type I. $\mathfrak{S}_t$ is an elementary Smale cobordism for all $t \in [0, 1]$.
- Type IIb (birth). There is $t_0 \in (0, 1)$ such that for $t < t_0$ the function ϕ_t has no critical points, ϕ_{t_0} has a birth type critical point, and for $t > t_0$ the function ϕ_t has has two critical points p_t and q_t of index i and $i - 1$, respectively, connected by a unique X_t-trajectory.
- Type IId (death). There is $t_0 \in (0, 1)$ such that for $t > t_0$ the function ϕ_t has no critical points, ϕ_{t_0} has a death type critical point, and for $t < t_0$ the function ϕ_t has has two critical points p_t and q_t of index i and $i - 1$, respectively, connected by a unique X_t-trajectory.

We will also refer to an elementary Smale homotopy of Type IIb (resp. IId) as a *creation (resp. cancellation) family.*

Lemma 9.30 has the following parametric version which is proved similarly. We use the notation introduced before Lemma 9.30.

LEMMA 9.35. *Let (W, X_t, ϕ_t), $t \in [-1, 1]$, be an elementary Smale homotopy of Type IIb with birth-type critical point $p_0 = q_0$ at $t = 0$ and nondegenerate critical points p_t, q_t, $t \in (0, 1]$, with* $\mathrm{ind}(p_t) = k$ *and* $\mathrm{ind}(q_t) = k - 1$. *Suppose that near the critical points the family (X_t, ϕ_t) has the standard form from Proposition* 9.13. *Then the skeletons Δ_t, $t \in [0, 1]$, form a smooth family of embedded half-discs with upper boundaries $\partial_+ \Delta_t = D^-_{q_t}$ and lower boundaries $\partial_- \Delta_t = \Delta_t \cap \partial_- W$. More precisely, there exists a smooth family of embeddings $\alpha_t : D_- \hookrightarrow W$ such that*

- *$\alpha_t(D_- \setminus \partial_+ D_-) = W^-_{p_t}$, $\alpha_t(\partial_+ D_-) = W^-_{q_t}$, and $\alpha_t(\partial_- D_-) \subset \partial_- W$;*
- *$\alpha_t(0) = q_t$, $\alpha(\sqrt{t}\, e) = p_t$, and $\alpha(\sqrt{t}\, E) = W^-_{p_t} \cap W^+_{q_t}$.*

An analogous statement holds for an elementary homotopy of Type IId.

DEFINITION 9.36. An *admissible partition* of a Smale homotopy $\mathfrak{S}_t = (W, X_t, \phi_t)$, $t \in [0, 1]$, is a sequence $0 = t_0 < t_1 < \cdots < t_p = 1$ of parameter values, and for each $k = 1, \dots, p$ a finite sequence of functions

$$m(t) = c^k_0(t) < c^k_1(t) < \cdots < c^k_{N_k}(t) = M(t), \qquad t \in [t_{k-1}, t_k],$$

where $m(t) := \phi_t(\partial_- W)$ and $M(t) := \phi_t(\partial_+ W)$, such that $c^k_j(t)$, $j = 0, \dots, N_k$ are regular values of ϕ_t and each Smale homotopy

$$\mathfrak{S}^k_j := \Big(W^k_j(t) := \{c^k_{j-1}(t) \le \phi_t \le c^k_j(t)\}, X_t|_{W^k_j(t)}, \phi_t|_{W^k_j(t)}\Big)_{t \in [t_{k-1}, t_k]}$$

is elementary.

LEMMA 9.37. *Any Smale homotopy admits an admissible partition.*

PROOF. Let $A \subset (0, 1)$ be the finite subset in the definition of a Smale homotopy. Using Lemma 9.28, we now first construct an admissible partition on $\mathcal{O}p\, A$ and then extend it over $[0, 1] \setminus \mathcal{O}p\, A$. $\square$

Equivalence of elementary Smale homotopies. We define the *profile* of a Smale homotopy $\mathfrak{S}_t = (W, X_t, \phi_t)$, $t \in [0,1]$, as the profile $C(\{\phi_t\}) \subset [0,1] \times \mathbb{R}$ of the family of functions $\phi_t : W \to \mathbb{R}$ as in Section 9.1. We will use the notion of profile only for elementary homotopies.

LEMMA 9.38. *Let $\mathfrak{S}_t = (W, X_t, \phi_t)$ and $\widetilde{\mathfrak{S}}_t = (W, \widetilde{X}_t, \widetilde{\phi}_t)$, $t \in [0,1]$, be two elementary Smale homotopies with the same profile such that $\mathfrak{S}_0 = \widetilde{\mathfrak{S}}_0$. Then there exists a diffeotopy $h_t : W \to W$ with $h_0 = \mathrm{Id}$ such that $\phi_t = \widetilde{\phi}_t \circ h_t$ for all $t \in [0,1]$. Moreover, if $\phi_t = \widetilde{\phi}_t$ near $\partial_+ W$ and/or $\partial_- W$ we can arrange $h_t = \mathrm{Id}$ near $\partial_+ W$ and/or $\partial_- W$.*

PROOF. Denote by $C_t, \widetilde{C}_t$ the critical point sets and by $\Delta_t, \widetilde{\Delta}_t$ the skeletons of $\mathfrak{S}_t, \widetilde{\mathfrak{S}}_t$. We first construct a family of diffeomorphisms $f_t : \mathcal{O}p\, C_t \to \mathcal{O}p\, \widetilde{C}_t$ with $f_0 = \mathrm{Id}$ and $\widetilde{\phi}_t \circ f_t = \phi_t$. For this, we first use Theorem 9.4 to construct f_t near the birth-death points, and then the Morse Lemma 9.1 to extend it over the Morse critical points.

Next we canonically extend the maps $f_t : \mathcal{O}p\, C_t \to \mathcal{O}p\, \widetilde{C}_t$ to diffeomorphisms $f_t : U_t \to \widetilde{U}_t$ between neighborhoods of $\Delta_t, \widetilde{\Delta}_t$ mapping ϕ_t to $\widetilde{\phi}_t$ and trajectories of X_t to trajectories of $\widetilde{X}_t$.

Note that $U_t^- := \partial_- W \cap U_t$ is a neighborhood of the submanifold $\Delta_t \cap \partial_- W$, and each restriction $f_t|_{U_t^-}$ is isotopic to the identity by following trajectories for shorter times. Hence by the smooth isotopy extension theorem, after shrinking U_t, the maps $f_t|_{U_t^-}$ extend to diffeomorphisms $g_t : \partial_- W \to \partial_- W$. Moreover, since $f_0 = \mathrm{Id}$ we can arrange $g_0 = \mathrm{Id}$.

Now we extend the maps $U_t \cup \partial_- W \to \widetilde{U}_t \cup \partial_- W$ given by f_t and g_t canonically to diffeomorphisms $h_t : W \to W$ mapping ϕ_t to $\widetilde{\phi}_t$ and trajectories of X_t to trajectories of $\widetilde{X}_t$. Note that $h_0 = \mathrm{Id}$.

Finally, if $\phi_t = \widetilde{\phi}_t$ near $\partial_\pm W$ we undo the diffeotopy h_t on level sets near $\partial_\pm W$ to arrange $h_t = \mathrm{Id}$ on $\mathcal{O}p\,(\partial_\pm W)$. Note that in this last step we destroy the property that h_t maps trajectories of X_t to trajectories of $\widetilde{X}_t$. □

LEMMA 9.39. *Let (W, X, ϕ) be an elementary Smale cobordism with $\phi|_{\partial_\pm W} = a_\pm$ and critical points $p_1, \dots, p_n$ of values $\phi(p_i) = c_i$. For $i = 1, \dots, n$ let $c_i : [0,1] \to (a_-, a_+)$ be smooth functions with $c_i(0) = c_i$. Then there exists a smooth family ϕ_t, $t \in [0,1]$, of Lyapunov functions for X with $\phi_0 = \phi$ and $\phi_t = \phi$ on $\mathcal{O}p\, \partial W$ such that $\phi_t(p_i) = c_i(t)$.*

PROOF. By hypothesis there are no X-trajectories between different critical points, so the stable manifolds are disjoint discs. Denote by $S_i \subset \partial_- W$ the stable sphere of p_i. Pick disjoint tubular neighborhoods of the S_i in $\partial_- W$ and denote by V_i the closures of their forward images under the flow of X. Define $U_i \setminus V_i$ analogously for slightly smaller neighborhoods. The flow of X induces diffeomorphisms $V_i \setminus \mathrm{Int}\, U_i \cong S^{k_i - 1} \times S^{m - k_i - 1} \times [0,1] \times [a_-, a_+]$ in which $\phi(u, v, x, y) = y$ and $X = \partial_y$. Here $m = \dim W$ and $k_i = \mathrm{ind}(p_i)$.

Fix a cutoff function $\rho : [0,1] \to [0,1]$ which equals 1 near 0 and 0 near 1. For diffeomorphisms $\sigma_i : [a_-, a_+] \to [a_-, a_+]$ with $\sigma_i = \mathrm{Id}$ near $a_\pm$ define a function

$\psi_\sigma : W \to [a_-, a_+]$ by

$$\psi_\sigma := \begin{cases} (1-\rho(x))y + \rho(x)\sigma_i(y) & \text{on } V_i \setminus \operatorname{Int} U_i, \\ \sigma_i \circ \phi & \text{on } U_i, \\ \phi & \text{on } W \setminus \bigcup_i V_i. \end{cases}$$

On $V_i \setminus \operatorname{Int} U_i$ we have $\frac{\partial \psi_\sigma}{\partial y} = \big(1-\rho(x)\big) + \rho(x)\sigma_i'(y) > 0$, so ψ_σ is a Lyapunov function for X. Moreover, $\psi_\sigma = \phi$ near ∂W, $\psi_\sigma(p_i) = \sigma_i(c_i)$, and ψ_σ depends smoothly on $\sigma_1, \dots, \sigma_n$. Now pick smooth families of diffeomorphisms $\sigma_{i,t}$ with $\sigma_{i,0} = \mathrm{Id}$ and $\sigma_{i,t}(c_i) = c_i(t)$ and set $\phi_t := \psi_{\sigma_t}$. □

Holonomy of Smale cobordisms.

DEFINITION 9.40. Let (W, X, ϕ) be a Smale cobordism such that the function ϕ has no critical points. The *holonomy* of X is the diffeomorphism

$$\Gamma_X : \partial_+ W \to \partial_- W,$$

which maps $x \in \partial_+ W$ to the intersection of its trajectory under the flow of $-X$ with $\partial_- W$.

Consider now a function $\phi : W \to \mathbb{R}$ without critical points and constant on $\partial_- W$ and $\partial_+ W$. Denote by $\mathcal{X}(W, \phi)$ the space of all gradient-like vector fields for ϕ. Note that the holonomy maps of all $X \in \mathcal{X}(W, \phi)$ are isotopic. We denote by $\mathcal{D}(\partial_+ W, \partial_- W)$ the corresponding path component in the space of diffeomorphisms from $\partial_+ W$ to $\partial_- W$. All spaces are equipped with the C^∞-topology.

Recall that a continuous map $p : E \to B$ is a *Serre fibration* if it has the homotopy lifting property for all closed discs D^k, i.e., given a homotopy $h_t : D^k \to B$, $t \in [0,1]$, and a lift $\widetilde{h}_0 : D^k \to E$ with $p \circ \widetilde{h}_0 = h_0$, there exists a homotopy $\widetilde{h}_t : D^k \to E$ with $p \circ \widetilde{h}_t = h_t$. For more background, see [**91**] or Appendix A.1.

We omit the proof of the following easy lemma.

LEMMA 9.41. *Let (W, ϕ) be a Morse cobordism without critical points. Then the map $\mathcal{X}(W, \phi) \to \mathcal{D}(\partial_+ W, \partial_- W)$ that assigns to X its holonomy Γ_X is a Serre fibration. In particular:*

(i) Given $X \in \mathcal{X}(W, \phi)$ and an isotopy $h_t \in \mathcal{D}(\partial_+ W, \partial_- W)$, $t \in [0,1]$, with $h_0 = \Gamma_X$, there exists a path $X_t \in \mathcal{X}(W, \phi)$ with $X_0 = X$ such that $\Gamma_{X_t} = h_t$ for all $t \in [0,1]$.

(ii) Given a path $X_t \in \mathcal{X}(W, \phi)$, $t \in [0,1]$, and a path $h_t \in \mathcal{D}(\partial_+ W, \partial_- W)$ which is homotopic to Γ_{X_t} with fixed endpoints, there exists a path $\widetilde{X}_t \in \mathcal{X}(W, \phi)$ with $\widetilde{X}_0 = X_0$ and $\widetilde{X}_1 = X_1$ such that $\Gamma_{\widetilde{X}_t} = h_t$ for all $t \in [0,1]$. □

As a consequence, we obtain

LEMMA 9.42. *Let X_t, Y_t be two paths in $\mathcal{X}(W, \phi)$ starting at the same point $X_0 = Y_0$. Suppose that for a subset $A \subset \partial_+ W$, one has $\Gamma_{X_t}(A) = \Gamma_{Y_t}(A)$ for all $t \in [0,1]$. Then there exists a path $\widehat{X}_t \in \mathcal{X}(W, \phi)$ such that*

(i) *$\widehat{X}_t = X_{2t}$ for $t \in [0, \frac{1}{2}]$;*
(ii) *$\widehat{X}_1 = Y_1$;*
(iii) *$\Gamma_{\widehat{X}_t}(A) = \Gamma_{Y_1}(A)$ for $t \in [\frac{1}{2}, 1]$.*

PROOF. Consider the path $\gamma : [0,1] \to \mathcal{D}(\partial_+ W, \partial_- W)$ given by the formula

$$\gamma(t) := \Gamma_{X_1} \circ \Gamma_{X_t}^{-1} \circ \Gamma_{Y_t}.$$

We have $\gamma(0) = \Gamma_{X_1}$ and $\gamma(1) = \Gamma_{Y_1}$. The path γ is homotopic with fixed endpoints to the concatenation of the paths $\Gamma_{X_{1-t}}$ and Γ_{Y_t}. Hence by Lemma 9.41 we conclude that there exists a path $X_t' \in \mathcal{X}(W, \phi)$ such that $X_0' = X_1$, $X_1' = Y_1$, and $\Gamma_{X_t'} = \gamma(t)$ for all $t \in [0,1]$. Since

$$\Gamma_{X_t'}(A) = \Gamma_{X_1}\left(\Gamma_{X_t}^{-1}(\Gamma_{Y_t}(A))\right) = \Gamma_{X_1}(A) = \Gamma_{Y_1}(A),$$

the concatenation $\widehat{X}_t$ of the paths X_t and X_t' has the required properties. □

9.8. The h-cobordism theorem

The topology of a manifold provides a number of constraints on the critical points of a Morse function on it. For instance, the *Morse inequalities* (see [**139**]) assert that the number of index k critical points of a Morse function on a manifold V or cobordism W (exhausting in the manifold case, with regular level sets $\partial_\pm W$ in the cobordism case) is bounded below by the rank of the homology group $H_k(V;\mathbb{Z})$ (or by the rank of the relative homology group $H_k(W, \partial_- W; \mathbb{Z})$ in the case of a cobordism). *Morse–Smale theory* deals with the problem of simplification of a Morse function on a manifold, as much as the topology allows. In particular, one has the celebrated

THEOREM 9.43 (h-cobordism theorem, Smale [**173**]). *Let W be a cobordism of dimension* $\dim W \geq 6$ *such that W and $\partial_\pm W$ are simply connected and* $H_*(W, \partial_- W; \mathbb{Z}) = 0$. *Then W carries a Morse function without critical points and constant on $\partial_\pm W$.*

More generally, a Morse function on a cobordism or a manifold is called *perfect* if it has the minimal number of critical points compatible with the Morse inequalities. Then one has

THEOREM 9.44 (Smale [**173**]). *Let W be a compact manifold with boundary of dimension* $\dim W \geq 6$ *such that W and ∂W are simply connected. Then W carries a perfect Morse function with regular level set ∂W.*

If W is not simply connected, one has a further obstruction to the cancellation of critical points called *Whitehead torsion.* An analogous result in this case, called the *s-cobordism theorem*, was proved by Barden, Mazur, and Stallings (see [**112**]).

The key geometric ingredients in the proof of the h-cobordism and s-cobordism theorem are the following four geometric lemmas about modifications of Smale cobordisms (see [**140**]).

The first lemma is an immediate consequence of Lemma 9.39.

LEMMA 9.45 (Moving critical levels). *Let (W, X, ϕ) be an elementary Smale cobordism. Then there is a homotopy (W, X, ϕ_t)* rel $\mathcal{O}p\,\partial W$ *of elementary Smale cobordisms which arbitrarily changes the ordering of the critical values.*

The second lemma is an immediate consequence of Lemma 9.41 and the smooth isotopy extension theorem.

LEMMA 9.46 (Moving attaching spheres). *Let (W, X, ϕ) be a Smale cobordism and $p \in W$ a critical point whose stable manifold W_p^- intersects $\partial_- W$ along a* *sphere* $S \subset \partial_- W$. *Then given any isotopy* $S_t \subset \partial_- W$ *of the sphere* $S = S_0$, *there exists a homotopy* X_t rel ∂W *of gradient-like vector fields for* ϕ *such that* $X_0 = X$ *and the stable manifold* $W_p^-(X_t)$ *intersects* $\partial_- W$ *along* S_t.

The third lemma is proved by simply implanting a model creation family near a regular point; see [**140**, Lemma 8.2].

LEMMA 9.47 (Creation of critical points). *Let* (W, X, ϕ) *be a Smale cobordism without critical points. Then for any* $1 \le k \le \dim W$ *and any* $p \in \mathrm{Int}\, W$ *there exists a birth type Smale homotopy* (W, X_t, ϕ_t), $t \in [0,1]$, *fixed outside a neighborhood of* p *with* $(X_0, \phi_0) = (X, \phi)$, *which creates a pair of critical points of index* $k-1$ *and* k *connected by a unique trajectory of* X_1 *along which the stable and unstable manifolds intersect transversely.*

The converse lemma is more difficult; see [**140**, Theorem 5.4].

LEMMA 9.48 (Cancellation of critical points). *Suppose that a Smale cobordism* (W, X, ϕ) *contains exactly two critical points of index* $k-1$ *and* k *which are connected by a unique trajectory of* X *along which the stable and unstable manifolds intersect transversely. Then there exists a death type Smale homotopy* (W, X_t, ϕ_t), $t \in [0,1]$, *fixed on* $\mathcal{O}p\, \partial W$ *with* $(X_0, \phi_0) = (X, \phi)$ *which kills the critical points, so the cobordism* (W, X_1, ϕ_1) *has no critical points.*

Using the smooth surroundings provided by Proposition 9.19, one can in fact deduce Lemma 9.48 from the following more elementary lemma (which is a special case of [**140**, Theorem 5.4]). This deduction will be carried out in Section 10.7 in the more difficult context of J-convex functions, using the J-convex surroundings from Chapter 4.

LEMMA 9.49. *Suppose that a Lyapunov pair* (X, ϕ) *on the* k*-dimensional disc* D^k *contains exactly two critical points of index* $k-1$ *and* k *in* $\mathrm{Int}\, D^k$ *which are connected by a unique trajectory of* X *along which the stable and unstable manifolds intersect transversely. Then there exists a family* (X_t, ϕ_t), $t \in [0,1]$, *of Lyapunov pairs on* D^k, *fixed on* $\mathcal{O}p\, \partial D^k$ *with* $(X_0, \phi_0) = (X, \phi)$, *which kills the critical points, so the pair* (X_1, ϕ_1) *has no critical points.*

Here is a sketch of the proof of the h-cobordism theorem based on Lemmas 9.45 to 9.48; see [**140**] for details. For W as in Theorem 9.43 pick any Morse function $\phi : W \to \mathbb{R}$ having $\partial_\pm W$ as regular level sets. Using Lemma 9.45 and a transversality argument, ϕ can be made *self-indexing*, i.e., such that the value of each critical point equals its Morse index.

For $k \in \mathbb{N}$ consider the regular level set $\Sigma = \phi^{-1}(k - \frac{1}{2})$. Let $p_1, \dots, p_s$ be the critical points on level k and $q_1, \dots, q_t$ those on level $k-1$. Denote by $S_i^- \subset \Sigma$ the stable sphere of p_i and by $S_j^+ \subset \Sigma$ the unstable sphere of q_j, and consider the matrix A of homological intersection numbers $S_i^- \cdot S_j^+$.

Next we modify the matrix A using *handle slides*. Namely, consider two critical points p_i and p_j on level k. We first apply Lemma 9.45 to raise p_i to a slightly higher level, and then use Lemma 9.46 to deform the vector field X so that at one moment during the deformation there appears a trajectory connecting p_i and p_j (= the handle slide). As a result, the stable manifold of p_i slides over the stable manifold of p_j, so that after the handle slide the homology class $[S_i^-]$ is replaced by $[S_i^-] + [S_j^-]$. Thus the handle slide has the effect of adding the j-th row to the i-th row in A. Using such elementary row operations and the hypothesis $H_*(W, \partial_- W; \mathbb{Z}) = 0$, we can modify the matrix A such that $S_i^- \cdot S_i^+ = 1$ for $i = 1, \ldots, r \leq \min\{s, t\}$ and all other intersection numbers are zero.

The next task is to get rid of homologically unnecessary intersections between S_i^- and S_j^+ in Σ. For this, consider two transverse intersection points $z_\pm$ with local intersection indices ± 1. Connect them by paths in S_i^- and S_j^+ to obtain an embedded loop γ in Σ. Suppose for the moment that there are no critical points of indices $0, 1, m-1, m$, where $m = \dim W$. Then the hypotheses $\pi_1(W) = 0$ and $\dim W \geq 6$ allow us to apply Whitney's theorem [**191**] and find a Whitney disc $\Delta \subset \Sigma$ with boundary γ meeting $S_i^- \cup S_j^+$ only transversely along the boundary. Then we can eliminate the intersection points $z_\pm$ by pushing S_i^- over Δ using Lemma 9.46.

After this elimination procedure, we are left with S_i^- and S_i^+ for $i = 1, \ldots, r$ intersecting in a unique point. This means that the critical points q_i and p_i are connected by a unique X-trajectory, so we can eliminate them by Lemma 9.48. Performing these steps on all levels, we end up with a Morse function without critical points. This concludes the proof provided there were no critical points of indices $0, 1, m-1, m$. To arrange this, we first use Lemma 9.48 to cancel critical points of index 0 and m. To get rid of a critical point p of index 1 (and similarly for $m-1$), one uses the so-called *Smale trick* (see [**173**]) to create with the use of Lemma 9.47 a pair of critical points q, r of indices 2, 3 in such a way that p and q can be cancelled using Lemma 9.48. This finishes the sketch of the proof of the h-cobordism theorem.

In Chapter 10 we will prove analogues of the four Lemmas 9.45–9.48 for J-convex functions. These will then be used to derive h-cobordism type results for J-convex functions as well as results on deformation of Stein structures in Chapter 15.

9.9. The two-index theorem

The following so-called "two-index theorem" of Hatcher and Wagoner ([**92**], see also [**107**]) will be important for our applications. This theorem is a 1-parametric version of the Smale trick which we mentioned above in our sketch of the proof of the h-cobordism theorem.

PROPOSITION 9.50 ([**92**, Chapter V Prop. 3.5], [**107**, Chapter VI Thm. 1.1]). *Let $f_t : W^m \to [0,1]$ be a generic 1-parameter family of functions on the cobordism W with regular level sets $\partial_- W = f_t^{-1}(0)$ and $\partial_+ W = f_t^{-1}(1)$. Let $i < m-3$ be the lowest index of critical points in this family. Suppose that $(W, \partial_- W)$ is i-connected and f_0, f_1 are Morse without critical points of index i. Then, by introducing new critical points of index $i+1$ and $i+2$, f_t can be deformed rel f_0, f_1 to a family without critical points of index i.*

PROOF. The statement is identical with Proposition 3.5 in Chapter V of [**92**], except that their hypothesis that W is an h-cobordism has been replaced by i-connectivity of the pair $(W, \partial_- W)$. Now the only place in the proof where the hypothesis of an h-cobordism is used is the first step in the proof of Lemma 3.3 in [**92**] where they consider the homotopy exact sequence of a certain triple $\partial_- W \subset W_1 \subset W$,

$$\cdots \to \pi_i(W, \partial_- W) \to \pi_i(W, W_1) \to \pi_{i-1}(W_1, \partial_- W) \to \cdots.$$

Here i-connectivity of $(W, \partial_- W)$ and $(i-1)$-connectivity of $(W_1, \partial_- W)$ together imply i-connectivity of (W, W_1), which is the only conclusion that is needed in the rest of the proof. □

COROLLARY 9.51. *Let $f_t : W^m \to [0,1]$ be a 1-parameter family of functions on the cobordism W with regular level sets $\partial_- W = f_t^{-1}(0)$ and $\partial_+ W = f_t^{-1}(1)$. For some $i < m-3$, suppose that f_0, f_1 are Morse without critical points of index $\leq i$. Then f_t can be deformed rel f_0, f_1 to a family without critical points of index $\leq i$.*

PROOF. The existence of a Morse function f_0 without critical points of index $\leq i$ implies that W is i-connected. Now the corollary follows from the preceding proposition by induction over i. □

COROLLARY 9.52. *Let $f_t : W^{2n} \to [0,1]$ be a 1-parameter family of functions on the cobordism W with regular level sets $\partial_- W = f_t^{-1}(0)$ and $\partial_+ W = f_t^{-1}(1)$.*

(i) *Suppose that $n > 2$ and f_0, f_1 are Morse without critical points of index $> n$. Then f_t can be deformed rel f_0, f_1 to a family without critical points of index $> n$.*
(ii) *Suppose $n > 3$ and f_0, f_1 are Morse without critical points of index $\geq n$. Then f_t can be deformed rel f_0, f_1 to a family without critical points of index $\geq n$.*

PROOF. Consider the cobordism $\overline{W}$ with reversed orientation and the family of functions $\bar{f}_t : \overline{W} \to [0,1]$, $\bar{f}_t(x) := 1 - f_t(x)$. In case (i) the Morse functions $\bar{f}_0, \bar{f}_1$ are without critical points of index $< n$, and in case (ii) without critical points of index $\leq n$. Hence the statement follows from the preceding corollary applied to $\bar{f}_t$ with $m = 2n$ and $i = n-1$ in case (i), and with $i = n$ in case (ii). The necessary inequality reduces to $n - 1 = i < m - 3 = 2n - 3$ or equivalently $n > 2$ in case (i), and to $n = i < m - 3 = 2n - 3$ or equivalently $n > 3$ in case (ii). □

9.10. Pseudo-isotopies

Let us recall the basic notions of pseudo-isotopy theory; see [**30**] and [**92**]. For a manifold W (possibly with boundary) and a closed subset $A \subset W$, we denote by $\mathrm{Diff}(W, A)$ the space of diffeomorphisms of W fixed on $\mathcal{O}p\,(A)$, equipped with the C^∞-topology. For a cobordism W the restriction map to $\partial_+ W$ defines a fibration

$$\mathrm{Diff}(W, \partial W) \to \mathrm{Diff}(W, \partial_- W) \to \mathrm{Diff}_{\mathcal{P}}(\partial_+ W),$$

where $\mathrm{Diff}_{\mathcal{P}}(\partial_+ W)$ denotes the image of the restriction map $\mathrm{Diff}(W, \partial_- W) \to \mathrm{Diff}(\partial_+ W)$. For the product cobordism $I \times M$, $I = [0,1]$, $\partial M = \varnothing$,

$$\mathcal{P}(M) := \mathrm{Diff}(I \times M, 0 \times M)$$

is the group of *pseudo-isotopies* of M. Denote by $\mathrm{Diff}_{\mathcal{P}}(M)$ the group of diffeomorphisms of M that are *pseudo-isotopic to the identity*, i.e., that appear as the restriction to $1 \times M$ of an element in $\mathcal{P}(M)$. Restriction to $1 \times M$ defines the fibration

$$\mathrm{Diff}(I \times M, \partial I \times M) \to \mathcal{P}(M) \to \mathrm{Diff}_{\mathcal{P}}(M),$$

and thus a homotopy exact sequence

$$\cdots \to \pi_0 \mathrm{Diff}(I \times M, \partial I \times M) \to \pi_0 \mathcal{P}(M) \to \pi_0 \mathrm{Diff}_{\mathcal{P}}(M) \to 0.$$

We will use the following alternative description of $\mathcal{P}(M)$; see [**30**]. Denote by $\mathcal{E}(M)$ the space of all smooth functions $f: I \times M \to I$ without critical points and satisfying $f(r,x) = r$ on $\mathcal{O}p\,(\partial I \times M)$. We have a homotopy equivalence

$$\mathcal{P}(M) \to \mathcal{E}(M), \qquad F \mapsto p \circ F,$$

where $p: I \times M \to I$ is the projection. A homotopy inverse is given fixing a metric and sending $f \in \mathcal{E}(M)$ to the unique diffeomorphism F mapping levels of f to levels of p and gradient trajectories of f to straight lines $I \times \{x\}$. Note that the last map in the homotopy exact sequence

$$\cdots \to \pi_0 \mathrm{Diff}(I \times M, \partial I \times M) \to \pi_0 \mathcal{E}(M) \to \pi_0 \mathrm{Diff}_{\mathcal{P}}(M)$$

associates to $f \in \mathcal{E}(M)$ the flow from $0 \times M$ to $1 \times M$ along trajectories of a gradient-like vector field (whose isotopy class does not depend on the gradient-like vector field).

We will discuss in Chapters 14 and 15 symplectic and Stein versions of these notions. For the symplectic version, it will be convenient to replace $I \times M$ by $\mathbb{R} \times M$ as follows: We replace $\mathcal{E}(M)$ by the space of functions $f : \mathbb{R} \times M \to \mathbb{R}$ without critical points and satisfying $f(r,x) = r$ outside a compact set; $\mathrm{Diff}(I \times M, \partial I \times M)$ by the space $\mathrm{Diff}_c(\mathbb{R} \times M)$ of diffeomorphisms that equal the identity outside a compact set; and $\mathcal{P}(M)$ by the space of diffeomorphisms of $\mathbb{R} \times M$ that equal the identity near $\{-\infty\} \times M$ and have the form $(r,x) \mapsto (r + f(x), g(x))$ near $\{+\infty\} \times M$. The last map in the exact sequence

$$\cdots \to \pi_0 \mathrm{Diff}_c(\mathbb{R} \times M) \to \pi_0 \mathcal{E}(M) \to \pi_0 \mathrm{Diff}_{\mathcal{P}}(M)$$

then associates to $f \in \mathcal{E}(M)$ the flow from $\{-\infty\} \times M$ to $\{+\infty\} \times M$ along trajectories of a gradient-like vector field which equals ∂_r outside a compact set.

We endow the spaces $\mathcal{P}(M)$, $\mathcal{E}(M)$ and $\mathrm{Diff}_c(\mathbb{R} \times M)$ with the topology of uniform C^∞-convergence on $\mathbb{R} \times M$ (and *not* the topology of uniform C^∞-convergence on compact sets), with respect to the product of the Euclidean metric on $\mathbb{R}$ and any Riemannian metric on M. In other words, a sequence $F_n \in \mathcal{P}(M)$ converges to $F \in \mathcal{P}(M)$ if and only if $\|F_n - F\|_{C^k(\mathbb{R} \times M)} \to 0$ for every $k = 0, 1, \dots$. For example, consider any non-identity element $F \in \mathcal{P}(M)$ and the translations $\tau_c(r,x) = (r + c, x)$, $c \in \mathbb{R}$, on $\mathbb{R} \times M$. Then the sequence $F_n := \tau_n \circ F \circ \tau_{-n}$ *does not converge* as $n \to \infty$ to the identity in $\mathcal{P}(M)$, although it does converge uniformly on compact sets. With this topology, the obvious inclusion maps from the spaces on $I \times M$ to the corresponding spaces on $\mathbb{R} \times M$ are weak homotopy equivalences.

REMARK 9.53. It was proven by Cerf in [**30**] that $\pi_0 \mathcal{P}(M)$ is trivial if $\dim M \geq 5$ and M is simply connected. In the non-simply connected case and for $\dim M \geq 6$, Hatcher and Wagoner ([**92**], see also [**107**]) have expressed $\pi_0 \mathcal{P}(M)$ in terms of algebraic K-theory of the group ring of $\pi_1(M)$. In particular, there are many fundamental groups for which $\pi_1 \mathcal{P}(M)$ is not trivial.

CHAPTER 10

Modifications of J-Convex Morse Functions

In this chapter we discuss modifications of J-convex Morse functions on a given complex manifold. This parallels the h-cobordism theory for ordinary Morse functions in Section 9.8. More precisely, we show how to perform the following operations:

- moving attaching spheres by isotropic isotopies (Section 10.1);
- moving critical levels (Section 10.3);
- creation and cancellation of critical points (Sections 10.4–10.8).

Section 10.2 is an aside on the J-orthogonality condition used in Chapter 8.

The proofs in this chapter rely on the techniques developed in Chapters 3 and 4.

10.1. Moving attaching spheres by isotropic isotopies

For a function $\phi : V \to \mathbb{R}$ we will use the notations

$$V^b := \phi^{-1}(b), \qquad V^{[a,b]} := \phi^{-1}([a,b]).$$

The goal in this section is to prove the following result:

PROPOSITION 10.1. *Consider a complex manifold (V, J) and a proper J-convex function $\phi : V \to \mathbb{R}$ without critical values in the interval $[a, b]$. Let $\Lambda \subset V^b$ be a closed isotropic submanifold and $L \subset V$ its image under the flow of $-\nabla_\phi \phi$. Let $(\Lambda_t)_{t\in[0,1]}$ be an isotropic isotopy of $\Lambda_0 := L \cap V^a$ in V^a.*

Then, after composing ϕ with a sufficiently convex increasing function f : $[a, b] \to \mathbb{R}$, there exists a diffeotopy $h_t : V \to V$ with the following properties for all $t \in [0, 1]$ (see Figure 10.1):

(i) $h_t = \mathrm{Id}$ outside $V^{[a,b]}$;

(ii) $\phi_t := \phi \circ h_t$ is J-convex;

(iii) the image L_t of Λ under the flow of $-\nabla_{\phi_t}\phi_t$ intersects V^a in Λ_t.

REMARK 10.2. The corresponding result for ordinary functions ϕ is very easy: It just states that one can realize a smooth isotopy of spheres Λ_t as descending spheres for a homotopy of gradient-like vector fields, keeping the function ϕ fixed. In contrast, Proposition 10.1 is more subtle because the gradient vector fields $\nabla_{\phi_t}\phi_t$ are determined by the functions ϕ_t themselves.

The proof requires some preparation. The following lemma is the main technical ingredient.

LEMMA 10.3. *Let Σ be a J-convex hypersurface with field of complex tangencies ξ in a complex manifold (V, J). Let $X^\perp$ be a transverse vector field along Σ with $JX^\perp \in T\Sigma$. Let $\Lambda \subset \Sigma$ be an isotropic submanifold, and let X be a vector field along Λ that is transverse to Σ. Then for any compact subset $K \subset \Lambda$ there exists a J-convex hypersurface Σ' with the following properties (see Figure 10.2):*

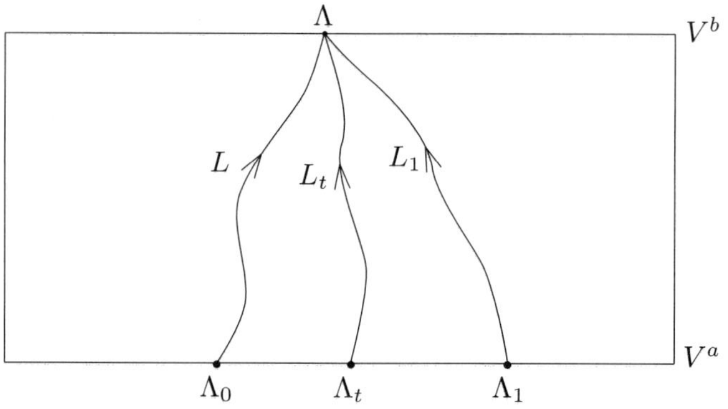

FIGURE 10.1. Moving attaching spheres by isotropic isotopies.

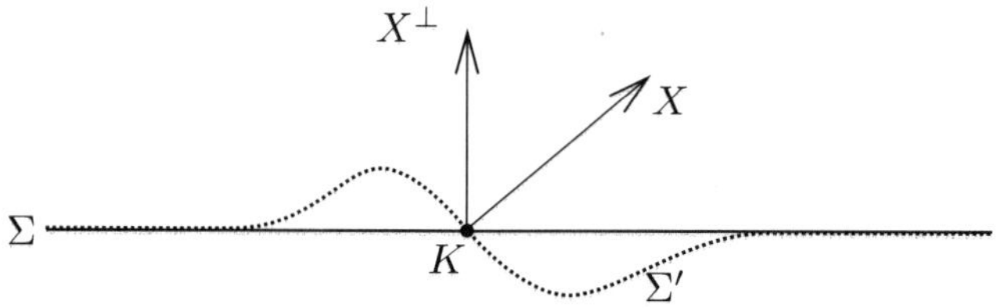

FIGURE 10.2. Turning a J-convex hypersurface along an isotropic submanifold.

(i) $K \subset \Sigma'$ *and* $\xi \subset T\Sigma'$ *along* K;
(ii) Σ' *is transverse to* $X^\perp$ *and* $\Sigma' = \Sigma$ *outside a neighborhood of* K;
(iii) $JX(x) \in T_x\Sigma'$ *for all* $x \in K$.

PROOF. Let $n = \dim_{\mathbb{C}} V$ and $k - 1 = \dim \Lambda$. We will only carry out the proof in the Legendrian case $k = n$, the case $k < n$ being analogous but notationally more involved. Note that the case $k < n$ formally follows from the Legendrian case provided that the symplectic normal bundle $(T\Lambda)^\omega/T\Lambda$ of Λ in the field of complex tangencies $\xi \subset T\Sigma$ is trivial. Indeed, in this case a neighborhood of Λ (after shrinking it) in Σ is contactomorphic to a neighborhood of the zero section in $J^1\Lambda \oplus \mathbb{C}^{n-k}$ (see Chapter 6). So we can extend Λ to a Legendrian submanifold $\widetilde{\Lambda} \cong \Lambda \times \mathbb{R}^{n-k} \subset \Sigma$ and X to a vector field $\widetilde{X}$ along $\widetilde{\Lambda}$ transverse to Σ.

After possibly changing its sign, we may assume that $X^\perp$ is opposite to the coorientation of Σ. The flow of $X^\perp$ extends Λ (after shrinking Λ) to a totally real submanifold $\Lambda \times [-1, 1] \subset V$. By Proposition 5.55 the inclusion $\Lambda \times [-1, 1] \hookrightarrow V$ extends to a diffeomorphism of a neighborhood of $\Lambda \times [-1, 1]$ in $\Lambda^{\mathbb{C}} \oplus \mathbb{C}$ onto a neighborhood of $\Lambda \times [-1, 1]$ in V such that the pullback of J (still denoted by J) and the standard structure $J_{\rm st}$ on $\Lambda^{\mathbb{C}} \oplus \mathbb{C}$ coincide along $\Lambda \times [-1, 1]$ together with their 7-jets. Here $\Lambda^{\mathbb{C}}$ is the complexification of Λ (for some real analytic structure on Λ) and $X^\perp$ generates the real line $0 \oplus i\mathbb{R}$. This implies that $T\Sigma = T\Lambda^{\mathbb{C}} \oplus \mathbb{R}$ with field of complex tangencies $\xi = T\Lambda^{\mathbb{C}}$ along Λ. Denote coordinates on $\Lambda^{\mathbb{C}} \oplus \mathbb{C}$ by $(z, w) = (x, y, u + iv)$, where y are coordinates on Λ and x coordinates in the fibers of $\Lambda^{\mathbb{C}}$. In these coordinates, Σ can be written near Λ as the graph

$$\Sigma = \{v = \phi(x, y, u)\}$$

of a function ϕ with $\phi(0, y, 0) = 0$ and $d\phi(0, y, 0) = 0$. The choice of $X^\perp$ implies that Σ is J-convex cooriented from above. We will find Σ' as the graph $\Sigma' = \{v = \widetilde{\phi}(x, y, u)\}$ of a function $\widetilde{\phi}$ with $\widetilde{\phi} = \phi$ outside a neighborhood of $K \oplus 0$ in $\Lambda^{\mathbb{C}} \oplus \mathbb{R}$. Then Σ' is transverse to $X^\perp = \partial_v$. The conditions $K \subset \Sigma'$ and $\xi \subset T\Sigma$ along K are equivalent to $\widetilde{\phi}(0, y, 0) = 0$ and $d_z\widetilde{\phi}(0, y, 0) = 0$ for $y \in K$. After rescaling and possibly changing its sign, we can write the given vector field X as $X = \partial_v - \tau(y)\partial_u + Y$ with Y tangent to $\Lambda^{\mathbb{C}}$ and τ some given function on Λ. Then $JX \in T\Sigma'$ along K is equivalent to $\widetilde{\phi}_u(0, y, 0) = \tau(y)$ for $y \in K$.

Let $Q := \mathrm{dist}^2_\Lambda$ be the squared distance (with respect to some Hermitian metric for J_{st}) from the zero section in $\Lambda^{\mathbb{C}}$. By Proposition 2.15, Q is a J_{st}-convex function. Note that the hypersurface $\{v = Q(x, y)\}$ is tangent to Σ along Λ. Its Levi form at points of Λ is given by $-dd^{\mathbb{C}}\big(Q(x, y) - v\big)|_{\xi = T\Lambda^{\mathbb{C}}} = -dd^{\mathbb{C}}Q$, so $\{v = Q(x, y)\}$ is J_{st}-convex along Λ cooriented from above. Since the Levi forms with respect to J and J_{st} agree along Λ, the hypersurfaces Σ and $\{v = Q(x, y)\}$ are also J-convex near Λ. Thus by Corollary 3.31 we can modify Σ near K, preserving J-convexity and the condition $\Lambda \subset \Sigma$, such that $\Sigma = \{v = Q(x, y)\}$ near K.

Now let a function $\tau(y)$ be given as above. Our task is to find a smooth function $\widetilde{\phi}$ with J-convex graph such that

$$\widetilde{\phi}(0, y, 0) = 0, \qquad d_z\widetilde{\phi}(0, y, 0) = 0, \qquad \widetilde{\phi}_u(0, y, 0) = \tau(y)$$

for $y \in K$ and $\widetilde{\phi}(x, y, u) = Q(x, y)$ outside a neighborhood of K.

Pick a function $g(y, u)$ on $\Lambda \oplus \mathbb{R}$ with $g(y, 0) = 0$ for all $y \in \Lambda$ and $g_u(y, 0) = \tau(y)$ for $y \in K$, and such that $g(y, u) < -1$ outside $K' \times [-1, 1]$ for some compact neighborhood K' of K in Λ. For any $\varepsilon > 0$ let $g^\varepsilon(y, u) := \varepsilon g(y, u/\varepsilon)$. These functions satisfy $g^\varepsilon(y, 0) = 0$, $g^\varepsilon_y(y, 0) = 0$ and $g^\varepsilon_u(y, 0) = \tau(y)$ for all $y \in K$, and $g^\varepsilon(y, u) < -\varepsilon$ outside $K' \times [-\varepsilon, \varepsilon]$. Moreover, we have

$$|g^\varepsilon(y, u)| \le C_0|u| \le C_0\varepsilon, \quad |g^\varepsilon_y|, |g^\varepsilon_{yy}| \le C_0\varepsilon, \quad |g^\varepsilon_u|, |g^\varepsilon_{yu}| \le C_0, \quad |g^\varepsilon_{uu}| \le C_0/\varepsilon$$

for $(y, u) \in K' \times [-\varepsilon, \varepsilon]$ with a constant $C_0 \ge 1$ not depending on ε. For $0 < a \le 1/2$ and $\varepsilon > 0$ consider the function

$$\psi(x, y, u) := aQ(x, y) + g^\varepsilon(y, u).$$

Our desired function $\widetilde{\phi}$ will be a smoothing of

$$\widetilde{\psi} := \max\,(Q - \varepsilon, \psi).$$

Let us first determine the region where $\psi < Q - \varepsilon$, or equivalently,

$$g^\varepsilon(y, u) + \varepsilon < (1 - a)Q(x, y). \tag{10.1}$$

For $|u| > \varepsilon$ or $y \notin K'$ this inequality holds because the left hand side is negative and the right hand side is nonnegative. Moreover, $1 - a \ge 1/2$ implies

$$g^\varepsilon(y, u) + \varepsilon \le (C_0 + 1)\varepsilon \le 2(C_0 + 1)\varepsilon(1 - a),$$

so inequality (10.1) holds if $Q(x, y) > C_1\varepsilon$ with the constant $C_1 := 2(C_0 + 1)$ not depending on ε and a. So we have $\psi < Q - \varepsilon$ outside the compact region

$$W' := \{(x, y, u) \mid y \in K', |u| \le \varepsilon, Q(x, y) \le C_1\varepsilon\}.$$

On the other hand, in the region

$$W := \{(x,y,u) \mid y \in K', Q(x,y) + C_0|u| \le \varepsilon\} \subset W',$$

we have the reverse estimate

$$g^\varepsilon(y,u) + \varepsilon \ge \varepsilon - C_0|u| \ge Q(x,y) \ge (1-a)Q(x,y).$$

Hence $\psi \ge Q - \varepsilon$ on the neighborhood W of K.

We will show below that for $\varepsilon = a^2$ sufficiently small the graph of ψ is J-convex on W'. Assuming this for the moment, note that the graph of $Q-\varepsilon$ is also J-convex. Thus by Corollary 3.23, we can C^0-approximate $\widetilde{\psi}$ by a smooth function with J-convex graph $\widetilde{\Sigma}$ which agrees with ψ on W and $Q - \varepsilon$ outside W'. (Note that in Corollary 3.23 the minimum appears rather than the maximum because the graphs are cooriented from below rather than above.) Now on any fixed (i.e., independent of a, ε) compact neighborhood U of K', the function $Q - \varepsilon$ C^2-approaches Q as $\varepsilon \to 0$. Hence for small ε we can modify $\widetilde{\Sigma}$ outside W' so that it agrees with Σ outside U. This yields the desired hypersurface Σ'.

It remains to prove J-convexity of the hypersurface $\Sigma_\psi = \{v = \psi(z,u)\}$ over W' for small a and ε. For this, cover K' by finitely many $J_{\rm st}$-holomorphic coordinate charts in which Λ corresponds to $i\mathbb{R}^{n-1}$. Choose ε so small that the coordinate charts cover the region $\{(x,y) \mid y \in K', Q(x,y) \le C_1\varepsilon\}$. We will show that the normalized modulus of $J_{\rm st}$-convexity of Σ_ψ satisfies $\mu(\Sigma_\psi) \ge \varepsilon^2$ for $\varepsilon = a^2$ sufficiently small. On the other hand, the definition of W' shows that the distance to Λ is bounded above by $C_2\sqrt{\varepsilon}$ on the graph of ψ over W', for some constant C_2 independent of ε. Since J and $J_{\rm st}$ coincide with their 7-jets along Λ, it follows that $\|J - J_{\rm st}\|_{C^2} \le C_3\varepsilon^{5/2}$ on the graph of ψ over W'. Thus by Corollary 3.37, the graph of ψ over W' is J-convex for sufficiently small $\varepsilon > 0$.

So it remains to prove the estimate $\mu(\Sigma_\psi) \ge \varepsilon^2$. We write Σ_ψ as the zero set of the function $\Psi(x,y,u,v) := \psi(x,y,u) - v$, whose gradient is given by $|\nabla\Psi|^2 = 1 + |\nabla\psi|^2 = 1 + \psi_u^2 + |d_z\psi|^2$. Using the definition (3.11) of the normalized modulus of convexity and Lemma 2.24 we find

$$\mu(\Sigma_\psi) = \frac{m(\mathbb{L}_{\Sigma_\psi})}{\max\{M(II_{\Sigma_\psi}), 1\}} = \frac{m(H_\Psi)}{\max\{M(\mathrm{Hess}_\Psi), |\nabla\Psi|\}}$$
$$= \frac{m(H_\psi)}{\max\{M(\mathrm{Hess}_\psi), \sqrt{1+|\nabla\psi|^2}\}}.$$

By Lemma 2.26, $m(H_\psi) = |\nabla\Psi| m(\mathbb{L}_{\Sigma_\psi})$ satisfies

$$m(H_\psi) \ge \frac{H_\psi^{\min}(1+\psi_u^2) - |\psi_{uu}|\,|d_z\psi|^2 - 2|d_z\psi_u|\,|d_z\psi|\sqrt{1+\psi_u^2}}{1+\psi_u^2+|d_z\psi|^2}$$
$$\ge \frac{H_\psi^{\min} - |\psi_{uu}|\,|d_z\psi|^2 - 2|d_z\psi_u|\,|d_z\psi|\,(1+|\psi_u|)}{1+\psi_u^2+|d_z\psi|^2}.$$

So for $\mu(\Sigma_\psi) \ge \varepsilon^2$ it suffices to show

$$\begin{aligned} &H_\psi^{\min} - |\psi_{uu}|\,|d_z\psi|^2 - 2|d_z\psi_u|\,|d_z\psi|\,(1+|\psi_u|) \\ &\quad \ge \varepsilon^2(1+\psi_u^2+|d_z\psi|^2)\max\{M(\mathrm{Hess}_\psi), \sqrt{1+\psi_u^2+|d_z\psi|^2}\}. \end{aligned} \tag{10.2}$$

By J-convexity of the function Q, we have $H_Q^{\min} \ge \gamma$ for some constant $\gamma > 0$. Moreover, $|Q_z| \le C|x|$ and all derivatives of Q involving a u-derivative vanish.

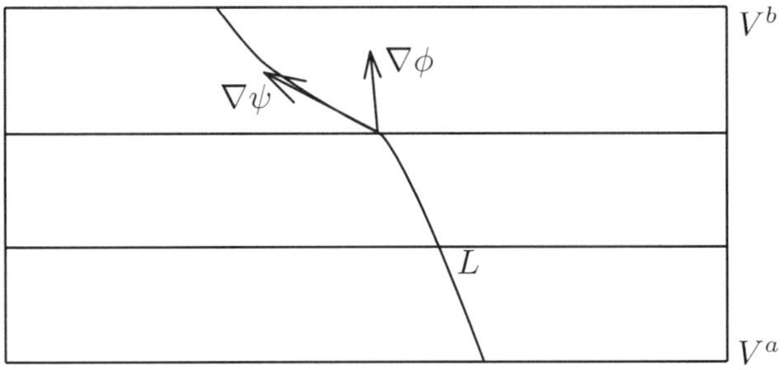

FIGURE 10.3. Making the gradient of a J-convex function tangent to a totally real submanifold J-orthogonal to its level sets.

Here and in the following C denotes a generic constant that depends on C_0, C_1, γ but not on a, ε. The estimates for g^ε yield

$$H_\psi^{\min} \geq \gamma a - C\varepsilon, \quad |\psi_z| \leq Ca|x| + C\varepsilon, \quad |\psi_u|, |\psi_{zu}| \leq C, \quad |\psi_{uu}| \leq C/\varepsilon$$

for $(y, u) \in K' \times [-\varepsilon, \varepsilon]$. It follows that the left hand side in (10.2) is estimated from below by

$$A := \gamma a - C\varepsilon - Ca|x| - Ca^2|x|^2/\varepsilon.$$

Now on W' we have $\gamma|x|^2 \leq Q(x, y) \leq C_1\varepsilon$, and hence

$$A \geq \gamma a - C\varepsilon - Ca\sqrt{\varepsilon} - Ca^2.$$

To estimate the right hand side from above, we first compute

$$\begin{aligned} M(\mathrm{Hess}_\psi) &\leq C(aM(Q) + |g_{yy}^\varepsilon| + |g_{yu}^\varepsilon| + |g_{uu}^\varepsilon|) \\ &\leq C(a\gamma + \varepsilon + 1 + 1/\varepsilon) \leq C/\varepsilon, \end{aligned}$$

since γ is fixed and $a \leq 1/2$. As $1 + \psi_u^2 + |d_z\psi|^2 \leq C$, we see that the right hand side of (10.2) is bounded from above by $C\varepsilon$, and therefore (10.2) is implied by

$$\gamma a - C\varepsilon - Ca\sqrt{\varepsilon} - Ca^2 \geq 0.$$

Choosing $\varepsilon = a^2$, this becomes

$$\gamma a - Ca^2 \geq 0$$

which is satisfied for $\varepsilon = a^2 > 0$ sufficiently small. This proves the estimate $\mu(\Sigma_\psi) \geq \varepsilon^2$ and hence Lemma 10.3. □

LEMMA 10.4. *Let ϕ be a proper J-convex function on the complex manifold V without critical values in $[a, b]$. Let $L \subset V^{[a,b]}$ be a totally real submanifold that intersects each level set J-orthogonally in a compact manifold (possibly with boundary). Then there exists a J-convex function ψ, C^1-close to ϕ, such that $\psi = \phi$ on L and $\nabla_\psi\psi$ is tangent to L.*

Moreover, if $\nabla_\phi\phi$ is already tangent to L near $V^{[a,a']} \cup V^{[b',b]}$ for some $[a', b'] \subset (a, b)$, then we can choose $\psi = \phi$ on $V^{[a,a']} \cup V^{[b',b]}$ (see Figure 10.3).

REMARK 10.5. The gradient $\nabla_\psi\psi$ will in general not be C^0-close to $\nabla_\phi\phi$. This is possible despite ψ being C^1-close to ϕ because the metric g_ψ need not be C^0-close to g_ϕ.

PROOF. If $\dim L < \dim_{\mathbb{C}} V$, we extend L to a totally real submanifold $L' \subset V^{[a,b]}$ of dimension $\dim L' = \dim_{\mathbb{C}} V$, still intersecting all level sets J-orthogonally. Hence it suffices to consider the case $\dim L = \dim_{\mathbb{C}} V$.

Let X be the unique vector field tangent to L, orthogonal (with respect to the metric g_ϕ) to the intersection of L with level sets of ϕ, with $d\phi(X) \equiv 1$. (However, X need not be orthogonal to the level sets of ϕ). By J-orthogonality, JX is tangent to the level sets of ϕ. The flow of X defines a diffeomorphism $\Lambda \times i[a,b] \cong L$, where $\Lambda := L \cap V^a$. By Proposition 5.55, this diffeomorphism extends to a diffeomorphism from a neighborhood of $\Lambda \times i[a,b]$ in $\Lambda^{\mathbb{C}} \times \mathbb{C}$ (where $\Lambda^{\mathbb{C}}$ is a complexification of Λ) onto a neighborhood of L in V, such that the pullback of J (still denoted by J) agrees with the standard complex structure $J_{\rm st}$ on $\Lambda^{\mathbb{C}} \times \mathbb{C}$ up to second order along $\Lambda \times i[a,b]$ (which we will again denote by L). Then the Levi forms of functions with respect to J and $J_{\rm st}$ agree along L, so we can compute Levi forms with respect to $J_{\rm st}$.

Denote coordinates on $\Lambda^{\mathbb{C}}$ by z and on $\mathbb{C}$ by $u + iv$. Under this identification L corresponds to $\Lambda \times i[a,b]$, and $X = \partial_v$, $\phi = v$ along L. Since the level sets of ϕ are J-orthogonal to L, they are tangent to $T\Lambda^{\mathbb{C}} \oplus \mathbb{R}$ along L. Define the function

$$\psi(z,u,v) := v + Q(z) + \frac{1}{2} f(z,v) u^2$$

on $\Lambda^{\mathbb{C}} \times \mathbb{C}$, where $Q := \mathrm{dist}^2_\Lambda$ for some Hermitian metric on $\Lambda^{\mathbb{C}}$ and f is a positive function. We compute

$$\begin{aligned} d\psi &= dv + dQ + f(z,v)u\,du + \frac{1}{2}u^2 df, \\ d^{\mathbb{C}}\psi &= du + dQ \circ J_{\Lambda^{\mathbb{C}}} - f(z,v)u\,dv + \frac{1}{2}u^2 d^{\mathbb{C}} f, \\ \omega_\psi &= -dd^{\mathbb{C}}\psi = \omega_Q + f(z,v) du \wedge dv \text{ along } L. \end{aligned}$$

In particular, ψ is J-convex and $d\psi = dv = d\phi$ along L. Hence by Proposition 3.26, ψ can be extended to a J-convex function on V which agrees with ϕ outside a neighborhood of L. Moreover, ψ can be chosen arbitrarily C^1-close to ϕ with modulus of J-convexity m_ψ bounded from below. The gradient of ψ is determined by the equation

$$\omega_\psi(\nabla_\psi \psi, Y) = -d^{\mathbb{C}}\psi(Y)$$

for all $Y \in TV$. Now $d^{\mathbb{C}}\psi = du$ along L implies $\nabla_\psi \psi = f(z,v)^{-1}\partial_v$ along L, so $\nabla_\psi \psi$ is tangent to L.

Finally, suppose that $\nabla_\phi \phi$ is already tangent to L near $V^{[a,a']} \cup V^{[b',b]}$. Pick a cutoff function $\beta : V \to [0,1]$ which equals 0 outside $V^{[a',b']}$ and 1 where $\nabla_\phi \phi$ is not tangent to L. Construct ψ as above and set

$$\theta := (1-\beta)\phi + \beta\psi.$$

This function agrees with ϕ on $V^{[a,a']} \cup V^{[b',b]}$, and by Lemma 3.28 (since m_ψ is bounded from below), θ is J-convex for ψ sufficiently C^1-close to ϕ.

It remains to show that $\nabla_\theta \theta$ is tangent to L at points x with $0 < \beta(x) < 1$. By construction, we have $\phi(x) = \psi(x)$ and $d\phi(x) = d\psi(x)$. Moreover, since $\nabla_\phi \phi(x)$ is tangent to L, the choice of X implies that X is proportional to $\nabla_\phi \phi$ at x. So the three vector fields X, $\nabla_\phi \phi$, and $\nabla_\psi \psi$ are positively proportional at x. Since $\nabla_\psi \psi = f(z,v)^{-1} X$ along L, we can therefore choose the positive function f in the construction of ψ to arrange $\nabla_\phi\phi = \nabla_\psi\psi$ along $L \cap \{0 < \beta < 1\}$. Since ϕ and ψ agree to first order along L, we have

$$d^{\mathbb{C}}\theta = (1-\beta)d^{\mathbb{C}}\phi + \beta d^{\mathbb{C}}\psi, \qquad \omega_\theta = (1-\beta)\omega_\phi + \beta\omega_\psi$$

at the point x. Hence for any $Y \in T_xV$,

$$\begin{aligned}\omega_\theta(\nabla_\theta\theta, Y) &= -d^{\mathbb{C}}\theta(Y) = -(1-\beta)d^{\mathbb{C}}\phi(Y) - \beta d^{\mathbb{C}}\psi(Y)\\ &= (1-\beta)\omega_\phi(\nabla_\phi\phi, Y) + \beta\omega_\psi(\nabla_\psi\psi, Y)\\ &= (1-\beta)\omega_\phi(\nabla_\phi\phi, Y) + \beta\omega_\psi(\nabla_\phi\phi, Y)\\ &= \omega_\theta(\nabla_\phi\phi, Y).\end{aligned}$$

This shows $\nabla_\theta\theta = \nabla_\phi\phi$ along L. In particular, $\nabla_\theta\theta$ is tangent to L, so θ is the desired function. □

PROOF OF PROPOSITION 10.1. Let $\Sigma := V^a$. The flow of the vector field $\nabla\phi/|\nabla\phi|$ defines a diffeomorphism

$$\Sigma \times [a,b] \cong V^{[a,b]}.$$

Under this identification, ϕ corresponds to the function $(x,r) \mapsto r$, $\nabla\phi/|\nabla\phi|$ to the vector field ∂_r, L to $\Lambda \times [a,b]$, and Λ_t to $\Lambda_t \times \{a\}$. In view of Lemma 11.13, $\Lambda_t \times \{r\}$ is isotropic for the contact structure ξ_r on $\Sigma \times \{r\}$ for all $r \in [a,b]$.

Pick a C^2-function $g : [a,b] \to [0,1]$ which equals 1 on $[a,a']$ and 0 on $[b',b]$, for some interval $[a',b'] \subset (a,b)$. For $t \in [0,1]$ define

$$L_t := \bigcup_{r\in[a,b]} \Lambda_{tg(r)} \times \{r\} \subset \Sigma \times [a,b].$$

This is a totally real submanifold which intersects each level set $\Sigma \times \{r\}$ in the isotropic submanifold

$$\Lambda_{t,r} := \Lambda_{tg(r)} \times \{r\}.$$

Let $X_{t,r}$ be the unique vector field tangent to L_t along $\Lambda_{t,r}$ and orthogonal to $\Lambda_{t,r}$ (with respect to the metric g_ϕ) with $dr(X_{t,r}) = 1$. In particular, $X_{t,r}$ is transverse to the level sets $\Sigma \times \{r\}$. Hence by Lemma 10.3 there exist J-convex hypersurfaces $\Sigma_{t,r}$ transverse to ∂_r such that $\Lambda_{t,r} \subset \Sigma_{t,r}$, the contact structure ξ_r is contained in $T\Sigma_{t,r}$ along $\Lambda_{t,r}$, and $JX_{t,r} \in T\Sigma_{t,r}$. Note that the last two conditions say that L_t intersects $\Sigma_{t,r}$ J-orthogonally for all r. Moreover, we may choose $\Sigma_{t,r} = \Sigma \times \{r\}$ for r outside $[a',b']$.

By construction, the $\Sigma_{t,r}$ for fixed t and varying r form a foliation near L_t. Thus by Proposition 3.25, we can modify the $\Sigma_{t,r}$ to a J-convex foliation, keeping them fixed near L_t and for r outside $[a',b']$. Let ψ_t be the function which equals r on the new hypersurfaces $\widetilde{\Sigma}_{r,t}$. Pick a sufficiently convex increasing function $f : \mathbb{R} \to \mathbb{R}$ such that $f \circ \psi_t$ is J-convex for all $t \in [0,1]$. Now we apply Lemma 10.4 to the functions $f \circ \psi_t$ and the totally real submanifolds L_t. We find J-convex functions ϕ_t, C^1-close to $f \circ \psi_t$ and agreeing with $f \circ \psi_t$ on L_t and for r outside $[a',b']$, such that $\nabla_{\phi_t}\phi_t$ is tangent to L_t. Thus L_t is the image of $\Lambda_{t,b} = \Lambda_0 \times \{b\} = \Lambda$ under the flow of $-\nabla_{\phi_t}\phi_t$, and by construction L_t intersects $\Sigma \times \{a\}$ in $\Lambda_{t,a} = \Lambda_t \times \{a\}$. This proves property (iii).

By construction, ϕ_t agrees with $f \circ \phi$ for r outside $[a',b']$. Moreover, since $L_0 = L$, we can arrange $\phi_0 = f \circ \phi$. It remains to find an isotopy h_t such that $\phi_t = f \circ \phi \circ h_t$. Define diffeomorphisms $g_t : V^{[a,b]} \to V^{[a,b]}$ on the level $\phi^{-1}(r)$ by following the flow of $-\nabla_\phi \phi$ down to V^a and then the flow of $\nabla_{\phi_t}\phi_t$ up to the level $\phi_t^{-1}(r)$. Then $g_0 = \mathrm{Id}$ and $\phi_t = f \circ \phi \circ g_t$. Moreover, $g_t = \mathrm{Id}$ on $V^{[a,a']}$ and

$$g_t : V^{[b',b]} \cong \Sigma \times [b', b] \to \Sigma \times [b', b], \qquad (x, r) \mapsto \big(\gamma_t(x), r\big),$$

with $\gamma_t := g_t|_{V^{b'}}$. Define h_t on the level $\phi^{-1}(r)$ as $g_{t\rho(r)}$ with a smooth function $\rho : [a, b] \to [0, 1]$ which equals 1 on $[a, b']$ and 0 near b. Then $h_t = \mathrm{Id}$ near b and h_t is the desired isotopy. This concludes the proof of Proposition 10.1. □

10.2. Relaxing the J-orthogonality condition

This section is an aside on the J-orthogonality condition used in the surrounding results in Chapter 8. Namely, Lemma 10.3 allows us to relax the J-orthogonality condition in Theorem 8.4 to the weaker condition that $\partial\Delta \subset \partial_- W$ is isotropic (i.e., tangent to the field of complex tangencies):

COROLLARY 10.6. *Let (W, J) be a complex cobordism. Suppose that $\partial_- W$ is J-concave as a boundary component of W. Let $(\Delta, \partial\Delta) \subset (W, \partial_- W)$ be a totally real disc transverse to $\partial_- W$ and such that $JT\partial\Delta \subset T\partial_- W$. Then $\partial_- W \cup \Delta$ can be surrounded by J-convex hypersurfaces.*

PROOF. Let $\Sigma := \partial_- W$, $S := \partial\Delta$. Consider a collar $G = S \times [0, \varepsilon] \subset \Delta$, $S \times 0 = \partial\Delta$. It can be extended to a collar $\widehat{G} = \Sigma \times [0, \varepsilon]$, $\Sigma \times 0 = \partial_- W$ such that $(\Sigma \times t) \cap \Delta = S \times t$ and $JT(S \times t) \subset T(\Sigma \times t), t \in [0, \varepsilon)$. By continuity there exists $\delta \in (0, \varepsilon)$, such that the hypersurfaces Σ_t are J-convex for $t \in [0, \delta]$. According to Lemma 10.3, we can modify the family $\Sigma \times t$ by a C^0-small isotopy fixed on Δ to arrange for the hypersurface $\Sigma \times \delta$ to be J-orthogonal to Δ. Set $\widetilde{\Delta} := \Delta \setminus \big(S \times [0, \delta)\big)$. Now let U be any neighborhood of $\partial_- W \cup \Delta$. We can assume δ so small that $\Sigma \times \delta \subset U$. Hence, we can apply Theorem 8.4 to find a hypersurface $\widetilde{\Sigma}$ which surrounds $(\Sigma \times \delta) \cup \widetilde{\Delta}$ in $\widetilde{W} := W \setminus \big(\Sigma \times [0, \delta)\big)$. Then it also surrounds $\partial_- W \cup \Delta$ in W. □

The same argument allows us to relax the J-orthogonality condition in Theorem 8.23:

COROLLARY 10.7. *Let (W, J) be a complex cobordism and $(\Delta, \partial\Delta) \subset (W, \partial_- W)$ a totally real disc transverse to $\partial_- W$ and such that $JT\partial_-\Delta \subset T\partial_- W$. Then $\partial_- W \cup \Delta$ is a local J-convex retract.*

Similarly, in Corollary 8.26 one can relax the condition of J-orthogonality of L to W by requiring instead that $JT\partial L \subset T\partial W$.

On the other hand, the condition that $\partial\Delta \subset \partial_- W$ is isotropic is also *necessary* for existence of a J-convex surrounding:

PROPOSITION 10.8. *Let $W \subset V$ be a compact domain with smooth J-convex boundary in a complex manifold (V, J). Let $L \subset V \setminus \mathrm{Int}\, W$ be a totally real submanifold transversely attached to ∂W along a submanifold $\partial L \subset \partial W$ that is somewhere not tangent to the field of complex tangencies $\xi \subset T\partial W$. Suppose that L and ∂W are real analytic. Then $W \cup L$ is not holomorphically convex, and therefore cannot be surrounded by J-convex hypersurfaces.*

PROOF. Let us extend L to a larger real analytic totally real submanifold $\widetilde{L} \supset L$ such that $\partial\widetilde{L} \subset \mathrm{Int}\, W$ and $\widetilde{L} \subset W \cup L$. By assumption, there exists a point $p \in \partial L$

and a real line $\lambda \subset T_p\partial L$ such that $J\lambda$ is transverse to $T_p\partial W$. There exists a real analytic family of embeddings $h_s : [-\varepsilon_1, \varepsilon_1] \to \widetilde{L}$, $s \in [-\tau, \tau]$, such that

- $h_0(0) = p$;
- $h_0'(0) \in \lambda$ and $Jh_0'(x)$ is inward transverse to the boundary ∂W for all $x \in [-\varepsilon, \varepsilon]$;
- $h_s([-\varepsilon, \varepsilon]) \subset W$ for $s \leq 0$;
- $h_s(0) \notin W$ for $s > 0$.

We complexify the family h_s for some $\delta > 0$ to a real analytic family of holomorphic embeddings $H_s : P := \{z = x + iy; \mid |x| \leq \varepsilon,\ |y| \leq \delta\} \hookrightarrow V$, $s \in [-\tau, \tau]$. Set $P_+ := P \cap \{y \geq 0\}$. Then for sufficiently small $\sigma < \tau$ we have

- $H_s(P_+) \subset W$ for $s \in [-\sigma, 0]$;
- $H_s(\partial P_+) \subset \widetilde{L} \subset W \cup L$ for $s \in [-\sigma, \sigma]$;
- $H_s(P_+) \not\subset W \cup L$ for $s \in (0, \sigma]$.

By Example 5.3, this implies that $W \cup L$ is not holomorphically convex. □

REMARK 10.9. Proposition 10.8 should remain true without the real analyticity hypothesis.

10.3. Moving critical levels

In this section we prove the following analogue of Lemma 9.39 for J-convex functions. Recall that a Stein cobordism (W, J, ϕ) is a Morse cobordism (W, ϕ) with a complex structure J for which ϕ is J-convex, equipped with the gradient vector field $\nabla_\phi \phi$.

PROPOSITION 10.10. *Let (W, J, ϕ) be an elementary Stein cobordism with $\phi|_{\partial_\pm W} = a_\pm$ and critical points $p_1, \dots, p_n$ of values $\phi(p_i) = c_i$. For $i = 1, \dots, n$ let $c_i : [0,1] \to (a_-, a_+)$ be smooth functions with $c_i(0) = c_i$. Then there exists a smooth family of J-lc Morse functions ϕ_t, $t \in [0,1]$, with $\phi_0 = \phi$ and $\phi_t = \phi$ on $\mathcal{O}p\,\partial W$, such that all ϕ_t have the same critical points and stable discs and $\phi_t(p_i) = c_i(t)$.*

PROOF. **Step 1.** Pick a family of diffeomorphisms $f_t : [a_-, a_+] \to [a_-, a_+]$ such that $f_0 = \mathrm{Id}$ and $f_t \circ c_i(t) \leq c_i$ for all i and t. If we can find a family of i-lc functions ψ_t starting at ϕ with critical values $\psi_t(p_i) = f_t \circ c_i(t)$, then the functions $\phi_t = f_t^{-1} \circ \psi_t$ will have the desired critical values $c_i(t)$. So we may assume without loss of generality that $c_i(t) \leq c_i$ for all i and t.

Moreover, as we will construct the functions ϕ_t to agree with ϕ outside a neighborhood of the stable discs, it suffices to consider the case with a unique critical point p of index k. Let $c(t) \leq c = \phi(p)$, $t \in [0,1]$, be the given function and denote by Δ the stable k-disc of p. Pick a value a with $a_- < a < \min_t c(t)$. We may assume without loss of generality $a = -1$ and $c = 0$; the general case then follows by composing all functions with the affine function $x \mapsto (c-a)x + c$.

Consider the standard handle $H_\varepsilon = D^k_{1+\varepsilon} \times D^{2n-k}_\varepsilon \subset \mathbb{C}^n$. Here $z_j = x_j + iy_j$ are complex coordinates such that $(y_1, \dots, y_k)$ are coordinates on $D^k_{1+\varepsilon}$ and $(x_1, \dots, x_n, y_{k+1}, \dots, y_n)$ on D^{2n-k}_ε. As in Chapter 4, we introduce the functions

$$r := \sqrt{x_1^2 + \cdots + x_n^2 + y_{k+1}^2 + \cdots + y_n^2}, \qquad R := \sqrt{y_1^2 + \cdots + y_k^2}.$$

After these preparations, we now turn to the actual proof.

Step 2. Fix some $a > 1$. As in the proof of Theorem 8.5, after C^1-perturbing ϕ near Δ, there exists an embedding $F : H_\gamma \hookrightarrow W$ mapping $D^k_{1+\gamma}$ to Δ such that $F^*\phi = ar^2 - R^2$, and F^*J satisfies the estimate $\|F^*J - i\|_{C^2} \le c(a,n)\gamma^{12}$ in Theorem 4.1. Let $\Psi = \Psi_1 : \mathbb{C}^n \to \mathbb{R}$ be the J-lc function provided by Corollary 4.4. It agrees with $F^*\phi$ near ∂H_γ and up to target reparametrization near $D^k_{1+\gamma}$, and it satisfies $\Psi(0) < -1$.

The homotopy $\mathrm{smooth}\max(F^*\phi, \Psi - t)$ is fixed near $\partial H_\gamma \cup D^k_{1+\gamma}$, agrees with $F^*\phi$ for large t and with $\mathrm{smooth}\max(F^*\phi, \Psi)$ at $t = 0$. After applying this homotopy we may thus assume that $F^*\phi = \mathrm{smooth}\max(F^*\phi, \Psi)$. Now the functions $\Phi_s := \mathrm{smooth}\max(F^*\phi + s, \Psi)$ agree with $F^*\phi$ near ∂H_γ and have critical values $\Phi_s(0,0) = s$ at the origin. Hence the J-lc functions $\phi_t := F_*\Phi_{c(t)}$ (extended by ϕ outside $F(H_\gamma)$) have critical values $\phi_t(p) = c(t)$. □

10.4. Creation and cancellation of critical points

In this section we state our two main results concerning the creation and cancellation of critical points of J-convex functions. For the relevant concepts in Morse theory, we refer to Chapter 9. Recall in particular that a family (X_t, ϕ_t), $t \in [0,1]$, of functions and gradient-like vector fields is called a *cancellation (resp. creation) family* if there is a $t_0 \in (0,1)$ such that the following holds:

(i) for $t > t_0$ (resp. $t < t_0$) the function ϕ_t has no critical points;
(ii) for $t < t_0$ (resp. $t > t_0$) it has exactly two critical points of index k and $k-1$ connected by a unique trajectory of X_t along which the stable and unstable manifolds intersect transversely;
(iii) for $t = t_0$ it has a unique embryonic critical point.

For J-convex functions we always assume in addition that $X_t = \nabla_{\phi_t}\phi_t$. We will say that a deformation of functions $\phi_t : W \to \mathbb{R}$, $t \in [0,1]$, is *weakly supported* in $U \subset W$ if there exists an isotopy $\alpha_t : \mathbb{R} \to \mathbb{R}$ such that on $W \setminus U$ we have $\phi_t = \alpha_t \circ \phi_0$.

The following theorem describes the creation of critical points of J-convex functions.

THEOREM 10.11 (Creation theorem). *Let (W, J, ϕ) be a Stein cobordism such that the J-convex function ϕ has no critical points. Then given any point $p \in \mathrm{Int}\, W$ and an integer $k = 1, \dots, n$, there is a creation family ϕ_t of J-convex functions, weakly supported in $\mathcal{O}p\, p$, such that $\phi_0 = \phi$ and ϕ_1 has a pair of critical points of index k and $k-1$.*

Note that in ordinary Morse theory the analogue of Theorem 10.11 is rather trivial: using an appropriate cut-off construction, any local creation family can be implanted into a globally defined family; see Lemma 9.47 above. However, in the context of J-convex functions this scheme does not seem to work. In fact, we do not know whether the statement remains true if one drops the word "weakly" and tries to construct a locally supported creation family.

The following theorem describes the cancellation of critical points of J-convex functions.

THEOREM 10.12 (Cancellation theorem). *Let (W, J, ϕ) be a Stein cobordism such that the J-convex function ϕ has exactly two critical points p, q of index k and $k-1$, respectively, which are connected by a unique gradient trajectory along* *which the stable and unstable manifolds intersect transversely. Set* $a_- := \phi|_{\partial_- W}$*,* $b := \phi(q)$*,* $c := \phi(p)$*. Choose a regular value* $a \in (a_-, b)$*. Let* Δ *be the closure of the stable disc of the critical point* p *in* $\{\phi \geq a\}$*. Then there exists a cancellation family* $\phi_t : W \to \mathbb{R}$*,* $t \in [0,1]$*, of* J*-convex functions, weakly supported in* $\mathcal{O}p\,\Delta$*, such that* $\phi_0 = \phi$ *and* ϕ_1 *has no critical points.*

The proof of these two theorems will occupy the remainder of this chapter.

10.5. Carving one J-convex function with another one

As preparation for the proofs, we describe in this section a general way to modify one J-convex function with the help of another one.

Let $\phi : U \to \mathbb{R}$ be a J-convex function on an open set U and $\Sigma = \{\phi = a\}$ be a regular level set. Let us denote by U_- and U_+ the domains $\{\phi \leq a\}$ and $\{\phi \geq a\}$, respectively. Let $\psi : \Omega \to [c_-, c_+]$ be another J-convex function defined on a compact subdomain $\Omega \subset U$ with boundary $\partial\Omega = \partial_+\Omega \cup \partial_-\Omega \cup \partial_v\Omega$ such that $\psi|_{\partial_\pm\Omega} = c_\pm$ and $\partial_+\Omega \cup \partial_v\Omega \subset \operatorname{Int} U_+$; see Figure 10.4(a).

For a small $\varepsilon > 0$ let us denote by Ω^ε the domain $\{c_- + \varepsilon \leq \psi \leq c_+ - \varepsilon\} \subset \Omega$, and by U_-^ε the domain $\{\phi \leq a - \varepsilon\} \subset U_-$. By composing ϕ, ψ with increasing weakly convex diffeomorphisms $g, h : \mathbb{R} \to \mathbb{R}$, we can arrange that the functions $\widetilde{\phi} = g \circ \phi$ and $\widetilde{\psi} = h \circ \psi$ satisfy the following conditions:

- $\widetilde{\psi} > \widetilde{\phi}$ on $U_-^\varepsilon \cap \Omega^\varepsilon$;
- $\widetilde{\phi} > \widetilde{\psi}$ on $(U_+ \cap \Omega) \cup \partial_-\Omega$.

To see this, first compose ψ with h such that $h(c_-) < \min_\Omega \phi$ and $h(c_- + \varepsilon) > \max_{\Omega^\varepsilon} \phi$; thus, $\widetilde{\psi} > \phi$ on Ω^ε and $\widetilde{\psi} < \phi$ on $\partial_-\Omega$. Then compose ϕ with g such that $g(x) = x$ for $x \leq a - \varepsilon$ and $g(a) > \max_{U_+ \cap \Omega} \widetilde{\psi}$, thus $\widetilde{\phi} > \widetilde{\psi}$ on $(U_+ \cap \Omega) \cup \partial_-\Omega$ and $\widetilde{\psi} > \widetilde{\phi}$ on $U_-^\varepsilon \cap \Omega^\varepsilon$.

Take the function $\max(\widetilde{\phi}, \widetilde{\psi})$ and apply to it the smoothing procedure from Section 3.2. We will call this operation *carving the level set* Σ *of* ϕ *with the function* ψ. The resulting function is shown in Figure 10.4(b); it will be denoted by

$$\mathfrak{carv}_\psi(\phi, \Sigma).$$

Though there are numerous ambiguities in the definition of this operation, it is important that it can be done for families of functions smoothly dependent on parameters, that ε can be chosen arbitrarily small, and that the smoothing can be chosen C^0-close to $\max(\widetilde{\phi}, \widetilde{\psi})$ in the sense of Corollary 3.15. In particular, everywhere below where we use the notation $\mathfrak{carv}_\psi(\phi, \Sigma)$, we assume that ε is chosen sufficiently small and the approximation is good enough.

Note that if both functions ϕ, ψ are transverse to the same vector field, then so is $\mathfrak{carv}_\psi(\phi, \Sigma)$ (see Corollary 3.20), hence the carving operation does not create new critical points. It follows that carving is well-defined in the class of J-lc functions without critical points; note that in this case we can rescale the functions to make $\widetilde{\phi} = \phi$.

10.6. Surrounding a stable half-disc

The main ingredient for cancellation of critical points are J-convex surroundings for a stable half-disc which we construct in this section. For this, consider the following setup as in Section 9.6.

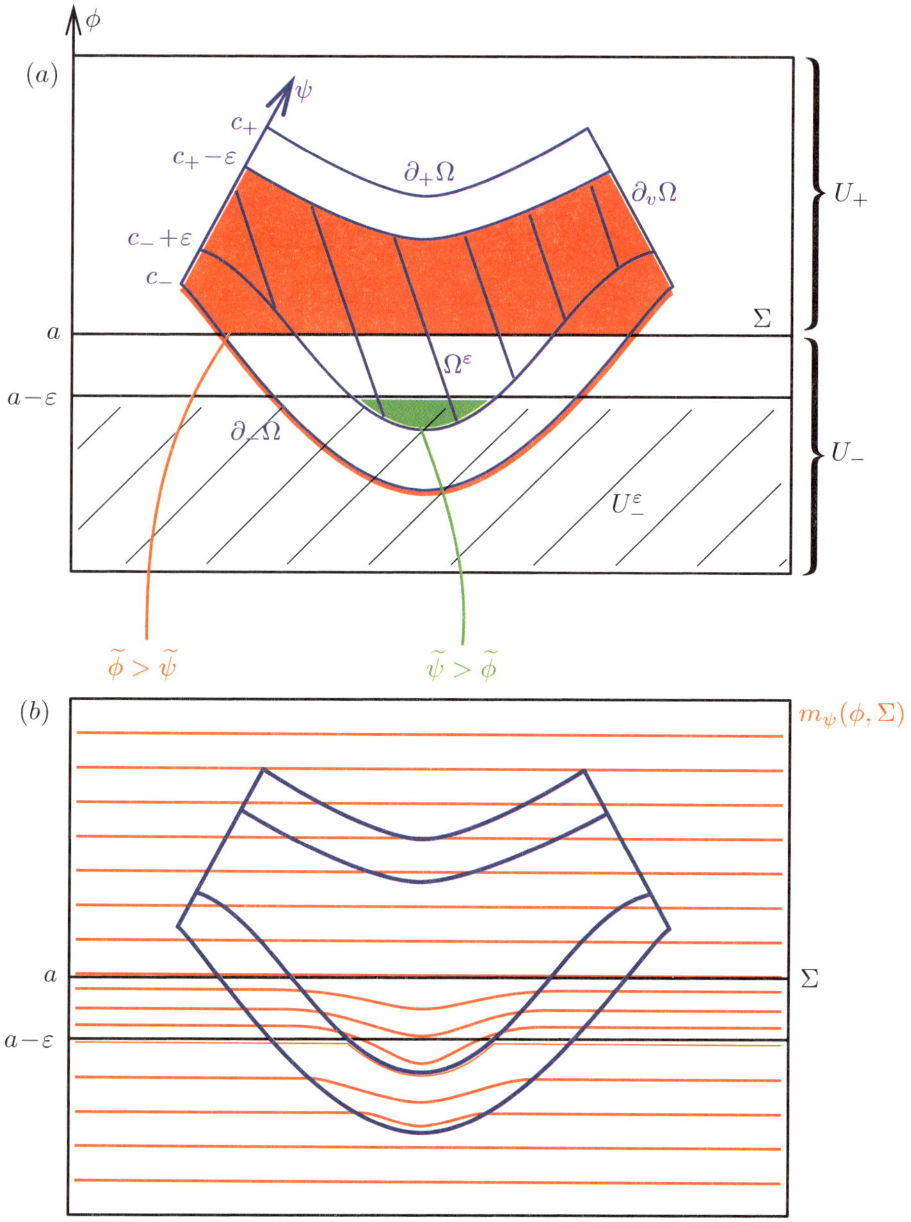

FIGURE 10.4. Carving the level set Σ of ϕ with the function ψ.

Let $\mathbb{R}^k = \mathbb{R}^{k-1} \times \mathbb{R}$ be the space with coordinates $(x_1, \dots, x_k)$. Let D_t, $t > 0$, denote the disc $\{\sum_{j=1}^k x_j^2 \le t^2\}$ of radius t, and we write D instead of D_1. We further denote by D_- the lower half-disc $D \cap \{x_k \le 0\}$, and set $\partial_+ D_- = D_- \cap \{x_k = 0\}$ and $\partial_- D_- = \partial D \cap D_-$, so that we have $\partial D_- = \partial_- D_- \cup \partial_+ D_-$; see Figure 9.4.

Viewing $\mathbb{R}^k$ as a coordinate subspace of $\mathbb{C}^n$ with complex coordinates $(x_1 + iy_1, \dots, x_n + iy_n)$, we will consider the splitting $\mathbb{C}^n = \mathbb{R}^k \times \mathbb{R}^{2n-k}$ and write $z \in \mathbb{C}^n$ as $z = (x, u)$, where $x = (x_1, \dots, x_k)$ and $u = (x_{k+1}, \dots, x_n, y_1, \dots, y_n)$.

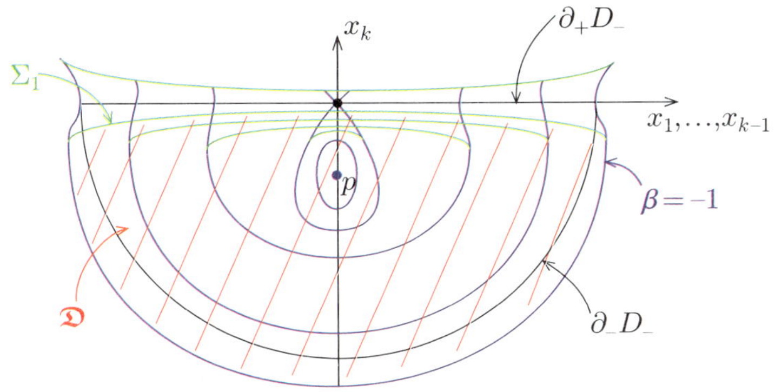

FIGURE 10.5. The function β near the lower half-disc.

Set

$$R = ||x|| = \sqrt{\sum_1^k x_j^2}, \qquad r = ||u|| = \sqrt{\sum_{k+1}^n x_j^2 + \sum_1^n y_j^2},$$

$$R' = \sqrt{\sum_1^{k-1} x_j^2}, \qquad r' = \sqrt{\sum_k^n x_j^2 + \sum_1^n y_j^2}.$$

For a constant $A > 1$, we introduce the vector fields

$$\vec{v} = A\frac{\partial}{\partial x_k} - \sum_1^{k-1} x_j \frac{\partial}{\partial x_j}, \qquad \vec{u} = \sum_{k+1}^n x_j \frac{\partial}{\partial x_j} + \sum_1^n y_j \frac{\partial}{\partial y_j}.$$

Given a compact subset $K \subset \mathbb{C}^n$ and $\sigma > 0$, we denote by $U_\sigma(K)$ its open σ-neighborhood in $\mathbb{C}^n$.

Suppose we are given an i-convex function $\phi_0 : \mathbb{C}^n \to \mathbb{R}$ of the form

$$\phi_0(x,u) = \beta(x) + Ar^2,$$

where the function $\beta : \mathbb{R}^k \to \mathbb{R}$ satisfies the following conditions (see Figure 10.5):

(i) β has exactly two nondegenerate critical points, an index $k-1$ critical point of value 0 at 0 and an index k critical point $p \in \operatorname{Int} D_-$ of value $0 < c < A - 1$;

(ii) $\beta > -1$ on $D_- \setminus \partial_+ D_-$ and $\beta(x) = Ax_k^2 - R'^2$ on $\mathcal{O}p\,(\partial_+ D_-)$.

Let us fix a neighborhood $U \subset \mathbb{C}^n$ of the disc $\{x_k \le 0,\ \beta \ge -1\} \subset \mathbb{R}^k$. We will deform the function ϕ_0 through i-lc functions $\phi_t : \mathbb{C}^n \to \mathbb{R}$ satisfying the following conditions:

- ϕ_t has exactly two nondegenerate critical points at 0 and p;
- $\phi_t = \phi_0$ outside U.

We will call such functions and deformations *admissible*. The desired surrounding will be a combination of three surroundings. Set $\Sigma_0 := \{\phi_0 = -1\}$.

First surrounding. Note that $\phi_0 = -R'^2 + Ar'^2$ on a neighborhood $U_1 \subset U$ of the $(k-1)$-disc $\partial_+ D_-$. So we can apply Corollary 4.4 to construct an admissible deformation ϕ_t, $t \in [0,1]$, with the following properties (see Figures 10.5 and 10.6):

- $\phi_t = \phi_0$ outside U_1;

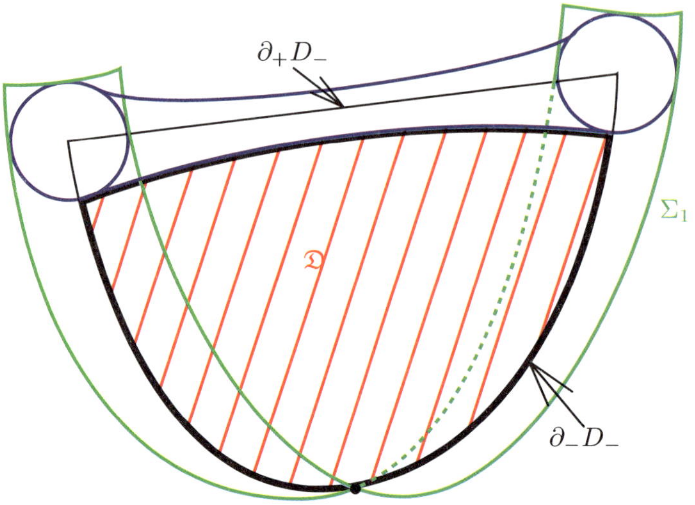

FIGURE 10.6. The first surrounding hypersurface Σ_1 and the disc $\mathfrak{D}$.

- the regular level set $\Sigma_1 = \phi_1^{-1}(-1)$ agrees with Σ_0 outside U_1 and with $\{r' = \delta\}$ in a smaller neighborhood of $\partial_+ D_-$, for some $\delta > 0$;
- the k-disc $\mathfrak{D} := \{u = 0,\ \phi_1 \geq -1\ , x_k < 0\}$ is attached i-orthogonally to Σ_1.

Here the last property follows from Lemma 4.14.

Second surrounding. The function $\phi_1|_{\mathfrak{D}}$ has a unique nondegenerate maximum at p of value c and $\phi_1|_{\partial\mathfrak{D}} \equiv -1$. Hence there exists a diffeomorphism $f : D \to \mathfrak{D}$ such that

$$\phi_1 \circ f(x) = c - (c+1)R^2.$$

Using Proposition 5.55, we extend f to a diffeomorphism $F : \mathcal{O}p\, D \to \mathcal{O}p\, \mathfrak{D}$ such that the pullback complex structure F^*i agrees with i to order 7 along D. Using Proposition 3.26, we adjust $\widetilde{\phi}_1 = F^*\phi_1$ via a C^1-small admissible deformation to make it equal to $c - (c+1)R^2 + Ar^2$ in a neighborhood $\widetilde{U}_2$ of D. Next we apply Corollary 4.4 to construct an F^*i-lc deformation $\widetilde{\phi}_t$, $t \in [1,2]$, supported in $\widetilde{U}_2$ such that the regular level set $\widetilde{\Sigma}_2 = \widetilde{\phi}_2^{-1}(-1)$ surrounds D. So ϕ_t, $t \in [1,2]$, is an admissible i-lc deformation with the following properties:

- $\phi_t = \phi_1$ outside a neighborhood U_2 of $\mathfrak{D}$;
- the regular level set $\Sigma_2 = \phi_2^{-1}(-1)$ agrees with Σ_1 outside U_2 and surrounds D_-.

See Figure 10.7 for a cross section of Σ_2. Before proceeding further, consider the following property of a function $\phi : \mathbb{C}^n \to \mathbb{R}$:

$$\vec{u} \cdot \phi \geq \mu r^2 \text{ for some constant } \mu > 0. \tag{10.3}$$

LEMMA 10.13. *The above functions ϕ_t, $t \in [0,2]$ satisfy property* (10.3).

PROOF. Note first that the function ϕ_0 satisfies (10.3) with constant $\mu = 2A$. The functions ϕ_t, $t \in [0,1]$ satisfy (10.3) because on U_1 they are given by shapes of the form $\Psi_t(R', r')$ with $\frac{\partial \Psi_t}{\partial r'} > 0$.

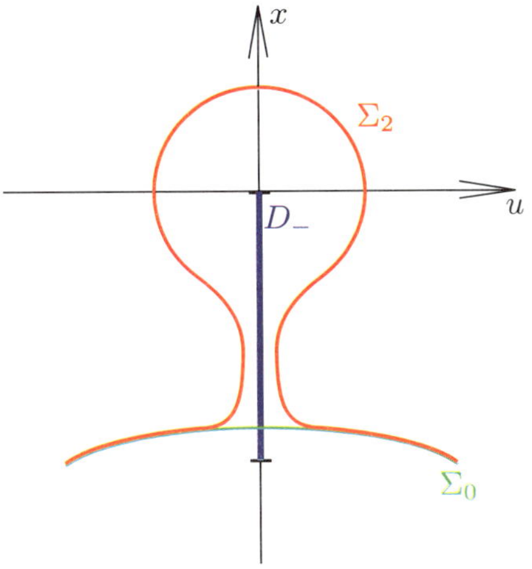

FIGURE 10.7. The dumbbell-shaped cross section of the second surrounding hypersurface Σ_2.

Next note that the map F above has complex linear differential along D, so it is of the form

$$F(x,y,z) = \Big(f(x), Df(x)y, B(x)z\Big) + O(r^2),$$

where $y = (y_1, \dots, y_k)$, $z = (z_{k+1}, \dots, z_n)$, and $B(x)$ is complex linear for each $x \in D$. Since the canonical vector field $\vec{u}$ is preserved by any linear map in the u-variables, it follows that

$$F^*\vec{u} = \vec{u} + O(r^2),$$

and hence the pullback function $\widetilde{\phi}_1 = F^*\phi_1$ satisfies (10.3) in a sufficiently small neighborhood of D. By Proposition 3.26(f), the ensuing C^1-small adjustment of $\widetilde{\phi}_1$ near D also preserves property (10.3). Now the functions $\widetilde{\phi}_t$, $t \in [1,2]$ satisfy (10.3) because on $\widetilde{U}_2$ they are given by shapes of the form $\Psi_t(R,r)$ with $\frac{\partial \Psi_t}{\partial r} > 0$. Finally, by the argument above, the functions $\phi_t = F_*\widetilde{\phi}_t$, $t \in [1,2]$, still satisfy (10.3). $\square$

Third surrounding. We will now construct another i-lc function Ψ and use the carving construction from Section 10.5 to construct our third surrounding. For this, let us rename ϕ_2 to ϕ. Pick a value a slightly below -1 such that the level set $\Sigma := \phi^{-1}(a)$ surrounds D_- and its intersection with the set $\{R \leq 1\}$ is contained in $\{r < \delta\}$.

By Corollary 4.3 (with $a = -1 - \sigma$ and after a target reparametrization), there exists an i-lc shape function $\Psi : \mathbb{C}^n \supset \Omega \to (0, \delta)$ without critical points and with the following properties (see Figure 10.8):

- $\Psi(r,R) = r$ for $r \in (0, \delta]$, $R \leq 1$;
- $\Psi(r,R) = f(Ar^2 - R^2)$ for $r \geq \gamma$, where $f : (a, -1] \to (0, \delta]$ is an increasing diffeomorphism.

Figure 10.9 shows the hypersurface Σ and the level sets of Ψ. So for any $t \in (0, \delta]$ the restriction Ψ_t of Ψ to the set $\Omega_t := \Psi^{-1}([t, \delta])$ together with ϕ, Σ satisfies the hypotheses of Section 10.5. Let

$$\phi_t := \mathfrak{carv}_{\Psi_t}(\phi, \Sigma)$$

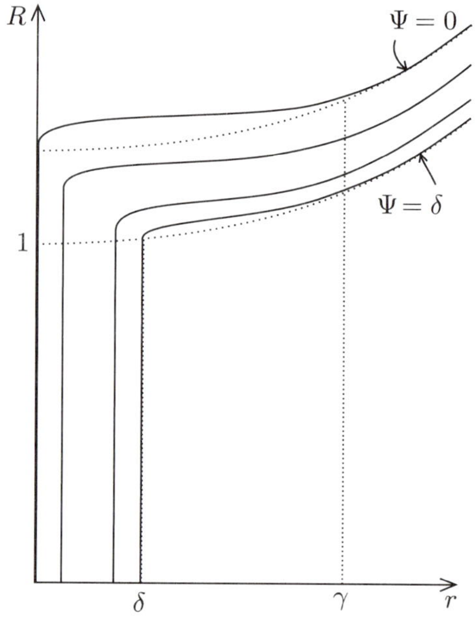

FIGURE 10.8. The shape function Ψ.

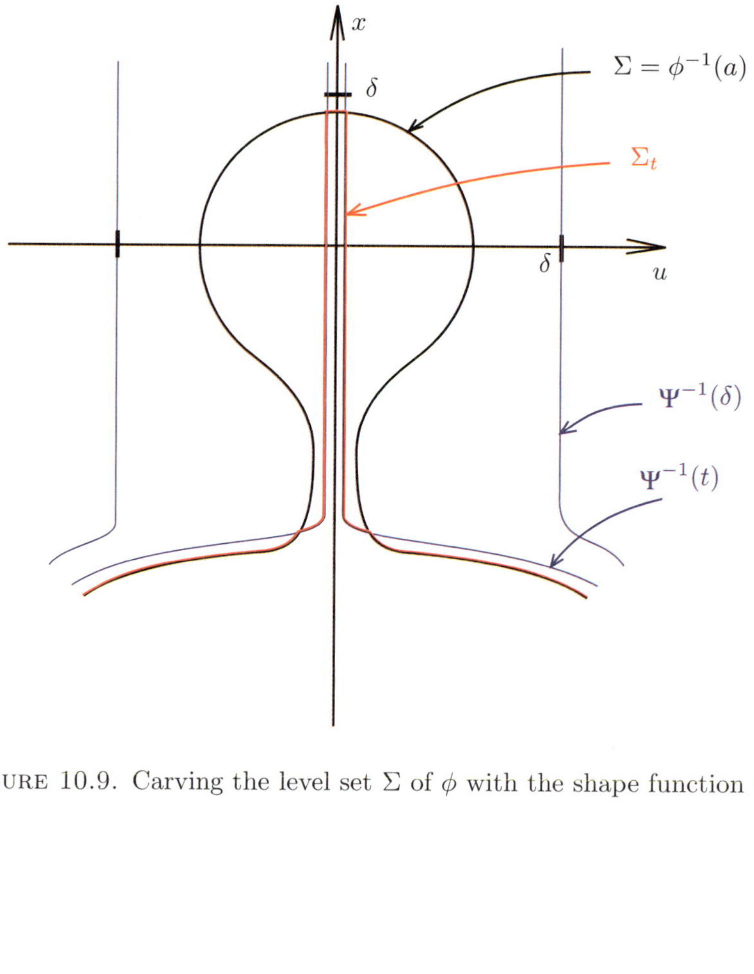

FIGURE 10.9. Carving the level set Σ of ϕ with the shape function Ψ_t.

be the carving of the level set Σ of ϕ with the function Ψ_t. Note that $\phi_\delta = \phi$. We claim that ϕ_t, $t \in (0,\delta]$, is an admissible deformation of i-lc functions with the following properties (see Figure 10.9):

(j) $\vec{u} \cdot \phi_t > 0$ for $u \neq 0$, and $\vec{v} \cdot \phi_t > 0$ on the set $\{u = 0,\ R \leq 1,\ 0 < x_k < \delta\}$;

(jj) for t sufficiently small, some level set Σ_t of ϕ_t surrounds the disc D_- in $U_\delta(D_-)$ and contains the set $\{r = t,\ R \leq 1,\ x_k \leq \delta/2\}$.

Property (jj) is clear from the construction. For property (j), note first that the functions ϕ (by Lemma 10.13) and Ψ (since it is a shape) are both transverse to the vector field $\vec{u}$ on $\Omega \cap \{u \neq 0\}$. On the other hand, the property $\vec{v} \cdot \phi_t > 0$ on the set $\{u = 0,\ R \leq 1,\ 0 < x_k < \delta\}$ holds during the first surrounding deformation above because shapes $\Psi(r', R')$ have this property, and the following deformations are fixed on this set. In view of property (j), the carving construction does not create new critical points, so the deformation ϕ_t is admissible.

10.7. Proof of the cancellation theorem

After these preparations, we now prove Theorem 10.12 in three steps. The first two steps contain preliminary deformations not affecting the critical points; the actual cancellation happens in Step 3.

Step 1. Let (W, J, ϕ) and Δ be as in Theorem 10.12. After a C^2-small perturbation of ϕ near p and q, we may assume that it agrees with a standard quadratic function as in Lemma 2.2 in suitable holomorphic coordinates near p, q. Then Lemma 9.30 shows that Δ is an embedded half-disc.

After rescaling ϕ, we may assume that $a = -1$ and $b = 0$, so we have $\phi(p) = c$, $\phi(q) = 0$ and $\phi|_{\partial_-\Delta} \equiv -1$. Pick any $A > c+1$. Then there exists a diffeomorphism $f : \widetilde{D}_- \to \Delta$ from a half-disc $\widetilde{D}_- \subset \mathbb{R}^k$ containing D_- such that the pullback function $\beta := \phi \circ f$ satisfies conditions (i)–(ii) in Section 10.6. Here and in the following we use the notation from Section 10.6.

Using Proposition 5.55, we extend f to a diffeomorphism $F : \mathcal{O}p\, \widetilde{D}_- \to \mathcal{O}p\, \Delta$ such that the pullback complex structure F^*i agrees with i to order 7 along $\widetilde{D}_-$. Let $\widetilde{A} \geq A$ be so large that the function $\beta(x) + \widetilde{A}r^2$ on $\mathbb{C}^n$ is i-convex. Using Proposition 3.26, we adjust $\phi \circ F$ via a C^1-small admissible deformation to make it equal to $\beta(x) + \widetilde{A}r^2$ in a neighborhood of $\widetilde{D}_-$. Now pick an increasing convex function $\xi : \mathbb{R}_+ \to \mathbb{R}_+$ which agrees with Ax^2 near $x = 0$ and with $\widetilde{A}x^2$ for $x \geq \varepsilon$, and modify β near $\{x_k = 0\}$ to $\widetilde{\beta}(x) = \xi(x_k) - R'^2$. Then the function $\widetilde{\beta}(x) + \widetilde{A}r^2$ is still i-convex. After renaming $\widetilde{A}, \widetilde{\beta}$ back to A, β, the function $\phi_0(x, u) = \beta(x) + Ar^2$ thus satisfies the conditions in Section 10.6.

In the following steps we will modify ϕ_0 through i-convex functions ϕ_t on a neighborhood of the half-disc $\widetilde{D}_-$. The tangency condition on F^*i and Proposition 4.34 ensure that the resulting functions $F_*\phi_t$ on W will be J-convex.

Step 2. After the three admissible deformations in Section 10.6, we may replace ϕ_0 by the function ϕ_t satisfying conditions (j)–(jj) at the end of Section 10.6, for some arbitrarily small $t > 0$. Let $\Sigma_t = \phi_t^{-1}(c_t)$ be the level set from condition (jj). Note that $c_t > c = \max_{D_-} \phi_t$. Let $\psi_t := g \circ \phi_t$ with an increasing function $g : \mathbb{R} \to \mathbb{R}$ satisfying $g(x) \leq x$, $g(x) = x$ for $x \geq c_t$, and $g(x) < b = \min_{U_\delta(D_-)} \phi_t$ for $x \leq c_t'$, for some $c_t' \in (c, c_t)$. Then $\phi_t = \max(\phi_t, \psi_t)$.

More generally, suppose that ϕ is an i-lc function on $U_\delta(D_-)$ satisfying the following conditions:

(i) $\vec{u} \cdot \phi > 0$ for $u \neq 0$;
(ii) $\phi = \phi_t$ on $U_\delta(\partial_- D_-) \cup \{u = 0,\ R \leq 1,\ \delta/2 < x_k < \delta\}$;
(iii) $b \leq \phi \leq c$.

Then smooth $\max(\phi, \psi_t)$ is an i-lc function which coincides with ϕ_t outside $U_\delta(D_-)$ and with ϕ near D_-. Moreover, conditions (i)–(ii) for ϕ together with condition (j) for ϕ_t ensure that the critical points of smooth $\max(\phi, \psi)$ coincide with the critical points of ϕ on the half-disc $D_-^\delta := \{u = 0,\ R \leq 1,\ x_k \leq \delta/2\}$.

Step 3. Recall that the restriction $\beta_t := \phi_t|_{D_-^\delta}$ is a Morse function with two critical points p, q of indices $k, k-1$ and values $c, 0$ connected by a unique trajectory of the vector field $X_t = \nabla_{\phi_t} \phi_t$. Moreover, X_t is inward pointing along $\partial_- D_-^\delta$ and outward pointing along $\partial_+ D_-^\delta$. Hence by Lemma 9.49 there exists a cancellation family β_s, $s \in [t, 1]$, fixed near ∂D_-^δ, such that β_1 has no critical points. We extend β_s by β_t to $\{R \geq 1\}$.

Pick a constant $B \geq A$ so large that the function $\beta_s(x) + Br^2$ is i-convex in $U_\delta(D_-)$ for all $s \in [t, 1]$. Using Proposition 3.26, we modify ϕ_t such that it agrees with $\beta_t(x) + Br^2$ near D_-^δ. Proposition 3.26(f) ensures that this can be done preserving the condition $\vec{u} \cdot \phi_t > 0$ for $u \neq 0$. After replacing ϕ_t by smooth $\max(\phi_t, \psi_{t'})$ as in Step 2, for sufficiently small $t' \in (0, t)$, and renaming t' back to t, we may hence assume that $\phi_t = \beta_t(x) + Br^2$ on the set $\{x \in D_-^\delta, r \leq t\}$. According to Step 2, we now obtain the desired cancellation family as smooth $\max(\beta_s(x) + Br^2, \psi_t)$, $s \in [t, 1]$. This completes the proof of Theorem 10.12. □

10.8. Proof of the creation theorem

The proof of Theorem 10.11 follows the same steps as the proof of Theorem 10.12, but it is much simpler and does not require the preparations in Section 10.6.

Step 1. Let (W, J, ϕ), $p \in \operatorname{Int} W$ and $1 \leq k \leq n$ be as in Theorem 10.11. After adding a constant to ϕ, we may assume that $\phi(p) = -1$. Pick an isotropic embedded $(k-1)$-sphere Λ through p in the level set $\{\phi = -1\}$. For some $d < 1$ close to 1 denote by $L \subset \{-1 \leq \phi \leq -d^2\} \subset \operatorname{Int} W$ the image of Λ under the gradient flow of ϕ.

We use the notation r, R from Section 10.6. There exists a diffeomorphism $f : \{d \leq R \leq 1,\ r = 0\} \to L$ such that $f^*\phi(R) = -R^2$. Using Proposition 5.55, we extend f for some $\gamma > 0$ to an embedding $F : U_\gamma := \{d - \gamma \leq R \leq 1 + \gamma,\ r \leq \gamma\} \hookrightarrow W$ such that the pullback complex structure F^*i agrees with i to order 7 along $\{r = 0\}$. After applying Proposition 3.26 and shrinking γ, we may assume that $F^*\phi(r, R) = Ar^2 - R^2 =: \Psi_{\rm st}(r, R)$ on U_γ, for any chosen constant $A > 1$. In the following we will deform the function $\Psi_{\rm st}$ on U_γ, keeping it fixed near ∂U_γ, and then implant it back into W by F to get the desired homotopy.

Step 2. A slight variation of Corollary 4.4 yields a smooth family of J-lc functions $\Psi_t : \mathbb{C}^n \to \mathbb{R}$, $\in [0, 1]$, with the following properties (see Figure 10.10):

(i) Ψ_t is of the form $\Psi_t(r, R)$ with $\frac{\partial \Psi_t}{\partial r} > 0$ and $\frac{\partial \Psi_t}{\partial R} \leq 0$;
(ii) $\Psi_0 = \Psi_{\rm st}$ and $\Psi_t = \Psi_{\rm st}$ outside U_γ;
(iii) Ψ_t is target equivalent to $\Psi_{\rm st}$ near $\{r- = 0\}$;

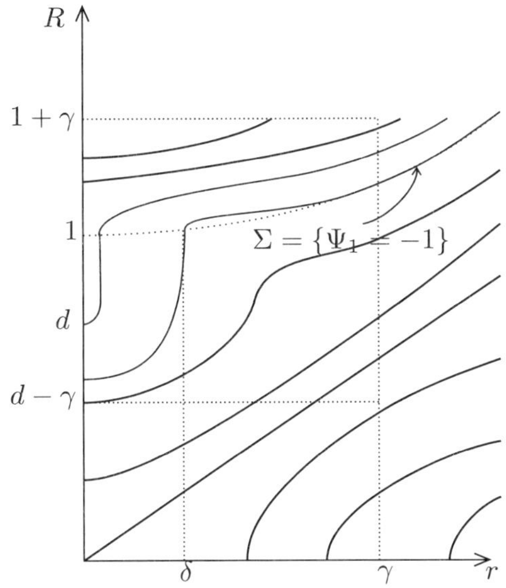

FIGURE 10.10. The modified shape function Ψ_1.

(iv) $\Psi_1 \equiv -1$ on the set $\{r = \delta,\ d \le R \le 1\}$, for some $\delta \in (0, \gamma)$.

To obtain this, we only need to replace the condition $g(a\gamma^2) < -1$ in the proof of Corollary 4.4 by the conditions $g(-d^2) < -1$ and $g\big(-(d-\gamma)^2\big) > -1$. Note that by construction we have $\Psi_1 = \max(\Psi_1, g \circ \Psi_{\rm st})$.

Step 3. After another application of Proposition 3.26 we may assume that $\Psi_1 = \text{smooth}\max(\Psi_1, \beta(x) + Ar^2)$ on $V_\gamma = \{r \le \gamma, d \le R \le 1\}$, where β is the restriction of Ψ to the cylinder $Z = \{r = 0, d \le R \le 1\}$. Note that β has no critical points and is constant on ∂Z. By Lemma 9.47 there exists a creation family β_t, $t \in [0,1]$, starting from $\beta_0 = \beta$ and creating a pair of critical points of index $k, k-1$ at some time $t_0 \in (0,1)$. Moreover, β_t is fixed on ∂Z and we make sure that $\max \beta_t = \max \beta$ and $\min \beta_t = \min \beta$. We choose the constant A so large that the functions $\beta_t(x) + Ar^2$ are F^*i-convex for all $t \in [0,1]$. Then the desired creation family is the push-forward under F of $\text{smooth}\max(\beta_t(x) + Ar^2, \Psi_1)$. Note that all the deformations can be made supported in an arbitrarily small neighborhood of the given point p. This completes the proof of Theorem 10.11. □

Part 4

From Stein to Weinstein and Back

CHAPTER 11

Weinstein Structures

In this chapter we introduce Weinstein cobordisms and manifolds and establish their basic properties. After a more general discussion of Liouville structures in the first three sections, we define Weinstein structures in Section 11.4.

In Section 11.5 we define the canonical map from Stein to Weinstein structures. The construction of a homotopy inverse to this map will be the main theme of Chapters 13 and 15.

In Section 11.6 we introduce Weinstein and Stein homotopies and show that different exhausting J-convex functions lead to homotopic Stein structures. In Section 11.7 we prove that every Stein structure on $\mathbb{C}^n$ with a unique critical point is homotopic to the standard structure. In the final Section 11.8 we introduce the classes of subcritical and flexible Weinstein structures, which will be extensively studied in Chapter 14.

11.1. Liouville cobordisms and manifolds

In this and the following two sections we discuss some basic properties of Liouville structures. See [**168, 169**] for more background.

A 1-form λ on a manifold V such that $d\lambda = \omega$ is symplectic is called a *Liouville form*. The vector field X that is ω-dual to λ, i.e., such that $i_X\omega = \lambda$, is called the *Liouville field* of λ. Note that the equation $d\lambda = \omega$ is equivalent to $L_X\omega = \omega$. If X integrates to a flow $X^t : V \to V$, then $(X^t)^*\omega = e^t\omega$, i.e., the Liouville field X is (symplectically) *expanding*, while $-X$ is *contracting*. By an *exact symplectic manifold* we will mean a pair (V, λ) where λ is a Liouville form, or equivalently, a triple (V, ω, X) where X is a Liouville field for the symplectic form ω, i.e., satisfying $L_X\omega = \omega$. Note that

$$i_X\lambda = 0, \qquad i_X d\lambda = \lambda, \qquad L_X\lambda = \lambda, \tag{11.1}$$

so the flow of X also expands the Liouville form, $(X^t)^*\lambda = e^t\lambda$. A map $\psi : (V_0, \lambda_0) \to (V_1, \lambda_1)$ between exact symplectic manifolds is called *exact symplectic* if $\psi^*\lambda_1 - \lambda_0$ is exact.

A *Liouville manifold* is an exact symplectic manifold (V, ω, X) such that

- the expanding vector field X is *complete*, and
- the manifold is *convex* (see [**49**]) in the sense that there exists an exhaustion $V = \bigcup_{k=1}^\infty V^k$ by compact domains $V^k \subset V$ with smooth boundaries along which X is outward pointing.[1]

[1]This notion of symplectic convexity is slightly more restrictive than one given in [**49**]. However, we do not know any examples of symplectic manifolds that are convex in one sense but not the other.

Note that the sets V^k are invariant under the contracting flow $X^{-t}, t > 0$. The set

$$\mathrm{Skel}(V, \omega, X) := \bigcup_{k=1}^{\infty} \bigcap_{t>0} X^{-t}(V^k)$$

is independent of the choice of the exhausting sequence of compact sets V^k and is called the *skeleton* of the Liouville manifold (V, ω, X). We have

LEMMA 11.1. $\mathrm{Int}\,\mathrm{Skel}(V, \omega, X) = \varnothing$.

PROOF. For each compact set V^k we have

$$\mathrm{Volume}\big(X^{-t}(V^k)\big) = e^{-t}\frac{1}{n!}\int_{V^k} \omega^n \underset{t\to\infty}{\longrightarrow} 0\,,$$

and hence $\mathrm{Volume}\big(\bigcap_{t>0} X^{-t}(V^k)\big) = 0$ for all $k \in \mathbb{N}$. □

We say that a Liouville manifold (V, ω, X) is of *finite type* if its skeleton is compact. In this case, let $W \subset V$ be a compact domain containing the skeleton with smooth boundary $\Sigma = \partial W$ along which X is outward pointing (e.g., $W = V^k$ for large k). Then the forward flow of X starting from Σ defines a diffeomorphism $V \setminus \mathrm{Int}\, W \cong \Sigma \times [0, \infty)$. (For this note that for every $p \in V$, $X^{-t}(p)$ gets close to the skeleton as $t \to \infty$ and thus in contained in W for large t.) Under this diffeomorphism the Liouville form $\lambda = i_X\omega$ corresponds to $e^t\alpha$, where $t \in \mathbb{R}$ is the parameter of the flow and $\alpha := \lambda|_\Sigma$. The form α is contact, and thus $(V \setminus \mathrm{Int}\, W, \omega)$ can be identified with the positive half of the symplectization of the contact manifold $(\Sigma, \xi = \ker \alpha)$. In fact, the whole symplectization of (Σ, ξ) sits in V as $\bigcup_{t\in\mathbb{R}} X^t(\Sigma)$ and this embedding is canonical in the sense that the image is independent of the choice of Σ: its complement $V \setminus \bigcup_{t\in\mathbb{R}} X^t(\Sigma)$ is exactly the skeleton $\mathrm{Skel}(V, \omega, X)$.

The following useful lemma shows that for finite type Liouville manifolds, we need not distinguish between symplectomorphisms and exact symplectomorphisms.

LEMMA 11.2. *Any symplectomorphism $f : (V, \lambda) \to (\widetilde{V}, \widetilde{\lambda})$ between finite type Liouville manifolds is diffeotopic to an exact symplectomorphism.*

PROOF. We have $f^*\widetilde{\lambda} = \lambda - \theta$ for a closed 1-form θ. Let Σ be a hypersurface in V transverse to the Liouville field of λ and bounding a compact domain W containing the skeleton, so the Liouville flow defines a splitting $V \setminus \mathrm{Int}\, W \cong \Sigma \times [0, \infty)$ as above. Since the projection $\pi : \Sigma \times [0, \infty) \to \Sigma$ induces an isomorphism on de Rham cohomology, we can write $\theta|_{\Sigma\times[0,\infty)} = \pi^*\beta + dF$ for a closed 1-form β on Σ and a smooth function F on $\Sigma \times [0, \infty)$. Pick a cutoff function $\rho : V \to [0, 1]$ which equals 0 on W and 1 on $\Sigma \times [1, \infty)$ and define the function $G := \rho F$ and the closed 1-form $\eta := \theta - dG$ on V. Since $\eta = \pi^*\beta$ on $\Sigma \times [1, \infty)$, the symplectic vector field Y on V defined by $i_Y\omega = \eta$ is complete. Let $h : V \to V$ be the time 1 map of its flow. Since $L_Y\eta = 0$ and $L_Y\lambda = \eta + di_Y\lambda$, it satisfies $h^*\eta = \eta$ and $h^*\lambda = \lambda + \eta + dH$ for some function H on V. Then the diffeomorphism $g := f \circ h : V \to \widetilde{V}$ is diffeotopic to f and satisfies

$$g^*\widetilde{\lambda} = h^*(\lambda - \theta) = h^*(\lambda - \eta - dG) = \lambda + d(H - h^*G).$$ □

REMARK 11.3. The contact manifold (Σ, ξ) above is canonically determined by the finite type Liouville manifold (V, ω, X). We do not know whether (Σ, ξ) actually depends on the Liouville form λ or only on the symplectic form $\omega = d\lambda$.

The answer depends on the following open problem: *Does symplectomorphism of symplectizations imply contactomorphism of contact manifolds?* We do not know how to distinguish contact manifolds with the same symplectization by currently known invariants.

A closely related concept is that of a *Liouville cobordism* (W, ω, X). This is a (compact) cobordism W with an exact symplectic structure (ω, X) such that X points outward along $\partial_+ W$ and inward along $\partial_- W$. A Liouville cobordism with $\partial_- W = \varnothing$ is called a *Liouville domain.*

For a Liouville domain (W, ω, X), the backward flow of X yields a collar neighborhood $(-\varepsilon, 0] \times \partial W$ on which λ corresponds to $e^t \alpha$, where $\alpha = \lambda|_{\partial W}$. So we can glue the semi-infinite cylinder $([0, \infty) \times \partial W$ to W and extend the Liouville form by $e^t \alpha$ to obtain a finite type Liouville manifold which we call the *completion* of (W, ω, X). Conversely, the discussion above shows that every finite type Liouville manifold is the completion of a Liouville domain.

We conclude this section with a brief discussion of holonomy in Liouville manifolds.

LEMMA 11.4. *Let $\Sigma, \widetilde{\Sigma}$ be hypersurfaces in a Liouville manifold (V, ω, X) such that following trajectories of X defines a diffeomorphism $\Gamma : \Sigma \to \widetilde{\Sigma}$. Then Γ is a contactomorphism for the contact structures induced by $i_X \omega$.*

We call Γ the *holonomy map* from Σ to $\widetilde{\Sigma}$.

PROOF. Use the Liouville flow to embed the symplectization $\mathbb{R} \times \Sigma$ into V such that Σ corresponds to $\{0\} \times \Sigma$, $\lambda = e^r \alpha$ and $X = \partial_r$, where $\alpha = \lambda|_\Sigma$ and r is the coordinate on $\mathbb{R}$. Then $\widetilde{\Sigma}$ is given as the graph $r = f(x)$ of a function $f : \Sigma \to \mathbb{R}$ and $\Gamma^*(\lambda|_{\widetilde{\Sigma}}) = e^f \alpha$. □

11.2. Liouville homotopies

In this section we introduce the notion of a homotopy of Liouville domains or manifolds. It has the important property (Proposition 11.8) that homotopic Liouville manifolds are symplectomorphic.

A *homotopy of Liouville cobordisms* is simply a smooth family of Liouville cobordisms (W, ω_s, X_s), $s \in [0, 1]$. However, the definition of a homotopy of Liouville manifolds requires some care.

DEFINITION 11.5. A smooth family (V, ω_s, X_s), $s \in [0, 1]$, of Liouville manifolds is called a *simple Liouville homotopy* if there exists a smooth family of exhaustions $V = \bigcup_{k=1}^{\infty} V_s^k$ by compact domains $V_s^k \subset V$ with smooth boundaries along which X_s is outward pointing. A smooth family (V, ω_s, X_s), $s \in [0, 1]$, of Liouville manifolds is called a *Liouville homotopy* if it is a composition of finitely many simple homotopies.

See Figure 11.1 in Section 11.6 illustrating a non-simple homotopy in the slightly more special case of Weinstein manifolds.

LEMMA 11.6. *A smooth family (V, ω_s, X_s), $s \in [0, 1]$, of Liouville manifolds of finite type is a Liouville homotopy if the closure $\overline{\bigcup_{t \in [0,1]} \mathrm{Skel}(V, \omega_s, X_s)}$ of the union of their skeletons is compact. In particular, the completions of a homotopy of Liouville domains define a homotopy of Liouville manifolds.*

PROOF. The compactness condition implies that around each $s \in [0,1]$ there exists an open interval I_s and a compact set $W_s \subset V$ such that $\mathrm{Skel}(V,\omega_t,X_t) \subset W_s$ and X_t points out of ∂W_s for all $t \in \bar{I}_s$. Finitely many such intervals cover $[0,1]$ and on each $\bar{I}_s$ we have a simple homotopy with exhaustion $V_s^k = X_s^k(W_s)$, $k \in \mathbb{N}$. □

The converse to Lemma 11.6 need not hold: For a homotopy of Liouville manifolds of finite type, the closure of the union of their skeletons need not be compact. We do not know (see the following example) whether the existence of a homotopy between the completions of two Liouville domains implies that the Liouville domains themselves are homotopic.

EXAMPLE 11.7. Let (V,ω,X) be a $2n$-dimensional Liouville manifold of finite type. Write $V = W \cup E$, where (W,ω,X) is a Liouville domain and $E = (\Sigma \times [0,\infty), d(e^t\alpha), \frac{\partial}{\partial t})$ a cylindrical end, and set $W_1 := W \cup (\Sigma \times [0,1])$. Suppose that there exists a diffeomorphism $f : \Sigma \times [0,1] \to \Sigma \times [0,1]$ with $f(x,t) = (x,t)$ near $t=0$ and $f(x,t) = (g(x),t)$ near $t=1$ representing a nontrivial pseudo-isotopy class; see Section 9.10. Let us extend f to the diffeomorphism $\widehat{f} : \Sigma \times \mathbb{R} \to \Sigma \times \mathbb{R}$ which maps (x,t) to $(g(x),t)$ for $t \geq 1$ and equals the identity on $\Sigma \times (-\infty, 0]$. Note that $\widehat{f}$ is isotopic to the identity via the isotopy

$$\widehat{f}_s := \begin{cases} \tau_{\sigma(s)} \circ \widehat{f} \circ \tau_{-\sigma(s)} & s \in (0,1], \\ \mathrm{Id} & s = 0, \end{cases}$$

where $\tau_c(x,t) = (x, t+c)$ is the translation by $c \in \mathbb{R}$ and $\sigma : (0,1] \to [0,\infty)$ is a decreasing diffeomorphism. Note that $\widehat{f}_1 = \widehat{f}$ and $\widehat{f}_s = \mathrm{Id}$ on $\Sigma \times (-\infty, 0]$ for all $s \in [0,1]$. Denote by $F : W_1 \to W_1$ the diffeomorphism equal to Id on W and f on $\Sigma \times [0,1]$. Similarly, we define $\widehat{F}_s : V \to V$ as equal to Id on W and $\widehat{f}_s$ on E. Let $(\widehat{\omega}_s := (\widehat{F}_s)_*\omega, \widehat{X}_s := (\widehat{F}_s)_*X)$, $s \in [0,1]$, be the push-forward Liouville manifold structure on V, and let $(\omega_1 := F_*\omega, X_1 := F_*X)$ be the push-forward Liouville domain structure on W_1. Note that $(V, \widehat{\omega}_1, \widehat{X}_1)$ is the completion of (W_1, ω_1, X_1). Then $(V, \widehat{\omega}_s, \widehat{X}_s)$ is a homotopy of Liouville manifolds connecting (V,ω,X) with $(V, \widehat{\omega}_1, \widehat{X}_1)$. On the other hand, there is no obvious homotopy of Liouville domains connecting (W_1,ω,X) and (W_1,ω_1,X_1). It follows from Theorem 14.3 below that the Liouville domains (W_1,ω,X) and (W_1,ω_1,X_1) are nevertheless homotopic if $n > 2$. The answer is unknown for $n = 2$.

PROPOSITION 11.8. *Let (V,ω_s,X_s), $s \in [0,1]$, be a homotopy of Liouville manifolds with Liouville forms λ_s. Then there exists a diffeotopy $h_s : V \to V$ with $h_0 = \mathrm{Id}$ such that $h_s^*\lambda_s - \lambda_0$ is exact for all $s \in [0,1]$. Moreover, if $\overline{\bigcup_{s\in[0,1]} \mathrm{Skel}(V,\omega_s,X_s)}$ is compact (e.g., for the completion of a homotopy of Liouville domains), then we can achieve $h_s^*\lambda_s - \lambda_0 = 0$ outside a compact set.*

PROOF. It suffices to consider the case of a simple homotopy (V,ω_s,X_s). Pick a family of exhaustions $V = \bigcup_{k=1}^{\infty} V_s^k$ as in Definition 11.5. Denote by Σ_s^k the boundary ∂V_s^k, by λ_s the Liouville form dual to X_s, and by ξ_s^k the contact structure induced on Σ_s^k by the contact form $\lambda_s|_{\Sigma_s^k}$, $s \in [0,1]$, $k \in \mathbb{N}$. By Gray's Stability Theorem 6.23 there are families of contactomorphisms

$$\psi_s^k : (\Sigma_0^k, \xi_0^k) \to (\Sigma_s^k, \xi_s^k),$$

so that $(\psi_s^k)^*\lambda_s = e^{f_s^k}\lambda_0$ for smooth families of functions $f_s^k : \Sigma_0^k \to \mathbb{R}$. (We denote the restrictions of λ_s to the various hypersurfaces by the same symbol.) For $c \in \mathbb{R}$

set $\Sigma_s^{k,c} := X_s^c(\Sigma_s^k)$ and define the diffeomorphisms

$$\psi_s^{k,c} := X_s^c \circ \psi_s^k \circ X_0^{-c} : \Sigma_0^{k,c} \to \Sigma_s^{k,c}.$$

By equation (11.1) we have $(\psi_s^{k,c})^*\lambda_s = e^{f_s^k \circ X_0^{-c}}\lambda_0$. For a sequence of real numbers d_k (which will be determined later) set

$$\widetilde{\Sigma}_s^k := \Sigma_s^{k,d_k}, \quad \widetilde{\psi}_s^k := \psi_s^{k,d_k}, \quad \widetilde{V}_s^k := X_s^{d_k}(V_s^k), \quad \widetilde{f}_s^k := f_s^k \circ X_0^{-d_k} \circ (\widetilde{\psi}_s^k)^{-1}.$$

A short computation using equation (11.1) shows that the map $\Psi_s^k := X_s^{-\widetilde{f}_s^k} \circ \widetilde{\psi}_s^k : \widetilde{\Sigma}_0^k \to V$ satisfies $(\Psi_s^k)^*\lambda_s = \lambda_0$ and hence canonically extends to a map (still denoted by the same symbol) $\Psi_s^k : \mathcal{O}p\,\widetilde{\Sigma}_0^k \to \mathcal{O}p\,(X_s^{-\widetilde{f}_s^k}\widetilde{\Sigma}_s^k)$, which maps trajectories of X_0 to trajectories of X_s and satisfies $(\Psi_s^k)^*\lambda_s = \lambda_0$.

Now we choose the constants d_k such that for each $s \in [0,1]$ the hypersurfaces $\widetilde{\Sigma}_s^k$, $k \in \mathbb{N}$, are mutually disjoint and the hypersurfaces $X_s^{-\widetilde{f}_s^k}(\widetilde{\Sigma}_s^k)$, $k \in \mathbb{N}$, are mutually disjoint. We achieve the first condition by choosing the d_k nondecreasing. The second condition holds if we have

$$\min_{x \in \widetilde{\Sigma}_s^k}\big(d_k - \widetilde{f}_s^k(x)\big) \geq \max_{x \in \widetilde{\Sigma}_s^{k-1}}\big(d_{k-1} - \widetilde{f}_s^{k-1}(x)\big)$$

for all $s \in [0,1]$ and $k \geq 2$. So we can achieve both conditions by defining the d_k inductively by $d_1 := 0$ and

$$d_k := d_{k-1} + \max\big\{0, \max_{s,x} f_s^k(x) - \min_{s,x} f_s^{k-1}(x)\big\}.$$

These conditions ensure that the Ψ_s^k induce a diffeomorphism

$$\Psi_s : \mathcal{O}p\left(\bigcup_{k=1}^{\infty}\widetilde{\Sigma}_0^k\right) \to \mathcal{O}p\left(\bigcup_{k=1}^{\infty} X_s^{-\widetilde{f}_s^k}\widetilde{\Sigma}_0^k\right)$$

satisfying $\Psi_s^*\lambda_s = \lambda_0$. Let us extend Ψ_s in any way to a diffeomorphism $\Psi_s : V \to V$. Now we apply Moser's Stability Theorem 6.8 to each of the open domains $\text{Int}\,\widetilde{V}_0^{k+1} \setminus \widetilde{V}_0^k$ and the family of exact symplectic forms $\Psi_s^*\omega_s = d(\Psi_s^*\lambda_s)$ whose primitives are s-independent near the boundary $\widetilde{\Sigma}_0^{k+1} \cup \widetilde{\Sigma}_0^k$. This yields a family of diffeomorphisms $\phi_s : V \to V$ which are the identity on $\mathcal{O}p\left(\bigcup_{k=1}^{\infty}\widetilde{\Sigma}_0^k\right)$ and such that the composition $h_s := \Psi_s \circ \phi_s$ is the required exact symplectomorphism $(V, \omega_0, X_0) \to (V, \omega_s, X_s)$.

If $K := \overline{\bigcup_{s\in[0,1]}\text{Skel}(V, \omega_s, X_s)}$ is compact we carry out an analogous proof with only one set V^1 containing K. □

11.3. Zeroes of Liouville fields

Here we study some local properties of Liouville fields near zeroes. Recall from Section 6.1 that a subspace W of a symplectic vector space (V,ω) (and similarly for manifolds) is called isotropic (resp. coisotropic) if $W \subset W^\omega$ (resp. $W^\omega \subset W$), where W^ω denotes the ω-orthogonal complement. Also recall from Section 9.2 the definitions of the various invariant manifolds $W_p^\pm, \dots$ near a zero p of a vector field and their tangent spaces $E_p^\pm, \dots$.

PROPOSITION 11.9. *Let (V,ω) be a symplectic manifold with Liouville field X, and let p be a (possibly degenerate) zero of X. Then:*

(a) *the center-stable space $E_p^- \oplus E_p^0 \subset T_pV$ is isotropic;*

(b) *the local stable manifold* W_p^- *is isotropic;*
(c) *the local unstable manifold* W_p^+ *is coisotropic;*
(d) *if* p *is embryonic, then the extended stable manifold* $\widehat{W}_p^-$ *is isotropic.*

In particular, $\dim W_p^- \leq n$ *and* $\dim W_p^+ \geq n$*, where* $2n = \dim V$*.*

PROOF. Let $\phi_t : V \to V$ be the flow of X. Recall that it expands the symplectic form as well as the Liouville form $\lambda = i_X\omega$, i.e., $\phi_t^*\omega = e^t\omega$ and $\phi_t^*\lambda = e^t\lambda$.

(a) The linearization $A := D_pX : T_pV \to T_pV$ preserves the splitting $T_pV = E_p^+ \oplus E_p^- \oplus E_p^0$ from Lemma 9.9(b) and its flow expands the symplectic form,

$$e^t\omega_p(v,w) = \omega_p(e^{tA}v, e^{tA}w).$$

For $v, w \in E_p^- \oplus E_p^0$ the right hand side is bounded for $t \geq 0$, so as $t \to \infty$ we find $\omega_p(v,w) = 0$.

For (b)–(d) abbreviate $W^\pm = W_p^\pm$, so $T_pV = T_pW^+ \oplus T_pW^- \oplus E_p^0$. All eigenvalues of the linearization of X at p have negative real part on T_pW^- and positive real part on T_pW^+. It follows that the differential $T_p\phi_t : T_xV \to T_{\phi_t(x)}V$ satisfies

$$\lim_{t\to\infty} T_x\phi_t(v) = 0 \text{ for } x \in W^-, v \in T_xW^-,$$
$$\lim_{t\to-\infty} T_x\phi_t(v) = 0 \text{ for } x \in W^+, v \in T_xW^+.$$

(b) Let $x \in W^-$ and $v \in T_xW^-$. Since $\phi_t(x) \to p$ as $t \to \infty$, the preceding discussion shows

$$e^t\lambda_x(v) = (\phi_t^*\lambda)(v) = \lambda_{\phi_t(x)}(T_x\phi_t \cdot v) \to 0$$

as $t \to \infty$. This implies $\lambda(v) = 0$, so λ and hence ω vanishes on W^-.

(c) Let $x \in W^+$ and $v \in (T_xW^+)^\omega \subset T_xV$. Suppose $v \notin T_xW^+$. Take a sequence $t_k \to -\infty$, and let $x_k := \phi_{t_k}(x)$. Pick $\lambda_k > 0$ such that $v_k := \lambda_k T_x\phi_{t_k} \cdot v$ has norm 1 with respect to some metric on V. Note that $v_k \in (T_{x_k}W^+)^\omega$ for all k. Pass to a subsequence so that $v_k \to v_\infty \in T_pV$. Since $T_x\phi_{t_k}$ (as $t_k \to -\infty$) exponentially contracts the component of v tangent to W^+ but not the transverse component (this follows, e.g., from the proof of the λ-lemma, see [**157**]), we find $0 \neq v_\infty \in T_pW^-$.

We claim that $v_\infty \in (T_pW^+)^\omega$. Otherwise, there would exist a $w_\infty \in T_pW^+$ with $\omega(v_\infty, w_\infty) \neq 0$. But then $\omega(v_k, w_k) \neq 0$ for k large and some $w_k \in T_{x_k}W^+$, contradicting $v_k \in (T_{x_k}W^+)^\omega$. Hence v_∞ is ω-orthogonal to T_pW^+. Since $T_pW^- \oplus E_p^0$ is isotropic by part (a), v_∞ is also ω-orthogonal to $T_pW^- \oplus E_p^0$. But this contradicts the nondegeneracy of ω because $T_pV = T_pW^+ \oplus T_pW^- \oplus E_p^0$.

(d) Suppose now that p is embryonic. The proof that $\widehat{W}_p^-$ is isotropic is similar to that of part (a). Let $x \in \widehat{W}^-$, so $\phi_t(x) \to p$ as $t \to \infty$. Since the eigenvalues on $T_p\widehat{W}^- = T_pW^- \oplus E_p^0$ have nonpositive real part, there exists a constant C such that

$$|T_x\phi_t(v)| \leq Ce^{t/2}|v| \text{ for all } v \in T_x\widehat{W}^-, \ t \geq 0.$$

It follows that

$$e^t|\lambda_x(v)| = |\lambda_{\phi_t(x)}(T_x\phi_t \cdot v)| \leq Ce^{t/2}|v|.$$

As $t \to \infty$ this implies $\lambda(v) = 0$, so λ and hence ω vanishes on $\widehat{W}^-$. □

11.4. Weinstein cobordisms and manifolds

DEFINITION 11.10. A *Weinstein manifold* (V, ω, X, ϕ) is a symplectic manifold (V, ω) with a complete Liouville field X which is gradient-like for an exhausting Morse function $\phi : V \to \mathbb{R}$. A *Weinstein cobordism* (W, ω, X, ϕ) is a Liouville cobordism (W, ω, X) whose Liouville field X is gradient-like for a Morse function $\phi : W \to \mathbb{R}$ which is constant on the boundary. In both cases the triple (ω, X, ϕ) is called a *Weinstein structure* on V (resp. W). A Weinstein cobordism with $\partial_- W = \varnothing$ is called a *Weinstein domain.*

Thus any Weinstein manifold (V, ω, X, ϕ) can be exhausted by Weinstein domains $W_k = \{\phi \leq d_k\}$, where $d_k \nearrow \infty$ is a sequence of regular values of the function ϕ.

A Weinstein manifold (V, ω, X, ϕ) is said to be of *finite type* if ϕ has only finitely many critical points. Note that by attaching a cylindrical end, any Weinstein domain (W, ω, X, ϕ) can be completed to a finite type Weinstein manifold, called its *completion.* Conversely, any finite type Weinstein manifold is the completion of a Weinstein domain.

REMARK 11.11. (i) Any Weinstein manifold (V, ω, X, ϕ) has the structure of a Liouville manifold (V, ω, X). However, not every Liouville manifold is diffeomorphic to a Weinstein manifold; see [**133, 64**].

(ii) Later on, in deformations of Weinstein structures, we will also allow ϕ to have embryonic (death-birth) singularities; in this section we restrict ourselves to the Morse case.

EXAMPLE 11.12. (1) $\mathbb{C}^n$ carries the canonical Weinstein structure

$$\omega_{\mathrm{st}} = \sum_{j=1}^n dx_j \wedge dy_j, \quad X_{\mathrm{st}} = \frac{1}{2}\sum_{j=1}^n \left(x_j \frac{\partial}{\partial x_j} + y_j \frac{\partial}{\partial y_j} \right), \quad \phi_{\mathrm{st}} = \frac{1}{4}\sum_{j=1}^n (x_j^2 + y_j^2).$$

(2) An important example of a Weinstein structure is provided by the cotangent bundle $V = T^*Q$ of a closed manifold Q with the standard symplectic form $\omega_{\mathrm{st}} = d\lambda_{\mathrm{st}}$, where $\lambda_{\mathrm{st}} = p\, dq$ is the standard Liouville form. The associated Liouville field $X_{\mathrm{st}} = p\frac{\partial}{\partial p}$ is gradient-like for the function $\phi_{\mathrm{st}}(q,p) = \frac{1}{2}|p|^2$. Since the function ϕ_{st} is not Morse, this does not yet define a Weinstein structure according to our definition (although the definition could be without difficulty relaxed to allow for Morse–Bott functions such as ϕ_{st}). To define a Weinstein structure, take any Riemannian metric on Q and a Morse function $f : Q \to \mathbb{R}$. Note that the Hamiltonian vector field X_F of the function $F(q,p) := \langle p, \nabla f(q) \rangle$ (or in more invariant notation $F = \lambda(\nabla f)$) coincides with ∇f along the zero section of T^*Q. Thus the vector field $X := p\frac{\partial}{\partial p} + X_F$ is Liouville and gradient-like for the Morse function $\phi(q,p) := \frac{1}{2}|p|^2 + f(q)$ if f is small enough.

(3) The product of two Weinstein manifolds $(V_1, \omega_1, X_1, \phi_1)$ and $(V_2, \omega_2, X_2, \phi_2)$ has a canonical Weinstein structure $(V_1 \times V_2, \omega_1 \oplus \omega_2, X_1 \oplus X_2, \phi_1 \oplus \phi_2)$. In particular, the product $(V, \omega, X, \phi) \times (\mathbb{R}^2, \omega_{\mathrm{st}}, X_{\mathrm{st}}, \phi_{\mathrm{st}})$ is called the *stabilization* of the Weinstein manifold (V, ω, X, ϕ).

Note that in a Weinstein manifold (V, ω, X, ϕ) any regular level set $\Sigma := \phi^{-1}(c)$ carries a canonical contact structure ξ defined by the contact form $\alpha := (i_X\omega)|_\Sigma$. In particular, this applies to the boundary of a Weinstein domain. Contact manifolds which appear as boundaries of Weinstein domains are called *Weinstein fillable.* We will see later (Theorem 13.5) that this is equivalent to being *Stein fillable.*

LEMMA 11.13. *Let (V,ω,X,ϕ) be a Weinstein manifold.*
(a) The stable manifold W_p^- of any critical point $p\in V$ of ϕ satisfies $\lambda|_{W_p^-}\equiv 0$. In particular, W_p^- is isotropic for the symplectic structure ω, and the intersection $W_p^-\cap\phi^{-1}(c)$ with any regular level set is isotropic for the contact structure induced by λ on $\phi^{-1}(c)$.
(b) Suppose ϕ has no critical values in $[a,b]$. Then the image of any isotropic submanifold $\Lambda^a\subset\phi^{-1}(a)$ under the flow of X intersects $\phi^{-1}(b)$ in an isotropic submanifold Λ^b.

PROOF. (a) Since X is tangent to W_p^- and W_p^- is isotropic by Proposition 11.9, the Liouville form $\lambda=i_X\omega$ vanishes on W_p^-. Part (b) is an immediate consequence of Lemma 11.4. □

In view of Lemma 9.9, every zero p of the Liouville field X in a Weinstein manifold (V,ω,X,ϕ) is hyperbolic. Thus the skeleton of (V,ω,X) is the union of all stable manifolds, which are isotropic by Proposition 11.9. Under suitable technical assumptions (X Morse–Smale and (X,ϕ) standard near critical points), the skeleton is in fact an isotropic embedded CW complex [**17**]. We will not use this fact, but rather the following interpretation of Lemma 11.13:

An exhaustion of a Weinstein manifold (V,ω,X,ϕ) by regular sublevel sets $\{\phi\le d_k\}$ such that each interval (d_{k-1},d_k) contains at most one critical value provides a handlebody decomposition of V whose core discs (the stable discs of critical points) are isotropic in the symplectic sense, and whose attaching spheres are isotropic in the contact sense.

11.5. From Stein to Weinstein

Until this point, by a Stein manifold we meant a complex manifold (V,J) which admits an exhausting J-convex function $\phi:V\to\mathbb{R}$. From now on, we will change our perspective and consider the function ϕ as part of the data. Moreover, we will require the function ϕ to be Morse, which can always be achieved by a C^2-small perturbation. The following analogue of Definition 11.10 in the Stein case will be relevant for the remainder of this book.

DEFINITION 11.14. A *Stein manifold* (V,J,ϕ) is a complex manifold (V,J) with an exhausting J-convex Morse function $\phi:V\to\mathbb{R}$. A *Stein cobordism* (W,J,ϕ) is a complex cobordism (W,J) with a J-convex Morse function $\phi:W\to\mathbb{R}$ having $\partial_\pm W$ as regular level sets. In both cases the triple (J,ϕ) is called a *Stein structure* on V (resp. W). A Stein cobordism with $\partial_-W=\varnothing$ is called a *Stein domain.*

Next recall that a J-convex function ϕ is called completely exhausting if it is exhausting and its gradient field $\nabla_\phi\phi$ is complete.

DEFINITION 11.15. To every Stein cobordism (W,J,ϕ) we associate the Weinstein cobordism structure

$$\mathfrak{W}(J,\phi):=(\omega_\phi,X_\phi,\phi):=(-dd^{\mathbb{C}}\phi,\nabla_\phi\phi,\phi)$$

on W. The same formula also associates a Weinstein manifold structure on V to every Stein manifold (V,J,ϕ) with ϕ a completely exhausting Morse function.

By Lemma 2.20, $\mathfrak{W}(J,\phi)$ defines indeed a Weinstein structure. Note that the contact structure ξ induced on a regular level set $\Sigma = \phi^{-1}(c)$ by the Liouville form $-d\phi^{\mathbb{C}}$ coincides with the field of complex tangencies on the J-convex hypersurface Σ.

REMARK 11.16. The completeness condition in Definition 11.15 is necessary because we require the Liouville field to be complete in our definition of a Weinstein manifold. According to Proposition 2.11, any exhausting J-convex function can be made completely exhausting by composing it with a sufficiently convex function $\mathbb{R} \to \mathbb{R}$. Subsequently, whenever we speak of the Weinstein manifold structure $\mathfrak{W}(J,\phi)$ associated to a Stein manifold, we will implicitly assume that ϕ is completely exhausting *without further mentioning it.*

REMARK 11.17. Let (V, J, ϕ) be an *almost* complex manifold with an exhausting J-convex Morse function $\phi : V \to \mathbb{R}$. Then even if the symplectic form $\omega_\phi = -dd^{\mathbb{C}}\phi$ is not compatible with J, one still gets a Weinstein structure $(\omega_\phi, X_\phi, \phi)$ on V similar to the one defined above. The only difference in this case is that the Liouville field X_ϕ should be defined directly as ω_ϕ-dual to $-d^{\mathbb{C}}\phi$, i.e., by

$$-d^{\mathbb{C}}\phi = i_{X_\phi}\omega_\phi.$$

Applying both sides to a tangent vector JZ, we find

$$d\phi(Z) = \omega_\phi(X_\phi, JZ),$$

so X_ϕ is gradient-like for ϕ with respect to the positive definite (but in general non-symmetric) $(2,0)$ tensor field $g_\phi := \omega_\phi(\cdot, J\cdot)$. Completeness of X_ϕ can be achieved as in the integrable case by composing ϕ with a sufficiently convex function.

Combined with Proposition 11.9, this yields another proof of the fact (Corollary 3.4) that the indices of critical points of a J-convex Morse function on a $2n$-dimensional almost complex manifold are $\leq n$.

EXAMPLE 11.18. Not every Weinstein structure equals $\mathfrak{W}(J,\phi)$ for some Stein structure (J,ϕ). Indeed, this fails already in a neighborhood of a critical point p: the linearization $D_pX : T_pV \to T_pV$ of the Liouville field is diagonalizable if $X = \nabla_\phi\phi$ for a J-convex function ϕ, while for a general Weinstein structure it need not be. For example, consider for $\varepsilon \in \mathbb{R}$ the Liouville 1-form

$$\lambda = \frac{1}{2}(x\,dy - y\,dx) + \varepsilon(x\,dx + y\,dy)$$

on $\mathbb{C}$ satisfying $d\lambda = dx \wedge dy$. The corresponding Liouville field

$$X = \frac{1}{2}(x\partial_x + y\partial_y) + \varepsilon(y\partial_x - x\partial_y)$$

has eigenvalues $1/2 \pm i\varepsilon$, so it is induced by a Stein structure if and only if $\varepsilon = 0$. A quadratic Lyapunov function for X is, e.g., given by $x^2 + y^2$.

11.6. Weinstein and Stein homotopies

Now we define Weinstein and Stein homotopies. They have the important property that homotopic Weinstein manifolds are symplectomorphic. Moreover, we prove that the Stein structures corresponding to two exhausting J-convex functions on the same complex manifold are homotopic.

DEFINITION 11.19. A *Weinstein homotopy* on a cobordism or manifold is a smooth family of Weinstein structures (ω_t, X_t, ϕ_t), $t \in [0,1]$, where we allow birth-death type degenerations, such that the associated Liouville structures (ω_t, X_t) form a Liouville homotopy. A *Stein homotopy* is a smooth family of Stein structures (J_t, ϕ_t), where we allow birth-death type degenerations, such that the associated Weinstein structures $\mathfrak{W}(J_t, \phi_t)$ form a Weinstein homotopy.

Since this definition is basic for all that follows, let us recall its main features. We begin with the case of a cobordism W. Then a Weinstein homotopy $(W, \omega_t, X_t, \phi_t)$ induces a Smale homotopy (W, X_t, ϕ_t) in the sense of Section 9.7. This means that the functions ϕ_t have $\partial_\pm W$ as regular level sets, and they are Morse except for finitely many $t \in (0,1)$ at which a birth-death type critical point occurs. Note again the slight abuse of language because (ω_t, X_t, ϕ_t) is *not* a Weinstein structure for such t.

In the case of a manifold V the conditions on the boundary are replaced by the existence of a smooth family of exhaustions as in Definition 11.5 which prevents critical points from escaping to infinity. Using the functions ϕ_t, this condition can be equivalently formulated as follows. Let $\phi_t : V \to \mathbb{R}$, $t \in [0,1]$, be a smooth family of exhausting functions on a manifold V having only Morse or birth-death type critical points. We call ϕ_t a *simple Morse homotopy* if there exists a sequence of smooth functions $c_1 < c_2 < \cdots$ on the interval $[0,1]$ such that for each $t \in [0,1]$, $c_i(t)$ is a regular value of the function ϕ_t and $\bigcup_k \{\phi_t \le c_k(t)\} = V$. A *Morse homotopy* is a composition of finitely many simple Morse homotopies. A *Smale homotopy* is a smooth family of Lyapunov pairs (X_t, ϕ_t) such that the associated functions ϕ_t form a Morse homotopy. Then a Weinstein homotopy is a family of Weinstein structures $(V, \omega_t, X_t, \phi_t)$ (again allowing birth-death type degenerations) such that the associated Lyapunov pairs (X_t, ϕ_t) form a Smale homotopy.

For Stein/Weinstein/Smale/Morse homotopies we will always use such exhaustions by sublevel sets $\{\phi_t \le c_k(t)\}$. Figure 11.1 shows the profile for a composition of two simple Morse homotopies which is not simple: the sublevel sets $\{\phi \le c_i\}$ (resp. $\{\phi \le c_i'\}$) provide exhaustions for the restrictions of the homotopy to the intervals $[0,1/2]$ (resp. $[1/2,1]$), while no such exhaustion exists over the whole interval $[0,1]$.

EXAMPLE 11.20. Consider an exhausting Morse function ϕ with gradient-like vector field X on a manifold V and a diffeotopy $h_t : V \to V$, $t \in [0,1]$. Then $(V, h_t^*X, h_t^*\phi)$ is a simple Smale homotopy. Indeed, in the definition we just take the constant functions $c_1 < c_2 < \cdots$ on the interval $[0,1]$ for a sequence $c_k \to \infty$ of regular values of ϕ.

Since a Weinstein homotopy (ω_t, X_t, ϕ_t) induces a Liouville homotopy (ω_t, X_t), Proposition 11.8 implies

COROLLARY 11.21. *If two Weinstein manifolds $\mathfrak{W}_0 = (V, \omega_0, X_0, \phi_0)$ and $\mathfrak{W}_1 = (V, \omega_1, X_1, \phi_1)$ are Weinstein homotopic, they are symplectomorphic. More precisely, there exists a diffeotopy $h_t : V \to V$ with $h_0 = \mathrm{Id}$ such that $h_1^*\lambda_1 - \lambda_0$ is exact, where $\lambda_i = i_{X_i}\omega_i$ are the Liouville forms. If $\mathfrak{W}_0$ and $\mathfrak{W}_1$ are the completions of homotopic Weinstein domains, then we can achieve $h_1^*\lambda_1 - \lambda_0 = 0$ outside a compact set.*

The following proposition shows that the existence of a Stein homotopy connecting two Stein structures (J_0, ϕ_0) and (J_1, ϕ_1) depends only on the Stein complex

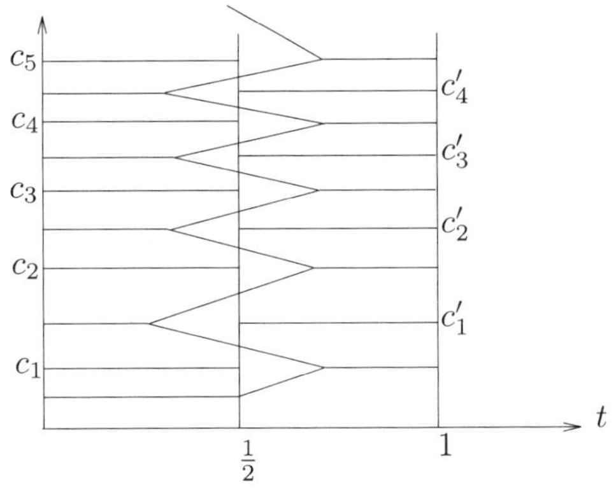

FIGURE 11.1. A composition of two simple homotopies which is not simple.

structures J_0, J_1 and not on the functions ϕ_0, ϕ_1. Moreover, the symplectic manifold (V, ω_ϕ) associated to a Stein manifold (V, J, ϕ) is independent, up to exact symplectomorphism isotopic to the identity, of the choice of a completely exhausting J-convex Morse function ϕ.

PROPOSITION 11.22 (see [**49**]). *Let $\phi_0, \phi_1 : V \to \mathbb{R}$ be two exhausting J-convex Morse functions on a complex manifold (V, J). Then (J, ϕ_0) and (J, ϕ_1) can be connected by a Stein homotopy (J, ϕ_t). In particular, if ϕ_0, ϕ_1 are completely exhausting Morse functions, then the corresponding Weinstein structures $(\omega_{\phi_0}, X_{\phi_0}, \phi_0)$ and $(\omega_{\phi_1}, X_{\phi_1}, \phi_1)$ are Weinstein homotopic.*

The proof of Proposition 11.22 is based on

LEMMA 11.23. *Let $\phi_0, \phi_1 : V \to \mathbb{R}$ be two exhausting J-convex functions on a complex manifold (V, J). Then there exist smooth functions $h_0, h_1 : \mathbb{R} \to \mathbb{R}$ with $h_0', h_1' \to \infty$ and $h_0'', h_1'' \geq 0$, a completely exhausting J-convex function $\psi : V \to \mathbb{R}_+$, and a sequence of compact domains V^k, $k = 1, \dots,$ with smooth boundaries $\Sigma^k = \partial V^k$, such that*

- *$V^k \subset \operatorname{Int} V^{k+1}$ for all $k \geq 1$ and $\bigcup_k V^k = V$;*
- *Σ^{2j-1} are level sets of the function ϕ_1 and Σ^{2j} are level sets of the function ϕ_0 for all $j \geq 1$;*
- *$\psi = h_1 \circ \phi_1$ on $\mathcal{O}p\left(\bigcup_{j=1}^\infty \Sigma^{2j-1}\right)$ and $\psi = h_0 \circ \phi_0$ on $\mathcal{O}p\left(\bigcup_{j=1}^\infty \Sigma^{2j}\right)$.*

PROOF. Let us call a diffeomorphism $h : \mathbb{R} \to \mathbb{R}$ an *admissible function* if $h'' \geq 0$ and $h' \to \infty$. Take any $c_1 > 0$ and define $V^1 := \{\phi_1 \leq c_1\}$, $\Sigma^1 := \partial V^1$. There exists an admissible function g_1 such that $\phi_0|_{\Sigma^1} < d_1 = g_1(c_1)$. Set $\psi_0 := \phi_0$, $\psi_1 := g_1 \circ \phi_1$. Next, take any $c_2 > d_1$ and define $V^2 := \{\psi_0 \leq c_2\}$, $\Sigma^2 := \partial V^2$. Then $V^1 \subset \operatorname{Int} V^2$. There exists an admissible function g_2 such that $g_2(x) = x$ for $x \leq d_1$ and $\psi_1|_{\Sigma^2} < d_2 = g_2(c_2)$. Set $\psi_2 := g_2 \circ \psi_0$. Continuing this process, we take $c_3 > d_2$ and define $V^3 := \{\psi_1 \leq c_3\}$, $\Sigma^3 := \partial V^3$. There exists an admissible function g_3 such that $g_3(x) = x$ for $x \leq d_2$ and $\psi_2|_{\Sigma^3} < d_3 = g_3(c_3)$. Set $\psi_3 := g_3 \circ \psi_1$, and so on. Continuing this process, we construct compact domains V^k with smooth boundaries $\partial V^k = \Sigma^k$, $k \geq 1$, and two sequences of admissible functions $\widetilde{g}_{2j} := g_{2j} \circ g_{2j-2} \circ \cdots \circ g_2$, $\widetilde{g}_{2j-1} := g_{2j-1} \circ g_{2j-3} \circ \cdots \circ g_1$. Since these sequences stabilize on compact sets, they converge to admissible functions $h_0 := \lim_{j\to\infty} \widetilde{g}_{2j}$ and $h_1 := \lim_{j\to\infty} \widetilde{g}_{2j-1}$. It follows that the sequences of functions $\psi_{2j} = \widetilde{g}_{2j} \circ \phi_0$ and $\psi_{2j-1} = \widetilde{g}_{2j-1} \circ \phi_1$ converge as $j \to \infty$ to exhausting J-convex functions $\psi_{\rm even}$ and $\psi_{\rm odd}$ on V. By construction, they have the following properties:

- $V^k \subset \operatorname{Int} V^{k+1}$ for all $k \geq 1$ and $\bigcup_k V^k = V$;
- ϕ_1 is constant on Σ^{2j-1} and ϕ_0 is constant on Σ^{2j} for all $j \geq 1$;
- $\psi_{\rm even} = h_0 \circ \phi_0$ and $\psi_{\rm odd} = h_1 \circ \phi_1$;
- $\psi_{\rm odd}|_{\Sigma^{2j-1}} > \psi_{\rm even}|_{\Sigma^{2j-1}}$ and $\psi_{\rm even}|_{\Sigma^{2j}} > \psi_{\rm odd}|_{\Sigma^{2j}}$ for all $j \geq 1$.

Smoothing the continuous J-convex function $\max(\psi_{\rm even}, \psi_{\rm odd})$ thus yields the required smooth J-convex function ψ. □

PROOF OF PROPOSITION 11.22. Let h_0, h_1 and ψ be the functions constructed in Lemma 11.23. Now the required Stein homotopy is constructed as a composition of four *elementary* homotopies. First, note that for any function $h : \mathbb{R}_+ \to \mathbb{R}_+$ with $h' > 0$ and $h'' \geq 0$ the linear combination $h^s(x) = (1-s)x + sh(x)$ has the same properties for any $s \in [0,1]$. Hence the exhausting J-convex functions $h_i^s \circ \phi_i$ provide elementary Stein homotopies $(J, h_i^s \circ \phi_i)$ between the Stein structures (J, ϕ_i) and $(J, h_i \circ \phi_i)$, $i = 0, 1$. On the other hand, for each $i = 0, 1$ the family $\phi_i^t = (1-t)h_i \circ \phi_i + t\psi$, $t \in [0,1]$, consists of exhausting J-convex functions which coincide near the boundaries of an exhausting sequence of compact domains. Hence they generate elementary Stein homotopies (J, ϕ_i^t) between $(J, h_i \circ \phi_i)$ and (J, ψ). Concatenating these four elementary homotopies yields the desired Stein homotopy (J, ϕ_t).

Now suppose ϕ_0 and ϕ_1 are completely exhausting Morse functions. In view of Proposition 2.11 (by choosing the h_i sufficiently convex) we can achieve that all the functions ϕ_t, $t \in [0,1]$, are completely exhausting. Moreover, we can perturb ϕ_t to a generic 1-parameter family of functions. Then $(\omega_{\phi_t}, X_{\phi_t}, \phi_t)$ provides a Weinstein homotopy between the Weinstein structures $(\omega_{\phi_0}, X_{\phi_0}, \phi_0)$ and $(\omega_{\phi_1}, X_{\phi_1}, \phi_1)$. This concludes the proof of Proposition 11.22. □

REMARK 11.24. Without the hypothesis on the functions $c_k(t)$, the notion of "Stein or Weinstein homotopy" would become rather trivial. For example, then all Stein structures on $\mathbb{C}^n$ would be "homotopic".

To see this, consider any Stein structure (J, ϕ) on $\mathbb{C}^n$. After a Stein homotopy, we may assume that (J, ϕ) agrees with the standard structure $(J_{\rm st} = i, \phi_{\rm st} = |z|^2)$ on the open unit ball B_1. Pick a smooth family of radial diffeomorphisms $h_t : \mathbb{C}^n \to \mathbb{C}^n$, $t \in [0,1)$, such that $h_0 = \mathrm{Id}$ and h_t converges as $t \to 1$ in $C^\infty_{\rm loc}$ to a radial diffeomorphism $h_1 : \mathbb{C}^n \to B_1$. Pick a smooth family of convex diffeomorphisms $g_t : \mathbb{R} \to \mathbb{R}$, $t \in [0,1)$, such that $g_0 = \mathrm{Id}$ and g_t converges in $C^\infty_{\rm loc}$ on $(-\infty, 1)$ as $t \to 1$ to a convex diffeomorphism $g_1 : (-\infty, 1) \to \mathbb{R}$. Then $(J_t, \phi_t) := (h_t^* J, g_t \circ \phi \circ h_t)$ would be a "Stein homotopy" from (J, ϕ) to a Stein structure (J_1, ϕ_1) which can be connected to the standard structure by another radial homotopy.

Since there exist Stein structures (J, ϕ) on $\mathbb{C}^n$ for which ω_ϕ is not symplectomorphic to the standard symplectic structure (see Chapter 17 below), this also shows that Corollary 11.21 would fail for this notion of "Weinstein homotopy".

REMARK 11.25. The proof of Proposition 11.22 (simply ignoring J-convexity) also shows that *any two exhausting Morse functions on the same manifold can be connected by a Morse homotopy.* Let us emphasize, however, that two exhausting Morse functions of finite type cannot in general be connected by a Morse homotopy during which all critical points remain in a fixed compact set. For example, let M_0, M_1 be two closed 4-manifolds that are homeomorphic but not diffeomorphic. Then the 5-manifolds $M_0 \times \mathbb{R}$ and $M_1 \times \mathbb{R}$ are diffeomorphic, so the functions $M_i \times \mathbb{R} \to \mathbb{R}$, $(x,t) \mapsto t^2$, can be perturbed to two finite type exhausting Morse functions ϕ_0, ϕ_1 on the same 5-manifold. Since high level sets of ϕ_0 and ϕ_1 are not diffeomorphic, the functions ϕ_0 and ϕ_1 cannot be connected by a Morse homotopy with critical points remaining in a compact set.

The notion of Weinstein (or Stein) homotopy can be formulated in more topological terms. Let us denote by $\mathfrak{Weinstein}$ the space of Weinstein structures on a fixed manifold V, where we allow the functions to have embryonic critical points. For any $\mathfrak{W}_0 \in \mathfrak{Weinstein}$, $\varepsilon > 0$, $A \subset V$ compact, $k \in \mathbb{N}$, and any unbounded sequence $c_1 < c_2 < \cdots$, we define the set

$$\mathcal{U}(\mathfrak{W}_0, \varepsilon, A, k, c) := \{\mathfrak{W} = (\omega, X, \phi) \in \mathfrak{Weinstein} \mid \|\mathfrak{W} - \mathfrak{W}_0\|_{C^k(A)} < \varepsilon, \\ c_i \text{ regular values of } \phi\}.$$

It is easy to see that these sets are the basis of a topology on $\mathfrak{Weinstein}$. Note that this topology is coarser that the $C^\infty_{\rm loc}$ topology, which we obtain by dropping the condition on the regular values c_i.

A smooth family of Weinstein structures $\mathfrak{W}_t = (\omega_t, X_t, \phi_t)$ defines a continuous path $[0,1] \to \mathfrak{Weinstein}$ with respect to this topology if and only if there exists a partition $0 = t_0 < t_1 < \cdots < t_N = 1$ and unbounded sequences $c_1^k < c_2^k < \cdots$, $k = 1, \dots, N$, such that c_i^k is a regular value of ϕ_t for all $t \in [t_{k-1}, t_k]$. Hence $\mathfrak{W}_t$ is a Weinstein homotopy according to our definition. Conversely, suppose that $\mathfrak{W}_t$ is a Weinstein homotopy. Then there exists a partition $0 = t_0 < t_1 < \cdots < t_N = 1$ and unbounded sequences of smooth *functions* $c_1^k(t) < c_2^k(t) < \cdots$, $t \in [t_{k-1}, t_k]$, $k = 1, \dots, N$, such that $c_i^k(t)$ is a regular value of ϕ_t for all $t \in [t_{k-1}, t_k]$. After a C^∞-small perturbation, we may assume that $c_i^k(t_k) \neq c_j^{k+1}(t_k)$ for all i, j. This allows us to pick a smooth family of diffeomorphisms $g_t : \mathbb{R} \to \mathbb{R}$ such that $g_0 = \mathrm{Id}$ and $g_t\big(c_i^k(t)\big)$ is constant in $t \in [t_{k-1}, t_k]$ for all $i \in \mathbb{N}$ and $k = 1, \dots, N$. Then $(\omega_t, X_t, g_t \circ \phi_t)$ defines a continuous path $[0,1] \to \mathfrak{Weinstein}$. Hence, up to target reparametrization $\phi_t \mapsto g_t \circ \phi_t$, continuous paths in $\mathfrak{Weinstein}$ correspond to Weinstein homotopies.

In view of the preceding discussion, we call a smooth k-parametric family of Weinstein structures $\mathfrak{W}_u = (\omega_u, X_u, \phi_u)$, $u \in D^k$, a *Weinstein family* if there exists a smooth family of diffeomorphisms $g_u : \mathbb{R} \to \mathbb{R}$ such that $(\omega_u, X_u, g_u \circ \phi_u)$ defines a continuous path $D^k \to \mathfrak{Weinstein}$.

The preceding discussion carries over to Stein structures with one minor modification: we require the target reparametrizations $g_u : \mathbb{R} \to \mathbb{R}$ to be weakly convex to ensure that $g_u \circ \phi_u$ remains J-convex. This can always be achieved by composing any family g_u with a sufficiently convex single function $f : \mathbb{R} \to \mathbb{R}$.

11.7. Weinstein structures with unique critical points

In this section we discuss Weinstein and Stein structures with a unique critical point.

PROPOSITION 11.26. *Let (V, J, ϕ) be a Stein manifold such that ϕ has a unique critical point, the minimum. Then there exists a diffeomorphism $h : \mathbb{C}^n \to V$ such that the Stein structure $(\mathbb{C}^n, h^*J, h^*\phi)$ is Stein homotopic to the standard structure* *on* $\mathbb{C}^n$. *Similarly, given a Stein domain* (W, J, ϕ) *such that* ϕ *has a unique critical point, there exists a diffeomorphism* $h : B^{2n} \to V$, *where* B^{2n} *is the closed unit ball in* $\mathbb{C}^n$, *such that the Stein structure* $(B^{2n}, h^*J, h^*\phi)$ *is Stein homotopic to the standard structure on* B^{2n}. *Analogous results hold for Weinstein structures.*

PROOF. We consider first the Stein manifold case. Assuming that the critical value of ϕ is 0, we first modify ϕ near the critical point so that $h_\varepsilon^*\phi = \phi_{\rm st} = |z|^2$ for some biholomorphic map h_ε from the open ε-ball $B_\varepsilon \subset \mathbb{C}^n$ onto a neighborhood of the minimum. Using gradient-like vector fields for ϕ and $\phi_{\rm st}$, we extend h_ε to a diffeomorphism $h : \mathbb{C}^n \to V$ with $h^*\phi = \phi_{\rm st}$. Define $\widetilde{J} := h^*J$, so $\widetilde{J}|_{B_\varepsilon} = i$. Pick a smooth family of maps $f_t : \mathbb{R}_+ \to \mathbb{R}_+$, $t \in [0,1]$, with the following properties:

- $f_0 = \mathrm{Id}$, and $f_t = \mathrm{Id}$ near 0 for all $t \in [0,1]$;
- f_t defines a diffeomorphism $\mathbb{R}_+ \to [0, \varepsilon/t)$ for $t \in (0,1]$.

Then the smooth family of maps $g_t : \mathbb{C}^n \to \mathbb{C}^n$, $g_t(z) := f_t(|z|)\frac{z}{|z|}$ satisfies:

- $g_0 = \mathrm{Id}$, and $g_t = \mathrm{Id}$ near $z = 0$ for all $t \in [0,1]$;
- g_t defines a diffeomorphism $\mathbb{C}^n \to B_{\varepsilon/t}$ for $t \in (0,1]$.

Since $\phi_{\rm st} \circ g_t(z) = f_t(|z|)^2$ is $g_t^*\widetilde{J}$-convex, the function $\phi_{\rm st}$ is $g_t^*\widetilde{J}$-lc for all $t \in [0,1]$. Hence (after a target reparametrization which we suppress) we can connect $(\widetilde{J}, \phi_{\rm st})$ on $\mathbb{C}^n$ by the Stein homotopy $(g_t^*\widetilde{J}, \phi_{\rm st})$ to $(g_1^*J = g_1^*i, \phi_{\rm st})$. Since $\phi_{\rm st}$ is also g_t^*i-lc, the Stein homotopy $(g_t^*i, \phi_{\rm st})$ on $\mathbb{C}^n$ connects $(g_1^*i, \phi_{\rm st})$ with the standard structure $(i, \phi_{\rm st})$.

The case of a Weinstein manifold is analogous. In the case of a Stein or Weinstein domain, we only need to replace $\mathbb{C}^n$ and B_ε by the closed balls B^{2n} and $\bar{B}_\varepsilon$. □

COROLLARY 11.27. *Every Stein (resp. Weinstein) structure on* $\mathbb{C}^n$ *with a unique critical point is Stein (resp. Weinstein) homotopic to the standard structure on* $\mathbb{C}^n$. *An analogous statement holds for Stein (resp. Weinstein) structures on the closed ball* B^{2n} *provided that* $n > 2$.

PROOF. Any orientation preserving diffeomorphism $h : \mathbb{C}^n \to \mathbb{C}^n$ is diffeotopic to the identity via the Alexander trick

$$h_t(z) := \begin{cases} \frac{1}{t}h(tz) & t \in (0,1], \\ z & t = 0, \end{cases}$$

after adjusting h to equal the identity near 0. Hence the claim for $\mathbb{C}^n$ follows from Proposition 11.26 and Example 11.20.

For the case of the closed ball and $n > 2$ the claim follows from Cerf's theorem [**30**] that the group $\mathrm{Diff}_+(B^{2n})$ of orientation preserving diffeomorphisms of B^{2n} is connected for $n > 2$. □

REMARK 11.28. Nothing, however, is known about the topology of $\mathrm{Diff}_+(B^4)$. We will encounter this phenomenon again in Chapter 16.

11.8. Subcritical and flexible Weinstein structures

A $2n$-dimensional Weinstein cobordism or manifold (W, ω, X, ϕ) is called *subcritical* if all critical points of the function ϕ have index $< n$. Similarly, one defines subcritical Stein cobordisms and manifolds. Clearly, the stabilization (see Section 11.4 above) of any Weinstein manifold is subcritical. The converse is also true due to the following theorem from [**33**] which we will prove in Section 14.4: *Every subcritical Weinstein manifold is symplectomorphic to a stabilization.*

In Section 7.7 we introduced a class of *loose* Legendrian links in contact manifolds of dimension ≥ 5. In the 3-dimensional case a Legendrian link is called *loose* if its complement is overtwisted. Recall from Sections 7.6 and 7.7 that loose Legendrian links satisfy an h-principle. The following definition was motivated by a talk of E. Giroux at ETH Zürich on November 9, 2010.

DEFINITION 11.29. An elementary $2n$-dimensional Weinstein cobordism (W, ω, X, ϕ) is called *flexible* if the attaching spheres of all index n handles form in $\partial_- W$ a *loose* Legendrian link. A Weinstein cobordism or manifold structure (W, ω, X, ϕ) is called flexible if it can be decomposed into elementary flexible cobordisms. A Stein structure is called *flexible* if the underlying Weinstein structure is flexible.

REMARK 11.30. (1) In particular, *any subcritical Weinstein cobordism is flexible.*

(2) Note that a 4-dimensional Weinstein cobordism can only be flexible if it is subcritical, or if the contact structure on $\partial_- W$ is overtwisted. In particular, *a 4-dimensional Weinstein manifold is flexible if and only if it is subcritical.*

(3) The property of a Weinstein structure being subcritical is clearly not preserved under Weinstein homotopies because one can always create index n critical points. We do not know whether flexibility is preserved under Weinstein homotopies. In fact, it is not even clear to us whether every decomposition of a flexible Weinstein cobordism W into elementary cobordisms consists of flexible elementary cobordisms. Indeed, if $\mathcal{P}_1$ and $\mathcal{P}_2$ are two partitions of W into elementary cobordisms and $\mathcal{P}_2$ is finer than $\mathcal{P}_1$, then flexibility of $\mathcal{P}_1$ implies flexibility of $\mathcal{P}_2$ (in particular the partition for which each elementary cobordism contains only one critical value is then flexible), but we do not know whether flexibility of $\mathcal{P}_2$ implies flexibility of $\mathcal{P}_1$.

We will see in Chapter 14 that, as the name suggests, flexible Weinstein manifolds indeed exhibit a good deal of flexibility. In particular, we will prove:

Two flexible Weinstein structures on the same manifold whose symplectic forms are homotopic as nondegenerate 2-forms are Weinstein homotopic (Theorem 14.5).

Every diffeomorphism $f : V_1 \to V_2$ between two flexible Weinstein manifolds $(V_i, \omega_i, X_i, \phi_i)$, $i = 1, 2$, such that $f^*\omega_2$ is homotopic to ω_1 as nondegenerate 2-forms, is diffeotopic to a symplectomorphism (Theorem 14.7).

CHAPTER 12

Modifications of Weinstein Structures

In this chapter we carry over various constructions for Morse cobordisms in Chapter 9 to Weinstein cobordisms. In particular, we discuss holonomy of Weinstein cobordisms (Section 12.2), modifications of Weinstein structures near critical points (Section 12.4) and stable discs, and equivalence of elementary Weinstein homotopies (Section 12.7). In Section 12.6 we prove the (easier) Weinstein analogues of the modifications of Stein structures in Chapter 10.

12.1. Weinstein structures with given functions

Given a Weinstein (manifold or cobordism) structure $\mathfrak{W} = (\omega, X, \phi)$ with Liouville form $\lambda = i_X\omega$, we denote by $\mathcal{C}(\mathfrak{W})$ the space of all Weinstein structures on the same manifold with the same function ϕ and with Liouville form

$$\widetilde{\lambda} = f\lambda + g\,d\phi \tag{12.1}$$

for smooth functions $f, g : W \to \mathbb{R}$ with $f > 0$.

Note that all Weinstein structures $\widetilde{\mathfrak{W}} \in \mathcal{C}(\mathfrak{W})$ induce the same contact structures on all level sets of ϕ. Conversely, if this is the case, then the Liouville form $\widetilde{\lambda}$ has the form (12.1) *outside the critical points.*

Let us first find the conditions on f, g under which the 1-form $\widetilde{\lambda}$ defined by (12.1) defines again a Weinstein structure with function ϕ.

LEMMA 12.1. *Let (W, ω, X, ϕ) be a Weinstein cobordism with Liouville form λ. Then for functions $f, g : W \to \mathbb{R}$ the following holds:*

(i) The 1-form $f\lambda$ defines a Weinstein structure if and only if $f > 0$ and $k := f + df(X) > 0$; in that case, it has Lyapunov function ϕ and Liouville field $\frac{f}{k}X$.

(ii) The 1-form $\lambda + g\,d\phi$ defines a Weinstein structure if and only if $k := 1 - dg(X_\phi) > 0$, where X_ϕ is the Hamiltonian vector field of ϕ; in that case, it has Lyapunov function ϕ and Liouville field $\frac{1}{k}X - \frac{g}{k}X_\phi + Z$ with $d\phi(Z) = \lambda(Z) = 0$.

REMARK 12.2. (a) Lemma 12.1 remains true for Weinstein manifolds instead of cobordisms if one additionally requires that the new Liouville field is complete. Note that this is automatic in the special case $df(X) \geq 0$ in (i), and it is implied by completeness of the vector field $\frac{1}{k}X$ in (ii).

(b) The proof of Lemma 12.1 shows that the Liouville fields of λ and $\widetilde{\lambda}$ are proportional if and only if g is constant on level sets of ϕ.

PROOF. Recall that the Liouville field X and the Hamiltonian vector field X_ϕ satisfy

$$\begin{gathered} i_X\omega = \lambda, \quad i_{X_\phi}\omega = -d\phi, \\ d\phi(X_\phi) = \lambda(X) = 0, \quad d\phi(X) = \lambda(X_\phi) =: h. \end{gathered} \tag{12.2}$$

Note that we have $X_\phi = hR$, where R is the Reeb vector field of the form λ restricted to the level sets of ϕ, i.e., $i_R\omega|_{\{\phi=\text{const}\}} = 0$, $d\phi(R) = 0$ and $\lambda(R) = 1$.

Consider a 1-form

$$\widetilde{\lambda} = f\lambda + g\,d\phi$$

as in (12.1). Let us derive the conditions for the form

$$\widetilde{\omega} = d\widetilde{\lambda} = f\,\omega + df \wedge \lambda + dg \wedge d\phi$$

to be symplectic. First note that at a critical point p of ϕ the form equals $\widetilde{\omega}_p = f(p)\omega$, so $\widetilde{\omega}$ is symplectic near p if and only if $f(p) > 0$. Hence, in the rest of the proof we will assume $f > 0$ and work in the complement of the critical locus of ϕ.

Consider any vector field Y, and write it in the form

$$Y = aX_\phi + bX + Z, \qquad Z \in \xi,$$

where $\xi = \ker d\phi \cap \ker\lambda$ is the contact structure on level sets of ϕ. A short computation using the relations (12.2) yields

$$\begin{aligned}
\beta &:= i_Y\widetilde{\omega} \\
&= a\Big[df(X_\phi)\lambda - h\,df - f\,d\phi + dg(X_\phi)d\phi\Big] \\
&\quad + b\Big[df(X)\lambda - h\,dg + f\,\lambda + dg(X)d\phi\Big] \\
&\quad + \Big[df(Z)\lambda + f\,i_Z\omega + dg(Z)d\phi\Big].
\end{aligned}$$

Thus $Y \in \ker\widetilde{\omega}$ is equivalent to the three equations

$$\begin{aligned}
\beta|_\xi &= (f\,i_Z\omega - ah\,df - bh\,dg)|_\xi = 0, \\
\beta(X_\phi)/h &= bk + df(Z) = 0, \\
\beta(X)/h &= -ak + dg(Z) = 0,
\end{aligned}$$

where we have set $k := f + df(X) - dg(X_\phi)$. For $k > 0$ one easily sees that in both cases (i) and (ii) these equations imply $Y = 0$, so $\widetilde{\omega}$ is symplectic. The necessity of the conditions $f > 0$ and $k > 0$ follows in case (i) from the nonvanishing of

$$i_X(\widetilde{\omega}^n) = n\,f^{n-1}\big(f + df(X)\big)\lambda \wedge \omega^{n-1},$$

and in case (ii) from the nonvanishing of

$$i_{X_\phi}(\widetilde{\omega}^n) = -n\big(1 - dg(X_\phi)\big)d\phi \wedge \omega^{n-1}.$$

Finally, we compute the Liouville field of $\widetilde{\lambda}$. We again write it in the form $\widetilde{X} = aX_\phi + bX + Z$ with $Z \in \xi$. Then the equation $i_{\widetilde{X}}\widetilde{\omega} = \widetilde{\lambda}$ is equivalent to the three equations

$$\begin{aligned}
\beta|_\xi &= (f\,i_Z\omega - ah\,df - bh\,dg)|_\xi = 0, \\
\beta(X_\phi)/h &= bk + df(Z) = f, \\
\beta(X)/h &= -ak + dg(Z) = g.
\end{aligned}$$

In both cases (i) and (ii) these equations imply $df(Z) = dg(Z) = 0$ (actually $Z = 0$ in case (i)), and we conclude

$$\widetilde{X} = \frac{f}{k}X - \frac{g}{k}X_\phi + Z, \qquad (fk\,i_Z\omega + gh\,df - fh\,dg)|_\xi = 0.$$

In particular, we see that $d\phi(\widetilde{X}) = f\,d\phi(X)/k$, so $\widetilde{X}$ is gradient-like for ϕ. $\square$

COROLLARY 12.3. *Let* $\overline{\mathfrak{W}} = (W, \overline{\omega}, \overline{X}, \phi)$ *be a Weinstein cobordism with Liouville form* $\overline{\lambda}$. *Then the space* $\mathcal{C}(\overline{\mathfrak{W}})$ *of Weinstein structures* (W, ω, X, ϕ) *with Liouville forms* $\lambda = f\overline{\lambda} + g\,d\phi$, $f > 0$, *has the following properties.*

(i) If λ *and* $f\lambda$ *belong to* $\mathcal{C}(\overline{\mathfrak{W}})$, *then so does* $(1-t+tf)\lambda$ *for all* $t \in [0,1]$.

(ii) If λ *and* $\lambda + g\,d\phi$ *belong to* $\mathcal{C}(\overline{\mathfrak{W}})$, *then so does* $\lambda + \rho \circ \phi\, g\, d\phi$ *for each function* $\rho : \mathbb{R} \to [0,1]$.

(iii) The space $\mathcal{C}(\overline{\mathfrak{W}})$ *is weakly contractible, and so is its subspace of* 1*-forms that equal* $\overline{\lambda}$ *near* $\partial_- W$ *and a positive constant multiple of* $\overline{\lambda}$ *near* $\partial_+ W$.

PROOF. (i) By Lemma 12.1(i), the 1-forms λ and $f\lambda$ both belong to $\mathcal{C}(\overline{\mathfrak{W}})$ if and only if $f + df(X) > 0$, where X is the Liouville field of λ. Then $1-t+tf+tdf(X) > 0$ for all $t \in [0,1]$, so $(1-t+tf)\lambda$ belongs to $\mathcal{C}(\overline{\mathfrak{W}})$.

(ii) By Lemma 12.1(ii), the 1-forms λ and $\lambda + g\,d\phi$ both belong to $\mathcal{C}(\overline{\mathfrak{W}})$ if and only if $dg(X_\phi) < 1$, where X_ϕ is the Hamiltonian vector field of ϕ with respect to $d\lambda$. Since $d(\rho \circ \phi)(X_\phi) = 0$, this implies $d(\rho \circ \phi\, g)(X_\phi) = \rho \circ \phi\, dg(X_\phi) < 1$ for all $\rho : \mathbb{R} \to [0,1]$, so $\lambda + \rho \circ \phi\, g\, d\phi$ belongs to $\mathcal{C}(\overline{\mathfrak{W}})$.

(iii) The proof of weak contractibility is based on the following observation: If (λ, ϕ) is a Liouville structure on a $2n$-dimensional cobordism, then

$$d\phi \wedge \lambda \wedge (d\lambda)^{n-1} > 0. \tag{12.3}$$

To see this, evaluate this $2n$-form at a point on the basis $(X, X_\phi, Z_1, \dots, Z_{2n-2})$, where X is the Liouville field of λ, X_ϕ is the Hamiltonian vector field with respect to $d\lambda$, and $Z_1, \dots, Z_{2n-2}$ is a symplectic basis of $\ker d\phi \cap \ker \lambda$. Conversely, if (λ, ϕ) satisfies (12.3), then $(e^{\rho \circ \phi}\lambda, \phi)$ is a Liouville structure for each sufficiently increasing function $\rho : \mathbb{R} \to \mathbb{R}$. For this, set $\widetilde{\lambda} := e^{\rho\circ\phi}\lambda$ and note that

$$(d\widetilde{\lambda})^n = e^{n\rho\circ\phi}\big(d\lambda^n + n\rho' \circ \phi\, d\phi \wedge \lambda \wedge (d\lambda)^{n-1}\big) > 0$$

for $\rho' > 0$ sufficiently large. Finally, we observe that if (λ, ϕ) satisfies (12.3), then so does $\widetilde{\lambda} = f\lambda + g\,d\phi$ for all functions $f, g : W \to \mathbb{R}$ with $f > 0$. Indeed, $d\widetilde{\lambda} = f\,d\lambda + df \wedge \lambda + dg \wedge d\phi$ implies

$$d\phi \wedge \widetilde{\lambda} \wedge (d \wedge \lambda)^{n-1} = f^n d\phi \wedge \lambda \wedge (d\lambda)^{n-1} > 0.$$

The last observation shows that the space $\widetilde{\mathcal{C}}(\overline{\mathfrak{W}})$ of 1-forms $\lambda = f\overline{\lambda} + g\,d\phi$ satisfying (12.3) with the given function ϕ is convex and thus contractible. Since any compact family in $\widetilde{\mathcal{C}}(\overline{\mathfrak{W}})$ can be lifted to a family in $\mathcal{C}(\overline{\mathfrak{W}})$ by multiplying the 1-forms with $e^{\rho]\circ\phi}$ for a sufficiently increasing function $\rho : \mathbb{R} \to \mathbb{R}$, this implies weak contractibility on $\mathcal{C}(\overline{\mathfrak{W}})$. For 1-forms that agree with λ near $\partial_- W$ and with $C\lambda$ near $\partial_+ W$ for constants $C > 0$, we can choose ρ to be zero near $\partial_- W$ and constant near $\partial_+ W$. □

COROLLARY 12.4. *For any Weinstein manifold* $\mathfrak{W} = (V, \omega, X, \phi)$ *and any* $\widetilde{\mathfrak{W}} = (V, \widetilde{\omega}, \widetilde{X}, \phi) \in \mathcal{C}(\mathfrak{W})$ *the manifolds* (V, ω) *and* $(V, \widetilde{\omega})$ *are symplectomorphic.*

PROOF. Corollary 12.3 provides a Weinstein homotopy $\mathfrak{W}_t$ from $\mathfrak{W}$ to $\widetilde{\mathfrak{W}}$ with common Lyapunov function ϕ. This homotopy defines a Liouville homotopy in the sense of Section 11.2, with an exhaustion given by regular sublevel sets of ϕ. So Proposition 11.8 yields a family of symplectomorphisms from $\mathfrak{W}_0$ to $\mathfrak{W}_t$. □

12.2. Holonomy of Weinstein cobordisms

In this section we consider Weinstein cobordisms $\mathfrak{W} = (W, \omega, X, \phi)$ *without critical points* (of the function ϕ). We denote by $\Gamma_{\mathfrak{W}} : \partial_+ W \to \partial_- W$ the *holonomy* diffeomorphism along trajectories of X. According to Lemma 11.4, it defines a contactomorphism

$$\Gamma_{\mathfrak{W}} : (\partial_+ W, \xi_+) \to (\partial_- W, \xi_-)$$

for the contact structures $\xi_\pm$ on $\partial_\pm W$ induced by the Liouville form $\lambda = i_X \omega$.

We say that two Weinstein structures $\mathfrak{W} = (\omega, X, \phi)$ and $\widetilde{\mathfrak{W}}$ agree *up to scaling* on a subset $A \subset W$ if $\widetilde{\mathfrak{W}}|_A = (C\omega, X, \phi)$ for a constant $C > 0$. Note that in this case $\widetilde{\mathfrak{W}}|_A$ has Liouville form $C\lambda$.

Let us fix a Weinstein cobordism $\overline{\mathfrak{W}} = (W, \overline{\omega}, \overline{X}, \phi)$ without critical points. We denote by $\mathcal{W}(\overline{\mathfrak{W}})$ the space of all Weinstein structures $\mathfrak{W} = (W, \omega, X, \phi)$ with the same function ϕ such that

- $\mathfrak{W}$ coincides with $\overline{\mathfrak{W}}$ on $\mathcal{O}p\, \partial_- W$ and up to scaling on $\mathcal{O}p\, \partial_+ W$;
- $\mathfrak{W} \in \mathcal{C}(\overline{\mathfrak{W}})$, i.e., $\mathfrak{W}$ and $\overline{\mathfrak{W}}$ induce the same contact structures on level sets of ϕ.

Equivalently, $\mathcal{W}(\overline{\mathfrak{W}})$ can be viewed as the space of Liouville forms $\lambda = f\overline{\lambda} + g\, d\phi$ with $f \equiv 1$ near $\partial_- W$, $f \equiv C$ near $\partial_+ W$, and $g \equiv 0$ near ∂W, where $\overline{\lambda}$ denotes the Liouville form of $\overline{\mathfrak{W}}$.

Denote by $\mathcal{D}(\overline{\mathfrak{W}})$ the space of contactomorphisms $(\partial_+ W, \xi_+) \to (\partial_- W, \xi_-)$, where $\xi_\pm$ is the contact structure induced on $\partial_\pm W$ by $\overline{\mathfrak{W}}$. Note that $\Gamma_{\mathfrak{W}} \in \mathcal{D}(\overline{\mathfrak{W}})$ for any $\mathfrak{W} \in \mathcal{W}(\overline{\mathfrak{W}})$. The following two lemmas are analogues of Lemmas 9.41 and 9.42 in the context of Weinstein cobordisms.

LEMMA 12.5. *Let $\overline{\mathfrak{W}}$ be a Weinstein cobordism without critical points. Then the map $\mathcal{W}(\overline{\mathfrak{W}}) \to \mathcal{D}(\overline{\mathfrak{W}})$ that assigns to $\mathfrak{W}$ its holonomy $\Gamma_{\mathfrak{W}}$ is a Serre fibration. In particular:*

(i) Given $\mathfrak{W} \in \mathcal{W}(\overline{\mathfrak{W}})$ and an isotopy $h_t \in \mathcal{D}(\overline{\mathfrak{W}})$, $t \in [0,1]$, with $h_0 = \Gamma_{\mathfrak{W}}$, there exists a path $\mathfrak{W}_t \in \mathcal{W}(\overline{\mathfrak{W}})$ with $\mathfrak{W}_0 = \mathfrak{W}$ such that $\Gamma_{\mathfrak{W}_t} = h_t$ for all $t \in [0,1]$.

(ii) Given a path $\mathfrak{W}_t \in \mathcal{W}(\overline{\mathfrak{W}})$, $t \in [0,1]$, and a path $h_t \in \mathcal{D}(\overline{\mathfrak{W}})$ which is homotopic to $\Gamma_{\mathfrak{W}_t}$ with fixed endpoints, there exists a path $\widetilde{\mathfrak{W}}_t \in \mathcal{W}(\overline{\mathfrak{W}})$ with $\widetilde{\mathfrak{W}}_0 = \mathfrak{W}_0$ and $\widetilde{\mathfrak{W}}_1 = \mathfrak{W}_1$ such that $\Gamma_{\widetilde{\mathfrak{W}}_t} = h_t$ for all $t \in [0,1]$.

PROOF. Following the flowlines of $\overline{X}$, we find a diffeomorphism $W \cong [a,c] \times \Sigma$ under which $\phi(r,x) = r$ and $\overline{X}$ is a positive multiple of ∂_r, hence $\overline{\lambda} = \bar{g}\, \alpha$ for the contact form $\alpha = \overline{\lambda}|_{\partial_- W}$ and a function $\bar{g} : W \to \mathbb{R}_+$. In particular, $\overline{\lambda}$ defines the same contact structure ξ on each level set $\{r\} \times \Sigma$. Let us fix a cutoff function $\tau : [a,c] \to [0,1]$ which equals 0 near a and c, and 1 near a point $b \in (a,c)$.

We will identify elements in $\mathcal{W}(\overline{\mathfrak{W}})$ with their Liouville forms λ and denote by Γ_λ their holonomy. Suppose we are given $\lambda \in \mathcal{W}(\overline{\mathfrak{W}})$ and an isotopy $h_t \in \mathcal{D}(\overline{\mathfrak{W}})$, $t \in [0,1]$, with $h_0 = \Gamma_\lambda$. For $t \in [0,1]$ we push forward λ to a 1-form $\lambda_t := (H_t)_* \lambda$ on $[a,b] \times \Sigma$ under the diffeomorphism $H_t(r,x) := h_{t\tau(r)}^{-1}(x)$. Since H_t induces a contactomorphism on each level $\{r\} \times \Sigma$, the form λ_t defines the contact structure ξ on each level set in $[a,b] \times \Sigma$. By construction, λ_t has holonomy $h_t : \{b\} \times \Sigma \to \{a\} \times \Sigma$.

Near $\{b\} \times \Sigma$ we have $\lambda_t = h_t^* \lambda = g_t \lambda$ for positive functions $g_t : \Sigma \to \mathbb{R}_+$ with $g_0 \equiv 1$. Pick a family of nondecreasing functions $\rho_t : [b,c] \to \mathbb{R}$ which equal

1 near b and constants $C_t \geq 1$ near c. Using these functions, we extend λ_t over $[b,c] \times \Sigma$ by the formula $\lambda_t := \rho_t(r) g_{t\tau(r)} \lambda$. Here we choose $\dot{\rho}_t$ sufficiently large so that the function $f_t(r,x) := \rho_t(r) g_{t\tau(r)}(x)$ satisfies $\frac{\partial}{\partial r} f_t \geq 0$, and hence $df_t(X) \geq 0$. Since $g_0 \equiv 1$, we may choose $\rho_0 \equiv C_0 = 1$. It follows from Lemma 12.1(i) that (λ_t, ϕ) defines a Weinstein structure on W whose holonomy over $[b,c] \times \Sigma$ equals the identity. Hence λ_t defines a path of Weinstein structures in $\mathcal{W}(\overline{\mathfrak{W}})$ starting at $\lambda_0 = \lambda$ and with holonomy h_t.

Since the above construction can be done smoothly with respect to a parameter in D^k, the general homotopy lifting property follows. $\square$

The proof of the following lemma is now analogous to that of Lemma 9.42, using Lemma 12.5 instead of Lemma 9.41.

LEMMA 12.6. *Let $\mathfrak{W}_t$, $\mathfrak{W}'_t$ be two paths in $\mathcal{W}(\overline{\mathfrak{W}})$ starting at the same point $\mathfrak{W}_0 = \mathfrak{W}'_0$. Suppose that for a subset $A \subset \partial_+ W$ one has $\Gamma_{\mathfrak{W}_t}(A) = \Gamma_{\mathfrak{W}'_t}(A)$ for all $t \in [0,1]$. Then there exists a path $\widehat{\mathfrak{W}}_t \in \mathcal{W}(\overline{\mathfrak{W}})$ such that*

(i) $\widehat{\mathfrak{W}}_t = \mathfrak{W}_{2t}$ *for* $t \in [0, \frac{1}{2}]$;

(ii) $\widehat{\mathfrak{W}}_1 = \mathfrak{W}'_1$;

(iii) $\Gamma_{\widehat{\mathfrak{W}}_t}(A) = \Gamma_{\mathfrak{W}'_1}(A)$ *for* $t \in [\frac{1}{2}, 1]$. $\square$

Finally, we discuss how to interpolate between Weinstein cobordisms without critical points. Let us fix a product cobordism $W \cong [a,d] \times \Sigma$ with function $\phi(r,x) = r$. We denote by $\mathcal{W}(W,\phi)$ the space of Weinstein structures on W with function ϕ. For a contact structure ξ on Σ we denote by $\mathcal{W}(W,\xi,\phi) \subset \mathcal{W}(W,\phi)$ the subspace of Weinstein structures inducing the contact structure ξ on each level set $r \times \Sigma$. For $a < b < c < d$, we set $W' := ([a,b] \cup [c,d]) \times \Sigma \subset W$.

LEMMA 12.7. *The restriction maps*

$$\mathcal{W}(W,\phi) \to \mathcal{W}(W',\phi)\big/\text{scaling},$$

$$\mathcal{W}(W,\xi,\phi) \to \mathcal{W}(W',\xi,\phi)\big/\text{scaling}$$

are Serre fibrations. In particular:

(i) Given $\mathfrak{W} \in \mathcal{W}(W,\phi)$ and a path $\mathfrak{W}'_t \in \mathcal{W}(W',\phi)$, $t \in [0,1]$, with $\mathfrak{W}'_0 = \mathfrak{W}|_{W'}$ there exists a path $\mathfrak{W}_t \in \mathcal{W}(W,\phi)$ with $\mathfrak{W}_0 = \mathfrak{W}$ such that $\mathfrak{W}_t|_{W'} = \mathfrak{W}'_t$ up to scaling for all $t \in [0,1]$.

(ii) Given a path $\mathfrak{W}_t \in \mathcal{W}(W,\phi)$, $t \in [0,1]$, and a path $\mathfrak{W}'_t \in \mathcal{W}(W',\phi)$ which is homotopic to $\mathfrak{W}_t|_{W'}$ with fixed endpoints, there exists a path $\widetilde{\mathfrak{W}}_t \in \mathcal{W}(W,\phi)$ which is homotopic to $\mathfrak{W}_t$ with fixed endpoints such that $\widetilde{\mathfrak{W}}_t|_{W'} = \mathfrak{W}'_t$ up to scaling for all $t \in [0,1]$.

Analogous statements hold with fixed contact structure ξ.

PROOF. Let us first consider the fibration $\mathcal{W}(W,\xi,\phi) \to \mathcal{W}(W',\xi,\phi)/\text{scaling}$. We again denote elements in $\mathcal{W}(W,\xi,\phi)$ just by their Liouville forms. Consider $\lambda \in \mathcal{W}(W,\xi,\phi)$ with Liouville field X and Hamiltonian vector field X_ϕ, and a path $\lambda'_t = f'_t \lambda + g'_t\, d\phi \in \mathcal{W}(W',\xi,\phi)$ with $\lambda'_0 = \lambda|_{W'}$. After multiplying λ'_t with $e^{\sigma \circ \phi}$ for a sufficiently increasing function $\sigma : \mathbb{R} \to \mathbb{R}$, we may assume that $df'_t(X'_t) \geq 0$, where X'_t denotes the Liouville field of λ'_t. We extend λ'_t to Weinstein structures (still denoted by the same letter) $\lambda'_t = f'_t \lambda + g'_t\, d\phi \in \mathcal{W}(\widetilde{W},\xi,\phi)$, where $\widetilde{W} = ([a,\widetilde{b}] \cup [\widetilde{c},d]) \times \Sigma$ for some $b < \widetilde{b} < \widetilde{c} < c$. Fix a function $\rho : [b,c] \to [0,1]$ which equals 0 on $[\widetilde{b},\widetilde{c}]$ and 1 near b,c and pick functions $f_t : W \to \mathbb{R}$ with $f_0 \equiv 1$ which agree with f_t' on $\widetilde{W}$ and satisfy $df_t(X) \geq 0$ (this is possible after choosing the function σ above sufficiently increasing). We claim that

$$\lambda_t := f_t\lambda + \rho \circ \phi\, g_t'\, d\phi$$

defines an extension of λ_t' to a path in $\mathcal{W}(W,\xi,\phi)$. Indeed, on $[\widetilde{b},\widetilde{c}] \times \Sigma$ the forms λ_t agree with $f_t\lambda$ and thus define Weinstein structures by Lemma 12.1(i). On $\widetilde{W}$ we have $\lambda_t = f_t'\lambda + \rho \circ \phi\, g_t'\, d\phi$, where $f_t'\lambda$ (by the preceding argument) and $f_t'\lambda + g_t'\, d\phi = \lambda_t'$ (by hypothesis) both belong to $\mathcal{W}(\widetilde{W},\xi,\phi)$ and $\rho : [b,c] \to [0,1]$, hence $\lambda_t \in \mathcal{W}(\widetilde{W},\xi,\phi)$ by Corollary 12.3(ii).

Since this construction works smoothly for families, it proves the Serre fibration property for $\mathcal{W}(W,\xi,\phi) \to \mathcal{W}(W',\xi,\phi)/\text{scaling}$.

The case $\mathcal{W}(W,\phi) \to \mathcal{W}(W',\phi)/\text{scaling}$ reduces to the case with fixed ξ by Gray's stability theorem: Consider families $\mathfrak{W}'_{\lambda,t} \in \mathcal{W}(W',\phi)$ and $\mathfrak{W}_\lambda \in \mathcal{W}(W,\phi)$ with $\mathfrak{W}'_{\lambda,0} = \mathfrak{W}_\lambda|_{W'}$ for $\lambda \in D^k$, $t \in [0,1]$. Let ξ be the contact structure induced by $\mathfrak{W}_{0,0}$ on $\{a\} \times \Sigma$. By Gray's Theorem 6.23 there exists a family of diffeomorphisms $h_{\lambda,t} : W \to W$, $\lambda \in D^k$, $t \in [0,1]$, such that the pullbacks $h^*_{\lambda,t}\mathfrak{W}'_{\lambda,t}$ and $h^*_{\lambda,0}\mathfrak{W}_\lambda$ induce ξ on all level sets. Let $\widetilde{\mathfrak{W}}_{\lambda,t} \in \mathcal{W}(W,\xi,\phi)$ be the lift of $h^*_{\lambda,t}\mathfrak{W}'_{\lambda,t}$ with $\widetilde{\mathfrak{W}}_{\lambda,0} = h^*_{\lambda,0}\mathfrak{W}_\lambda$. Then $\mathfrak{W}_{\lambda,t} := (h_{\lambda,t})_*\widetilde{\mathfrak{W}}_{\lambda,t}$ is the desired lift of $\mathfrak{W}'_{\lambda,t}$. □

12.3. Liouville fields near isotropic submanifolds

In this section we discuss the construction and modification of Liouville fields near isotropic submanifolds. We begin with a construction to extend a vector field on an isotropic submanifold to a Liouville field on a neighborhood.

Consider an isotropic submanifold L of a symplectic manifold (V,ω) and a compact subset $K \subset L$. Assume for simplicity that the symplectic normal bundle $(TL)^\omega/TL$ is trivial (this assumption is not necessary but will be satisfied in our applications). Then by the isotropic neighborhood theorem (Corollary 6.13), a neighborhood of K in (V,ω) is symplectomorphic to a neighborhood of K in $T^*L \times \mathbb{C}^\ell$ with coordinates $(q,p,z=x+iy)$ and the symplectic form

$$\omega_{\rm st} = \sum_{i=1}^{k} dp_i \wedge dq_i + \sum_{j=1}^{\ell} dx_j \wedge dy_j.$$

It has the canonical Liouville field

$$p\partial_p + \frac{1}{2}z\partial_z = \sum_i p_i\partial_{p_i} + \frac{1}{2}\sum_j (x_j\partial_{x_j} + y_j\partial_{y_j}).$$

To each tangent vector field Y on L, we associate the Liouville vector field

$$\widehat{Y} := p\partial_p + \frac{1}{2}z\partial_z + X_H$$

on $T^*L \times \mathbb{C}^\ell$, where X_H is the Hamiltonian vector field of the Hamiltonian function

$$H(q,p,z) := \langle p, Y(q)\rangle.$$

We extend each smooth function $\psi : L \to \mathbb{R}$ to a function

$$\widehat{\psi}(q,p,z) := \psi(q) + \rho(q,p,z), \qquad \rho(q,p,z) := |p|^2 + |z|^2$$

on $T^*L \times \mathbb{C}^\ell$, for some Riemannian metric on L.

LEMMA 12.8. *Suppose that all eigenvalues at zeroes of Y have real part < 1. Then the pair $(\widehat{Y}, \widehat{\psi})$ has the following properties.*

(i) *$\widehat{Y}$ is a Liouville field for $\omega_{\rm st}$ which coincides with Y along L and satisfies $\widehat{Y} \cdot \rho \geq \varepsilon\rho$ near K for some $\varepsilon > 0$.*
(ii) *The zeroes of $\widehat{Y}$ agree with the zeroes of Y and have the same nullity and Morse index.*
(iii) *If Y is gradient-like for ψ, then $\widehat{Y}$ is gradient-like for $\widehat{\psi}$ near K.*
(iv) *Suppose that Y is the restriction of a vector field X defined on a neighborhood of L which is gradient-like for a function $\phi : \mathcal{O}p\, L \to \mathbb{R}$. Suppose that the zeroes of X are isolated and at each zero $q \in L$ the center-stable space $E_q^0 \oplus E_q^-$ equals T_qL and the unstable space E_q^+ is coisotropic. Then we can arrange that $\widehat{Y}$ is gradient-like for the given function ϕ.*
(v) *The construction of $(\widehat{Y}, \widehat{\psi})$ also works if L has nonempty smooth boundary.*

PROOF. (i) Let us write $Y(q) = \sum_i Y_i(q)\partial_{q_i}$ in local coordinates (q_i, p_i), hence $H(q, p, z) = \sum_i p_iY_i(q)$, and we compute

$$\begin{aligned} dH &= \sum_i Y_i(q)dp_i + \sum_{i,j} \frac{\partial Y_j}{\partial q_i} p_j dq_i, \\ X_H &= \sum_i Y_i(q)\partial_{q_i} - \sum_{i,j} \frac{\partial Y_j}{\partial q_i} p_j \partial_{p_i}, \\ \widehat{Y} &= \sum_i Y_i(q)\partial_{q_i} + \sum_{i,j} \left(\delta_{ij} - \frac{\partial Y_j}{\partial q_i}\right) p_j\partial_{p_i} + \frac{1}{2} z\partial_z. \end{aligned}$$

This shows that $\widehat{Y} = Y$ along L. Moreover, as the real parts of eigenvalues of $\mathrm{Id} - D_qY$ are bounded below by some $\varepsilon \in (0, 1)$ near K, we compute in geodesic normal coordinates at $q \in L$:

$$\widehat{Y} \cdot \rho = 2\langle p, (\mathrm{Id} - D_qY)p\rangle + |z|^2 \geq 2\varepsilon|p|^2 + |z|^2 \geq \varepsilon\rho.$$

(ii) The formula for $\widehat{Y}$ shows that $\widehat{Y}(q, p, z) = 0$ if and only if $z = 0$, $Y(q) = 0$, and $p \in \ker(\mathrm{Id} - D_qY) = \{0\}$, so the zeroes of $\widehat{Y}$ coincide with the zeroes of Y on L. If q is a zero for Y, then positivity of the real parts of all eigenvalues of $\mathrm{Id} - D_qY$ implies that $(q, 0, 0)$ is a zero for $\widehat{Y}$ with the same nullity and Morse index.

(iii) Suppose that $Y \cdot \psi \geq \delta(|Y|^2 + |d\psi|^2)$ for some $\delta > 0$. Using $\widehat{Y} \cdot \rho \geq \varepsilon\rho$ and $|D_qY|^2 \leq C$ near K for some $C \geq 1$, we estimate at points of K:

$$\begin{aligned} \widehat{Y} \cdot \widehat{\psi} &\geq Y \cdot \psi + \varepsilon(|p|^2 + |z|^2) \\ &\geq \delta(|Y|^2 + |d\psi|^2) + \frac{\varepsilon}{4C}(|p|^2 + |(\mathrm{Id} - D_qY)p|^2 + |z|^2 + |z\partial_z|^2) \\ &\geq \min\{\delta, \varepsilon/4C\}(|\widehat{Y}|^2 + |d\widehat{\psi}|^2). \end{aligned}$$

(iv) Under the hypotheses of (iv), we can choose the identification with $\mathcal{O}p\, K \subset T^*L \times \mathbb{C}^\ell$ such that $E_q^+ = T_q^*L \times \mathbb{C}^\ell$ at each critical point $q \in L$. Since this also equals the unstable space with respect to $\widehat{Y}$, and the Hessian of ϕ is positive definite on E_p^+ by Lemma 9.9, it follows that $\widehat{Y}$ is gradient-like for ϕ near critical points and hence near K.

(v) If L has nonempty smooth boundary, we extend L to a slightly larger isotropic submanifold L' with coordinates $(q_1, \dots, q_k)$ such that $L = \{q_1 \leq 0\}$. Now we extend a vector field Y on L to Y' on L' and define the Liouville field $\widehat{Y}'$ as above, and similarly for a function ψ. Properties (i)–(iv) follow. $\square$

The second result concerns interpolation between two Liouville fields tangent to an isotropic submanifold with common Lyapunov function. Recall that a Lyapunov pair (X, ϕ) satisfies in particular an estimate

$$|X| \leq C|d\phi|. \tag{12.4}$$

In suitable coordinates Z near a nondegenerate critical point, we have $|d\phi(Z)| = |Z|$, so (12.4) (for some constant C) is equivalent to $X(0) = 0$. In suitable coordinates (x, y, z) near an embryonic point, we have $|d\phi(x, y, z)| = |x| + |y| + |z^2|$, so (12.4) is equivalent to $X(0,0,0) = \frac{\partial X}{\partial z}(0,0,0) = 0$. This implies that in both cases (12.4) carries over to families as follows: Let (X_t, ϕ) be a smooth family of Lyapunov pairs with ϕ having nondegenerate or embryonic critical points. Then there exists a constant C such that

$$|X_s - X_t| \leq C\,|d\phi|\,|s - t| \quad \text{for all } t \in [0, 1]. \tag{12.5}$$

We say that a function $\phi : V \to \mathbb{R}$ is *transversely nondegenerate* along a submanifold L if at each critical point $x \in L$ the Hessian satisfies $\ker \mathrm{Hess}_x \phi \subset T_x L$. By Lemma 9.3, this implies that near each critical point $x \in L$ there exist coordinates (q, p) in which

$$\phi(q, p) = \psi(q) + \frac{1}{2}\sum_{i=1}^{\ell} p_i^2 - \frac{1}{2}\sum_{i=\ell+1}^{m} p_i^2,$$

where q is the coordinate along L. It follows that $|d\phi(q, p)|^2 = |d\psi(q)|^2 + |p|^2$, so ϕ satisfies the estimate

$$|p| \leq |d\phi(q, p)| \leq |d\phi(q, sp)| \quad \text{for all } s \geq 1. \tag{12.6}$$

This estimate globalizes to a tubular neighborhood of L for a suitable bundle metric on the normal bundle.

Now we can state the second interpolation result.

LEMMA 12.9. *Let (V, ω) be a symplectic manifold. Let $L \subset V$ be an isotropic submanifold, possibly with nonempty smooth boundary, and $K \subset L$ a compact subset. Let $\phi : V \to \mathbb{R}$ be a function with nondegenerate or embryonic critical points which is transversely nondegenerate along L. Let X and $X_{\rm loc}$ be Liouville fields for ω on V (resp. on a neighborhood $V_{\rm loc} \subset V$ of K). Assume that both X and $X_{\rm loc}$ are tangent to L and gradient-like for ϕ. Then there exists a homotopy of Liouville fields X_t, $t \in [0, 1]$, on V with the following properties:*

(i) *X_t is tangent to L and gradient-like for ϕ for all t;*
(ii) *$X_0 = X$, $X_t = X$ outside $V_{\rm loc}$ and on the set $L \cap \{X_{\rm loc} = X\}$;*
(iii) *$X_t = (1 - t)X + tX_{\rm loc}$ on $\mathcal{O}p\, K$.*

PROOF. Suppose first that L has no boundary. Let us identify (after shrinking) $V_{\rm loc}$ with a subset of the normal bundle to L such that the estimate (12.6) holds. Let $\delta > 0$ be a constant such that

$$X \cdot \phi \geq \delta |d\phi|^2, \qquad X_{\rm loc} \cdot \phi \geq \delta |d\phi|^2.$$

Step 1. We first prove the assertion under the additional hypothesis

$$|X - X_{\rm loc}| \le \alpha |d\phi|, \qquad \alpha := \delta/4. \tag{12.7}$$

The 1-form $\lambda := i_{X-X_{\rm loc}}\omega$ on $V_{\rm loc}$ satisfies $d\lambda = 0$, and $\lambda|_L = 0$ because $X - X_{\rm loc}$ is tangent to L and L is isotropic. By the relative Poincaré lemma (see [**87**]), there exists a function $H : V_{\rm loc} \to \mathbb{R}$ with $H \equiv 0$ on $L \cap V_{\rm loc}$ and $dH = \lambda$. It follows that $X_{\rm loc} - X = X_H$. For each $\varepsilon \in (0,1)$ choose a cutoff function $g : [0,\varepsilon] \to [0,1]$ with $g \equiv 1$ near 0, $g \equiv 0$ near ε, and $|g'(t)| \le \frac{2}{\varepsilon}$ for all t. Fix another cutoff function $h : L \to [0,1]$ with $h \equiv 1$ near K and $h \equiv 0$ outside a larger neighborhood of K in $L \cap V_{\rm loc}$. Then the function $f(q,p) := h(q)g(|p|)$ satisfies $f \equiv 1$ near K, $f \equiv 0$ outside $V_{\rm loc}$, and $|df(q,p)| \le 2/\varepsilon$ for sufficiently small ε. Define

$$H_t := tfH, \qquad X_t := X + X_{H_t}.$$

Note that

$$X_t = X + tfX_H + tHX_f = (1-tf)X + tfX_{\rm loc} + tHX_f.$$

So the vector fields X_t are tangent to L (where $H = 0$) and satisfy $L_{X_t}\omega = \omega$, $X_0 = X$, $X_t = X$ outside $V_{\rm loc}$ (where $f \equiv 0$) and on the set $L \cap \{X = X_{\rm loc}\}$ (where $H \equiv 0$), and $X_1 = X_{\rm loc}$ on $\mathcal{O}p\,K$ (where $f \equiv 1$). By (12.6) and (12.7) we have

$$\begin{aligned}
|H(q,p)| &= \left|\int_0^1 \frac{d}{ds} H(q,sp)ds\right| \\
&\le \int_0^1 |p|\, \left|\frac{\partial H}{\partial p}(q,sp)\right| ds \\
&\le \int_0^1 |p|\, |X_{\rm loc}(q,sp) - X(q,sp)| ds \\
&\le \alpha \int_0^1 |p|\, |d\phi(q,sp)| ds \\
&\le \alpha \int_0^1 |p|\, |d\phi(q,p)| ds \\
&\le \alpha |p|\, |d\phi(q,p)|.
\end{aligned}$$

Using this and hypothesis $\alpha = \delta/4$, we obtain

$$\begin{aligned}
X_t \cdot \phi &= \big[(X + tfX_H) + tHX_f\big] \cdot \phi \\
&= \big[(1-tf)X + tfX_{\rm loc} + tHX_f\big] \cdot \phi \\
&\ge \delta |d\phi|^2 - |H|\, |df|\, |d\phi| \\
&\ge \left(\delta - \alpha |p| \frac{2}{\varepsilon}\right) |d\phi|^2 \\
&\ge (\delta - 2\alpha)\, |d\phi|^2 \\
&\ge \frac{\delta}{2} |d\phi|^2.
\end{aligned}$$

Here we have used that the term involving $|p|$ is not present for $|p| > \varepsilon$ because then df vanishes. This proves gradient-likeness and thus the assertion under the additional hypothesis (12.7).

Step 2. For the general case (still assuming $\partial L = \varnothing$), consider the vector fields $\bar{X}_t := (1-t)X+tX_{\rm loc}$ on $V_{\rm loc}$. They all satisfy $L_{\bar{X}_t}\omega = \omega$ and $\bar{X}_t\cdot\phi \geq \delta|d\phi|^2$. In view of the estimate (12.5), we can pick a partition $0 = t_0 < t_1 < \cdots < t_N = 1$ such that $|\bar{X}_{t_i} - \bar{X}_{t_{i-1}}| \leq \alpha|d\phi|$ for all $i = 1,\dots,N$. Apply Step 1 with $(X, X_{\rm loc}) = (X_0, X_{t_1})$ and some $\varepsilon = \varepsilon_0$ to find a vector field $\widetilde{X}_1$ which equals $X_0 = X$ for $|p| \geq \varepsilon_0$ and X_{t_1} for $|p| \leq 2\varepsilon_1$ with some $\varepsilon_1 < \varepsilon_0$. Next apply Step 1 with $(X, X_{\rm loc}) = (\widetilde{X}_1, X_{t_2})$ and $\varepsilon = \varepsilon_1$ to find a vector field $\widetilde{X}_2$ which equals $\widetilde{X}_1$ for $|p| \geq \varepsilon_1$ and X_{t_2} for $|p| \leq 2\varepsilon_2$ with some $\varepsilon_2 < \varepsilon_1$. Continuing this way, we find a Liouville field $\widetilde{X}_N$ with $\widetilde{X}_N = X$ outside $V_{\rm loc}$ and $\widetilde{X}_N = X_{\rm loc}$ near K. Now $X_t := (1-t)X + t\widetilde{X}_N$ is the desired homotopy.

Step 3. Finally, consider the case that L has nonempty boundary. Extend L to a slightly larger isotropic submanifold L' with coordinates $(q_1,\dots,q_k)$ such that $L = \{q_1 \leq 0\}$. Note that the Liouville fields $X, X_{\rm loc}$ are tangent to L but not necessarily to L'. However, their components normal to L' vanish to infinite order as $q_1 \to 0$. It follows that the closed 1-form $\lambda = i_{X-X_{\rm loc}}\omega$ and its primitive H vanish to infinite order along L' as $q_1 \to 0$. In particular, we have $H(q,0) = O(q_1^6)$ and the estimate above yields

$$|H(q,p)| \leq \alpha|p|\;|d\phi(q,p)| + O(q_1^6).$$

For $\varepsilon \in (0,1)$ we pick a cutoff function $f(q,p) := h(q)g(|p|)$ such that $f \equiv 1$ near K, $f \equiv 0$ outside $V_{\rm loc} \cap \{q_1 \leq \varepsilon, |p| \leq \varepsilon\}$, and $|df(q,p)| \leq 2/\varepsilon$. Then the last estimate in Step 1 gets modified to

$$X_t\cdot\phi \geq \frac{\delta}{2}|d\phi|^2 - \frac{2}{\varepsilon}|d\phi|O(q_1^6),$$

where the last term is only present if $q_1 \leq \varepsilon$ because otherwise $df = 0$. So we can estimate the term $O(q_1^6)$ by $c\varepsilon^2 q_1^4$ with a constant $c > 0$ independent of ε. On the other hand, since all critical points of ϕ at $q_1 = 0$ are nondegenerate or embryonic, it satisfies near $\{q_1 = 0\}$ an estimate $|d\phi|^2 \geq \gamma q_1^4$ with a constant $\gamma > 0$ independent of ε. It follows that

$$X_t\cdot\phi \geq \frac{\delta}{4}|d\phi|^2 + \left(\frac{\delta\gamma}{4} - 2c\varepsilon|d\phi|\right)q_1^4 \geq \frac{\delta}{4}|d\phi|^2$$

for ε sufficiently small, and the proof is concluded as before. □

We conclude this section with another interpolation result which will be used in Section 12.6 for cancellation and creation of critical points of Weinstein structures.

LEMMA 12.10. *Let (W,ω,X,ϕ) be a Weinstein cobordism and $L \subset W$ an isotropic submanifold with boundary such that L is invariant under the forward flow of $-X$. Let $(\omega, X_{\rm loc}, \phi_{\rm loc})$ be a Weinstein structure on a subset $W_{\rm loc} \subset W$ containing L which coincides with (ω,X,ϕ) near $L \cap \partial W_{\rm loc} = \partial L$ and such that $X_{\rm loc}$ is tangent to L. Suppose that X, $X_{\rm loc}$ satisfy estimates*

$$\varepsilon\rho \leq X\cdot\rho,\; X_{\rm loc}\cdot\rho \leq \varepsilon^{-1}\rho$$

for some $\varepsilon > 0$ and a smooth function $\rho : W_{\rm loc} \to [0,\infty)$ with $\rho^{-1}(0) = L$. Then there exists a Weinstein structure (ω, Y, ψ) on W which agrees with (ω,X,ϕ) outside $W_{\rm loc}$ and with $(\omega, X_{\rm loc}, \phi_{\rm loc})$ near L, and which has no critical points in $W_{\rm loc} \setminus L$.

REMARK 12.11. Note that the restrictions of X and $X_{\rm loc}$ to L are completely unrelated, in particular they can have different zero sets. Note also that Lemma 12.10 does not provide a Weinstein homotopy between (ω, X, ϕ) and (ω, Y, ψ).

PROOF. As in the proof of Lemma 12.9, we find that $X_{\rm loc} - X = X_H$ for a Hamiltonian function H on $W_{\rm loc}$ which vanishes on L. Choose $\gamma > 0$ such that $(X_{\rm loc}, \phi_{\rm loc}) = (X, \phi)$ on $\{\rho \le \gamma\} \cap \partial W_{\rm loc}$. Pick a cutoff function $f = f(\rho)$ which equals 1 near $\rho = 0$ and 0 for $\rho \ge \gamma$ and define $Y := X + X_{fH}$. Since $X_f \cdot \rho = 0$, we obtain

$$Y \cdot \rho = (1 - f)X \cdot \rho + fX_{\rm loc} \cdot \rho \ge \varepsilon\rho.$$

Let $g = g(\rho)$ be another cutoff function with support in the set $\{f = 1\}$. Then on $\operatorname{supp} g$ we have $Y = X_{\rm loc}$ and thus

$$Y \cdot \big(g(\rho)\phi_{\rm loc}\big) = g(\rho)X_{\rm loc} \cdot \phi_{\rm loc} + \phi_{\rm loc}g'(\rho)X_{\rm loc} \cdot \rho \ge g\delta|X_{\rm loc}|^2 - C|g'(\rho)|\,\rho$$

for some constants $\delta, C > 0$. Choose g such that $C|g'| \le \varepsilon/2$ and $g(0) > 0$. Then

$$Y \cdot \big(g(\rho)\phi_{\rm loc} + \rho\big) \ge g(\rho)\delta|X_{\rm loc}|^2 + \varepsilon\rho/2 > 0,$$

so $\psi_{\rm loc} := g(\rho)\phi_{\rm loc} + \rho$ is a Lyapunov function for Y on $W_{\rm loc}$. We can adjust $\psi_{\rm loc}$ to make it agree with ϕ near $W_{\rm loc} \cap \partial_- W$.

Finally, we interpolate between the Lyapunov functions $\psi_{\rm loc}$ and ϕ for Y near $\partial W_{\rm loc}$ as follows. After adding a constant to ϕ we may assume that $\phi|_{\partial_- W} = 0$. Pick $b > 0$ so small that $\phi = \phi_{\rm loc}$ on the set $L \cap \{g(0)\phi < b\}$. By Corollary 9.21 there exists a Lyapunov function $\vartheta : W \to \mathbb{R}$ for X with the following properties:

- $\vartheta = \phi$ on $\mathcal{O}p\,\partial W$ and outside $W_{\rm loc}$;
- $\vartheta|_L < b$.

We claim that the function $\psi := \operatorname{smooth\,max}(\vartheta, \psi_{\rm loc})$ has the desired properties. Indeed, for b and γ sufficiently small, we have $\psi = \psi_{\rm loc}$ on $\operatorname{supp} f$, so ψ is a Lyapunov function for Y on $\operatorname{supp} f$. On $W_{\rm loc} \setminus \operatorname{supp} f$ the functions ϑ and $\psi_{\rm loc} = \rho$ are both Lyapunov for the vector field $Y = X$, hence so is ψ. Near $\partial_- W$ we have $\psi = \phi$, and outside $W_{\rm loc} \cap \mathcal{O}p\,\partial_- W$ we have $\vartheta = \phi > \psi_{\rm loc}$ and thus $\psi = \phi$. This concludes the proof of Lemma 12.10. □

12.4. Weinstein structures near critical points

In this section we prove that a Weinstein structure can be arbitrarily altered near a hyperbolic or embryonic critical point. The precise formulation is given in the following proposition, which is a Weinstein version of Corollary 9.14. We refer to Chapter 9 for the relevant notions concerning hyperbolic and embryonic critical points.

PROPOSITION 12.12. *Let p be a hyperbolic (resp. embryonic) critical point of ϕ in a Weinstein manifold $\mathfrak{W} = (V, \omega, X, \phi)$. Let $\mathfrak{W}_{\rm loc} = (\omega_{\rm loc}, X_{\rm loc}, \phi_{\rm loc})$ be a Weinstein structure on a neighborhood $V_{\rm loc}$ of p such that p is a hyperbolic (resp. embryonic) critical point of $\phi_{\rm loc}$ of value $\phi_{\rm loc}(p) = \phi(p)$ and Morse index $\operatorname{ind}_p(\phi_{\rm loc}) = \operatorname{ind}_p(\phi)$. Then there exists a homotopy of Weinstein structures $\mathfrak{W}_t = (\omega_t, X_t, \phi_t)$ on V with the following properties:*

(i) *$\mathfrak{W}_0 = \mathfrak{W}$ and $\mathfrak{W}_t = \mathfrak{W}$ outside $V_{\rm loc}$;*
(ii) *X_t has a unique hyperbolic (resp. embryonic) zero at p in $V_{\rm loc}$ for all t;*
(iii) *$\mathfrak{W}_1 = \mathfrak{W}_{\rm loc}$ near p;*

(iv) *if* $W_p^-(X_{\rm loc}) = W_p^-(X)$ *(resp.* $\widehat{W}_p^-(X_{\rm loc}) = \widehat{W}_p^-(X)$*), then* $W_p^-(X_t) = W_p^-(X)$ *(resp.* $\widehat{W}_p^-(X_t) = \widehat{W}_p^-(X)$*) for all* t.

Moreover, if $\omega_{\rm loc} = \omega$ *we can arrange* $\omega_t = \omega$ *for all* t*, and if* $\phi_{\rm loc} = \phi$ *we can arrange* $\phi_t = \phi$ *for all* t.

Since there exist Stein models for hyperbolic and embryonic critical points, we obtain in particular

COROLLARY 12.13. *A Weinstein structure with hyperbolic and embryonic critical points is homotopic to one which is Stein for a given complex structure near the critical points.*

PROOF OF PROPOSITION 12.12. By Darboux's theorem (Proposition 6.5), after moving $\mathfrak{W}$ by a diffeotopy near p we may assume that $\omega_{\rm loc} = \omega$. After this, we will keep ω fixed and modify (X, ϕ) near p in three steps.

Step 1. Denote by $W_p^\pm$ (resp. $\widehat{W}_p^-$) the stable and unstable manifolds with respect to X. Let $L := W_p^-$ in the hyperbolic and $L := \widehat{W}_p^-$ in the embryonic case. By Proposition 11.9, L is isotropic (with $\partial L \neq \varnothing$ in the embryonic case). Denote by $L_{\rm loc}$ the corresponding isotropic submanifold for $X_{\rm loc}$. After pulling back $(X_{\rm loc}, \phi_{\rm loc})$ by a symplectic isotopy supported near p, we may assume that $L_{\rm loc} = L$ near p and the unstable spaces $E_p^+(X_{\rm loc}) = E_p^+(X) =: E_p^+$ agree.

So it suffices to consider the case that X and $X_{\rm loc}$ have a common stable manifold L and unstable space E_p^+. During the following modifications L and E_p^+ will remain fixed. Let us call a homotopy of Weinstein structures (ω, X_t, ϕ_t) *admissible* if it has properties (i)–(ii) of the proposition, X_t is tangent to L and has unstable space E_p^+ for all t.

Step 2. Note that the quadruple $(L, K := \{p\}, Y := X|_L, \phi)$ satisfies the hypotheses of Lemma 12.8(iv). Let $\widehat{Y}$ be the new Liouville field obtained by Lemma 12.8. Thus $\widehat{Y}$ is gradient-like for ϕ and agrees with X on L. Since ϕ is transversely nondegenerate along L, Lemma 12.9 (with $K = \{p\}$) provides an admissible homotopy (ω, X_t, ϕ) from $X_0 = X$ to $X_1 = \widehat{Y}$. After renaming X_1 back to X, we may hence assume that $X|_{\mathcal{O}p\,p} = \widehat{Y}$ is obtained by the construction of Lemma 12.8 from its restriction $Y := X|_L$. After applying Proposition 9.23, we may further assume that $\phi|_{\mathcal{O}p\,p} = \widehat{\psi}$ is obtained by the construction of Lemma 12.8 from its restriction $\psi := \phi|_L$.

Step 3. By Corollary 9.14, there exists a homotopy of Lyapunov pairs (Y_t, ψ_t) on $L \cap \mathcal{O}p\,p$ having a hyperbolic (resp. embryonic) critical point p from $(Y_0, \psi_0) = (Y, \psi)$ to $(Y_1, \psi_1) = (Y_{\rm loc}, \psi_{\rm loc}) := (X_{\rm loc}|_L, \phi_{\rm loc}|_L)$. Lemma 12.8 (which works smoothly in the parameter t) provides an extension to a homotopy of Weinstein structures $(\omega, \widehat{Y}_t, \widehat{\psi}_t)$ on $\mathcal{O}p\,p$ from $(\widehat{Y}_0, \widehat{\psi}_0) = (X, \phi)|_{\mathcal{O}p\,p}$ to $(\widehat{Y}_1, \widehat{\psi}_1) = (\widehat{Y}_{\rm loc}, \widehat{\psi}_{\rm loc})$.

By Lemma 9.12 there exists a partition $0 = t_0 < t_1 < \cdots < t_N = 1$ such that for all i the following hold:

- $\widehat{\psi}_{t_i}$ is a Lyapunov function for $\widehat{Y}_t$ for all $t \in [t_i, t_{i+1}]$;
- $\widehat{\psi}_t$ is a Lyapunov function for $\widehat{Y}_{t_{i+1}}$ for all $t \in [t_i, t_{i+1}]$.

Therefore, for each i we can inductively apply Lemma 12.9 to change $\widehat{Y}_{t_i}$ to $\widehat{Y}_{t_{i+1}}$ near p (fixing $\widehat{\psi}_{t_i}$), and then Proposition 9.23 to change $\widehat{\psi}_{t_i}$ to $\widehat{\psi}_{t_i+1}$ near p (fixing $\widehat{Y}_{t_{i+1}}$).

Renaming the new Weinstein structure resulting from this construction back to (X, ϕ), we have thus achieved that $(X, \phi) = (\widehat{Y}_{\rm loc}, \widehat{\psi}_{\rm loc})$ near p. Since $\phi_{\rm loc}$ is a Lyapunov function for both $X_{\rm loc}$ and $\widehat{Y}_{\rm loc}$ by Lemma 12.8, we can use Lemma 12.9 to arrange $X = X_{\rm loc}$ near p. Finally, we apply once again Proposition 9.23 to arrange $\phi = \phi_{\rm loc}$ near p.

The proof shows that $\omega_{\rm loc} = \omega$ implies $\omega_t = \omega$ for all t. If $\phi_{\rm loc} = \phi$, then Lemma 9.38 yields a diffeotopy $h_t : V \to V$ with $h_0 = \mathrm{Id}$ such that $\phi_t \circ h_t = \phi$ for all $t \in [0, 1]$. Moreover, $h_t = \mathrm{Id}$ outside $V_{\rm loc}$ and $h_1 = \mathrm{Id}$ on $\mathcal{O}p\, p$, so $h_t^* \mathfrak{W}_t$ is the desired Weinstein homotopy with fixed function ϕ. This concludes the proof of Proposition 12.12. □

12.5. Weinstein structures near stable discs

Now we apply the results of the previous two sections to construct and deform Weinstein structures near stable discs. Our first result concerns deformations of a given Weinstein structure.

PROPOSITION 12.14. *Consider a Weinstein manifold (V, ω, X, ϕ) with a nondegenerate critical point p of index k and an embedded k-disc $\Delta \subset W_p^-$ containing p. Let $(\omega_{\rm loc}, X_{\rm loc}, \phi_{\rm loc})$ be a Weinstein structure on a neighborhood $V_{\rm loc}$ of Δ which coincides with (ω, X, ϕ) on $\Delta \cup \mathcal{O}p\, \partial\Delta$ (so in particular $X_{\rm loc}$ is tangent to Δ). Then there exists a homotopy of Weinstein structures (ω_t, X_t, ϕ_t) on V such that $(\omega_t, X_t, \phi_t) = (\omega, X, \phi)$ outside $V_{\rm loc}$ and on the region where $(\omega_{\rm loc}, X_{\rm loc}, \phi_{\rm loc}) = (\omega, X, \phi)$, and $(\omega_1, X_1, \phi_1) = (\omega_{\rm loc}, X_{\rm loc}, \phi_{\rm loc})$ on $\mathcal{O}p\, \Delta$.*

If $\omega_{\rm loc} = \omega|_{V_{\rm loc}}$, we can achieve $\omega_t = \omega$ for all $t \in [0, 1]$.

PROOF. After an application of the isotropic neighborhood theorem (Corollary 6.13) and shrinking $V_{\rm loc}$, we may assume that $\omega_{\rm loc} = \omega|_{V_{\rm loc}}$. In the following argument all Weinstein structures will have symplectic form ω. Let $(\widehat{Y}, \widehat{\psi})$ be the Weinstein structure obtained by Lemma 12.8 from the restriction $(Y, \psi) := (X|_\Delta, \phi|_\Delta)$. After applying Lemma 12.9 (with Lyapunov function ϕ), we may assume that $X = \widehat{Y}$ on $\mathcal{O}p\, \Delta$. Next we deform ϕ to $\widehat{\psi}$ (through Lyapunov functions for X fixed outside $V_{\rm loc}$) first near p using Proposition 9.23, and then by interpolation on $\mathcal{O}p\, \Delta \setminus \mathcal{O}p\, p$ using $\phi|_\Delta = \widehat{\psi}|_\Delta$. After these deformations we may hence assume that $(X, \phi) = (\widehat{Y}, \widehat{\psi})$ on $\mathcal{O}p\, \Delta$. Since $(Y, \psi) = (X_{\rm loc}|_\Delta, \phi_{\rm loc}|_\Delta)$, we can now reverse the preceding argument to deform $(\widehat{Y}, \widehat{\psi})$ to $(X_{\rm loc}, \phi_{\rm loc})$ near Δ. □

REMARK 12.15. One can similarly prove a parametric version of Proposition 12.14.

The following two lemmas concern the construction of Weinstein structures near stable discs of Smale cobordisms. They will be used in Chapters 13 and 14 to upgrade Morse cobordisms and homotopies to Weinstein cobordisms and homotopies.

LEMMA 12.16. *Let $\mathfrak{S} = (W, X, \phi)$ be an elementary Smale cobordism and ω a nondegenerate 2-form on W. Let $D_1, \dots, D_k$ be the stable discs of critical points of ϕ, and set $\Delta := \bigcup_{j=1}^k D_j$. Suppose that the discs $D_1, \dots, D_k$ are ω-isotropic and the pair (ω, X) is Liouville on $\mathcal{O}p\,(\partial_- W)$. Then for any neighborhood U of $\partial_- W \cup \Delta$ there exists a homotopy (ω_t, X_t), $t \in [0, 1]$, with the following properties:*

(i) *X_t is a gradient-like vector field for ϕ and ω_t is a nondegenerate 2-form on W for all $t \in [0,1]$;*

(ii) *$(\omega_0, X_0) = (\omega, X)$, and $(\omega_t, X_t) = (\omega, X)$ outside U and on $\Delta \cup \mathcal{O}p\,(\partial_- W)$ for all $t \in [0,1]$;*

(iii) *(ω_1, X_1) is a Liouville structure on $\mathcal{O}p\,(\partial_- W \cup \Delta)$.*

PROOF. To simplify the notation, assume that ϕ has a unique critical point p of index k with stable disc Δ. Let $D \subset \mathbb{R}^n \subset \mathbb{C}^n$ be the unit k-disc and $\omega_{\rm st}$ the standard symplectic structure on $\mathbb{C}^n$. Since $\omega|_\Delta = 0$, there exists an embedding $f : \mathcal{O}p\, D \hookrightarrow W$ mapping D onto Δ such that $f^*\omega = \omega_{\rm st}$ along D. According to Proposition 6.22 (since $\lambda = i_X\omega$ vanishes on Δ), we can modify f such that in addition $f^*\omega = \omega_{\rm st}$ on $\mathcal{O}p\,(\partial D)$. Thus $f_*\omega_{\rm st}$ and $\omega|_{\mathcal{O}p\,(\partial_- W)}$ fit together to a symplectic form $\widetilde{\omega}$ on $\mathcal{O}p\,(\partial_- W \cup \Delta)$ which agrees with ω along $\Delta \cup \mathcal{O}p\,(\partial_- W)$. The condition $\widetilde{\omega} = \omega$ along $\Delta \cup \mathcal{O}p\,(\partial_- W)$ allows us to find a homotopy ω_t of nondegenerate 2-forms on W, fixed on $\Delta \cup \mathcal{O}p\,(\partial_- W)$ and outside a neighborhood of Δ, such that $\omega_0 = \omega$ and $\omega_1|_{\mathcal{O}p\,\Delta} = \widetilde{\omega}$.

Note that the stable space E_p^- equals $T_p\Delta$, but the unstable space E_p^+ need not be ω_1-coisotropic. After a further homotopy of symplectic 2-forms, supported near p and keeping Δ isotropic (but changing on T_pW), we may assume that E_p^+ is ω_1-coisotropic.

Next we apply Lemma 12.8 (the hypothesis on the eigenvalues is satisfied since $E_p^- = T_p\Delta$) to find a Liouville field X' for ω_1 on $\mathcal{O}p\,\Delta$ which agrees with X on Δ and is gradient-like for ϕ. On $\mathcal{O}p\,(\partial\Delta)$ we have $X' = X + X_H$ for a function H that vanishes together with its differential along Δ. So $\widetilde{X} := X + X_{fH}$ for a cutoff function f yields a Liouville field for ω_1 on $\mathcal{O}p\,(\partial_- W \cup \Delta)$ which is gradient-like for ϕ and coincides with X on $\Delta \cup \mathcal{O}p\,(\partial_- W)$. Now we use Lemma 9.8 to extend $\widetilde{X}$ to a gradient-like vector field for ϕ on W and set $X_t := (1-t)X + t\widetilde{X}$. □

Lemma 12.16 has the following version for homotopies.

LEMMA 12.17. *Let $\mathfrak{S}_t = (W, X_t, \phi_t)$, $t \in [0,1]$, be an elementary Smale homotopy and ω_t, $t \in [0,1]$, a family of nondegenerate 2-forms on W. Let Δ_t be the skeleton of X_t and set*

$$\Delta := \bigcup_{t\in[0,1]} \{t\} \times \Delta_t \subset [0,1] \times W.$$

Suppose that Δ_t is ω_t-isotropic for all $t \in [0,1]$, the pair (ω_t, X_t) is Liouville on $\mathcal{O}p\,(\partial_- W)$ for all $t \in [0,1]$, and (ω_0, X_0) and (ω_1, X_1) are Liouville on all of W. Then for any open neighborhood $V = \bigcup_{t\in[0,1]}\{t\} \times V_t$ of Δ there exists an open neighborhood $U = \bigcup_{t\in[0,1]}\{t\} \times U_t \subset V$ of Δ and a 2-parameter family (ω_t^s, X_t^s), $s,t \in [0,1]$, with the following properties:

(i) *X_t^s is a gradient-like vector field for ϕ_t and ω_t^s is a nondegenerate 2-form on W for all $s,t \in [0,1]$;*

(ii) *$(\omega_t^0, X_t^0) = (\omega_t, X_t)$ for all $t \in [0,1]$, $(\omega_0^s, X_0^s) = (\omega_0, X_0)$ and $(\omega_1^s, X_1^s) = (\omega_1, X_1)$ for all $s \in [0,1]$, and $(\omega_t^s, X_t^s) = (\omega_t, X_t)$ outside V_t and on $\Delta_t \cup \mathcal{O}p\,(\partial_- W)$ for all $s,t \in [0,1]$;*

(iii) *(ω_t^1, X_t^1) is a Liouville structure on U_t for all $t \in [0,1]$.*

PROOF. For a Type I homotopy the proof is just a 1-parametric version of the proof of Lemma 12.16. For an elementary homotopy of Type IIb (the Type IId case is analogous), the proof is again analogous to the one of Lemma 12.16 with the following modifications.

Let us parametrize the homotopy over $t \in [-1, 1]$ with an embryonic critical point at $t = 0$. By Lemma 9.35 (cf. Figure 9.5), the skeletons Δ_t, $t \in [0, 1]$, form a smooth family of embedded half-discs with upper boundaries $\partial_+\Delta_t = D^-_{q_t}$ and lower boundaries $\partial_-\Delta_t = \Delta_t \cap \partial_- W$. Arguing as in the proof of Lemma 12.16, we find homotopies of nondegenerate 2-forms ω^s_t, fixed at $s = 0$, $t = 0$, $t = 1$ and on $\Delta_t \cup \mathcal{O}p\,(\partial_- W)$, such that ω^1_t are symplectic on $\mathcal{O}p\,\Delta$ and the unstable spaces $E^+_{p_t}, E^+_{q_t}$ are ω_t-coisotropic. Now a 1-parametric version of Lemma 12.8 yields the desired homotopy of gradient-like vector fields X^s_t. □

12.6. Morse–Smale theory for Weinstein structures

In this section we consider modifications of Weinstein structures analogous to those considered in Chapter 10 for Stein structures.

The first two lemmas have been proved in [**32**]; we include their easy proofs for the sake of completeness.

LEMMA 12.18. *Let $\mathfrak{W} = (W, \omega, X, \phi)$ be a Weinstein cobordism structure such that ϕ has no critical points. Let $\Lambda \subset \partial_+ W$ be an isotropic submanifold and $L \subset W$ its image under the flow of $-X$. Let $(\Lambda_t)_{t\in[0,1]}$ be an isotropic isotopy of $\Lambda_0 := L \cap \partial_- W$ in $\partial_- W$. Then there exists a family of Weinstein structures $\mathfrak{W}_t = (\omega_t, X_t, \phi)$, $t \in [0, 1]$, with the following properties (see Figure 10.1):*

(i) *$\mathfrak{W}_0 = \mathfrak{W}$, and $\mathfrak{W}_t$ is fixed on $\mathcal{O}p\,\partial_- W$ and up to scaling on $\mathcal{O}p\,\partial_+ W$;*
(ii) *the $\mathfrak{W}_t$, $t \in [0, 1]$, induce the same contact structures on level sets of ϕ;*
(iii) *the image L_t of Λ under the flow of $-X_t$ intersects $\partial_- W$ in Λ_t.*

PROOF. By the contact isotopy extension theorem (Proposition 6.24), there exists a contact diffeotopy $h_t : \partial_- W \to \partial_- W$ with $h_0 = \mathrm{Id}$ and $h_t(\Lambda_0) = \Lambda_t$. Apply Lemma 12.5 to obtain a homotopy of Weinstein structures $\mathfrak{W}_t$ with properties (i) and (ii) and whose holonomy equals $h_t \circ \Gamma : \partial_+ W \to \partial_- W$, where Γ is the holonomy of $\mathfrak{W}$. Hence the image L_t of Λ under the flow of $-X_t$ intersects $\partial_- W$ in $h_t(\Lambda_0) = \Lambda_t$. □

REMARK 12.19. If the isotopy Λ_t in Lemma 12.18 is sufficiently C^1-small, we can keep $\mathfrak{W}_t$ fixed near $\partial_+ W$. We do not know whether in general the rescaling near $\partial_+ W$ is actually needed.

The second lemma is just a restatement of Lemma 9.45.

LEMMA 12.20. *Let (W, ω, X, ϕ) be an elementary Weinstein cobordism. Then there is a homotopy (W, ω, X, ϕ_t) rel $\mathcal{O}p\,\partial W$ of elementary Weinstein cobordisms which arbitrarily changes the ordering of the critical values.*

The following two propositions are the Weinstein analogues of Theorems 10.11 and 10.12 in the Stein case.

PROPOSITION 12.21. *Let (W, ω, X, ϕ) be a Weinstein cobordism without critical points. Then given any point $p \in \mathrm{Int}\,W$ and integer $k = 1, \dots, n$ there exists a Weinstein homotopy (ω, X_t, ϕ_t) with the following properties:*

(i) *$(X_0, \phi_0) = (X, \phi)$ and $(X_t, \phi_t) = (X, \phi)$ outside a neighborhood of p;*
(ii) *ϕ_t is a creation family such that ϕ_1 has a pair of critical points of index k and $k - 1$.*

PROPOSITION 12.22. *Let (W, ω, X, ϕ) be a Weinstein cobordism with exactly two critical points p, q of index k and $k-1$, respectively, which are connected by a unique gradient trajectory along which the stable and unstable manifolds intersect transversely. Let Δ be the skeleton of (W, X), i.e., the closure of the stable manifold of the critical point p. Then there exists a Weinstein homotopy (ω, X_t, ϕ_t) with the following properties:*

(i) $(X_0, \phi_0) = (X, \phi)$, *and* $(X_t, \phi_t) = (X, \phi)$ *near* ∂W *and outside a neighborhood of* Δ;

(ii) ϕ_t *has no critical points outside* Δ;

(iii) ϕ_t *is a cancellation family such that* ϕ_1 *has no critical points.*

PROOF OF PROPOSITION 12.22. Pick a slightly larger embedded isotropic half-disc Δ' containing Δ in its interior. Pick a gradient-like vector field Y on Δ' for $\psi := \phi|_{\Delta'}$ with $Y|_\Delta = X|_\Delta$. Let $(\omega, \widehat{Y}, \widehat{\psi})$ be the Weinstein structure on $\mathcal{O}p\,\Delta'$ provided by Lemma 12.8. After applying Proposition 12.14 twice and shrinking Δ', we may assume that $(X, \phi) = (\widehat{Y}, \widehat{\psi})$ on $\mathcal{O}p\,\Delta'$.

Note that Y is inward pointing along $\partial_-\Delta'$ and outward pointing along $\partial_+\Delta'$. Hence Lemma 9.49 provides a cancellation family (Y_t, ψ_t) on Δ' which agrees with (Y, ψ) at $t = 0$ and near $\partial\Delta'$. Using Lemma 12.8 we extend it to a Weinstein homotopy $(\omega, \widehat{Y}_t, \widehat{\psi}_t)$ on $\mathcal{O}p\,\Delta'$ which agrees with (X, ϕ) at $t = 0$ and on $\mathcal{O}p\,\partial\Delta'$. Finally, we apply Lemma 12.10 to the pairs (X, ϕ) and $(X_{\rm loc}, \phi_{\rm loc}) = (\widehat{Y}_t, \widehat{\psi}_t)$, $t \in [0, 1]$, to obtain a cancellation type Weinstein homotopy (ω, X_t, ϕ_t) on W which agrees with (X, ϕ) at $t = 0$ and outside a neighborhood of Δ', and with $(\omega, \widehat{Y}_t, \widehat{\psi}_t)$ on a smaller neighborhood of Δ'. □

PROOF OF PROPOSITION 12.21. The proof is similar to that of Proposition 12.22. Define the vector field $Y(x) = \partial_{x_k}$ and the function $\psi(x) = x_k$ on $\mathbb{R}^k$ and their extensions to a Liouville field $\widehat{Y}$ and a function $\widehat{\psi}$ on $\mathbb{C}^n$ as in Lemma 12.8. Proposition 6.22 provides an isomorphism of isotropic setups

$$F : (\mathcal{O}p\,0 \subset \mathbb{C}^n, \omega_{\rm st}, \widehat{Y}, \{\widehat{\psi} = 0\}, 0) \cong (\mathcal{O}p\,p \subset W, \omega, X, \{\phi = a\}, p),$$

where $a = \phi(p)$. We will suppress the diffeomorphism F and just identify the corresponding objects.

After a homotopy of ϕ we may assume that $\phi = \widehat{\psi}$ on a smaller neighborhood U of p. Lemma 9.47 provides a creation family (Y_t, ψ_t) on the disc $\Delta := U \cap \mathbb{R}^k$ which agrees with (Y, ψ) at $t = 0$ and near $\partial\Delta$. Moreover, we can arrange that all eigenvalues of DY_t at critical points have real part < 1. Lemma 12.8 provides an extension of (Y_t, ψ_t) to a Weinstein homotopy $(\widehat{Y}_t, \widehat{\psi}_t)$ on $\mathcal{O}p\,\Delta$ which agrees with (X, ϕ) at $t = 0$ and on $\mathcal{O}p\,\partial\Delta$. Finally, we apply Lemma 12.10 to obtain a creation type Weinstein homotopy (ω, X_t, ϕ_t) on W which agrees with (X, ϕ) at $t = 0$ and outside a neighborhood of Δ, and with $(\omega, \widehat{Y}_t, \widehat{\psi}_t)$ on a smaller neighborhood of Δ. □

12.7. Elementary Weinstein homotopies

A Weinstein homotopy $(W, \omega_t, X_t, \phi_t)$ is called *elementary* if the underlying Smale homotopy (W, X_t, ϕ_t) is elementary. An *admissible partition* for a Weinstein homotopy is an admissible partition for the underlying Smale homotopy. According to Lemma 9.37, any Weinstein homotopy admits an admissible partition.

Lemma 9.38 has the following analogue for elementary Weinstein homotopies.

LEMMA 12.23. *Let* $\mathfrak{W}_t = (W, \omega_t, X_t, \phi_t)$ *and* $\widetilde{\mathfrak{W}}_t = (W, \widetilde{\omega}_t, \widetilde{X}_t, \widetilde{\phi}_t)$, $t \in [0,1]$, *be two elementary Weinstein homotopies with the same profile such that* $\mathfrak{W}_0 = \widetilde{\mathfrak{W}}_0$. *Then there exists a diffeotopy* $h_t : W \to W$ *with* $h_0 = \mathrm{Id}$ *such that* $\phi_t = \widetilde{\phi}_t \circ h_t$, *and the paths of Weinstein structures* $\mathfrak{W}_t$ *and* $h_t^*\widetilde{\mathfrak{W}}_t$ *are homotopic with fixed functions* ϕ_t *and fixed at* $t = 0$.

Moreover, if $\mathfrak{W}_t = \widetilde{\mathfrak{W}}_t$ *up to scaling near* $\partial_\pm W$, *we can arrange that* $h_t = \mathrm{Id}$ *near* $\partial_\pm W$ *and the homotopies between* $\mathfrak{W}_t$ *and* $h_t^*\widetilde{\mathfrak{W}}_t$ *are fixed up to scaling near* $\partial_\pm W$.

PROOF. The proof follows the same scheme as that of Lemma 9.38, keeping track of the contact structures on level sets.

Denote by $C_t, \widetilde{C}_t$ the critical point sets and by $\Delta_t, \widetilde{\Delta}_t$ the skeletons of $\mathfrak{W}_t, \widetilde{\mathfrak{W}}_t$. We first use Theorem 9.4 and the Morse Lemma 9.1 to construct a family of diffeomorphisms $f_t : \mathcal{O}p\, C_t \to \mathcal{O}p\, \widetilde{C}_t$ with $f_0 = \mathrm{Id}$ and $\widetilde{\phi}_t \circ f_t = \phi_t$. By Proposition 12.12, the path of Weinstein structures $\mathfrak{W}_t$ is homotopic with fixed functions ϕ_t and fixed at $t = 0$ to one which agrees with $f_t^*\widetilde{\mathfrak{W}}_t$ on $\mathcal{O}p\, C_t$. After replacing $\mathfrak{W}_t$ with this new path, we may hence assume that $\mathfrak{W}_t = f_t^*\widetilde{\mathfrak{W}}_t$ on $\mathcal{O}p\, C_t$.

Next we canonically extend the maps $f_t : \mathcal{O}p\, C_t \to \mathcal{O}p\, \widetilde{C}_t$ to diffeomorphisms $f_t : U_t \to \widetilde{U}_t$ between neighborhoods of Δ_t mapping ϕ_t to $\widetilde{\phi}_t$ and trajectories of X_t to trajectories of $\widetilde{X}_t$.

By Lemma 11.4, f_t induces contactomorphisms on all level sets. Note that $U_t^- := \partial_- W \cap U_t$ is a neighborhood of the isotropic submanifold $\Delta_t \cap \partial_- W$, and each restriction $f_t|_{U_t^-}$ is contact isotopic to the identity by following trajectories for shorter times. Hence by the contact isotopy extension theorem (Proposition 6.24), after shrinking U_t, the maps $f_t|_{U_t^-}$ extend to contactomorphisms $g_t : (\partial_- W, \xi_-) \to (\partial_- W, \widetilde{\xi}_-)$. Moreover, since $f_0 = \mathrm{Id}$ we can arrange $g_0 = \mathrm{Id}$.

Now we extend the maps $U_t \cup \partial_- W \to \widetilde{U}_t \cup \partial_- W$ given by f_t and g_t canonically to diffeomorphisms $h_t : W \to W$ mapping ϕ_t to $\widetilde{\phi}_t$ and trajectories of X_t to trajectories of $\widetilde{X}_t$. We have $h_0 = \mathrm{Id}$ and, again by Lemma 11.4, h_t induces contactomorphisms on all level sets. Hence according to Corollary 12.3, the paths of Weinstein structures $\mathfrak{W}_t$ and $h_t^*\widetilde{\mathfrak{W}}_t$ are homotopic with fixed functions ϕ_t and fixed at $t = 0$.

Finally, if $\mathfrak{W}_t = \widetilde{\mathfrak{W}}_t$ up to scaling near $\partial_\pm W$, we undo the contact diffeotopy h_t on level sets near $\partial_\pm W$ to arrange $h_t = \mathrm{Id}$ on $\mathcal{O}p\, \partial_\pm W$. Then Corollary 12.3 shows that the homotopies between $\mathfrak{W}_t$ and $h_t^*\widetilde{\mathfrak{W}}_t$ can be chosen fixed up to scaling near $\partial_\pm W$. Note that in this last step we destroy the property that h_t maps trajectories of X_t to trajectories of $\widetilde{X}_t$. ☐

CHAPTER 13

Existence Revisited

In this chapter we prove a more precise version of the Stein Existence Theorem 1.5 by splitting it into two theorems: Theorem 13.1 on the existence of Weinstein structures and Theorem 13.9 on upgrading a Weinstein structure to a Stein structure. Moreover, we establish the homotopy equivalence between the spaces of Stein and Weinstein structures with a given function (Theorem 1.2 from the Introduction).

13.1. Existence of Weinstein structures

The following is the analogue of Theorem 8.17 for Weinstein cobordisms.

THEOREM 13.1 (Weinstein existence theorem). *Let (W, ϕ) be a $2n$-dimensional Morse cobordism such that ϕ has no critical points of index $> n$. Let η be a nondegenerate (not necessarily closed) 2-form on W and Y a vector field near $\partial_- W$ such that (η, Y, ϕ) defines a Weinstein structure on $\mathcal{O}p\, \partial_- W$. Suppose that either $n > 2$, or $n = 2$ and the contact structure induced by the Liouville form $\lambda = i_Y \eta$ on $\partial_- W$ is overtwisted. Then there exists a Weinstein structure (ω, X, ϕ) on W with the following properties:*

(i) *$(\omega, X) = (\eta, Y)$ on $\mathcal{O}p\, \partial_- W$;*

(ii) *the nondegenerate 2-forms ω and η on W are homotopic* rel $\mathcal{O}p\, \partial_- W$.

Moreover, we can arrange that (ω, X, ϕ) is flexible.

Let us point out that Theorem 13.1 does not follow from the Stein Existence Theorem 8.17 in the case $n > 2$ because the given Weinstein structure on $\mathcal{O}p\, \partial_- W$ need not be Stein. For example, if $n > 2$ and the induced contact structure on $\partial_- W$ is not symplectically fillable, then by Theorem 5.60 the Weinstein structure on $\mathcal{O}p\, \partial_- W$ cannot be deformed to a Stein structure.

The following version for manifolds follows directly from Theorem 13.1. Note that it is also a formal consequence of the Stein Existence Theorem 1.5.

THEOREM 13.2. *Let (V, ϕ) be a $2n$-dimensional manifold with an exhausting Morse function ϕ that has no critical points of index $> n$. Let η be a nondegenerate (not necessarily closed) 2-form on V. Suppose that $n > 2$. Then there exists a Weinstein structure (ω, X, ϕ) on V such that the nondegenerate 2-forms ω and η on V are homotopic. Moreover, we can arrange that (ω, X, ϕ) is flexible.*

The proof of Theorem 13.1 is based on the following special case.

LEMMA 13.3. *Theorem 13.1 holds for an elementary cobordism.*

PROOF. The proof follows the same scheme as the proof of Lemma 8.20 in the Stein case. To simplify the notation, we will assume that ϕ has a unique critical

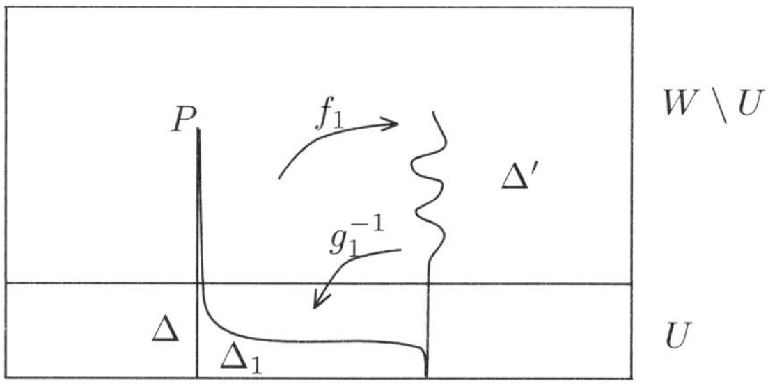

FIGURE 13.1. Deforming the disc Δ to one which is totally real and J-orthogonally attached.

point p. The general case is similar. Let us extend Y to a gradient-like vector field for ϕ on W and denote by Δ the stable disc of p.

Step 1. We first show that, after a homotopy of (η, Y) fixed on $\mathcal{O}p\,\partial_-W$, we may assume that Δ is η-isotropic.

The Liouville form $\lambda = i_Y\eta$ on $\mathcal{O}p\,\partial_-W$ defines a contact structure $\xi := \ker(\lambda|_{\partial_-W})$ on ∂_-W. We choose an auxiliary η-compatible almost complex structure J on W which preserves ξ and maps Y along ∂_-W to the Reeb vector field R of $\lambda|_{\partial_-W}$. We apply Theorem 7.34 to find a diffeotopy $f_t : W \to W$ such that the disc $\Delta' = f_1(\Delta)$ is J-totally real and J-orthogonally attached to ∂_-W. This is the only point in the proof where the overtwistedness assumption for $n = 2$ is needed. Moreover, according to Theorem 7.34, in the case $\dim \Delta = n$ we can arrange that the Legendrian sphere $\partial\Delta'$ in (∂_-W, ξ) is loose (meaning that $\partial_-W \setminus \partial\Delta'$ is overtwisted in the case $n = 2$).

Next we modify the homotopy f_t^*J to keep it fixed near ∂_-W. J-orthogonality implies that $\partial\Delta'$ is tangent to the maximal J-invariant subspace $\xi \subset T(\partial_-W)$ and thus $\lambda|_{\partial\Delta'} = 0$. Since the spaces $T\Delta'$ and $\mathrm{span}\{T\partial\Delta', Y\}$ are both totally real and J-orthogonal to $T(\partial_-W)$, we can further adjust the disc Δ' (keeping $\partial\Delta'$ fixed) to make it tangent to Y in a neighborhood of $\partial\Delta'$. It follows that we can modify f_t such that it preserves the function ϕ and the vector field Y on a neighborhood U of ∂_-W (extend f_t from ∂_-W to U using the flow of Y).

Hence, there exists a diffeotopy $g_t : W \to W$, $t \in [0, 1]$, which equals f_t on $W \setminus U$, the identity on $\mathcal{O}p\,\partial_-W$, and preserves ϕ (but not Y!) on U; see Figure 13.1. Then the diffeotopy $k_t := f_t^{-1} \circ g_t$ equals the identity on $W \setminus U$, f_t^{-1} on $\mathcal{O}p\,\partial_-W$, and preserves ϕ on all of W. Thus the vector fields $Y_t := k_t^*Y$ are gradient-like for $\phi = k_t^*\phi$ and coincide with Y on $(W \setminus U) \cup \mathcal{O}p\,\partial_-W$. The nondegenerate 2-forms $\eta_t := g_t^*\eta$ are compatible with $J_t := g_t^*J$ and coincide with η on $\mathcal{O}p\,\partial_-W$. Moreover, since Δ' is J-totally real, the stable disc $\Delta_1 := k_1^{-1}(\Delta) = g_1^{-1}(\Delta')$ of p with respect to Y_1 is J_1-totally real and J_1-orthogonally attached to ∂_-W.

After renaming (η_1, Y_1, Δ_1) back to (η, Y, Δ), we may hence assume that Δ is J-totally real and J-orthogonally attached to ∂_-W for some η-compatible almost complex structure J on W which preserves ξ and maps Y to the Reeb vector field R along ∂_-W. In particular, $\partial\Delta$ is λ-isotropic and $\Delta \cap \mathcal{O}p\,\partial_-W$ is η-isotropic. Since the space of nondegenerate 2-forms compatible with J is contractible, after a further homotopy of η fixed on $\mathcal{O}p\,\partial_-W$ and outside a neighborhood of Δ, we may assume that Δ is η-isotropic.

Step 2. By Lemma 12.16 there exists a homotopy (η_t, Y_t), $t \in [0,1]$, of gradient-like vector fields for ϕ and nondegenerate 2-forms on W, fixed on $\Delta \cup \mathcal{O}p\,\partial_- W$ and outside a neighborhood of Δ, such that $(\eta_0, Y_0) = (\eta, Y)$ and (η_1, Y_1) is Liouville on $\mathcal{O}p\,(\partial_- W \cup \Delta)$. After renaming (η_1, Y_1) back to (η, Y), we may hence assume that (η, Y) is Liouville on a neighborhood U of $\partial_- W \cup \Delta$.

Step 3. Using Proposition 9.19 (pushing down along trajectories of Y), we construct an isotopy of embeddings $h_t : W \hookrightarrow W$, $t \in [0,1]$, with $h_0 = \mathrm{Id}$ and $h_t = \mathrm{Id}$ on $\mathcal{O}p\,(\partial_- W \cup \Delta)$, which preserves trajectories of Y and such that $h_1(W) \subset U$. Then $(\eta_t, Y_t) := (h_t^*\eta, h_t^*Y)$ defines a homotopy of nondegenerate 2-forms and vector fields on W, fixed on $\mathcal{O}p\,(\partial_- W \cup \Delta)$, from $(\eta_0, Y_0) = (\eta, Y)$ to the Liouville structure $(\eta_1, Y_1) =: (\omega, X)$. Since the Y_t are proportional to Y, they are gradient-like for ϕ for all $t \in [0,1]$.

The Weinstein structure (ω, X, ϕ) will be flexible if we choose the stable sphere $\partial\Delta$ in Step 1 to be loose, so Lemma 13.3 is proved. □

PROOF OF THEOREM 13.1. We decompose the Morse cobordism $\mathfrak{M} = (W, \phi)$ into elementary ones, $W = W_1 \cup \cdots \cup W_N$, and inductively apply Lemma 13.3 to extend the Weinstein structure over $W_1, \dots, W_N$. □

13.2. From Weinstein to Stein: existence

In this section we formulate various theorems about passing from Weinstein to Stein structures. The proofs of the two main results, Theorems 13.4 and 13.6, are postponed to Section 13.3 below. Let us point out that all the results in this section also hold in dimension 4 without further hypotheses.

We begin with the case of cobordisms. Our first theorem concerns the passing from Weinstein to Stein within an ambient complex cobordism.

THEOREM 13.4 (Ambient Stein existence theorem). *Let $\mathfrak{W} = (W, \omega, X, \phi)$ be a Weinstein cobordism and J an integrable complex structure on W. Suppose that on $\mathcal{O}p\,\partial_- W$ the function ϕ is J-convex and $\mathfrak{W}$ coincides with $\mathfrak{W}(J, \phi)$. Suppose moreover that J is homotopic* rel $\partial_- W$ *to an almost complex structure compatible with ω. Then, after target reparametrizing ϕ, there exists an isotopy $h_t : W \hookrightarrow W$* rel $\mathcal{O}p\,\partial_- W$ *with $h_0 = \mathrm{Id}$ such that the function $h_{1*}\phi$ is J-convex, and the Weinstein structures $\mathfrak{W}(h_1^*J, \phi)$ and $\mathfrak{W}$ on W are homotopic rel $\mathcal{O}p\,\partial_- W$ with fixed function ϕ.*

Combining this theorem with the existence of complex structures, we obtain

THEOREM 13.5 (Stein existence theorem). *Let $\mathfrak{W} = (W, \omega, X, \phi)$ be a Weinstein cobordism which is Stein near $\partial_- W$. Then, after target reparametrizing ϕ, the Stein structure on $\mathcal{O}p\,\partial_- W$ extends to a Stein structure (J, ϕ) on W such that the Weinstein structures $\mathfrak{W}$ and $\mathfrak{W}(J, \phi)$ are homotopic* rel $\mathcal{O}p\,\partial_- W$ *and with fixed function ϕ.*

PROOF. By assumption $\mathfrak{W} = \mathfrak{W}(\widetilde{J}, \phi)$ for a Stein structure $(\widetilde{J}, \phi)$ on $\mathcal{O}p\,\partial_- W$. We extend $\widetilde{J}$ to an almost complex structure on W compatible with ω. Let L be the skeleton of $\mathfrak{W}$.

By Theorem 8.11, $\widetilde{J}$ is homotopic rel $\mathcal{O}p\,\partial_- W$ to an almost complex structure J' which is integrable on a neighborhood U of $\partial_- W \cup L$. By Proposition 9.19, there exists an isotopy $g_t : W \hookrightarrow W$ rel $\mathcal{O}p\,(\partial_- W \cup L)$ with $g_0 = \mathrm{Id}$ and $g_1(W) \subset U$.

Then $J'' := g_1^* J'$ is a complex structure on W which is homotopic rel $\mathcal{O}p\,\partial_- W$ to $\widetilde{J}$. Thus we can apply Theorem 13.4 to find a Stein structure $(h_1^* J'', \phi)$ on W such that the Weinstein structures $\mathfrak{W}(h_1^* J'', \phi)$ and $\mathfrak{W}$ are homotopic rel $\mathcal{O}p\,\partial_- W$ and with fixed function ϕ. □

This theorem has the following multiparametric version, where D^k denotes the closed k-disc.

THEOREM 13.6 (Parametric Stein existence theorem). *Let $\mathfrak{W}_u = (W, \omega_u, X_u, \phi)$, $u \in D^k$, $k \geq 0$, be a family of Weinstein cobordism structures which share the same Morse function ϕ. Suppose $\mathfrak{W}_u$ is Stein near $\partial_- W$ for all u, and on W for $u \in \partial D^k$. Then, after target reparametrizing ϕ, there exist a family of Stein structures (J_u, ϕ), $u \in D^k$, extending the given structures near $\partial_- W$ and for $u \in \partial D^k$, and a homotopy of Weinstein structures $\mathfrak{W}_{t,u} = (\omega_{t,u}, X_{t,u}, \phi)$, $(t,u) \in [0,1] \times D^k$, such that $\mathfrak{W}_{0,u} = \mathfrak{W}(J_u, \phi)$ and $\mathfrak{W}_{1,u} = \mathfrak{W}_u$ for all $u \in D^k$, $\mathfrak{W}_{t,u} = \mathfrak{W}_u$ near $\partial_- W$, and $\mathfrak{W}_{t,u} = \mathfrak{W}_u$ for $u \in \partial D^k$ and all $t \in [0,1]$.*

In order to rephrase this theorem in a more topological way, let us fix a Morse function $\phi : W \to \mathbb{R}$ which has $\partial_\pm W$ as regular level sets. We denote by $\mathfrak{Stein}(W, \phi)$ the space of Stein structures on W with J-lc function ϕ, and by $\mathfrak{Weinstein}(W, \phi)$ the space of Weinstein structures on W with function ϕ which are Stein near $\partial_- W$. Then Theorem 13.6 implies

COROLLARY 13.7. *The map $\mathfrak{W} : \mathfrak{Stein}(W, \phi) \to \mathfrak{Weinstein}(W, \phi)$ is a weak homotopy equivalence.*

See Corollary A.4 for the purely topological argument. In fact, Corollary 13.7 is equivalent to Theorem 13.6 if we drop the condition that homotopies are fixed near $\partial_- W$.

Corollary 13.7 continues to hold if ϕ is a generalized Morse function (using our results about half-discs at embryonic critical points). The case $\partial_- W = \varnothing$ is then Theorem 1.2 from the Introduction.

The preceding theorems have the following analogues for Weinstein/Stein *manifolds*, which are derived from the cobordism versions by induction over sublevel sets as in the proof of Theorem 8.16.

THEOREM 13.8. *Let $\mathfrak{W} = (V, \omega, X, \phi)$ be a Weinstein manifold. Let J be an integrable complex structure on V which is homotopic to an almost complex structure compatible with ω. Then there exists an isotopy $h_t : V \hookrightarrow V$ with $h_0 = \mathrm{Id}$ such that the function $h_{1*}\phi$ is J-convex, and the Weinstein structures $\mathfrak{W}(h_1^* J, \phi)$ and $\mathfrak{W}$ on V are homotopic with fixed function ϕ.*

The following result is Theorem 1.1(a) from the Introduction.

THEOREM 13.9. *Let $\mathfrak{W} = (V, \omega, X, \phi)$ be a Weinstein manifold. Then there exists a Stein structure (J, ϕ) on W such that the Weinstein structures $\mathfrak{W}$ and $\mathfrak{W}(J, \phi)$ are homotopic with fixed function ϕ.*

THEOREM 13.10. *Let $\mathfrak{W}_u = (V, \omega_u, X_u, \phi)$, $u \in D^k$, $k \geq 0$, be a family of Weinstein manifolds which share the same Morse function ϕ. Suppose $\mathfrak{W}_u$ is Stein for $u \in \partial D^k$. Then there exist a family of Stein structures (J_u, ϕ), $u \in D^k$, extending the given structures for $u \in \partial D^k$, and a homotopy of Weinstein structures $\mathfrak{W}_{t,u} = (\omega_{t,u}, X_{t,u}, \phi)$, $(t,u) \in [0,1] \times D^k$, such that $\mathfrak{W}_{0,u} = \mathfrak{W}(J_u, \phi)$ and $\mathfrak{W}_{1,u} = \mathfrak{W}_u$ for all $u \in D^k$, and $\mathfrak{W}_{t,u} = \mathfrak{W}_u$ for $u \in \partial D^k$ and all $t \in [0,1]$.*

The last theorem can again be formulated in a more topological way. We fix an exhausting Morse function $\phi : V \to \mathbb{R}$ and denote by $\mathfrak{Stein}(V,\phi)$ and $\mathfrak{Weinstein}(V,\phi)$ the spaces of Stein (resp. Weinstein) structures on V with function ϕ. These spaces are equipped with the topologies explained in Section 11.6. Then, by Corollary A.4, Theorem 13.10 is equivalent to

COROLLARY 13.11. *The map* $\mathfrak{W} : \mathfrak{Stein}(V,\phi) \to \mathfrak{Weinstein}(V,\phi)$ *is a weak homotopy equivalence.*

13.3. Proof of the Stein existence theorems

In this section we prove Theorems 13.4 and 13.6. The first one will be an easy consequence of the following proposition.

PROPOSITION 13.12. *Under the hypotheses of Theorem* 13.4 *there exists a homotopy of Weinstein structures* $\mathfrak{W}_t = (\omega_t, X_t, \phi_t)$ *on* W, $t \in [0,1]$, *and a regular value* c *of the function* ϕ_1 *with the following properties:*

(i) $\mathfrak{W}_0 = \mathfrak{W}$, *and* $\mathfrak{W}_t$ *agrees with* $\mathfrak{W}$ *on* $\mathcal{O}p\,\partial_- W$ *and up to scaling on* $\mathcal{O}p\,\partial_+ W$;
(ii) *on* $W' = \{\phi_1 \le c\}$ *the function* ϕ_1 *is* J*-convex and* $\mathfrak{W}_1|_{W'} = \mathfrak{W}(J,\phi_1)$;
(iii) *on* $\{\phi_1 \ge c\}$ *the function* ϕ_1 *has no critical points;*
(iv) $\phi_t = \phi \circ f_t$ *for a diffeotopy* $f_t : W \to W$ *fixed on* $\mathcal{O}p\,\partial W$ *with* $f_0 = \mathrm{Id}$.

PROOF. We first consider the case of an elementary cobordism. To simplify the notation, we will assume that there is only one critical point $p \in W$; the general case differs only in notation. Let Δ be the stable disc of p. We will construct in four steps an elementary Weinstein homotopy $\mathfrak{W}_t$ and a regular value with properties (i)–(iii) such that the ϕ_t have fixed values on $\partial_\pm W$ and p. Lemma 9.38 then ensures the existence of a diffeotopy f_t as in (iv).

Step 1. After applying Corollary 12.13, we may assume that near p the function ϕ is J-convex and $\mathfrak{W} = \mathfrak{W}(J,\phi)$. In the following steps we will keep $\mathfrak{W}$ fixed on $\mathcal{O}p\,p$.

Step 2. Choose a homotopy J_t rel $\mathcal{O}p\,(\partial_- W \cup p)$ of almost complex structures such that $J_1 = J$ and J_0 is compatible with ω. Then the disc Δ is totally real for J_0. Hence, according to Corollary 7.31, there exists a C^0-small isotopy of J_t-totally real discs Δ_t, starting with $\Delta_0 = \Delta$ and fixed on $\Delta \cap \mathcal{O}p\,(\partial_- W \cup p)$. We extend this isotopy to a global diffeotopy $g_t : W \to W$ fixed on $\mathcal{O}p\,(\partial W \cup p)$. After replacing $\mathfrak{W}$ by $(g_1)_*\mathfrak{W}$, we may hence assume that Δ is totally real for J. Note that Δ is also J-orthogonally attached to $\partial_- W$.

Step 3. According to Lemma 8.7, there exists a J-convex function $\widetilde{\phi}$ on $\mathcal{O}p\,(\partial_- W \cup \Delta)$ which agrees with ϕ on $\Delta \cup \mathcal{O}p\,(\partial_- W \cup \{p\})$ and such that along Δ the gradient vector field $\nabla_{J,\widetilde{\phi}}\widetilde{\phi}$ equals X. Next, we use Proposition 12.14 to deform $\mathfrak{W}$ rel $\mathcal{O}p\,(\partial W \cup p) \cup \Delta$ to a Weinstein structure $\widetilde{\mathfrak{W}}$ which coincides with $\mathfrak{W}(J,\widetilde{\phi})$ on $\mathcal{O}p\,\Delta$. After replacing $\mathfrak{W}$ by $\widetilde{\mathfrak{W}}$, we may hence assume that ϕ is J-convex and $\mathfrak{W} = \mathfrak{W}(J,\phi)$ on a neighborhood U of $\partial_- W \cup \Delta$.

Step 4. Finally, we use Theorem 8.5 to construct a deformation ϕ_t of J-convex functions on U with the following properties:

- $\phi_0 = \phi|_U$;

- ϕ_t is target equivalent to ϕ near ∂U, and equal to ϕ on a smaller neighborhood $N \subset U$ of $\partial_- W \cup \Delta$;
- ϕ_t has no critical points besides p;
- some level set $\{\phi_1 = c\}$ surrounds $\partial_- W \cup \Delta$ in U.

By the second property, near ∂U we have $\phi_t = g_t \circ \phi$ for some increasing function $g_t : \mathbb{R} \to \mathbb{R}$. After composing g_t with a sufficiently convex function, we may assume that $g_t'' \geq 0$. Moreover, we can arrange $g_t(x) = c_t x + d_t$ for $x \geq \max_U \phi$, for some smooth families of constants $c_t > 0$ and d_t. Then near ∂U the Weinstein structure $\mathfrak{W}(J, \phi_t)$ has Liouville form

$$\lambda_t = -d\phi_t \circ J = -g_t' \circ \phi \, d\phi \circ J = f_t \lambda, \qquad f_t := g_t' \circ \phi.$$

Note that f_t satisfies

$$f_t + df_t(X) = g_t' \circ \phi + g_t'' \circ \phi \, d\phi(X) > 0.$$

According to Lemma 12.1, this ensures that $(f_t\lambda, g_t \circ \phi)$ defines an extension $\mathfrak{W}_t$ of the Weinstein structure $\mathfrak{W}(J, \phi_t)$ from U to the whole cobordism W. Near $\partial_+ W$ we have $f_t\lambda = c_t\lambda$, so (ω_t, X_t) is fixed up to scaling near $\partial_+ W$. Finally, we target reparametrize ϕ_t to make it equal to ϕ on $\mathcal{O}p\,(\partial W \cup \Delta)$. This concludes the proof for the case of an elementary cobordism.

Consider now the case of a general cobordism W. Take an admissible partition $\min \phi = c_0 < c_1 < \cdots < c_N = \max \phi$. First we apply the above construction to deform the Weinstein structure on the elementary cobordism $W_1 = \{\phi \leq c_1\}$, keeping it fixed up to scaling on $W \setminus W_1$, such that $\mathfrak{W}_1 = \mathfrak{W}(J, \phi_1)$ on $W_1' = \{\phi_1 \leq c\}$, and the function ϕ_1 has no critical points on $W_1 \setminus W_1'$. Then we can apply again the same construction to the restriction of $\mathfrak{W}_1$ to the elementary cobordism $W_2 = \{\phi \leq c_2\} \setminus W_1'$. Continuing this process we construct the required deformation on the whole cobordism W. □

Proof of Theorem 13.4. Let $\mathfrak{W}_t = (W, \omega_t, X_t, \phi_t = \phi \circ f_t)$ and $W' = \{\phi_1 \leq c\}$ be as in Proposition 13.12. Since ϕ_1 has no critical points outside W', pushing down along trajectories of X_1 we find an isotopy $g_t : W \hookrightarrow W$ rel $\mathcal{O}p\,\partial_- W$ satisfying $g_0 = \mathrm{Id}$, $g_1(W) = W'$, and $\phi_1 \circ g_t^{-1} = \alpha_t \circ \phi_1|_{g_t(W)}$ for convex increasing functions $\alpha_t : \mathbb{R} \to \mathbb{R}$. Set $h_t := g_t \circ f_t^{-1} : W \hookrightarrow W$. Then J-convexity of $\phi_1|_{W'}$ implies J-convexity of $h_{1*}\phi = \phi_1 \circ g_1^{-1} = \alpha_1 \circ \phi_1|_{W'}$.

It remains to show that $\mathfrak{W}(h_1^* J, \phi)$ is homotopic rel $\partial_- W$ with fixed function ϕ to $\mathfrak{W}$. To see this, let us write $\alpha \circ \mathfrak{W} := (\omega, X, \alpha \circ \phi)$ for a Weinstein structure $\mathfrak{W} = (\omega, X, \phi)$. Then $\mathfrak{W}(h_1^* J, \phi)$ is connected to $\mathfrak{W}$ by the following chain of homotopies rel $\partial_- W$ with fixed function ϕ:

$$\begin{aligned}\mathfrak{W}(h_1^* J, \phi) &= f_{1*} g_1^* \mathfrak{W}(J, \alpha_1 \circ \phi_1) \sim f_{1*} g_1^* \alpha_1 \circ \mathfrak{W}(J, \phi_1) \\ &= f_{1*} g_1^* \alpha_1 \circ \mathfrak{W}_1 \overset{f_{1*} g_t^* \alpha_t \circ \mathfrak{W}_1}{\sim} f_1^* \mathfrak{W}_1 \overset{f_{t*}\mathfrak{W}_t}{\sim} \mathfrak{W}.\end{aligned}$$ □

Proof of Theorem 13.6. The proof is essentially a k-parametric version of the proof of Theorem 13.5. However, some care must be taken to make the Stein family match the given Stein structures on ∂D^k.

Assume first that all the $\mathfrak{W}_u$ are elementary. We reparametrize the family $\mathfrak{W}_u$ to make it Stein for $u \in \mathcal{O}p\,\partial D^k$. To simplify notation, let us assume that there is exactly one critical point $p \in W$ of ϕ with X_u-stable discs Δ_u. We construct the desired Weinstein family $\mathfrak{W}_u$ in the following three steps that are similar to Steps 2–4 in the proof of Proposition 13.12.

Step 1. Pick a family $\widetilde{J}_u$, $u \in D^k$, of almost complex structures on W that are compatible with ω_u and agree with the given integrable structures on $\mathcal{O}p\,\partial_- W$ and for $u \in \mathcal{O}p\,\partial D^k$. Note that the discs Δ_u are $\widetilde{J}_u$-totally real and $\widetilde{J}_u$-orthogonally attached. Hence a k-parametric version of Proposition 8.12 yields a family of integrable complex structures J_u on $\mathcal{O}p\,(\partial_- W \cup \Delta_u)$, restricting to the given structures on $\mathcal{O}p\,\partial_- W$ and for $u \in \mathcal{O}p\,\partial D^k$, such that Δ_u is J_u-totally real and J_u-orthogonally attached.

Step 2. Next, we use k-parametric versions of Lemma 8.7 and Proposition 12.14 to deform $\mathfrak{W}_u$ to a Weinstein family $\widetilde{\mathfrak{W}}_u = (W, \widetilde{\omega}_u, \widetilde{X}_u, \widetilde{\phi}_u)$, coinciding with $\mathfrak{W}_u$ on $\mathcal{O}p\,(\partial W) \cup \Delta_u$ and for $u \in \mathcal{O}p\,\partial D^k$, such that on $\mathcal{O}p\,\Delta_u$ the function $\widetilde{\phi}_u$ is J_u-convex and $\widetilde{\mathfrak{W}}_u = \mathfrak{W}(J_u, \widetilde{\phi}_u)$. By Lemma 9.29 we have $\widetilde{\phi}_u \circ h_u = \phi$ for diffeomorphisms $h_u : W \to W$ fixed on ∂W. After pulling back $\widetilde{\mathfrak{W}}_u$ under h_u and renaming it back to $\mathfrak{W}_u$, we may hence assume that $\mathfrak{W}_u = \mathfrak{W}(J_u, \phi)$ on a neighborhood U_u of $\partial_- W \cup \Delta_u$. Since (J_u, ϕ) is already Stein for $u \in \mathcal{O}p\,\partial D^k$, we may choose $U_u = W$ for $u \in \mathcal{O}p\,\partial D^k$.

Step 3. Finally, we use a k-parametric version of Theorem 8.5 to construct a family ψ_u, $u \in D^k$, of J_u-convex functions on U_u with the following properties:

- ψ_u is target equivalent to ϕ for $u \in \partial D^k$;
- ψ_u is target equivalent to ϕ near ∂U_u, and equal to ϕ on a smaller neighborhood $N_u \subset U_u$ of $\partial_- W \cup \Delta_u$;
- ψ_u has no critical points besides p;
- the regular level sets $\{\psi_u = c\}$ surround $\partial_- W \cup \Delta_u$ in U_u and agree with $\partial_+ W$ for $u \in \partial D^k$.

By Lemma 9.29 there exists a family of diffeomorphisms $h_u : W \to \{\psi_u \leq c\}$, fixed on N_u with $h_u = \mathrm{Id}$ for $u \in \partial D^k$, such that $\psi_u \circ h_u$ is target equivalent to ϕ. After a target reparametrization of ϕ the desired Stein family is thus $(h_u^* J_u, \phi)$. Since the families of Weinstein structures $\mathfrak{W}_u$ and $\mathfrak{W}(h_u^* J_u, \phi)$ agree on $\mathcal{O}p\,(\partial_- W \cup \Delta_u)$, the homotopy rel $\mathcal{O}p\,\partial_- W$ with fixed function ϕ between them follows from Gray's Stability Theorem 6.23 and Corollary 12.3. This concludes the proof for an elementary family.

If the $\mathfrak{W}_u$ are not elementary, we pick regular values

$$\phi|_{\partial_- W} = c_0 < c_1 < \cdots < c_N = \phi|_{\partial_+ W}$$

of ϕ such that each (c_{k-1}, c_k) contains at most one critical value. Then the restriction of $\mathfrak{W}_u$ to each cobordism $W^k := \{c_{k-1} \leq \phi \leq c_k\}$ is elementary. We apply Steps 1–3 to the restriction of the family $\mathfrak{W}_u$ to W^1 to construct a Stein family (W^1, J_u, ϕ) extending the given Stein structures on $\mathcal{O}p\,\partial_- W$ and for $u \in \partial D^k$ such that the Weinstein families $\mathfrak{W}_u|_{W^1}$ and $\mathfrak{W}(W^1, J_u, \phi)$ are connected by a homotopy $\mathfrak{W}_{t,u}$, $(t, u) \in [0,1] \times D^k$, rel $\mathcal{O}p\,\partial_- W$ with fixed function ϕ and fixed on ∂D^k. Using the homotopy $\mathfrak{W}_{t,u}$ and Lemma 12.7, we extend $\mathfrak{W}(W^1, J_u, \phi)$ to a Weinstein homotopy on W with the same function ϕ and continue inductively with $W^2, \ldots, W^N$. $\square$

CHAPTER 14

Deformations of Flexible Weinstein Structures

In this chapter we show that flexible Weinstein structures in dimension > 4 are indeed "flexible": any Morse homotopy can be followed by a flexible Weinstein homotopy, and two flexible Weinstein structures on the same manifold whose symplectic forms are homotopic as nondegenerate 2-forms are Weinstein homotopic. As applications we obtain a Weinstein version of the h-cobordism theorem (Corollary 14.2), a realization result of isotopy classes of diffeomorphisms by symplectomorphisms (Theorem 14.7), and a realization result of pseudo-isotopies by symplectic pseudo-isotopies (Theorem 14.23). Moreover, in Section 14.4 we prove the result from [**33**] that subcritical Weinstein manifolds split as a product with $\mathbb{C}$.

Combining Theorems 13.1, 14.5, and 14.7, we obtain Theorem 1.8 from the Introduction.

14.1. Homotopies of flexible Weinstein cobordisms

The following Theorems 14.1 and 14.3 are our main results concerning deformations of flexible Weinstein structures.

THEOREM 14.1 (First Weinstein deformation theorem). *Let $\mathfrak{W} = (W, \omega, X, \phi)$ be a flexible Weinstein cobordism of dimension $2n > 4$. Let $\psi : W \to \mathbb{R}$ be a Morse function without critical points of index $> n$. Then there exists a homotopy $\mathfrak{W}_t = (W, \omega_t, X_t, \phi_t)$, $t \in [0,1]$, of flexible Weinstein structures, fixed on $\mathcal{O}p\, \partial_- W$ and fixed up to scaling on $\mathcal{O}p\, \partial_+ W$, such that $\mathfrak{W}_0 = \mathfrak{W}$ and $\phi_1 = \psi$.*

In particular, we have the following Weinstein version of the h-cobordism theorem.

COROLLARY 14.2 (Weinstein h-cobordism theorem). *Any flexible Weinstein structure on a product cobordism $W = Y \times [0,1]$ of dimension $2n > 4$ is homotopic to a Weinstein structure (W, ω, X, ϕ), where $\phi : W \to [0,1]$ is a function without critical points.* □

THEOREM 14.3 (Second Weinstein deformation theorem). *Let $\mathfrak{W}_0 = (\omega_0, X_0, \phi_0)$ and $\mathfrak{W}_1 = (\omega_1, X_1, \phi_1)$ be two flexible Weinstein structures on a cobordism W of dimension $2n > 4$ which coincide on $\mathcal{O}p\, \partial_- W$. Let η_t be a homotopy rel $\mathcal{O}p\, \partial_- W$ of nondegenerate 2-forms on W connecting ω_0 and ω_1. Then there exists a homotopy $\mathfrak{W}_t = (\omega_t, X_t, \phi_t)$ of flexible Weinstein structures connecting $\mathfrak{W}_0$ and $\mathfrak{W}_1$, fixed on $\mathcal{O}p\, \partial_- W$, such that the paths ω_t and η_t of nondegenerate 2-forms are homotopic* rel *$\mathcal{O}p\, \partial_- W$ with fixed endpoints.*

Theorems 14.1 and 14.3 will be proved in Sections 14.2 and 14.3. They have the following analogues for deformations of flexible Weinstein *manifolds*, which are derived from the cobordism versions by induction over sublevel sets as in the proof of Theorem 8.16 (using Remark 11.25 as the starting point).

THEOREM 14.4. *Let $\mathfrak{W} = (V, \omega, X, \phi)$ be a flexible Weinstein manifold of dimension $2n > 4$. Let $\psi : V \to \mathbb{R}$ be a Morse function without critical points of index $> n$. Then there exists a homotopy $\mathfrak{W}_t = (V, \omega_t, X_t, \phi_t)$, $t \in [0,1]$, of flexible Weinstein structures such that $\mathfrak{W}_0 = \mathfrak{W}$ and $\phi_1 = \psi$.*

If the Morse functions ϕ and ψ agree outside a compact set, then the Weinstein homotopy $\mathfrak{W}_t$ can be chosen fixed outside a compact set.

THEOREM 14.5. *Let $\mathfrak{W}_0 = (\omega_0, X_0, \phi_0)$ and $\mathfrak{W}_1 = (\omega_1, X_1, \phi_1)$ be two flexible Weinstein structures on the same manifold V of dimension $2n > 4$. Let η_t be a homotopy of nondegenerate 2-forms on V connecting ω_0 and ω_1. Then there exists a homotopy $\mathfrak{W}_t = (\omega_t, X_t, \phi_t)$ of flexible Weinstein structures connecting $\mathfrak{W}_0$ and $\mathfrak{W}_1$ such that the paths ω_t and η_t of nondegenerate 2-forms are homotopic with fixed endpoints.*

REMARK 14.6. Theorems 14.1, 14.3, 14.4, and 14.5 remain true in dimension $2n = 4$ if we assume the existence of a Morse homotopy ϕ_t connecting ϕ and ψ (resp. ϕ_0 and ϕ_1) *without critical points of index* > 1, or without critical points of index > 2 in the case that $\partial_- W \neq \varnothing$ is overtwisted in Theorems 14.1 and 14.3.

Theorem 14.5 has the following consequence for symplectomorphisms of flexible Weinstein manifolds.

THEOREM 14.7. *Let $\mathfrak{W} = (V, \omega, X, \phi)$ be a flexible Weinstein manifold of dimension $2n > 4$, and $f : V \to V$ a diffeomorphism such that $f^*\omega$ is homotopic to ω as nondegenerate 2-forms. Then there exists a diffeotopy $f_t : V \to V$, $t \in [0,1]$, such that $f_0 = f$, and f_1 is an exact symplectomorphism of (V, ω).*

PROOF. By Theorem 14.5, there exists a Weinstein homotopy $\mathfrak{W}_t$ connecting $\mathfrak{W}_0 = \mathfrak{W}$ and $\mathfrak{W}_1 = f^*\mathfrak{W}$. Thus Corollary 11.21 provides a diffeotopy $h_t : V \to V$ such that $h_0 = \mathrm{Id}$ and $h_1^* f^* \lambda - \lambda$ is exact, where λ is the Liouville form of $\mathfrak{W}$. Now $f_t = f \circ h_t$ is the desired diffeotopy. □

REMARK 14.8. Even if $\mathfrak{W}$ is of finite type and $f = \mathrm{Id}$ outside a compact set, the diffeotopy f_t provided by Theorem 14.7 will in general *not* equal the identity outside a compact set.

14.2. Proof of the first Weinstein deformation theorem

By Corollary 9.52, any two Morse functions without critical points of index $> n$ on a cobordism of dimension $2n > 4$ can be connected by a Morse homotopy without critical points of index $> n$. Hence Theorem 14.1 is an immediate consequence the following

THEOREM 14.9. *Let $\mathfrak{W} = (W, \omega, X, \phi)$ be a flexible Weinstein cobordism of dimension $2n$. Let ϕ_t, $t \in [0,1]$, be a Morse homotopy without critical points of index $> n$ with $\phi_0 = \phi$ and $\phi_t = \phi$ near ∂W. In the case $2n = 4$ assume that either $\partial_- W$ is overtwisted, or ϕ_t has no critical points of index > 1. Then there exists a homotopy $\mathfrak{W}_t = (W, \omega_t, X_t, \phi_t)$, $t \in [0,1]$, of flexible Weinstein structures, starting at $\mathfrak{W}_0 = \mathfrak{W}$, which is fixed near $\partial_- W$ and fixed up to scaling near $\partial_+ W$.*

The proof of Theorem 14.9 is based on the following three lemmas.

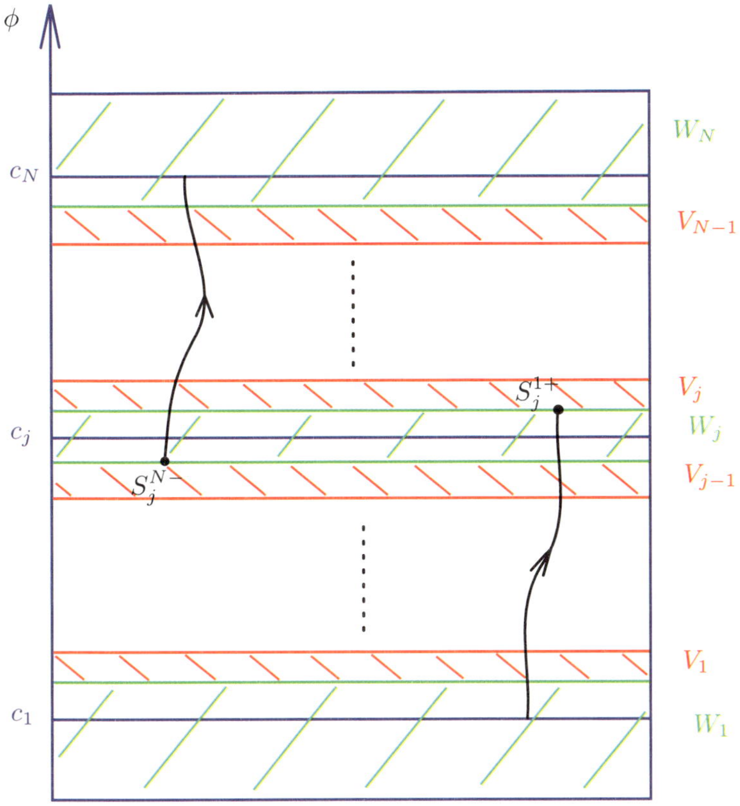

FIGURE 14.1. The partition of W into subcobordisms.

LEMMA 14.10. *Let $\mathfrak{W} = (W, \omega, X, \phi)$ be a flexible Weinstein cobordism and Y a gradient-like vector field for ϕ such that the Smale cobordism (W, Y, ϕ) is elementary. Then there exists a family X_t, $t \in [0,1]$, of gradient-like vector fields for ϕ and a family ω_t, $t \in [0, \frac{1}{2}]$, of symplectic forms on W such that*

- *$\mathfrak{W}_t = (W, \omega_t, X_t, \phi)$, $t \in [0, \frac{1}{2}]$, is a Weinstein homotopy with $\mathfrak{W}_0 = \mathfrak{W}$, fixed on $\mathcal{O}p\, \partial_- W$ and fixed up to scaling on $\mathcal{O}p\, \partial_+ W$;*
- *$X_1 = Y$ and the Smale cobordisms (W, X_t, ϕ), $t \in [\frac{1}{2}, 1]$, are elementary.*

PROOF. **Step 1.** Let $c_1 < \cdots < c_N$ be the critical values of the function ϕ. Set $c_0 := \phi|_{\partial_- W}$ and $c_{N+1} := \phi|_{\partial_+ W}$. Choose $\varepsilon \in (0, \min_{j=0,\dots,N} \frac{c_{j+1}-c_j}{2})$ and define

$$
\begin{aligned}
W_j &:= \{c_j - \varepsilon \le \phi \le c_j + \varepsilon\}, \quad j = 2, \dots, N-1, \\
W_1 &:= \{\phi \le c_1 + \varepsilon\}, \qquad W_N := \{\phi \ge c_N - \varepsilon\}, \\
V_j &:= \{c_j + \varepsilon \le \phi \le c_{j+1} - \varepsilon\}, \quad j = 1, \dots, N-1, \\
\Sigma_j^{\pm} &:= \{\phi = c_j \pm \varepsilon\}, \quad j = 1, \dots, N;
\end{aligned}
$$

see Figure 14.1.

Thus we have $\Sigma_j^+ = \partial_- V_j = \partial_+ W_j$ for $j = 1, \dots, N-1$ and $\Sigma_j^- = \partial_+ V_j = \partial_- W_j$ for $j = 2, \dots, N$. We denote by $\xi_j^{\pm}$ the contact structure induced by the Liouville form $i_X \omega$ on $\Sigma_j^{\pm}$, $j = 1, \dots, N$.

For $k \geq j$ we denote by S_j^{k-} the intersection of the union of the Y-stable manifolds of the critical points on level c_k with the hypersurface Σ_j^-. Similarly, for $i \leq j$ we denote by S_j^{i+} the intersection of the union of the Y-unstable manifolds of the critical points on level c_i with the hypersurface Σ_j^+; see Figure 14.1. Set

$$\mathbf{S}_j^- := \bigcup_{k \geq j} S_j^{k-}, \qquad \mathbf{S}_j^+ := \bigcup_{i \leq j} S_j^{i+}.$$

The assumption that the Smale cobordism (Y, ϕ) is elementary implies that $\mathbf{S}_j^\pm$ is a union of spheres in $\Sigma_j^\pm$.

Consider on $\bigcup_{j=1}^N W_j$ the gradient-like vector fields $Y_t := (1-t)Y + tX$, $t \in [0,1]$, for ϕ. Let us pick ε so small that for all $t \in [0,1]$ the Y_t-unstable spheres in Σ_j^+ of the critical points on level c_j do not intersect the Y-stable spheres in Σ_j^+ of any critical points on higher levels. By Lemma 9.8 we can extend the Y_t to gradient-like vector fields for ϕ on W such that $Y_0 = Y$ and $Y_t = Y$ outside $\mathcal{O}p \bigcup_{j=1}^N W_j$ for all $t \in [0,1]$. By Lemma 9.41, this can be done in such a way that the intersection of the Y_t-stable manifold of the critical point locus on level c_i with the hypersurface Σ_j^+ remains unchanged. This implies that the cobordisms (W, Y_t, ϕ) are elementary for all $t \in [0,1]$. After renaming Y_1 back to Y and shrinking the W_j, we may hence assume that $Y = X$ on $\mathcal{O}p \bigcup_{j=1}^N W_j$. Moreover, after modifying Y near ∂W we may assume that $Y = X$ on $\mathcal{O}p\, \partial W$.

We will construct the required homotopies X_t, $t \in [0,1]$, and ω_t, $t \in [0, \frac{1}{2}]$, separately on each V_j, $j = 1, \dots, N-1$, in such a way that X_t is fixed near ∂V_j for all $t \in [0,1]$ and ω_t is fixed up to scaling near ∂V_j for $t \in [0, \frac{1}{2}]$. This will allow us then to extend the homotopies X_t and ω_t to $\bigcup_{j=1}^N W_j$ as constant (resp. constant up to scaling).

Step 2. Consider V_j for $1 \leq j \leq N-1$. To simplify the notation, we denote the restriction of objects to V_j by the same symbol as the original objects, omitting the index j. Let us denote by $\mathcal{X}(V_j, \phi)$ the space of all gradient-like vector fields for ϕ on V_j that agree with X near ∂V_j. We connect X and Y by the path $Y_t := (1-t)X + tY$ in $\mathcal{X}(V_j, \phi)$.

Denote by $\Gamma_{Y_t} : \Sigma_{j+1}^- \to \Sigma_j^+$ the holonomy of the vector field Y_t on V_j and consider the isotopy $g_t := \Gamma_{Y_t}|_{\mathbf{S}_{j+1}^-} : \mathbf{S}_{j+1}^- \hookrightarrow \Sigma_j^+$. Suppose for the moment that $\mathbf{S}_{j+1}^- \subset \Sigma_{j+1}^-$ *is isotropic and loose* (this hypothesis will be satisfied below when we perform induction on descending values of j).

Since $\Gamma_{Y_0} = \Gamma_X$ is a contactomorphism, this implies that the embedding g_0 is loose isotropic. Hence, by the h-principles in Chapter 7 (Theorem 7.11 for the subcritical case, Theorem 7.19 for the Legendrian overtwisted case in dimension 4, and Theorem 7.25 in the Legendrian loose case in dimension $2n > 4$), the isotopy g_t can be C^0-approximated by an isotropic isotopy. More precisely, there exists a C^0-small diffeotopy $\delta_t : \Sigma_j^+ \to \Sigma_j^+$ with $\delta_0 = \mathrm{Id}$ such that $\delta_t \circ g_t$, $t \in [0,1]$, is loose isotropic with respect to the contact structure ξ_j^+.

The path Γ_{Y_t}, $t \in [0,1]$, in $\mathrm{Diff}(\Sigma_{j+1}^-, \Sigma_j^+)$ is homotopic with fixed endpoints to the concatenation of the paths $\delta_t \circ \Gamma_{Y_t}$ (from Γ_{Y_0} to $\delta_1 \circ \Gamma_{Y_1}$) and $\delta_t^{-1} \circ \delta_1 \circ \Gamma_{Y_1}$ (from $\delta_1 \circ \Gamma_{Y_1}$ to Γ_{Y_1}). Hence by Lemma 9.41 we find paths Y_t' and Y_t'', $t \in [0,1]$, in $\mathcal{X}(V_j, \phi)$ such that

- $Y'_0 = X$, $Y'_1 = Y''_0$ and $Y''_1 = Y$;
- $\Gamma_{Y'_t} = \delta_t \circ \Gamma_{Y_t}$ and $\Gamma_{Y''_t} = \delta_t^{-1} \circ \delta_1 \circ \Gamma_{Y_1}$, $t \in [0,1]$.

Note that $\Gamma_{Y'_t}|_{\mathbf{S}^-_{j+1}}$ is loose isotropic. Moreover, by choosing δ_t sufficiently C^0-small, we can ensure that $\Gamma_{Y''_t}(\mathbf{S}^-_{j+1}) \cap \mathbf{S}^+_j = \varnothing$ in Σ^+_j and $\Gamma_Y(\mathbf{S}^-_{j+1})$ is loose in $\Sigma^+_j \setminus \mathbf{S}^+_j$. We extend the vector fields Y'_t and Y''_t to W by setting $Y'_t := (1-t)X + tY$ and $Y''_t := Y$ on $W \setminus V_j$. The preceding discussion shows that the cobordisms (W, Y''_t, ϕ) are elementary for all $t \in [0,1]$. Hence it is sufficient to prove the lemma with the original vector field Y replaced by $Y'_1 = Y''_0$. To simplify the notation, we rename Y'_1 to Y and the homotopy Y'_t to Y_t. The new homotopy now has the property that the isotopy $\Gamma_{Y_t}|_{\mathbf{S}^-_{j+1}} : \mathbf{S}^-_{j+1} \hookrightarrow \Sigma^+_j$ is loose isotropic and $\Gamma_Y(\mathbf{S}^-_{j+1})$ is loose in $\Sigma^+_j \setminus \mathbf{S}^+_j$. So the image of $\Gamma_Y(\mathbf{S}^-_{j+1})$ under the holonomy of the elementary Weinstein cobordism $(W_j, \omega, X = Y, \phi)$ is loose isotropic in Σ^-_j. Since the union S^-_j of the stable spheres of (W_j, Y) are loose by the flexibility hypothesis on $\mathfrak{W}$, this implies that $\mathbf{S}^-_j \subset \Sigma^-_j$ is loose isotropic.

Now we perform this construction inductively in *descending* order over V_j for $j = N-1, N-2, \dots, 1$, always renaming the new vector fields back to Y. The resulting vector field Y is then connected to X by a homotopy Y_t such that the manifolds $\mathbf{S}^-_{j+1} \subset \Sigma^-_{j+1}$ and the isotopies $\Gamma_{Y_t}|_{\mathbf{S}^-_{j+1}} : \mathbf{S}^-_{j+1} \hookrightarrow \Sigma^+_j$, $t \in [0,1]$, are loose isotropic for all $j = 1, \dots, N-1$.

Step 3. Let Y and Y_t be as constructed in Step 2. Now we construct the desired homotopies X_t and ω_t separately on each V_j, $j = 1, \dots, N-1$, keeping them fixed near ∂V_j. We keep the notation from Step 2. By the contact isotopy extension theorem (Proposition 6.24), we can extend the isotropic isotopy $\Gamma_{Y_t}|_{\mathbf{S}^-_{j+1}} : \mathbf{S}^-_{j+1} \hookrightarrow \Sigma^+_j$ to a contact isotopy $G_t : (\Sigma^-_{j+1}, \xi^-_{j+1}) \to (\Sigma^+_j, \xi^+_j)$ starting at $G_0 = \Gamma_{Y_0} = \Gamma_X$. By Lemma 12.5 we find a Weinstein homotopy $\widetilde{\mathfrak{W}}_t = (V_j, \widetilde{\omega}_t, \widetilde{X}_t, \phi)$ beginning at $\widetilde{\mathfrak{W}}_0 = \mathfrak{W}$ with holonomy $\Gamma_{\widetilde{\mathfrak{W}}_t} = G_t$ for all $t \in [0,1]$. Now Lemma 9.42 provides a path $X_t \in \mathcal{X}(V_j, \phi)$ such that

(i) $X_t = \widetilde{X}_{2t}$ for $t \in [0, \frac{1}{2}]$;
(ii) $X_1 = Y_1 = Y$;
(iii) $\Gamma_{X_t}(\mathbf{S}^-_{j+1}) = \Gamma_Y(\mathbf{S}^-_{j+1})$ for $t \in [\frac{1}{2}, 1]$.

Over the interval $[0, \frac{1}{2}]$ the Smale homotopy $\mathfrak{S}_t = (V_j, X_t, \phi)$ can be lifted to the Weinstein homotopy $\mathfrak{W}_t = (V_j, \omega_t, X_t, \phi)$, where $\omega_t := \widetilde{\omega}_{2t}$.

Condition (iii) implies that $\Gamma_{X_t}(\mathbf{S}^-_{j+1}) \cap \mathbf{S}^+_j = \varnothing$ for all $t \in [\frac{1}{2}, 1]$, so the resulting Smale homotopy on W is elementary over the interval $[\frac{1}{2}, 1]$. □

LEMMA 14.11. *Let $\mathfrak{W} = (W, \omega, X, \phi)$ be a flexible Weinstein cobordism and Y a gradient-like vector field for ϕ. Suppose that the function ϕ has exactly two critical points transversely connected by a unique Y-trajectory. Then there exists a family X_t, $t \in [0,1]$, of gradient-like vector fields for ϕ and a family ω_t, $t \in [0, \frac{1}{2}]$, of symplectic forms on W such that*

- *$\mathfrak{W}_t = (W, \omega_t, X_t, \phi)$, $t \in [0, \frac{1}{2}]$, is a homotopy with $\mathfrak{W}_0 = \mathfrak{W}$, fixed on $\mathcal{O}p\, \partial_- W$ and fixed up to scaling on $\mathcal{O}p\, \partial_+ W$;*
- *$X_1 = Y$ and for $t \in [\frac{1}{2}, 1]$ the critical points of the function ϕ are connected by a unique X_t-trajectory.*

PROOF. Let us denote the critical points of the function ϕ by p_1 and p_2 and the corresponding critical values by $c_1 < c_2$. As in the proof of Lemma 14.10, for sufficiently small $\varepsilon > 0$, we split the cobordism W into three parts:

$$W_1 := \{\phi \le c_1 + \varepsilon\}, \quad V := \{c_1 + \varepsilon \le \phi \le c_2 - \varepsilon\}, \quad W_2 := \{\phi \ge c_2 - \varepsilon\}.$$

Arguing as in Step 1 of the proof of Lemma 14.10, we reduce to the case that $Y = X$ on $\mathcal{O}p\,(W_1 \cup W_2)$.

On V consider the gradient-like vector fields $Y_t := (1-t)X + tY$ for ϕ. Let $\Sigma := \{\phi = c_1 + \varepsilon\} = \partial_- V$. Denote by $S_t \subset \Sigma$ the Y_t-stable sphere of p_2 and by $S^+ \subset \Sigma$ the Y-unstable sphere of p_1. Note that S^+ is coisotropic, S_0 is isotropic, and S_1 intersects S^+ transversely in a unique point q. We deform S_1 to S_1' by a C^0-small deformation, keeping the unique transverse intersection point q with S^+, such that S_1' is isotropic near q. Connect S_0 to S_1' by an isotopy S_t' which is C^0-close to S_t. Due to the flexibility hypothesis on $\mathfrak{W}$, the isotropic sphere $S_0' = S_0$ is loose. Hence by Theorems 7.11, 7.19, and 7.25 we can C^0-approximate S_t' by an isotropic isotopy $\widetilde{S}_t$ such that $\widetilde{S}_0 = S_0' = S_0$, and $\widetilde{S}_1$ coincides with S_1' near q. In particular, $\widetilde{S}_1$ has q as the unique transverse intersection point with S^+. Arguing as in Steps 2 and 3 of the proof of Lemma 14.10, we now construct a Weinstein homotopy $\mathfrak{W}_t = (V, \omega_t, X_t, \phi)$, $t \in [0, \frac{1}{2}]$, fixed near $\partial_- V$ and fixed up to scaling near $\partial_+ V$, and Smale cobordisms (V, X_t, ϕ), $t \in [\frac{1}{2}, 1]$, fixed near ∂V, such that

- $\mathfrak{W}_0 = \mathfrak{W}|_V$ and $X_1 = Y|_V$;
- the X_t-stable sphere of p_2 in Σ equals $\widetilde{S}_{2t}$ for $t \in [0, \frac{1}{2}]$, and $\widetilde{S}_1$ for $t \in [\frac{1}{2}, 1]$.

In particular, for $t \in [\frac{1}{2}, 1]$ the X_t-stable sphere of p_2 in Σ intersects S^+ transversely in the unique point q, so the two critical points p_1, p_2 are connected by a unique X_t-trajectory for $t \in [\frac{1}{2}, 1]$. □

The following lemma will serve as induction step in proving Theorem 14.9.

LEMMA 14.12. *Let $\mathfrak{W} = (W, \omega, X, \phi)$ be a flexible Weinstein cobordism of dimension $2n$. Let $\mathfrak{S}_t = (W, Y_t, \phi_t)$, $t \in [0, 1]$, be an* elementary *Smale homotopy without critical points of index $> n$ such that $\phi_0 = \phi$ on W and $\phi_t = \phi$ near ∂W (but not necessarily $Y_0 = X$!). If $2n = 4$ and $\mathfrak{S}_t$ is of Type IIb assume that either $\partial_- W$ is overtwisted, or ϕ_t has no critical points of index > 1. Then there exists a homotopy $\mathfrak{W}_t = (W, \omega_t, X_t, \phi_t)$, $t \in [0, 1]$, of flexible Weinstein structures, starting at $\mathfrak{W}_0 = \mathfrak{W}$, which is fixed near $\partial_- W$ and fixed up to scaling near $\partial_+ W$.*

PROOF. **Type I.** Consider first the case when the homotopy $\mathfrak{S}_t$ is elementary of Type I. We point out that (W, X, ϕ) need not be elementary. To remedy this, we apply Lemma 14.10 to construct families X_t and ω_t such that

- $\mathfrak{W}_t = (W, \omega_t, X_t, \phi)$, $t \in [0, \frac{1}{2}]$, is a Weinstein homotopy with $\mathfrak{W}_0 = \mathfrak{W}$, fixed on $\mathcal{O}p\,\partial_- W$ and fixed up to scaling on $\mathcal{O}p\,\partial_+ W$;
- $X_1 = Y_0$ and the Smale cobordisms (W, X_t, ϕ), $t \in [\frac{1}{2}, 1]$, are elementary.

Thus it is sufficient to prove the lemma for the Weinstein cobordism $(\omega_{\frac{1}{2}}, X_{\frac{1}{2}}, \phi)$ instead of $\mathfrak{W}$, and the concatenation of the Smale homotopies $(X_t, \phi)_{t \in [\frac{1}{2}, 1]}$ and $(Y_t, \phi_t)_{t \in [0,1]}$ instead of (Y_t, ϕ_t). To simplify the notation, we rename the new Weinstein cobordism and Smale homotopy back to $\mathfrak{W} = (\omega, X, \phi)$ and (Y_t, ϕ_t). So in the new notation we now have $X = Y_0$.

According to Lemma 9.39 there exists a family $\widetilde{\phi}_t$, $t \in [0,1]$, of Lyapunov functions for X with the same profile as the family ϕ_t, and such that $\widetilde{\phi}_0 = \phi$ and $\widetilde{\phi}_t = \phi_t$ on $\mathcal{O}p\,\partial W$. Then Lemma 9.38 provides a diffeotopy $h_t : W \to W$, $t \in [0,1]$, such that $h_0 = \mathrm{Id}$, $h_t|_{\mathcal{O}p\,\partial W} = \mathrm{Id}$, and $\phi_t = \widetilde{\phi}_t \circ h_t$ for all $t \in [0,1]$. Thus the Weinstein homotopy $(W, \omega_t = h_t^*\omega, X_t = h_t^*X, \phi_t = h_t^*\widetilde{\phi}_t)$, $t \in [0,1]$, has the desired properties. It is flexible because the X_t-stable spheres in $\partial_- W$ are loose for $t = 0$ and moved by an isotropic isotopy, so they remain loose for all $t \in [0,1]$.

Type IId. Suppose now that the homotopy $\mathfrak{S}_t$ is of Type IId. Let $t_0 \in [0,1]$ be the parameter value for which the function ϕ_t has a death-type critical point. In this case the function ϕ has exactly two critical points p and q connected by a unique Y_0-trajectory. Arguing as in the Type I case, using Lemma 14.11 instead of Lemma 14.10, we can again reduce to the case that $X = Y_0$.

Then Proposition 12.22 provides an elementary Weinstein homotopy $(W, \omega, \widetilde{X}_t, \widetilde{\phi}_t)$ of Type IId starting from $\mathfrak{W}$ and killing the critical points p and q at time t_0. One can also arrange that $(\widetilde{X}_t, \widetilde{\phi}_t)$ coincides with (X, ϕ) on $\mathcal{O}p\,\partial W$, and (by composing $\widetilde{\phi}_t$ with suitable functions $\mathbb{R} \to \mathbb{R}$) that the homotopies $\widetilde{\phi}_t$ and ϕ_t have equal profiles. Then Lemma 9.38 provides a diffeotopy $h_t : W \to W$, $t \in [0,1]$, such that $h_0 = \mathrm{Id}$, $h_t|_{\mathcal{O}p\,\partial W} = \mathrm{Id}$, and $\phi_t = \widetilde{\phi}_t \circ h_t$ for all $t \in [0,1]$. Thus the Weinstein homotopy $(W, \omega_t = h_t^*\omega, X_t = h_t^*\widetilde{X}, \phi_t = h_t^*\widetilde{\phi}_t)$, $t \in [0,1]$, has the desired properties. It is flexible because the intersections of the X_t-stable manifolds with regular level sets remain loose for $t \in [0, t_0]$ and there are no critical points for $t > t_0$.

Type IIb. The argument in the case of Type IIb is similar, except that we use Proposition 12.21 instead of Proposition 12.22 and we do not need a preliminary homotopy. However, the flexibility of $\mathfrak{W}_t$ for $t \geq t_0$ requires an additional argument.

Consider first the case $2n > 4$. Suppose ϕ_1 has critical points p and q of index n and $n-1$, respectively (if they have smaller indices flexibility is automatic). Then the closure Δ of the X_1-stable manifold of the point p intersects $\partial_- W$ along a Legendrian disc $\partial_-\Delta$; see Figure 9.5. The boundary S_q^- of this disc is the intersection with $\partial_- W$ of the X_1-stable manifold D_q^- of q. According to Remark 7.22(2) all Legendrian discs are loose, or more precisely, $\partial_-\Delta \setminus S_q^-$ is loose in $\partial_- W \setminus S_q^-$. Let c be a regular value of ϕ_1 which separates $\phi_1(q)$ and $\phi_1(p)$ and consider the level set $\Sigma := \{\phi_1 = c\}$. Flowing along X_1-trajectories defines a contactomorphism $\partial_- W \setminus S_q^- \to \Sigma \setminus D_q^+$ mapping $\partial_-\Delta \setminus S_q^-$ onto $\Delta \cap \Sigma \setminus \{r\}$, where r is the unique intersection point of Δ and the X_1-unstable manifold D_q^+ in the level set Σ. It follows that $\Delta \cap \Sigma \setminus \{r\}$ is loose in $\Sigma \setminus \{r\}$, and hence $\Delta \cap \Sigma$ is loose in Σ. This proves flexibility of $\mathfrak{W}_1$, and thus of $\mathfrak{W}_t$ for $t \geq t_0$.

Finally, consider the case $2n = 4$. If the critical points have indices 1 and 0, flexibility is automatic. If they have indices 2 and 1 and $\partial_- W$ is overtwisted, we can arrange that $\partial_-\Delta \subset \partial_- W$ (in the notation above) has an overtwisted disc in its complement, hence so does the intersection of Δ with the regular level set $\{\phi = c\}$. □

PROOF OF THEOREM 14.9. Let us pick gradient-like vector fields Y_t for ϕ_t with $Y_0 = X$ and $Y_t = X$ near ∂W to get a Smale homotopy $\mathfrak{S}_t = (W, Y_t, \phi_t)$, $t \in [0,1]$. By Lemma 9.37 we find an admissible partition for the Smale homotopy $\mathfrak{S}_t$. Thus we get a sequence $0 = t_0 < t_1 < \cdots < t_p = 1$ of parameter values and smooth families of partitions

$$W = \bigcup_{j=1}^{N_k} W_j^k(t), \qquad W_j^k(t) := \{c_{j-1}^k(t) \le \phi_t \le c_j^k(t)\}, \qquad t \in [t_{k-1}, t_k],$$

such that each Smale homotopy

$$\mathfrak{S}_j^k := \Big(W_j^k(t), Y_t|_{W_j^k(t)}, \phi_t|_{W_j^k(t)}\Big)_{t\in[t_{k-1},t_k]}$$

is elementary. We will construct the Weinstein homotopy (ω_t, X_t, ϕ_t) on the cobordisms $\bigcup_{t\in[t_{k-1},t_k]} W_j^k(t)$ inductively over $k = 1, \dots, p$, and for fixed k over $j = 1, \dots, N_k$.

Suppose the required Weinstein homotopy is already constructed on W for $t \le t_{k-1}$. To simplify the notation we rename $\phi_{t_{k-1}}$ to ϕ, the vector fields X_{t_k} and Y_{t_k} to X and Y, and the symplectic form $\omega_{t_{k-1}}$ to ω. We also write N instead of N_k, W_j and $W_j(t)$ instead of $W_j^k(t_{k-1})$ and $W_j^k(t)$, and replace the interval $[t_{k-1}, t_k]$ by $[0, 1]$.

There exists a diffeotopy $f_t : W \to W$, fixed on $\mathcal{O}p\,\partial W$, with $f_0 = \mathrm{Id}$ and such that $f_t(W_j) = W_j(t)$ for all $t \in [0, 1]$. Moreover, we can choose f_t and a diffeotopy $g_t : \mathbb{R} \to \mathbb{R}$ with $g_0 = \mathrm{Id}$ such that the function $\widehat{\phi}_t := g_t \circ \phi_t \circ f_t$ coincides with ϕ on $\mathcal{O}p\,\partial W_j$ for all $t \in [0, 1]$, $j = 1, \dots, N$. Set $\widehat{Y}_t := f_t^* Y_t$. So we have a flexible Weinstein cobordism $\mathfrak{W} = (W = \bigcup_{j=1}^N W_j, \omega, X, \phi = \widehat{\phi}_0)$ and a Smale homotopy $(\widehat{Y}_t, \widehat{\phi}_t)$, $t \in [0, 1]$, whose restriction to each W_j is elementary. (But the restriction of $\mathfrak{W}$ to W_j need not be elementary.)

Now we apply Lemma 14.12 inductively for $j = 1, \dots, N$ to construct Weinstein homotopies $\widehat{\mathfrak{W}}_t^j = (W_j, \widehat{\omega}_t, \widehat{X}_t, \widehat{\phi}_t)$, fixed near $\partial_- W_j$ and fixed up to scaling near $\partial_+ W_j$, with $\widehat{\mathfrak{W}}_0^j = \mathfrak{W}|_{W_j}$. Thus the $\mathfrak{W}_t^j$ fit together to form a Weinstein homotopy $\widehat{\mathfrak{W}}_t = (\widehat{\omega}_t, \widehat{X}_t, \widehat{\phi}_t)$ on W. The desired Weinstein homotopy on W is now given by

$$\mathfrak{W}_t := \Big(f_{t*}\widehat{\omega}_t, f_{t*}\widehat{X}_t, g_t^{-1} \circ \widehat{\phi}_t \circ f_t^{-1}\Big). \qquad \square$$

14.3. Proof of the second Weinstein deformation theorem

Theorem 14.3 is an immediate consequence of Corollary 9.52 and

THEOREM 14.13. *Let $\mathfrak{W}_0 = (\omega_0, X_0, \phi_0)$ and $\mathfrak{W}_1 = (\omega_1, X_1, \phi_1)$ be two flexible Weinstein structures on a cobordism W of dimension $2n$. Let ϕ_t, $t \in [0, 1]$, be a Morse homotopy without critical points of index $> n$ connecting ϕ_0 and ϕ_1. In the case $2n = 4$, assume that either $\partial_- W$ is overtwisted or ϕ_t has no critical points of index > 1. Let η_t, $t \in [0, 1]$, be a homotopy of nondegenerate (not necessarily closed) 2-forms connecting ω_0 and ω_1 such that (η_t, Y_t, ϕ_t) is Weinstein near $\partial_- W$ for a homotopy of vector fields Y_t on $\mathcal{O}p\,\partial_- W$ connecting X_0 and X_1.*

Then $\mathfrak{W}_0$ and $\mathfrak{W}_1$ can be connected by a homotopy $\mathfrak{W}_t = (\omega_t, X_t, \phi_t)$, $t \in [0, 1]$, of flexible Weinstein structures, agreeing with (η_t, Y_t, ϕ_t) on $\mathcal{O}p\,\partial_- W$, such that the paths of nondegenerate 2-forms $t \mapsto \eta_t$ and $t \mapsto \omega_t$, $t \in [0, 1]$, are homotopic rel $\mathcal{O}p\,\partial_- W$ with fixed endpoints.

Let us extend the vector fields Y_t from $\mathcal{O}p\,\partial_- W$ to a path of gradient-like vector fields for ϕ_t on W connecting X_0 and X_1. We will deduce Theorem 14.13 from Theorem 14.9 and the following special case, which is just a 1-parametric version of the Weinstein Existence Theorem 13.1.

LEMMA 14.14. *Theorem 14.13 holds under the additional hypothesis that* $\phi_t = \phi$ *is independent of* $t \in [0,1]$ *and the Smale homotopy* (W, Y_t, ϕ) *is elementary.*

PROOF. The proof is just a 1-parametric version of the proof of Lemma 13.3, using Theorem 7.36 and Lemma 12.17 instead of Theorem 7.34 and Lemma 12.16. □

LEMMA 14.15. *Theorem 14.13 holds under the additional hypothesis that* $\phi_t = \phi$ *is independent of* $t \in [0,1]$.

PROOF. Let us pick regular values

$$\phi|_{\partial_- W} = c_0 < c_1 < \cdots < c_N = \phi|_{\partial_+ W}$$

such that each (c_{k-1}, c_k) contains at most one critical value. Then the restriction of the homotopy (Y_t, ϕ), $t \in [0,1]$, to each cobordism $W^k := \{c_{k-1} \leq \phi \leq c_k\}$ is elementary.

We apply Lemma 14.14 to the restriction of the homotopy (η_t, Y_t, ϕ) to W^1. Hence $\mathfrak{W}_0|_{W^1}$ and $\mathfrak{W}_1|_{W^1}$ are connected by a homotopy $\mathfrak{W}_t^1 = (\omega_t^1, X_t^1, \phi)$, $t \in [0,1]$, of flexible Weinstein structures on W^1, agreeing with (η_t, Y_t, ϕ_t) on $\mathcal{O}p\, \partial_- W$, such that the paths $t \mapsto \omega_t^1$ and $t \mapsto \eta_t$, $t \in [0,1]$, of nondegenerate 2-forms on W^1 are connected by a homotopy η_t^s, $s, t \in [0,1]$ rel $\mathcal{O}p\, \partial_- W$ with fixed endpoints. We use the homotopy ω_t^s to extend ω_t^1 to nondegenerate 2-forms η_t^1 on W such that $\eta_0^1 = \omega_0$, $\eta_1^1 = \omega_1$, $\eta_t^1 = \eta_t$ outside a neighborhood of W^1, and the paths $t \mapsto \eta_t^1$ and $t \mapsto \eta_t$, $t \in [0,1]$, of nondegenerate 2-forms on W are homotopic rel $\mathcal{O}p\, \partial_- W$ with fixed endpoints. By Lemma 9.8, we can extend X_t^1 to gradient-like vector fields Y_t^1 for ϕ on W such that $Y_0^1 = X_0$ and $Y_1^1 = X_1$. Now we can apply Lemma 14.14 to the restriction of the homotopy (η_t^1, Y_t^1, ϕ) to the elementary cobordism W^2 and continue inductively to construct homotopies (η_t^k, Y_t^k, ϕ) on W which are Weinstein on W^k, so (η_t^N, Y_t^N, ϕ) is the desired Weinstein homotopy. Note that (η_t^N, Y_t^N, ϕ) is flexible because its restriction to each W^k is flexible. □

PROOF OF THEOREM 14.13. Let us reparametrize the given homotopy (η_t, Y_t, ϕ_t), $t \in [0,1]$, to make it constant for $t \in [\frac{1}{2}, 1]$. After pulling back (η_t, Y_t, ϕ_t) by a diffeotopy and target reparametrizing ϕ_t, we may further assume that ϕ_t is independent of t on $\mathcal{O}p\, \partial W$.

By Theorem 14.9, $\mathfrak{W}_0$ can be extended to a homotopy $\mathfrak{W}_t = (\omega_t, X_t, \phi_t)$, $t \in [0, \frac{1}{2}]$, of flexible Weinstein structures on W, fixed on $\mathcal{O}p\, \partial_- W$. We modify $\mathfrak{W}_t$ using Lemma 12.7(i) to make it agree with (η_t, Y_t, ϕ_t) on $\mathcal{O}p\, \partial_- W$. Note that $\mathfrak{W}_{\frac{1}{2}}$ and $\mathfrak{W}_1$ share the same function $\phi_{\frac{1}{2}} = \phi_1$. We connect $\omega_{\frac{1}{2}}$ and ω_1 by a path η_t', $t \in [\frac{1}{2}, 1]$ of nondegenerate 2-forms by following the path ω_t backward and then η_t forward. Since $\omega_t = \eta_t$ on $\mathcal{O}p\, \partial_- W$ for $t \in [0, \frac{1}{2}]$, we can modify the path η_t' to make it constant equal to $\omega_{\frac{1}{2}} = \omega_1$ on $\mathcal{O}p\, \partial_- W$. By Lemma 9.8, we can connect $X_{\frac{1}{2}}$ and X_1 by a homotopy Y_t', $t \in [\frac{1}{2}, 1]$, of gradient-like vector fields for ϕ_1 which agree with $X_{\frac{1}{2}} = X_1$ on $\mathcal{O}p\, \partial_- W$.

So we can apply Lemma 14.15 to the homotopy (η_t', Y_t', ϕ_1), $t \in [\frac{1}{2}, 1]$. Hence $\mathfrak{W}_{\frac{1}{2}}$ and $\mathfrak{W}_1$ are connected by a homotopy $\mathfrak{W}_t = (\omega_t, X_t, \phi_1)$, $t \in [\frac{1}{2}, 1]$, of flexible Weinstein structures, agreeing with (ω_1, X_1, ϕ_1) on $\mathcal{O}p\, \partial_- W$, such that the paths of nondegenerate 2-forms $t \mapsto \omega_t$ and $t \mapsto \eta_t'$, $t \in [\frac{1}{2}, 1]$, are homotopic rel $\mathcal{O}p\, \partial_- W$ with fixed endpoints. It follows from the definition of η_t' that the concatenated path ω_t, $t \in [0,1]$, is homotopic to η_t, $t \in [0,1]$. Thus the concatenated Weinstein homotopy $\mathfrak{W}_t$, $t \in [0,1]$, has the desired properties. □

14.4. Subcritical Weinstein manifolds are split

In this section we prove the following theorem which asserts that subcritical Weinstein manifolds split as a product with $\mathbb{C}$ (see Section 11.8 for the definitions). We call two Weinstein manifolds (or cobordisms) $\mathfrak{W} = (V, \omega, X, \phi)$ and $\mathfrak{W}' = (V', \omega', X', \phi')$ *deformation equivalent* if there exists a diffeomorphism $h : V \to V'$ such that $h_*\mathfrak{W}'$ is homotopic to $\mathfrak{W}$. See Chapter 16 for more discussion of this notion.

THEOREM 14.16 ([**33**]). *Every subcritical Weinstein manifold (V, ω, X, ϕ) of dimension $2n$ is deformation equivalent to the stabilization of a Weinstein manifold (V', ω', X', ϕ') of dimension $2n - 2$.*

REMARK 14.17. Theorem 14.16 implies by induction: If a $2n$-dimensional Weinstein manifold (V, ω, X, ϕ) is *k-subcritical*, i.e., all critical points of ϕ have index $\leq n - k$, then it is deformation equivalent to the k-fold stabilization of a Weinstein manifold (V', ω', X', ϕ') of dimension $2(n - k)$.

EXAMPLE 14.18. Consider an oriented real plane bundle $V' \to S^2$ of even Euler number $e \in 2\mathbb{Z}$. Then $V := V' \times \mathbb{C} \cong S^2 \times \mathbb{C}^2$ carries a subcritical Weinstein structure with trivial first Chern class which is unique up to homotopy. On the other hand, V' carries a Weinstein structure (which can be chosen to have trivial first Chern class) if and only if $e \leq -2$ ([**125**], see also Section 16.3 below). This shows that not every smooth splitting $V = V' \times \mathbb{C}$ gives rise to a Weinstein splitting, and the diffeomorphism type of the manifold V' in a Weinstein splitting of V is not uniquely determined.

The proof of Theorem 14.16 uses the following lemma.

LEMMA 14.19. *Let V be a smooth orientable manifold of dimension $2n$ and $V' \subset V$ a codimension 2 submanifold with trivial normal bundle. Let $\phi : V \to \mathbb{R}$ be an exhausting Morse function and X a gradient-like vector field for ϕ such that the vector field X is tangent to V', and all critical points of ϕ and their stable manifolds are contained in V'.*

Then there exists a diffeomorphism $f : V' \times \mathbb{R}^2 \to V$ such that $f(x', 0) = x'$ for all $x' \in V'$, and $\phi \circ f(x', u) = \phi(x') + |u|^2$ for all $x' \in V'$, $u \in \mathbb{R}^2$.

PROOF. Since V' has trivial normal bundle, we can find an embedding $V' \times \mathbb{R}^2 \hookrightarrow V$ mapping $(x', 0)$ to x' for all $x' \in V'$. We will view $V' \times \mathbb{R}^2$ as a subset of V via this embedding. We can choose the embedding such the function $\phi' : V' \times \mathbb{R}^2 \to \mathbb{R}$, $\phi'(x', u) := \phi(x') + |u|^2$ satisfies $\phi'(x', u) \geq \phi(x', u)$ for all (x', u). Fix the gradient-like vector field $X' := X|_{V'} + u\partial_u$ for ϕ' on $V' \times \mathbb{R}^2$.

By assumption, the functions $\phi : V \to \mathbb{R}$ and $\phi' : V' \times \mathbb{R}^2 \to \mathbb{R}$ have the same critical points and stable manifolds (with respect to X, resp. X'), and the same values at critical points. Pick an unbounded sequence of regular values $c_0 < \min \phi < c_1 < \cdots$ such that the Smale cobordisms $(W_j := \{c_{j-1} \leq \phi \leq c_j\}, X, \phi)$ and $(W_j' := \{c_{j-1} \leq \phi' \leq c_j\}, X', \phi')$ are elementary. Since $\phi' \geq \phi$ we have $V_j' := \bigcup_{i=1}^j W_j' \subset V_j := \bigcup_{i=1}^j W_j$ for all j.

We will inductively modify the embedding $V' \times \mathbb{R}^2 \hookrightarrow V$ such that $\phi' = \phi$ on $V_j' = V_j$. For $j = 1$ this can be done by the Morse Lemma 9.1 and Remark 9.2.

Now suppose we already have $\phi' = \phi$ on $V'_{j-1} = V_{j-1}$. Applying Lemma 9.29 to the cobordisms $W'_j \subset W_j$, we find an isotopy $h_t : W_j \hookrightarrow W_j$, $t \in [0,1]$, with $h_0 = \mathrm{Id}$ and $h_t = \mathrm{Id}$ on $\mathcal{O}p\,(\partial_- W_j)$, such that $h_1(W_j) = W'_j$ and $\phi = \phi' \circ h_1$ on W_j. Moreover, since $X = X'$ on $V' \cap W_j$, the proof of Lemma 9.29 (which maps X-trajectories to X'-trajectories) yields $h_t = \mathrm{Id}$ on $V' \cap W_j$. We extend h_t to diffeomorphisms $h_t : V \to V$ which equal the identity on V_{j-1} and outside a neighborhood of V_j. Then $f_j := h_1^{-1}|_{V' \times \mathbb{R}^2} : V' \times \mathbb{R}^2 \hookrightarrow V$ is the desired new embedding satisfying $\phi' = \phi \circ f_j$ on $V_j = f_j(V'_j)$.

Since the sequence of embeddings stabilizes on each V'_j, it converges as $j \to \infty$ to a diffeomorphism $f : V' \times \mathbb{R}^2 \to V$ satisfying $f(x', 0) = x'$ for all $x' \in V'$ and $\phi \circ f = \phi'$. □

Using this lemma, we now prove the following purely topological analogue of Theorem 14.16.

PROPOSITION 14.20. *Let V be a smooth orientable manifold of dimension $2n$ which admits an exhausting Morse function ϕ without critical points of index $\geq n$. Then there exists a codimension 2 properly embedded submanifold $V' \subset V$ and a diffeomorphism $V' \times \mathbb{R}^2 \to V$ such that $f(x', 0) = x'$ for all $x' \in V'$, and $\phi \circ f(x', u) = \phi(x') + |u|^2$ for all $x' \in V'$, $u \in \mathbb{R}^2$.*

PROOF OF PROPOSITION 14.20. Let X_0 be any gradient-like vector field for the function ϕ. We slice V into elementary Smale cobordisms $(W_j := \{c_{j-1} \leq \phi \leq c_j\}, \phi|_{W_j}, X|_{W_j})$, $j \in \mathbb{N}$, where $c_0 < \min \phi < c_1 < \cdots$ are regular values of ϕ. We will inductively construct codimension 2 submanifolds $V'_j \subset V_j := \bigcup_{i=1}^{j} W_i$ and gradient-like vector fields X_j on V for ϕ satisfying the following conditions:

(i) $V'_j \subset V_j$ has trivial normal bundle;
(ii) the vector field X_j is tangent to V'_j;
(iii) all critical points of ϕ and their stable manifolds are contained in V'_j;
(iv) the pair $(\partial V_j, \partial V'_j)$ is $(n-2)$-connected;
(v) $V'_j \cap V_{j-1} = V'_{j-1}$ and $X_j|_{V_{j-1}} = X_{j-1}$ for all $j \geq 1$.

Then, by the last property, the V'_j and X_j stabilize on each compact set and thus converge to a codimension 2 submanifold $V' \subset V$ and a gradient-like vector field X for ϕ satisfying the hypotheses of Lemma 14.19, and the conclusion follows.

To simplify the notation, we will assume that each elementary cobordism W_j contains exactly one critical point p_j of the function ϕ.

By the Morse Lemma, the function ϕ has the form $\sum_{i=1}^n (x_i^2 + y_i^2)$ for some local coordinates near the minimum p_1. We deform X_0 to a gradient-like vector field X_1 for ϕ which agrees with X_0 outside V_1 and equals $\sum_{i=1}^n (x_i \frac{\partial}{\partial x_i} + y_i \frac{\partial}{\partial y_i})$ near p_1. We define V'_1 as the union of all trajectories of X_1 in V_1 which near p_1 lie in the subspace $\{x_n = y_n = 0\}$. Then V'_1 is a codimension 2 equatorial ball in the $2n$-dimensional ball V_1. In particular, the pair $(\partial V_1, \partial V'_1)$ is $(n-2)$-connected.

Now suppose we already have constructed $V'_{j-1} \subset V_{j-1}$ and X_{j-1} satisfying the above conditions. Pick a trivialization of the normal bundle of V'_{j-1} in V_{j-1}. Let k be the index of the critical point $p_j \in W_j$. By assumption, we have $k \leq n-1$. The stable manifold of p_j intersects ∂V_{j-1} along a sphere S of dimension $k-1 \leq n-2$. Since the pair $(\partial V_{j-1}, \partial V'_{j-1})$ is $(n-2)$-connected by the induction hypothesis, the sphere S is homotopic to a sphere in $\partial V'_{j-1}$. By a general position argument using the dimensional constraint $\dim S \leq n-2$ it is, in fact, *isotopic* to an embedded sphere $S' \subset \partial V'_{j-1}$.

In some local coordinate neighborhood U of p_j the function ϕ has the form $-\sum_1^k x_i^2 + \sum_{k+1}^n x_i^2 + \sum_1^n y_i^2$. Using Lemma 9.46, we deform X_{j-1} inside W_j to a gradient vector field X_j for ϕ which equals $-\sum_1^k x_i \frac{\partial}{\partial x_i} + \sum_{k+1}^n x_i \frac{\partial}{\partial x_i} + \sum_1^n y_i \frac{\partial}{\partial y_i}$ near p_j, and for which the stable disc of p_j is attached to ∂V_{j-1} along the sphere S'.

Next we adjust the normal framings. For small $\varepsilon > 0$ consider the local hypersurface $\Sigma_\varepsilon := \{\sum_1^k x_i^2 = \varepsilon,\ \sum_{k+1}^n x_i^2 + \sum_1^n y_i^2 < \varepsilon\}$. Following flow lines of X_j backward, we obtain an embedding $\Sigma \hookrightarrow \partial V_{j-1}$ mapping $S_\varepsilon := \{\sum_1^k x_i^2 = \varepsilon,\ \sum_{k+1}^n x_i^2 + \sum_1^n y_i^2 = 0\}$ onto S'. Its differential Φ maps the normal $(2n-k)$-frame $\partial_{x_{k+1}}, \dots, \partial_{x_n}, \partial_{y_1}, \dots, \partial_{y_n}$ to S_ε in Σ_ε onto a normal frame to S' in ∂V_{j-1}. Since by Corollary A.10 the homomorphism $\iota_* : \pi_{k-1}(SO_{2n-k-2}) \to \pi_{k-1}(SO_{2n-k})$ is surjective for $k \leq n-1$, we can deform Φ such that it maps $\partial_{x_{k+1}}, \dots, \partial_{x_{n-1}}, \partial_{y_1}, \dots, \partial_{y_{n-1}}$ to $T(\partial V'_{j-1})$ and $\partial_{x_n}, \partial_{y_n}$ to the given normal framing to V'_{j-1}. Note that here we are making a choice if the homomorphism ι_* is not injective: we can change the resulting normal framing of S' in $\partial V'_{j-1}$ by any element in $\ker \iota_*$.

After matching the normal framings, by further deforming X_j inside W_j, we can now arrange that the image U' of $U \cap \{x_n = y_n = 0\}$ under the backward flow of X_j intersects ∂V_{j-1} in $\partial V'_{j-1}$, and the given normal framing to V'_{j-1} extends over U'. Hence the union $V'_j \subset V_j$ of the unstable disc of p_j in W_j with the image of V'_{j-1} under the forward flow of X_j is a smooth codimension 2 submanifold satisfying conditions (i)–(iii) and (v).

It remains to verify that the pair $(\partial V_j, \partial V'_j)$ is $(n-2)$-connected. Recall that the pair $(\partial V_j, \partial V'_j)$ is obtained from the pair $(\partial V_{j-1}, \partial V'_{j-1})$ by a simultaneous surgery of index $k \leq n-1$ along a $(k-1)$-dimensional sphere $S' \subset \partial V'_{j-1}$. Pick a tubular neighborhood pair (N, N') for S in $(\partial V_{j-1}, \partial V'_{j-1})$ such that $N' = N \cap \partial V'_{j-1}$, thus (N, N') is diffeomorphic to $S^{k-1} \times (D^{2n-k}, D^{2n-k-2})$. We define the complement $(L, L') := (\partial V_{j-1}, \partial V'_{j-1}) \setminus \mathrm{Int}\,(N, N') \subset$ and denote its image in $(\partial V_j, \partial V'_j)$ under the forward flow of X_j also by (L, L'). Then $(M, M') := (\partial V_j, \partial V'_j) \setminus \mathrm{Int}\,(L, L')$ is a tubular neighborhood pair for the unstable sphere if p_j in $(\partial V_j, \partial V'_j)$ and thus diffeomorphic to $D^k \times (S^{2n-k-1}, S^{2n-k-3})$. Hence, $(M \cup L, M' \cup L') = (\partial V_j, \partial V'_j)$, and $(M \cap L, M' \cap L')$ is diffeomorphic to $S^{k-1} \times (S^{2n-k-1}, S^{2n-k-3})$. Next observe that the removal of (N, N') from $(\partial V_{j-1}, \partial V'_{j-1})$ did not affect its $(n-2)$-connectedness because the codimension of S' in ∂V_{j-1} and $\partial V'_{j-1}$ is $> n-1$. Therefore, the pair (L, L') is $(n-2)$-connected. Clearly, (M, M') and $(M \cap L, M' \cap L')$ are $(n-2)$-connected as well. Now the relative Mayer–Vietoris sequence (see [**91**]) implies that $(M \cup L, M' \cup L')$ has vanishing homology up to degree $n-2$, while van Kampen's theorem implies the triviality of $\pi_1(M \cup L, M' \cup L')$. Hence by the relative Hurewicz theorem (see [**91**]) we conclude that $(\partial V_j, \partial V'_j) = (M \cup L, M' \cup L')$ is $(n-2)$-connected.

This concludes the induction step and hence the proof of Proposition 14.20. □

The final ingredient in the proof of Theorem 14.16 is the following homotopical lemma.

LEMMA 14.21. *Let V be a noncompact $2n$-dimensional orientable manifold, and let $V' \subset V$ be a properly embedded orientable codimension 2 submanifold which* *has no closed components. Then every nondegenerate (not necessarily closed) 2-form ω on V is homotopic to a a nondegenerate 2-form ω' such that $\omega'|_{TV'}$ is nondegenerate.*

PROOF. A homotopically equivalent problem is making a codimension 2 orientable submanifold V' of an almost complex manifold an almost complex submanifold by deforming the almost complex structure. This is, in turn, equivalent to the problem of rotating the 2-dimensional normal bundle to V' in V to a complex 1-dimensional subbundle. Note that the assumption that V' has no closed components implies that it has the homotopy type of a $(2n-3)$-dimensional cell complex. Hence, arguing inductively over the cells of a cell decomposition of V', we come to the following problem. Given two vector fields e_1, e_2 normal to V' over a k-cell $D \subset V'$, $k \leq 2n-3$, such that $Je_1 = e_2$ over ∂D, we need to homotope the vector field e_2 relative to ∂D to the vector field Je_1, keeping it orthogonal to e_1. But the obstruction to doing this lies in $\pi_k S^{2n-2} = 0$ for $k \leq 2n-3$. □

PROOF OF THEOREM 14.16. Suppose first that $n > 3$. Let $\mathfrak{W} = (V, \omega, X, \phi)$ be a subcritical Weinstein manifold of dimension $2n$. According to Proposition 14.20, the manifold V is diffeomorphic to a product $V' \times \mathbb{R}^2$, where V' admits an exhausting function $\widetilde{\phi}$ without critical points of index $\geq n$. By Lemma 14.21 we find a nondegenerate 2-form ω' on $V = V' \times \mathbb{R}^2$ which is homotopic to ω through nondegenerate 2-forms such that $\omega'|_{TV'}$ is nondegenerate. Since $\dim V' = 2n-2 > 4$, we can use Theorem 13.2 to construct on V' a Weinstein structure $\widetilde{\mathfrak{W}} = (V', \widetilde{\omega}, \widetilde{X}, \widetilde{\phi})$ such that $\widetilde{\omega}$ and $\omega'|_{V'}$ are homotopic as nondegenerate 2-forms. Now Theorem 14.5 provides a Weinstein homotopy on V connecting the subcritical Weinstein structures $\mathfrak{W}$ and the stabilization of the Weinstein structure $\widetilde{\mathfrak{W}}$. Then, according to Proposition 11.8 the underlying symplectic manifolds are symplectomorphic.

The previous argument breaks down for $n = 3$ because then we cannot apply Theorem 13.2 to find a Weinstein structure on the 4-manifold V'. Indeed, the submanifold $V' \subset V$ provided by Proposition 14.20 may not carry a Weinstein structure; see Example 14.18. However, we can find in this case a *different* submanifold which carries a Weinstein structure. To do this, we inductively construct $V' \subset V$ *together with its Weinstein structure* as follows; see **[33]** for details. Consider the extension from V'_{j-1} to V'_j in the critical case $k = \mathrm{ind}(p_j) = 2$ in which we cannot apply Theorem 13.1 to extend the Weinstein structure to V'_j. The reason is that the stabilization construction in Proposition 7.12 only provides Legendrian regular homotopies of the attaching sphere S' with *positive* self-intersection index. Recall, however, that in the construction of V'_j in the proof of Proposition 14.20, we have the freedom to change the normal framing of S' in $\partial V'_{j-1}$ by an element in the kernel of the homomorphism $\iota_* : \pi_{k-1}(SO_{2n-k-2}) \to \pi_{k-1}(SO_{2n-k})$. In the present case $n = 3$, $k = 2$ this is the canonical projection $\pi_1(SO_2) \cong \mathbb{Z} \to \pi_1(SO_4) \cong \mathbb{Z}_2$ and thus $\ker \iota_* = 2\mathbb{Z}$. So we can change the normal framing of S', and hence the class of the formal Legendrian knot to which we want to apply Theorem 7.16, by an arbitrary even integer. By decreasing this class by a large even integer and then increasing it by stabilizations, we can thus make the obstruction in Theorem 7.16 vanish and continue as in the proof of Theorem 13.1 to extend the Weinstein structure over V'_j.

Note that when $n > 3$ the homotopy between the given and the split Weinstein structure can be made subcritical, while when $n = 3$ this cannot be guaranteed.

Finally, consider the case $n = 2$. Note that any two exhausting functions which have unique critical points of index 0 and the same number of critical points of index 1 are diffeomorphic. Hence, by pulling back the structure $\mathfrak{W}$ under this diffeomorphism, we can arrange that both Weinstein structures share the same Lyapunov function. According to Remark 14.6 this implies that the two Weinstein structures are homotopic. This concludes the proof of Theorem 14.16. □

14.5. Symplectic pseudo-isotopies

In this section we define and study symplectic analogues of the topological notions introduced in Section 9.10.

Let us fix a contact manifold (M^{2n-1}, ξ) and denote by $(SM, \lambda_{\rm st})$ its symplectization with its canonical Liouville structure $(\omega_{\rm st} = d\lambda_{\rm st}, X_{\rm st})$.

Any choice of a contact form α for ξ yields an identification of SM with $\mathbb{R}\times M$ and the Liouville structure $\lambda_{\rm st} = e^r\alpha$, $\omega_{\rm st} = d\lambda_{\rm st}$, $X_{\rm st} = \partial_r$. However, the following constructions do not require the choice of a contact form. We will refer to the two ends of SM as $\{\pm\infty\}\times M$.

We define the group of *symplectic pseudo-isotopies* of (M, ξ) as

$$\mathcal{P}(M,\xi) := \{F \in \mathrm{Diff}(SM) \mid F^*\omega_{\rm st} = \omega_{\rm st},\ F = \mathrm{Id} \text{ near } \{-\infty\}\times M,$$
$$F^*\lambda_{\rm st} = \lambda_{\rm st} \text{ near } \{+\infty\}\times M\}.$$

Moreover, we introduce the space

$$\mathcal{E}(M,\xi) := \{(\lambda,\phi) \text{ Weinstein structure on SM without critical points } \mid$$
$$d\lambda = \omega_{\rm st},\ (\lambda,\phi) = (\lambda_{\rm st},\phi_{\rm st}) \text{ outside a compact set}\}$$

and its image $\bar{\mathcal{E}}(M,\xi)$ under the projection $(\lambda,\phi)\mapsto\lambda$. We endow the spaces $\mathcal{P}(M,\xi)$, $\mathcal{E}(M,\xi)$, and $\bar{\mathcal{E}}(M,\xi)$ with the topology of uniform C^∞-convergence on $SM = \mathbb{R}\times M$ as explained in Section 9.10.

LEMMA 14.22. *The map*

$$\mathcal{E}(M,\xi) \to \bar{\mathcal{E}}(M,\xi), \qquad (\lambda,\phi)\mapsto\lambda,$$

is a homotopy equivalence and the map

$$\mathcal{P}(M,\xi) \to \bar{\mathcal{E}}(M,\xi), \qquad F\mapsto F^*\lambda_{\rm st},$$

is a homeomorphism.

PROOF. The first map defines a fibration whose fiber over λ is the contractible space of Lyapunov functions for X which are standard at infinity. The inverse of the second map associates to λ the unique $F\in\mathrm{Diff}(SM)$ satisfying $F_*X = X_{\rm st}$ on SM and $F = \mathrm{Id}$ near $\{-\infty\}\times M$ (which implies $F^*\lambda_{\rm st} = \lambda$ on SM). □

Since $F\in\mathcal{P}(M,\xi)$ satisfies $F^*\lambda_{\rm st} = \lambda_{\rm st}$ near $\{+\infty\}\times M$, it descends there to a contactomorphism $F_+ : M\to M$ (see Section 6.8). By construction, F_+ belongs to the group $\mathrm{Diff}_{\mathcal{P}}(M)$ of diffeomorphisms that are pseudo-isotopic to the identity, so it defines an element in

$$\mathrm{Diff}_{\mathcal{P}}(M,\xi) := \{F_+\in\mathrm{Diff}_{\mathcal{P}}(M) \mid F_+^*\xi = \xi\}.$$

Moreover, $F_+ = \mathrm{Id}$ if and only if F belongs to the space

$$\mathrm{Diff}_c(SM,\omega_{\rm st}) := \{F\in\mathrm{Diff}_c(SM) \mid F^*\omega_{\rm st} = \omega_{\rm st}\}$$

of compactly supported symplectomorphisms of $(SM, \omega_{\rm st})$. Thus we have a fibration

$$\mathrm{Diff}_c(SM, \omega_{\rm st}) \to \mathcal{P}(M, \xi) \to \mathrm{Diff}_{\mathcal{P}}(M, \xi).$$

The corresponding homotopy exact sequence fits into a commuting diagram

$$\begin{array}{ccccccc} \pi_0\mathrm{Diff}_c(SM, \omega_{\rm st}) & \longrightarrow & \pi_0\mathcal{P}(M, \xi) & \longrightarrow & \pi_0\mathrm{Diff}_{\mathcal{P}}(M, \xi) & \longrightarrow & 0 \\ \downarrow & & \downarrow & & \downarrow & & \\ \pi_0\mathrm{Diff}_c(\mathbb{R} \times M) & \longrightarrow & \pi_0\mathcal{P}(M) & \longrightarrow & \pi_0\mathrm{Diff}_{\mathcal{P}}(M) & \longrightarrow & 0, \end{array} \tag{14.1}$$

where the vertical maps are induced by the obvious inclusions.

The following is the main result of this section.

THEOREM 14.23. *For any closed contact manifold (M, ξ) of dimension $2n-1 \geq 5$, the map $\pi_0\mathcal{P}(M, \xi) \to \pi_0\mathcal{P}(M)$ is surjective.*

PROOF. By the discussion above and in Section 9.10, it suffices to show that the map $\pi_0\mathcal{E}(M, \xi) \to \pi_0\mathcal{E}(M)$ induced by the projection $(\lambda, \phi) \mapsto \phi$ is surjective. So let $\psi \in \mathcal{E}(M)$, i.e., $\psi : \mathbb{R} \times M \to \mathbb{R}$ is a function without critical points which agrees with $\phi_{\rm st}(r, x) = r$ outside a compact set $W = [a, b] \times M$. We apply Theorem 14.1 to the Weinstein cobordism $\mathfrak{W} = (W, \omega_{\rm st}, X_{\rm st}, \phi_{\rm st})$ and the function $\psi : W \to \mathbb{R}$. Hence there exists a Weinstein homotopy $\mathfrak{W}_t = (W, \omega_t, X_t, \phi_t)$, fixed on $\mathcal{O}p\, \partial_- W$ and fixed up to scaling on $\mathcal{O}p\, \partial_+ W$, such that $\mathfrak{W}_0 = \mathfrak{W}$ and $\phi_1 = \psi$. Note that $\lambda_t = c_t \lambda_{\rm st}$ on $\mathcal{O}p\, \partial_+ W$ for constants c_t with $c_0 = 1$. So we can extend $\mathfrak{W}_t$ over the rest of $\mathbb{R} \times M$ by the function $\phi_{\rm st}$ and Liouville forms of the form $f_t(r)\lambda_{\rm st}$ such that $\mathfrak{W}_t = \mathfrak{W}$ on $\{r \leq a\}$ and on $\{r \geq c\}$ for some sufficiently large $c > b$. By Moser's Stability Theorem 6.8, we find a diffeotopy $h_t : SM \to SM$ with $h_0 = \mathrm{Id}$, $h_t = \mathrm{Id}$ outside $[a, c] \times M$, and $h_t^* \mathfrak{W}_t = \mathfrak{W}$. Thus $h_1^* \mathfrak{W}_1 = (\lambda, \phi)$ with the function $\phi := \psi \circ h_1$ and a Liouville form λ which agrees with $\lambda_{\rm st}$ outside $[a, c] \times M$ and satisfies $d\lambda = \omega_{\rm st}$. Hence $(\lambda, \phi) \in \mathcal{E}(M, \xi)$ and ϕ is homotopic (via $\psi \circ h_t$) to ψ in $\mathcal{E}(M)$, i.e., $[\phi] = [\psi] \in \pi_0\mathcal{E}(M)$. □

Thus the second vertical map in the diagram (14.1) is surjective and we obtain

COROLLARY 14.24. *Let (M, ξ) be a closed contact manifold of dimension $2n - 1 \geq 5$. Then every diffeomorphism of M that is pseudo-isotopic to the identity is smoothly isotopic to a contactomorphism of (M, ξ).*

REMARK 14.25. Considering in the diagram (14.1) elements in $\pi_0\mathcal{P}(M)$ that map to $\mathrm{Id} \in \pi_0\mathrm{Diff}_{\mathcal{P}}(M)$, we obtain the following (non-exclusive) dichotomy for a contact manifold (M, ξ) of dimension ≥ 7 for which the map $\pi_0\mathrm{Diff}_c(\mathbb{R} \times M) \to \pi_0\mathcal{P}(M)$ is nontrivial: Either there exists a contactomorphism of (M, ξ) that is smoothly but not contactly isotopic to the identity; or there exists a compactly supported symplectomorphism of $(SM, \omega_{\rm st})$ which represents a nontrivial smooth pseudo-isotopy class in $\mathcal{P}(M)$. Unfortunately, we cannot decide which of the two cases occurs.

CHAPTER 15

Deformations of Stein Structures

In this chapter we show that Weinstein homotopies can be lifted to Stein homotopies, thus proving Theorem 1.1(b) and (c) from the Introduction. As a consequence, in Section 15.3 we carry over the flexibility results of Chapter 14 from Weinstein to Stein structures and deduce Theorems 1.9 and 1.10 from the Introduction.

15.1. From Weinstein to Stein: homotopies

The main results of this chapter are the following two theorems. Let us point out that all the results in this section also hold in dimension 4 without further hypotheses.

THEOREM 15.1 (First Stein deformation theorem). *Let* $\mathfrak{W}_t = (W, \omega_t, X_t, \phi_t)$ *be a homotopy of Weinstein cobordisms such that* $\mathfrak{W}_0 = \mathfrak{W}(J, \phi_0)$ *for a Stein structure* (J, ϕ_0) *on* W*. Then, after target reparametrizing the* ϕ_t*, there exists a diffeotopy* $h_t : W \to W$ rel $\mathcal{O}p\,\partial W$ *with* $h_0 = \mathrm{Id}$ *such that the functions* $h_{t*}\phi_t$ *are* J*-convex and the paths of Weinstein structures* $\mathfrak{W}_t$ *and* $\mathfrak{W}(h_t^* J, \phi_t)$ *are homotopic with fixed functions* ϕ_t *and fixed at* $t = 0$.

If $\mathfrak{W}_t$ *is fixed near* $\partial_- W$ *and/or fixed up to scaling near* $\partial_+ W$*, then the same can be arranged for the homotopy connecting the paths* $\mathfrak{W}_t$ *and* $\mathfrak{W}(h_t^* J, \phi_t)$.

Theorem 15.1 will be proved in the next section. Combined with Theorem 13.6, it implies

THEOREM 15.2 (Second Stein deformation theorem). *Let* (J_0, ϕ_0) *and* (J_1, ϕ_1) *be two Stein structures on the same cobordism* W*. Let* $\mathfrak{W}_t = (\omega_t, X_t, \phi_t)$ *be a Weinstein homotopy connecting* $\mathfrak{W}_0 = \mathfrak{W}(J_0, \phi_0)$ *and* $\mathfrak{W}_1 = \mathfrak{W}(J_1, \phi_1)$ *which is Stein near* $\partial_- W$*. Suppose that* $\mathfrak{W}_t = \mathfrak{W}_0$ *on* $\mathcal{O}p\,\partial_- W$ *for* $t \in [0, \frac{1}{2}]$*, and* $\phi_t = \phi_1$ *for* $t \in [\frac{1}{2}, 1]$*. Then, after target reparametrizing the* ϕ_t*, the Stein structures on* $\mathcal{O}p\,\partial_- W$ *extend to a Stein homotopy* (J_t, ϕ_t) *connecting* (J_0, ϕ_0) *and* (J_1, ϕ_1) *such that the paths of Weinstein structures* $\mathfrak{W}_t$ *and* $\mathfrak{W}(J_t, \phi_t)$ *are homotopic* rel $\mathcal{O}p\,\partial_- W$ *with fixed functions* ϕ_t *and fixed at* $t = 0, 1$.

PROOF. The proof of Theorem 15.2 follows the same scheme as that of Theorem 14.13. It is based on Theorem 15.1 and the 1-parametric case in Theorem 13.6.

We will construct the Stein/Weinstein homotopies as in Figure 15.1, where the vertical lines denote Weinstein homotopies with fixed functions.

First we apply Theorem 15.1 to the Weinstein homotopy $\mathfrak{W}_t$, $t \in [0, \frac{1}{2}]$, and the Stein structure (J_0, ϕ_0). Thus we find a diffeotopy $h_t : W \to W$, $t \in [0, \frac{1}{2}]$, rel $\mathcal{O}p\,\partial W$ with $h_0 = \mathrm{Id}$ such that the functions $h_{t*}\phi_t$ are J_0-convex and the paths of Weinstein structures $\mathfrak{W}_t$ and $\mathfrak{W}(J_t := h_t^* J_0, \phi_t)$, $t \in [0, \frac{1}{2}]$, are homotopic rel $\mathcal{O}p\,\partial_- W$ with fixed functions ϕ_t and fixed at $t = 0$.

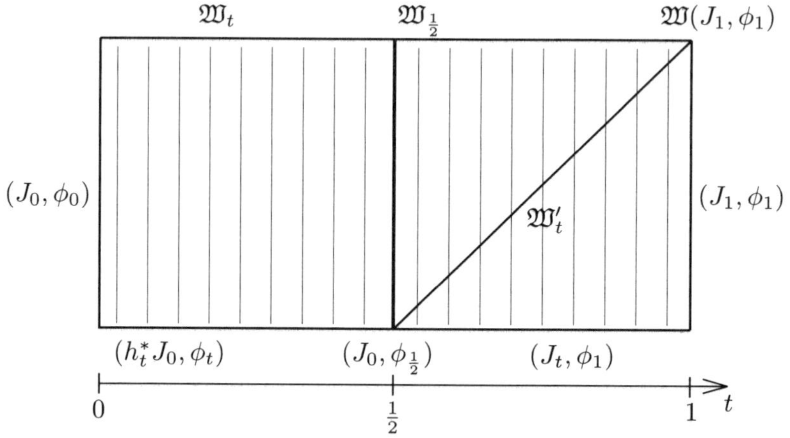

FIGURE 15.1. Proof of the second Stein deformation theorem.

We connect $\mathfrak{W}(J_{\frac{1}{2}},\phi_{\frac{1}{2}})$ to $\mathfrak{W}(J_1,\phi_1)$ by a Weinstein homotopy $\mathfrak{W}'_t$, $t\in[\frac{1}{2},1]$, with fixed function $\phi_{\frac{1}{2}}=\phi_1$ by concatenating the homotopy with fixed function from $\mathfrak{W}(J_{\frac{1}{2}},\phi_{\frac{1}{2}})$ to $\mathfrak{W}_{\frac{1}{2}}$ constructed in the preceding paragraph with the given homotopy $\mathfrak{W}_t$, $t\in[\frac{1}{2},1]$. Now we apply the 1-parametric case ($k=1$) in Theorem 13.6 to the homotopy $\mathfrak{W}'_t$, $t\in[\frac{1}{2},1]$, to find a Stein homotopy $(J_t,\phi_t\equiv\phi_1)$, $t\in[\frac{1}{2},1]$, connecting $(J_{\frac{1}{2}},\phi_{\frac{1}{2}})$ and (J_1,ϕ_1) such that the paths of Weinstein structures $\mathfrak{W}'_t$ and $\mathfrak{W}(J_t,\phi_t)$, $t\in[\frac{1}{2},1]$, are homotopic rel $\mathcal{O}p\,\partial_-W$ with fixed function $\phi_t\equiv\phi_1$ and fixed at $t=\frac{1}{2},1$. By construction, (J_t,ϕ_t) agrees with the Stein structure underlying $\mathfrak{W}_t$ on $\mathcal{O}p\,\partial_-W$ for all $t\in[0,1]$, and the paths $\mathfrak{W}(J_t,\phi_t)$ and $\mathfrak{W}_t$, $t\in[0,1]$, are homotopic rel $\mathcal{O}p\,\partial_-W$ with fixed functions ϕ and fixed endpoints. Hence (J_t,ϕ_t), $t\in[0,1]$, is the desired Stein homotopy. □

The same proofs also give the following versions of Theorems 15.1 and 15.2 for Weinstein/Stein *manifolds*, which correspond to Theorem 1.1(b) and (c) from the Introduction.

THEOREM 15.3. *Let* $\mathfrak{W}_t=(V,\omega_t,X_t,\phi_t)$ *be a homotopy of Weinstein manifolds such that* $\mathfrak{W}_0=\mathfrak{W}(J,\phi_0)$ *for a Stein structure* (J,ϕ_0) *on* V. *Then, after target reparametrizing the* ϕ_t, *there exists a diffeotopy* $h_t:V\to V$ *with* $h_0=\mathrm{Id}$ *such that the functions* $h_{t*}\phi_t$ *are* J*-convex and the paths of Weinstein structures* $\mathfrak{W}_t$ *and* $\mathfrak{W}(h_t^*J,\phi_t)$ *are homotopic with fixed functions* ϕ_t *and fixed at* $t=0$.

THEOREM 15.4. *Let* (J_0,ϕ_0) *and* (J_1,ϕ_1) *be two Stein structures on the same manifold* V. *Let* $\mathfrak{W}_t=(\omega_t,X_t,\phi_t)$ *be a Weinstein homotopy connecting* $\mathfrak{W}_0=\mathfrak{W}(J_0,\phi_0)$ *and* $\mathfrak{W}_1=\mathfrak{W}(J_1,\phi_1)$. *Then, after target reparametrizing the* ϕ_t, *there exists a Stein homotopy* (J_t,ϕ_t) *connecting* (J_0,ϕ_0) *and* (J_1,ϕ_1) *such that the paths of Weinstein structures* $\mathfrak{W}_t$ *and* $\mathfrak{W}(J_t,\phi_t)$ *are homotopic with fixed functions* ϕ_t *and fixed at* $t=0,1$.

REMARK 15.5. Theorem 15.3 has the following consequence. If (V,J_t,ϕ_t) is a homotopy of Stein manifolds, then there exist diffeomorphisms $h:V\to V$ and $g:\mathbb{R}\to\mathbb{R}$ isotopic to the identity such that $g\circ\phi_1\circ h^{-1}$ is J_0-convex. In view of Proposition 11.22, this shows that every exhausting J_1-convex function is equivalent to a J_0-convex function. So Morse theoretic properties of the space of exhausting J-convex functions, such as the minimal number of critical points of an exhausting J-convex function, are invariant under Stein homotopies. It would be interesting to further investigate such properties.

To put these theorems into a more topological context, we recall the setup from the Introduction; see Appendix A.1 for the topological notions. Let us fix a cobordism W. Denote by $\mathfrak{Stein}$ the space of Stein structures on W, by $\mathfrak{Weinstein}$ the space of Weinstein structures which are Stein near $\partial_- W$, and by $\mathfrak{Morse}$ the space of generalized Morse functions on W (as usual with regular level sets $\partial_\pm W$ and considered modulo target reparametrization). We have the commutative diagram

$$\begin{array}{ccc} \mathfrak{Stein} & \xrightarrow{\quad\mathfrak{W}\quad} & \mathfrak{Weinstein} \\ & {\scriptstyle\pi_{\mathfrak{S}}}\searrow \qquad \swarrow{\scriptstyle\pi_{\mathfrak{W}}} & \\ & \mathfrak{Morse} & \end{array}$$

where $\pi_{\mathfrak{W}}(\omega, X, \phi) := \phi$ and $\pi_{\mathfrak{S}}(J, \phi) := \phi$. Consider the fibers $\mathfrak{Stein}(\phi) := \pi_{\mathfrak{S}}^{-1}(\phi)$ and $\mathfrak{Weinstein}(\phi) := \pi_{\mathfrak{W}}^{-1}(\phi)$ of the projections $\pi_{\mathfrak{S}}$ and $\pi_{\mathfrak{W}}$ over $\phi \in \mathfrak{Morse}$. For a function $\phi \in \mathfrak{Morse}$ we introduce the spaces

$$\begin{aligned} \mathcal{P}(\phi) &:= \{(J,\gamma) \mid (J,\phi) \in \mathfrak{Stein},\ \gamma : [0,1] \to \mathfrak{Weinstein}(\phi) \text{ fixed near } \partial_- W, \\ &\qquad\qquad \gamma(0) = \mathfrak{W}(J,\phi)\}, \\ \mathcal{P} &:= \bigcup_{\phi \in \mathfrak{Morse}} \mathcal{P}(\phi). \end{aligned}$$

Theorem 15.2 asserts that the projection $\pi_{\mathcal{P}} : \mathcal{P} \to \mathfrak{Weinstein}$, $(h,\gamma) \mapsto \gamma(1)$ has the lift extension property for the pair $([0,1], \partial[0,1])$.

To rephrase Theorem 15.1, let us denote by $\mathcal{D}$ the identity component of the group of diffeomorphisms of W fixed near the boundary. For a Stein structure (J, ϕ_0) on W and a function $\phi \in \mathfrak{Morse}$, we introduce the spaces

$$\begin{aligned} \mathcal{D}_J(\phi) &:= \{h \in \mathcal{D} \mid \phi \text{ is } h^*J\text{-convex}\}, \\ \mathcal{P}_J(\phi) &:= \{(h,\gamma) \mid h \in \mathcal{D}_J(\phi),\ \gamma : [0,1] \to \mathfrak{Weinstein}(\phi) \text{ fixed near } \partial_- W, \\ &\qquad\qquad \gamma(0) = \mathfrak{W}(h^*J,\phi)\}, \\ \mathcal{P}_J &:= \bigcup_{\phi \in \mathfrak{Morse}} \mathcal{P}_J(\phi). \end{aligned}$$

We denote by $\mathfrak{Weinstein}_J \subset \mathfrak{Weinstein}$ the connected component of $\mathfrak{W}(J, \phi_0)$ in the space of Weinstein structures that agree with $\mathfrak{W}(J, \phi_0)$ near $\partial_- W$. Now Theorem 15.1 asserts the path lifting property for the projection $\pi_{\mathcal{P}_J} : \mathcal{P}_J \to \mathfrak{Weinstein}_J$, $(h,\gamma) \mapsto \gamma(1)$. Note that we have a commutative diagram

$$\begin{array}{ccc} \mathcal{P}_J & \xrightarrow{\qquad\qquad} & \mathcal{P} \\ & {\scriptstyle\pi_{\mathcal{P}_J}}\searrow \qquad \swarrow{\scriptstyle\pi_{\mathcal{P}}} & \\ & \mathfrak{Weinstein} & \end{array}$$

where the horizontal map sends (h,γ) to (h^*J,γ).

Now we see that the above proof of Theorem 15.2 just repeats the proof of Corollary A.5 from Appendix A.1 in the special case $k = 1$: We are given a path

$\mathfrak{W}_t$ in $\mathfrak{Weinstein}$ connecting $\mathfrak{W}(J_0,\phi_0)=\pi_{\mathcal{P}}(P_0)$ and $\mathfrak{W}(J_1,\phi_1)=\pi_{\mathcal{P}}(P_1)$ which belongs to $\mathfrak{Weinstein}_{J_0}$ for $t\in[0,\frac12]$ and to $\mathfrak{Weinstein}(\phi_1)$ for $t\in[\frac12,1]$. Theorem 15.1 and the last diagram provides a lift $P_t\in\mathcal{P}$, $t\in[0,\frac12]$. Now the projections of $P_{\frac12},P_1\in\mathcal{P}(\phi_1)$ are connected by the path $\mathfrak{W}_t$, $t\in[\frac12,1]$, in $\mathfrak{Weinstein}(\phi_1)$. Hence Theorem 13.6, which asserts that the projection $\mathcal{P}(\phi_1)\to\mathfrak{Weinstein}(\phi_1)$ is a weak homotopy equivalence, provides a path P_t, $t\in[\frac12,1]$, in $\mathcal{P}(\phi_1)$ connecting $P_{\frac12}$ and P_1.

We believe that Theorem 15.1 can be improved to the following

CONJECTURE 15.6. The projection $\pi_{\mathcal{P}_J}:\mathcal{P}_J\to\mathfrak{Weinstein}_J$, $(h,\gamma)\mapsto\gamma(1)$ is a Serre fibration.

According to Corollary A.6 in Appendix A.1, Conjecture 1.3 combined with Theorem 1.2 would imply

CONJECTURE 15.7. The map $\mathfrak{W}:\mathfrak{Stein}\to\mathfrak{Weinstein}$ is a weak homotopy equivalence.

15.2. Proof of the first Stein deformation theorem

The proof of Theorem 15.1 follows the same scheme as that of Theorem 14.9, based on the following three lemmas.

LEMMA 15.8. *Let (W,J,ϕ) be a Stein cobordism and $\mathfrak{W}=(W,\omega,X,\phi)$ an elementary Weinstein cobordism such that $\mathfrak{W}=\mathfrak{W}(J,\phi)$ on $\mathcal{O}p\,\partial W$. Suppose that $\mathfrak{W}$ and $\mathfrak{W}(J,\phi)$ are connected by a Weinstein homotopy with fixed function ϕ and fixed on $\mathcal{O}p\,\partial W$. Then, after target reparametrizing ϕ, there exists a Weinstein homotopy $\mathfrak{W}_t=(W,\omega_t,X_t,\phi)$, $t\in[0,1]$, such that*

- $\mathfrak{W}_0=\mathfrak{W}(J,\phi)$ *and* $\mathfrak{W}_1=\mathfrak{W}$;
- *the homotopy $\mathfrak{W}_t$ is fixed on $\mathcal{O}p\,\partial_-W$ and fixed up to scaling on $\mathcal{O}p\,\partial_+W$;*
- *for $t\in[0,\frac12]$ the function ϕ is h_t^*J-convex and $\mathfrak{W}_t=\mathfrak{W}(h_t^*J,\phi)$, for a diffeotopy $h_t:W\to W$, $t\in[0,\frac12]$, with $h_0=\mathrm{Id}$ and $h_t|_{\mathcal{O}p\,\partial W}=\mathrm{Id}$;*
- *for $t\in[\frac12,1]$ the Weinstein cobordisms $\mathfrak{W}_t$ are elementary and the attaching spheres in ∂_-W of all critical points of ϕ remain fixed for $t\in[\frac12,1]$.*

PROOF. **Step 0.** By Remark 9.2 there exists a diffeotopy $h_t:W\to W$, $t\in[0,1]$, with $h_0=\mathrm{Id}$, $h_t|_{\mathcal{O}p\,\partial W}=\mathrm{Id}$, and $\phi\circ h_t=\phi$, such that $h_1^*\mathfrak{W}(J,\phi)$ and $\mathfrak{W}$ have the same local stable and unstable manifolds near critical points. By Proposition 12.12 there exists a homotopy rel $\mathcal{O}p\,\partial W$ of Weinstein structures $\mathfrak{W}_t=(W,\omega_t,X_t,\phi)$ with $\mathfrak{W}_0=\mathfrak{W}$ such that $\mathfrak{W}_1$ agrees with $h_1^*\mathfrak{W}(J,\phi)$ near the critical points, and the stable and unstable manifolds of all $\mathfrak{W}_t$ agree with those of $\mathfrak{W}$. The last property ensures that all the $\mathfrak{W}_t$ are elementary cobordisms. After replacing $\mathfrak{W}$ by $\mathfrak{W}_1$ and J by h_1^*J, we may hence assume that $\mathfrak{W}$ agrees with $\mathfrak{W}(J,\phi)$ on $\mathcal{O}p\,(\partial W\cup\mathrm{Crit}\,\phi)$.

By a 1-parametric version of Proposition 12.12, we can connect $\mathfrak{W}$ and $\mathfrak{W}(J,\phi)$ by a Weinstein homotopy $\mathfrak{W}_t$ with fixed function ϕ and fixed on $\mathcal{O}p\,(\partial W\cup\mathrm{Crit}\,\phi)$. After applying Gray's Theorem 6.23 on each level set and pulling $\mathfrak{W}_t$ back by a diffeotopy, we can further arrange that the $\mathfrak{W}_t$, $t\in[0,1]$, induce the same contact structures on all level sets of ϕ.

After these preparations, the rest of the proof follows the same steps as that of Lemma 14.10, using the same notation.

Step 1. Define c_j, W_j, V_j and $\mathbf{S}_j^\pm \subset \Sigma_j^\pm$ (with respect to the Liouville field X) as in the proof of Lemma 14.10, see Figure 14.1.

By Step 0, the Weinstein structures $\mathfrak{W}$ and $\mathfrak{W}(J,\phi)$ induce the same contact structures $\xi_j^\pm$ on $\Sigma_j^\pm$. The assumption that the Weinstein cobordism $\mathfrak{W}$ is elementary implies that $\mathbf{S}_j^\pm$ is a union of isotropic (resp. coisotropic) spheres in the contact manifold $(\Sigma_j^\pm, \xi_j^\pm)$.

By Step 0, there exists a Weinstein homotopy $\mathfrak{W}_t := (\omega_t, X_t, \phi)$ from $\mathfrak{W}_0 = \mathfrak{W}$ to $\mathfrak{W}_1 = \mathfrak{W}(J,\phi)$ which is fixed on $\mathcal{O}p\,(\partial W \cup \operatorname{Crit}\phi)$ and induces the same contact structures on all level sets. Since $\mathfrak{W}_t$ is fixed near the critical points, after shrinking the W_j, we may assume that the X_t-unstable spheres in Σ_j^+ of the critical points on level c_j are fixed. Using Lemma 12.5 we can modify $\mathfrak{W}_t$ on $\bigcup_j V_j$ to a simple Weinstein homotopy $\widetilde{\mathfrak{W}}_t = (\widetilde{\omega}_t, \widetilde{X}_t, \phi)$ such that the intersections of the $\widetilde{X}_t$-stable manifolds of critical points on level $c_i > c_j$ with Σ_j^+ remain unchanged, and hence the Weinstein homotopy $\widetilde{\mathfrak{W}}_t$ is elementary. After renaming $\widetilde{\mathfrak{W}}_1$ back to $\mathfrak{W}$ we may thus assume that $\mathfrak{W} = \mathfrak{W}(J,\phi)$ on $\mathcal{O}p \bigcup_{j=1}^N W_j$.

We will construct the required homotopy $\mathfrak{W}_t = (\omega_t, X_t, \phi_t)$ separately on each V_j, fixed near $\partial_- V_j$ and fixed up to scaling near $\partial_+ V_j$. This will allow us to extend the homotopy to $\bigcup_{j=1}^N W_j$ as fixed up to scaling.

Step 2. Consider V_j for $1 \le j \le N-1$. To simplify the notation, we will omit the index j and denote the restriction of objects to V_j by the same symbol as the original objects.

By assumption we have a Weinstein homotopy $\mathfrak{W}_t = (V_j, \omega_t, X_t, \phi)$, $t \in [0,1]$, from $\mathfrak{W}_0 = \mathfrak{W}|_{V_j}$ to $\mathfrak{W}_1 = \mathfrak{W}(J,\phi)|_{V_j}$ which is fixed on $\mathcal{O}p\,(\partial W \cup \operatorname{Crit}\phi)$ and induces the same contact structures on all level sets. Recall that the holonomy along X_t defines contactomorphisms

$$\Gamma_{X_t} : (\Sigma_{j+1}^-, \xi_{j+1}^-) \to (\Sigma_j^+, \xi_j^+).$$

By Proposition 10.1, after target reparametrizing ϕ, we find a diffeotopy $h_t : V_j \to V_j$, $t \in [0,1]$, with $h_0 = \mathrm{Id}$ and $h_t = \mathrm{Id}$ near ∂V_j, such that the functions $h_{t*}\phi$ are J-convex and the holonomy of the cobordism $\mathfrak{W}'_t = (V_j, \omega'_t, X'_t, \phi) := \mathfrak{W}(V_j, h_t^* J, \phi)$ satisfies

$$\Gamma_{X'_t}(\mathbf{S}_{j+1}^-) = \Gamma_{X_t}(\mathbf{S}_{j+1}^-) \quad \text{for all } t \in [0,1].$$

Now Lemma 12.6 provides a Weinstein homotopy $\widetilde{\mathfrak{W}}_t = (V_j, \widetilde{\omega}_t, \widetilde{X}_t, \phi)$ such that

(i) $\widetilde{\mathfrak{W}}_t = \mathfrak{W}'_{2t} = \mathfrak{W}(h_{2t}^* J, \phi)$ for $t \in [0, \frac{1}{2}]$;

(ii) $\widetilde{\mathfrak{W}}_1 = \mathfrak{W}_1 = \mathfrak{W}$;

(iii) $\widetilde{\mathfrak{W}}_t$ coincides up to scaling with $\mathfrak{W}$ on $\mathcal{O}p\,\partial V_j$ and induces the same contact structures on level sets;

(iv) $\Gamma_{\widetilde{X}_t}(\mathbf{S}_{j+1}^-) = \Gamma_X(\mathbf{S}_{j+1}^-) = \mathbf{S}_j^+$ for $t \in [\frac{1}{2}, 1]$.

Condition (iv) implies that the resulting Weinstein homotopy $\widetilde{\mathfrak{W}}_t$ on W is elementary over the interval $[\frac{1}{2}, 1]$, and moreover, the intersection of the $\widetilde{X}_t$-stable manifolds of all critical points with $\partial_- W$ remains unchanged for $t \in [\frac{1}{2}, 1]$. □

LEMMA 15.9. *Let (W, J, ϕ) be a Stein cobordism and $\mathfrak{W} = (W, \omega, X, \phi)$ a Weinstein cobordism such that the function ϕ has exactly two critical points connected by a unique X-trajectory. Suppose that $\mathfrak{W}$ and $\mathfrak{W}(J,\phi)$ are connected by a Weinstein homotopy with fixed function ϕ and fixed on $\mathcal{O}p\,\partial W$. Then, after* *target reparametrizing* ϕ*, there exists a Weinstein homotopy* $\mathfrak{W}_t = (W, \omega_t, X_t, \phi)$*,* $t \in [0,1]$*, such that*

- $\mathfrak{W}_0 = \mathfrak{W}(J, \phi)$ *and* $\mathfrak{W}_1 = \mathfrak{W}$*;*
- *the homotopy* $\mathfrak{W}_t$ *is fixed on* $\mathcal{O}p\,\partial_- W$ *and fixed up to scaling on* $\mathcal{O}p\,\partial_+ W$*;*
- *for* $t \in [0, \frac{1}{2}]$ *the function* ϕ *is* $h_t^* J$*-convex and* $\mathfrak{W}_t = \mathfrak{W}(h_t^* J, \phi)$*, for a diffeotopy* $h_t : W \to W$*,* $t \in [0, \frac{1}{2}]$*, with* $h_0 = \mathrm{Id}$ *and* $h_t|_{\mathcal{O}p\,\partial W} = \mathrm{Id}$*;*
- *for* $t \in [\frac{1}{2}, 1]$*, the two critical points of the function* ϕ *are connected by a unique* X_t*-trajectory.*

PROOF. In this case the function ϕ has exactly two critical points $p_1, p_2 \in W$ of index $k-1, k$ and with critical values $c_1 < c_2$. For sufficiently small $\varepsilon > 0$, we split the cobordism W into two parts:

$$U := \{\phi \le c_1 + \varepsilon\}, \quad V := \{\phi \ge c_1 + \varepsilon\}.$$

Arguing as in Steps 0 and 1 of the proof of Lemma 15.8, we can reduce to the case that $\mathfrak{W} = \mathfrak{W}(J, \phi)$ on $\mathcal{O}p\,U$.

Now the restriction of $\mathfrak{W}$ to V is elementary of Type I. Hence by Lemma 15.8, after target reparametrizing ϕ, there exists a Weinstein homotopy $\mathfrak{W}_t = (V, \omega_t, X_t, \phi)$, $t \in [0,1]$, such that

- $\mathfrak{W}_0 = \mathfrak{W}(V, J, \phi)$ and $\mathfrak{W}_1 = \mathfrak{W}|_V$;
- the homotopy $\mathfrak{W}_t$ is fixed on $\mathcal{O}p\,\partial_- V$ and fixed up to scaling on $\mathcal{O}p\,\partial_+ V$;
- for $t \in [0, \frac{1}{2}]$ the function ϕ is $h_t^* J$-convex and $\mathfrak{W}_t = \mathfrak{W}(V, h_t^* J, \phi)$, for a diffeotopy $h_t : V \to V$, $t \in [0, \frac{1}{2}]$, with $h_0 = \mathrm{Id}$ and $h_t|_{\mathcal{O}p\,\partial V} = \mathrm{Id}$;
- for $t \in [\frac{1}{2}, 1]$ the Weinstein cobordisms $\mathfrak{W}_t$ are elementary and the attaching spheres in $\partial_- V$ of all critical points of ϕ remain fixed for $t \in [\frac{1}{2}, 1]$.

The homotopies $\mathfrak{W}_t$ and h_t extend canonically over U as a rescaling of $\mathfrak{W}$ (resp. the identity). The last property guarantees that for $t \in [\frac{1}{2}, 1]$ the two critical points are connected by a unique X_t-trajectory. □

The following lemma will serve as induction step in proving Theorem 15.1.

LEMMA 15.10. *Let* (W, J, ϕ) *be a Stein cobordism and* $\mathfrak{W}_t = (W, \omega_t, X_t, \phi_t)$*,* $t \in [0,1]$*, an* elementary *Weinstein homotopy such that* $\mathfrak{W}_t = \mathfrak{W}(J, \phi)$ *on* $\mathcal{O}p\,(\partial W)$*. Suppose that we are given a Weinstein homotopy* $\mathfrak{W}_0^s$*,* $s \in [0,1]$*, with fixed function* ϕ *and fixed on* $\mathcal{O}p\,\partial W$*, from* $\mathfrak{W}_0^0 = \mathfrak{W}_0$ *to* $\mathfrak{W}_0^1 = \mathfrak{W}(J, \phi)$*.*

Then, after target reparametrizing the ϕ_t*, there exists a diffeotopy* $h_t : W \to W$ *fixed on* $\mathcal{O}p\,\partial W$ *with* $h_0 = \mathrm{Id}$ *such that the functions* $h_{t*}\phi_t$ *are* J*-convex. Moreover, the Weinstein homotopy* $\mathfrak{W}_0^s$ *extends to a homotopy of paths of Weinstein structures* $\mathfrak{W}_t^s$*,* $s, t \in [0,1]$*, fixed on* $\mathcal{O}p\,\partial_- W$ *and up to scaling on* $\mathcal{O}p\,\partial_+ W$*, with fixed functions* ϕ_t*, from* $\mathfrak{W}_t^0 = \mathfrak{W}_t$ *to* $\mathfrak{W}_t^1 = \mathfrak{W}(h_t^* J, \phi_t)$*.*

PROOF. **Type I.** Consider first the case when the homotopy $\mathfrak{W}_t$ is elementary of Type I. We point out that $\mathfrak{W}(J, \phi)$ need not be elementary. To remedy this, we apply Lemma 15.8 to construct a Weinstein homotopy $\widetilde{\mathfrak{W}}_t = (W, \widetilde{\omega}_t, \widetilde{X}_t, \phi)$, $t \in [0,1]$, such that

- $\widetilde{\mathfrak{W}}_0 = \mathfrak{W}(J, \phi)$ and $\widetilde{\mathfrak{W}}_1 = \mathfrak{W}_0$;
- the homotopy $\widetilde{\mathfrak{W}}_t$ is fixed on $\mathcal{O}p\,\partial_- W$ and fixed up to scaling on $\mathcal{O}p\,\partial_+ W$;
- for $t \in [0, \frac{1}{2}]$ the function ϕ is $h_t^* J$-convex and $\widetilde{\mathfrak{W}}_t = \mathfrak{W}(h_t^* J, \phi)$, for a diffeotopy $h_t : W \to W$, $t \in [0, \frac{1}{2}]$, with $h_0 = \mathrm{Id}$ and $h_t|_{\mathcal{O}p\,\partial W} = \mathrm{Id}$;

- for $t \in [\frac{1}{2}, 1]$ the Weinstein cobordisms $\widetilde{\mathfrak{W}}_t$ are elementary.

Thus it is sufficient to prove the lemma for the Stein cobordism $(h^*_{\frac{1}{2}} J, \phi)$ instead of (J, ϕ), and the concatenation of the Weinstein homotopies $\widetilde{\mathfrak{W}}_{t\in[\frac{1}{2},1]}$ and $\mathfrak{W}_{t\in[0,1]}$ instead of $\mathfrak{W}_t$. To simplify the notation, we rename the new Stein cobordism and Weinstein homotopy back to (J, ϕ) and $\mathfrak{W}_t$. So in the new notation we have $\mathfrak{W}_0 = \mathfrak{W}(J, \phi)$.

According to Proposition 10.10, after target reparametrization of the ϕ_t, there exists a family of J-convex functions $\widetilde{\phi}_t$, $t \in [0, 1]$, on W with the same profile as the family ϕ_t and such that $\widetilde{\phi}_0 = \phi$ and $\widetilde{\phi}_t = \phi_t$ on $\mathcal{O}p\,(\partial W)$. Then Lemma 12.23 provides a diffeotopy $h_t : W \to W$ fixed on $\mathcal{O}p\,(\partial W \cup \operatorname{Crit} \phi_t)$ with $h_0 = \mathrm{Id}$ such that $\phi_t = \widetilde{\phi}_t \circ h_t$, and the paths of Weinstein structures $\mathfrak{W}_t$ and $\mathfrak{W}(h^*_t J, \phi_t)$ are homotopic rel $\mathcal{O}p\,\partial W$ with fixed functions ϕ_t and fixed at $t = 0$.

Types IId and IIb. Suppose now that the homotopy $\mathfrak{W}_t$ is of Type IId. Let $t_0 \in [0, 1]$ be the parameter value for which the function ϕ_t has a death-type critical point. In this case the function ϕ has exactly two critical points p and q connected by a unique X_0-trajectory. Arguing as in the Type I case, using Lemma 15.9 instead of Lemma 15.8, we can again reduce to the case that $\mathfrak{W}_0 = \mathfrak{W}(J, \phi)$.

Then Theorem 10.12 provides an elimination family of J-convex functions $\widetilde{\phi}_t : W \to \mathbb{R}$, $t \in [0, 1]$, starting from $\widetilde{\phi}_0 = \phi$ and killing the critical points p and q at time t_0. After target reparametrization of the ϕ_t, we can also arrange that $\widetilde{\phi}_t$ coincides with ϕ_t on $\mathcal{O}p\,(\partial W)$ and the homotopies $\widetilde{\phi}_t$ and ϕ_t have equal profiles. Now we again apply Lemma 12.23 to construct the required diffeotopy h_t and homotopy between the paths $\mathfrak{W}_t$ and $\mathfrak{W}(h^*_t J, \phi_t)$.

The argument in the case of Type IIb is similar, except that we use the Creation Theorem 10.11 instead of the Cancellation Theorem 10.12, and we do not need a preliminary homotopy. □

PROOF OF THEOREM 15.1. By Lemma 9.37 we find an admissible partition for the homotopy $\mathfrak{W}_t$:

$$0 = t_0 < t_1 < \cdots < t_p = 1, \qquad m(t) = c_0^k(t) < c_1^k(t) < \cdots < c_{N_k}^k(t) = M(t),$$

$t \in [t_{k-1}, t_k]$, $k = 1, \dots, p$. As in the proof of Theorem 14.9, by twisting the homotopy $\mathfrak{W}_t$ by a diffeotopy of W, we can arrange that the values $c_j^k = c_j^k(t)$ and the hypersurfaces $\Sigma_j^k = \{\phi_t = c_j^k(t)\}$ are independent of $t \in [t_{k-1}, t_k]$. We will extend the desired isotopy h_t and the homotopy $\mathfrak{W}_t^s$ between $\mathfrak{W}_t$ and $\mathfrak{W}(h^*_t J, \phi_t)$ inductively over the intervals $[t_{k-1}, t_k]$, $k = 1, \dots, p$, and for each k we extend them inductively over the elementary cobordisms W_j^k bounded by $\partial_- W_j^k = \Sigma_{j-1}^k$ and $\partial_+ W_j^k = \Sigma_j^k$.

Suppose h_t and $\mathfrak{W}_t^s$, $s \in [0, 1]$, are already constructed on all of W for $t \le t_{k-1}$. Recall that the restriction of the homotopy $\mathfrak{W}_t$, $t \in [t_{k-1}, t_k]$ to each cobordism W_j^k is elementary. Using Lemma 12.7, we can modify the family $\mathfrak{W}_t$, $t \in [t_{k-1}, t_k]$, near the Σ_j^k to make it agree with $\mathfrak{W}(h^*_{t_{k-1}} J, \phi_{t_{k-1}})$ on $\mathcal{O}p\,\Sigma_j^k$ for all j and $t \in [t_{k-1}, t_k]$. The resulting family, which we continue to denote by $\mathfrak{W}_t$, will still be elementary over each cobordism $\mathfrak{W}_j^k$. Hence we can apply Lemma 15.10 to each elementary homotopy $\mathfrak{W}_t|_{W_j^k}$, the complex structure $h^*_{t_{k-1}} J$, and the homotopy $\mathfrak{W}^s_{t_{k-1}}|_{W_j^k}$, $s \in [0, 1]$. For each j let $h_t^j : W_j^k \to W_j^k$, $t \in [t_{k-1}, t_k]$, be the diffeotopy provided by Lemma 15.10. The h_t^j fit together to form a diffeotopy $\widetilde{h}_t : W \to W$, $t \in [t_{k-1}, t_k]$.

Now $h_t := h_{t_{k-1}} \circ \widetilde{h}_t : W \to W$ is the desired extension of the diffeotopy to the interval $[t_{k-1}, t_k]$. Moreover, the 2-parametric Weinstein families on each W_j^k provided by Lemma 15.10 fit together (after rescaling) to the desired extension of the family $\mathfrak{W}_t^s$ over the interval $[t_{k-1}, t_k]$. □

15.3. Homotopies of flexible Stein structures

Using the results of Section 15.1, we can upgrade the results on flexible Weinstein homotopies in Chapter 14 to corresponding results on flexible Stein homotopies. Recall that a Stein cobordism or manifold structure (W, J, ϕ) is called subcritical (resp. flexible) if the corresponding Weinstein structure $\mathfrak{W}(W, J, \phi)$ is subcritical (resp. flexible).

Theorems 14.1 and 15.1 (resp. 14.4 and 15.3 in the manifold case) together with Remark 14.6 imply

THEOREM 15.11. *Let (W, J, ϕ) be a flexible Stein cobordism or manifold of real dimension $2n > 4$ and $\psi : W \to \mathbb{R}$, be a Morse function without critical points of index $> n$ which in the manifold case is exhausting, and in the cobordism case has $\partial_\pm W$ as its regular level sets. Then there exist diffeomorphisms $h : W \to W$ and $\alpha : \mathbb{R} \to \mathbb{R}$ diffeotopic to the identity such that the function $\alpha \circ \psi \circ h$ is J-convex.*

The same holds in dimension $2n = 4$ if we assume the existence of a Morse homotopy ϕ_t connecting ϕ and ψ without critical points of index > 1, or without critical points of index > 2 in the case that $\partial_- W \neq \varnothing$ is overtwisted.

In particular, we have the following Stein version of the h-cobordism theorem.

COROLLARY 15.12 (Stein h-cobordism theorem). *Any flexible Stein structure (J, ϕ) on a product cobordism $M \times [0, 1]$ of dimension $2n > 4$ admits a J-convex function without critical points.*

More generally, recall that a Morse function on a cobordism or manifold is called *perfect* if it has the minimal number of critical points compatible with the Morse inequalities.

COROLLARY 15.13. *(a) Let (W, J, ϕ) be a simply connected flexible Stein domain. Then there exists a perfect J-convex Morse function $\psi : W \to \mathbb{R}$ having ∂W as regular level set. In particular, the stabilization $V \times \mathbb{C}$ of any simply connected finite type Stein manifold V admits a perfect exhausting J-convex Morse function.*

(b) Let (J, ϕ) be a flexible Stein structure on $\mathbb{R}^{2n}$. Then there exists an exhausting J-convex function $\psi : \mathbb{R}^{2n} \to \mathbb{R}$ with a unique critical point, the minimum. In particular, such a function exists on the stabilization $V \times \mathbb{C}$ of any contractible Stein manifold V.

PROOF. The first statement in (a) for $\dim W = 2n \geq 6$ follows from Smale's Theorem 9.44 and Theorem 15.11. (Here simple connectedness of ∂W follows from that of W because W is obtained from ∂W by attaching handles of index ≥ 3.) In the case $2n = 4$ the Stein domain (W, J, ϕ) is subcritical, so ϕ has only critical points of index 0 and 1. Thus we just need to cancel all unnecessary minima of ϕ, which can be done in any dimension.

The second statement in (a) follows from this and the observation that the stabilization $V \times \mathbb{C}$ of a simply connected finite type Stein manifold V is the completion of a simply connected Stein domain W.

The first statement in (b) follows for $n \geq 3$ directly from Theorem 15.11, and for $n = 2$ from the argument for part (a).

The second statement in (b) follows from Stallings' theorem [**176**] which asserts that any product $V_1 \times V_2$ of two contractible manifolds with $\dim V_i \geq 1$ and $\dim(V_1 \times V_2) \geq 5$ is diffeomorphic to Euclidean space. (Actually Stallings' result is in the PL-category; the smooth case follows using [**142**].) □

The last statement in Corollary 15.13(b) is Theorem 1.9 from the Introduction. Theorems 14.3 and 15.2 (resp. 14.5 and 15.4 in the manifold case) together with Remark 14.6 imply

THEOREM 15.14. *Let (J_0, ϕ_0) and (J_1, ϕ_1) be two flexible Stein structures on a cobordism or manifold W of real dimension $2n > 4$. Suppose J_0 and J_1 are homotopic as almost complex structures. Then (J_0, ϕ_0) and (J_1, ϕ_1) are Stein homotopic.*

The same holds in dimension $2n = 4$ if we assume the existence of a Morse homotopy ϕ_t connecting ϕ_0 and ϕ_1 without critical points of index > 1, or without critical points of index > 2 in the case that $\partial_- W \neq \varnothing$ is overtwisted.

In particular, we have

COROLLARY 15.15. *Any two flexible Stein structures on $\mathbb{R}^{2n}$ are homotopic. In particular, the underlying Weinstein structures are exact symplectomorphic.*

PROOF. For $n > 2$ this follows directly from Theorem 15.14. Alternatively, we can use the following argument that works for any n: by Corollary 15.13(b) each flexible Stein structure on $\mathbb{R}^{2n}$ admits an exhausting J-convex function with a unique critical point, the minimum. Now the Stein homotopy is provided by Proposition 11.22 and Corollary 11.27. □

Without the flexibility hypothesis, Corollary 15.15 remains true for finite type Stein structures in dimension $2n = 4$ (see Chapter 16 below), while it becomes false in all dimensions $2n > 4$ (see Chapter 17).

REMARK 15.16. We do not know whether any two flexible (i.e., subcritical) Stein structures on a boundary connected sum (see Chapter 16) W of $k \geq 1$ copies of $B^3 \times S^1$ are homotopic. However, we will show in Chapter 16 that they become homotopic after applying a diffeomorphism of W.

Finally, let us consider the *J-convex pseudo-isotopy problem*, i.e., the study of the topology of the space of J-convex functions without critical points. Namely, let $(M \times [0,1], J, \phi)$ be a topologically trivial Stein cobordism. Let us denote by $\mathcal{E}(M \times [0,1], J)$ the space of J-convex functions $M \times [0,1] \to \mathbb{R}$ without critical points which are constant on $M \times 0$ and $M \times 1$. If $\dim M > 3$ and the Stein structure (J, ϕ) is flexible, then according to Corollary 15.12 the space $\mathcal{E}(M \times [0,1], J)$ is nonempty. It is interesting to study the canonical inclusion $\mathcal{I} : \mathcal{E}(M \times [0,1], J) \hookrightarrow \mathcal{E}(M)$ into the pseudo-isotopy space $\mathcal{E}(M)$ introduced in Section 9.10 of *all* smooth functions $M \times [0,1] \to \mathbb{R}$ without critical points which are constant on $M \times 0$ and $M \times 1$.

The following theorem corresponds to Theorem 1.10 in the Introduction.

THEOREM 15.17. *For any topologically trivial flexible Stein cobordism $(M \times [0,1], J, \phi)$ of dimension $2n > 4$, the induced homomorphism*

$$\mathcal{I}_* : \pi_0 \mathcal{E}(M \times [0,1], J) \to \pi_0 \mathcal{E}(M)$$

is surjective. □

PROOF. Let $\psi \in \mathcal{E}(M)$ be given. By Theorem 15.11 there exist diffeotopies $h_t : M \times [0,1] \to M \times [0,1]$ and $\alpha_t : \mathbb{R} \to \mathbb{R}$ with $h_0 = \mathrm{Id}$ and $\alpha_0 = \mathrm{Id}$ such that the function $\psi_1 := \alpha_1 \circ \psi \circ h_1$ is J-convex. Since ψ_1 is connected to ψ by the path $\alpha_t \circ \psi \circ h_t$ of functions without critical points, the functions ψ_1 and ψ belong to the same path connected component of $\mathcal{E}(M)$. $\square$

Part 5

Stein Manifolds and Symplectic Topology

The main tool in modern symplectic topology is the theory of J-holomorphic curves which was introduced by Gromov in 1985 ([**83**]). It was preceded by the method of filling by holomorphic discs introduced in the theory of functions of several complex variables by Bishop [**19**], and developed for global applications by Bedford and Gaveau [**15**]. In the final part of this book we discuss applications of this theory to the topology of Stein manifolds.

In Chapter 16 we outline how, in complex dimension 2, foliations by J-holomorphic curves give rise to various uniqueness results for Stein structures.

Chapter 17 starts with a review of symplectic homology, the main current tool for distinguishing Stein structures up to symplectomorphism. Then we outline how recent work by McLean and others leads to the existence of infinitely many pairwise non-symplectomorphic Stein structures on every smooth manifold that admits a Stein structure.

CHAPTER 16

Stein Manifolds of Complex Dimension Two

Foliations by J-holomorphic curves provide a powerful tool for the study of (almost) complex manifolds of complex dimension two. For example, Gromov used in [**83**] foliations by holomorphic spheres to prove that every symplectic filling (W, ω) of the standard 3-sphere with ω vanishing on $\pi_2(W)$ is symplectomorphic to the standard 4-ball. The results in the chapter are based on foliations by holomorphic *discs*, which were introduced in [**43**] and whose main properties (with sketches of proofs) we review in Section 16.1.

In Section 16.2 we derive uniqueness of Stein fillings up to deformation equivalence for S^3 and connected sums of copies of $S^2 \times S^1$. Along the way we prove that the property of having unique Stein fillings up to deformation equivalence is preserved under 0-surgery. In Section 16.3 we prove that certain 4-manifolds, including $\mathbb{R}^4$ and $\mathbb{R}^3 \times S^1$, have unique finite type Stein structures up to deformation equivalence, and that finite type Stein manifolds cannot be homeomorphic to $S^2 \times \mathbb{R}^2$.

16.1. Filling by holomorphic discs

Consider an embedded surface S in an almost complex 4-manifold (V, J) (i.e., V has real dimension 4 and S has real dimension 2). Generically, S has isolated points $p \in S$ where the tangent plane T_pS is a complex line. If the surface S is oriented, then a complex point is called *positive* or *negative* depending on whether the orientation of T_pS coincides with its orientation as a complex line, or is opposite to it. The complement of the complex points in S is a totally real surface. Generically, complex points can also be subdivided into *elliptic* and *hyperbolic* points; see [**43**]. An example one should have in mind is a surface $S \subset \mathbb{R}^3 = \{y_2 = 0\} \subset \mathbb{C}^2$. Then complex points of S are critical points of the function $x_2|_S$. A complex point is nondegenerate if and only if the corresponding critical point is nondegenerate; it is hyperbolic if the Morse index of this critical point is 1, and elliptic otherwise.

In this book we will deal only with surfaces S which are contained in a J-convex hypersurface $M \subset V$. Hence we will restrict our further discussion to this special case. See [**53**] for more detail.

Let us denote by ξ the induced contact structure, i.e., the field of complex tangencies to M. Given a surface $S \subset M \subset V$, its complex points are exactly the points where S is tangent to ξ. In other words, the complex points are singularities of the *characteristic foliation* generated by the line field $\xi \cap TS$ on S in the complement of complex points. Assuming that the surface S is oriented, the characteristic line field $\xi \cap TS$ inherits an orientation and hence can be generated by a vector field v. Generically, the index of the vector field v at complex points is equal to ± 1. We say that a complex point is *elliptic* if the index is $+1$, and *hyperbolic* if it is -1.

Let us denote by $e_\pm$ and $h_\pm$ the numbers of positive and negative elliptic and hyperbolic points, and set $d_\pm := e_\pm - h_\pm$.

If S is closed, the Euler characteristic χ of S and the value $c := e(\xi)[S]$ of the Euler class of ξ on $[S]$ can be computed from the singular points as

$$\begin{aligned} \chi &= d_+ + d_-, \\ c &= d_+ - d_-. \end{aligned} \tag{16.1}$$

Indeed, the first formula is just the Poincaré–Hopf index theorem (see [**87**]). To see the second one, note that c is the obstruction to constructing two $\mathbb{C}$-linearly independent vector fields tangent to W along S. Consider the pair of vector fields $(v, v^\perp)$ outside the complex points, where v is a vector field generating the characteristic foliation and $(v, v^\perp)$ is a basis of TS defining the orientation. These fields are $\mathbb{C}$-linearly independent away from the complex points, while positive elliptic and negative hyperbolic points contribute 1 to the total index c, and negative elliptic and positive hyperbolic points contribute -1.

The equations (16.1) can be rewritten in the form

$$d_\pm := e_\pm - h_\pm = \frac{1}{2}(\chi \pm c). \tag{16.2}$$

REMARK 16.1. For a general oriented closed surface in an almost complex 4-manifold, the corresponding formula for $d_\pm$ contains an extra term (see [**119, 51**]),

$$d_\pm = \frac{1}{2}(\chi + \nu \pm c). \tag{16.3}$$

Here ν is the normal Euler number, or equivalently, the self-intersection index, of the surface S. Note that ν vanishes in the special case considered above when S is contained in a J-convex hypersurface $M \subset V$, as well as when S is homologically trivial.

EXAMPLE 16.2. Consider the unit ball $B^4 = \{|z| \le 1\} \subset \mathbb{C}^2$ with complex coordinates $z = (z_1, z_2)$, $z_j = x_j + iy_j$. The field of complex tangencies on its i-convex boundary $S^3 = \partial B^4$ defines the standard contact structure on S^3. Let $p_\pm := (0, \pm i) \in S^3$. Then $S^3 \setminus \{p_+, p_-\}$ is foliated by the 2-spheres

$$S_t := \{|z| = 1,\ y_2 = t\} \subset S^3, \qquad t \in (-1, 1);$$

see Figure 16.1. Each S_t has precisely two complex points $q_t^\pm = (0, \pm\sqrt{1 - t^2} + it)$, with q_t^+ positive and q_t^- negative elliptic. Note that S_t bounds the Levi-flat 3-ball $\{|z| \le 1,\ y_2 = t\} \subset B^4$ which is foliated by the holomorphic discs

$$\Delta_{s,t} := \{x_2 = s,\ y_2 = t,\ |z_1|^2 \le 1 - s^2 - t^2\}, \qquad t \in (-1, 1),\ |s| < \sqrt{1 - t^2}.$$

The boundaries of the discs $\Delta_{s,t}$ foliate $S_t \setminus \{q_t^+, q_t^-\}$ by the circles $\{x_2 = s,\ y_2 = t,\ |z_1|^2 = 1 - s^2 - t^2\}$. For later reference, let us define the Levi-flat 3-ball

$$D := \{|z_1|^2 + x_2^2 \le 1\} \subset \mathbb{R}^3 = \{y_2 = 0\} \subset \mathbb{C}^2. \tag{16.4}$$

Let us call an almost complex structure J *tame* if it admits a symplectic form ω taming J, i.e., such that ω is positive on complex directions. An almost complex 4-manifold (W, J) is called *minimal* if it contains no embedded holomorphic spheres with self-intersection number -1. Any complex manifold can be blown down to a (not necessarily unique) minimal one, and the same holds for an almost complex manifold with taming symplectic form [**134**].

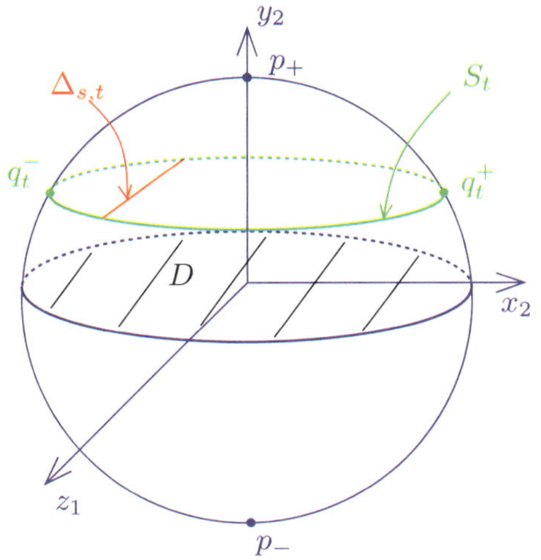

FIGURE 16.1. The foliation of the standard 4-ball by holomorphic discs.

Bishop proved in [**19**] that, as in Example 16.2 above, each elliptic complex point p on a surface S in a complex surface (V, J) has a neighborhood $U \subset S$ such that $U \setminus \{p\}$ is foliated by concentric circles which bound J-holomorphic discs in V. Such a family of J-holomorphic discs is called a *Bishop family*. Importantly, this result has the following global version; see [**15, 83, 43**].

THEOREM 16.3. *Let (W, J) be a tame compact almost complex 4-manifold with J-convex boundary. Suppose that W contains no nonconstant J-holomorphic spheres.*

(a) Let $S \subset \partial W$ be an embedded 2-sphere with exactly two complex points p_+, p_- which are both elliptic. Then $S \setminus \{p_-, p_+\}$ is foliated by circles which bound J-holomorphic discs inside W. All these discs are embedded, disjoint, and fill a Levi-flat embedded 3-ball $B \subset W$ bounded by $S = \partial B$. Moreover, there exists a diffeomorphism $F : D \to B$ which is holomorphic on the discs $\{x_2 = \text{const}\} \subset D$. If the sphere S is real analytic, then the diffeomorphism F can be chosen real analytic in the complement of the points $(z_1 = 0, x_2 = \pm 1)$.

(b) Let $f : \partial D \times [a, b] \hookrightarrow \partial W$ be an embedding such that each sphere $S_t = f(\partial D \times t) \subset \partial W$, $t \in [a, b]$, satisfies the hypotheses of (a). Then f extends to an embedding $F : D \times [a, b] \hookrightarrow W$ such that the balls $B_t = F(D \times t)$, $t \in [a, b]$, are Levi-flat and foliated by J-holomorphic discs. If the embedding f is real analytic, then the extension F can be chosen holomorphic along the discs $(D \cap \{x_2 = s\}) \times t$, $s \in (-1, 1)$, $t \in [a, b]$, and real analytic in the complement of the arcs $\gamma_\partial = \{z_1 = 0, x_2 = \pm 1\} \times [a, b]$.

Note that Theorem 16.3 is applicable, in particular, in the situation when (W, J) is a Stein domain of complex dimension 2.

SKETCH OF PROOF. For details of the following arguments, see [**99**]. Let us begin with (a). As already mentioned above, there exist Bishop families ([**19**], see [**195**] for the case of nonintegrable J) of J-holomorphic discs emanating from the elliptic points $p_\pm$. The problem is to extend them globally to fill the sphere S. Note that near $p_\pm$ Bishop's discs are embedded and disjoint. Positivity of intersections (see [**83, 135, 138**]) then implies that all the discs in the family are embedded and disjoint. If one can prove compactness for the moduli space of J-holomorphic discs with boundary on S this implies that the Bishop families emanating from $p_\pm$ are ends of the same 1-dimensional moduli space of embedded disjoint discs filling S. Note that Stokes' theorem implies that the symplectic area of all J-holomorphic discs with boundary on S is uniformly bounded by $\int_S |\omega|$. Hence, by Gromov's compactness theorem [**83**], compactness can only fail due to bubbling on the boundary or in the interior.

The boundaries of all holomorphic discs which fill S have to be transverse to the characteristic foliation on S. Indeed, tangency of the boundary of a disc to the characteristic foliation would imply the tangency of the disc itself to the J-convex boundary ∂W, which is impossible due to the maximum principle. Since $(\partial W, \xi)$ is fillable and hence tight, the characteristic foliation on S is homeomorphic to the foliation by meridians connecting elliptic points. An embedded boundary curve of a holomorphic disc thus has winding number ± 1 around the elliptic points, and hence it cannot split. This rules out bubbling at the boundary. On the other hand, bubbling in the interior is ruled out by the assumption that W has no nonconstant J-holomorphic spheres.

Part (b) can be proved similarly, taking into account that holomorphic discs filling different 2-spheres are disjoint due to positivity of intersections. □

The method of filling by holomorphic discs can also be used to prove

THEOREM 16.4 ([**43**]). *Let S be an embedded oriented closed surface contained in the J-convex boundary of a tame compact almost complex 4-manifold (W, J).*

(i) *If $S \not\cong S^2$, then $d_\pm \leq 0$, or equivalently (in view of (16.2)) $|c| \leq -\chi$.*
(ii) *If $S \cong S^2$, then $c = 0$ and hence (in view of (16.2)) $d_+ = d_- = 1$.*
(iii) *By a C^0-small isotopy of the surface S in ∂W, the nonnegative integers $e_\pm$ and $h_\pm$ can be arbitrarily changed as long as the differences $d_\pm$ are preserved.*

In particular, by a C^0-small isotopy of S in ∂W, one can get rid of all elliptic points in case (i) and kill all complex points except two elliptic points, one positive and one negative, in case (ii).

For a general surface S in a 4-manifold V the analogue of (iii) also holds, and in fact it is simpler because one is allowed an isotopy unconstrained by the condition $S \subset \partial W$; see [**84**] and [**51**].

Using the Giroux–Fuchs elimination lemma [**66**], Theorem 16.4 was extended in [**44**] to the more general case of a surface in an arbitrary tight contact 3-manifold.

16.2. Stein fillings

When complex analysts talk about *holomorphic fillings* they usually mean fillings of CR-manifolds. The existence of such a filling is a very delicate analytic question; see the discussion in Section 5.10 above. In this section we are interested in holomorphic fillings of smooth, or contact, manifolds.

A *Stein filling of a closed oriented 3-manifold M* is a Stein domain (W, J, ϕ) such that there exists an orientation preserving diffeomorphism between ∂W (with the boundary orientation) and M.

A *Stein filling of a closed contact* 3*-manifold* (M,ξ) is a Stein domain (W,J,ϕ) such that there exists an orientation preserving contactomorphism between ∂W with the field of complex tangencies and (M,ξ).

Two Stein cobordisms (W,J,ϕ) and (W',J',ϕ') are called *deformation equivalent* if there exists a diffeomorphism $h : W \to W'$ such that the Stein structures (J,ϕ) and $(h^*J',h^*\phi')$ on W are homotopic. An analogous definition applies to Weinstein cobordisms and to Stein/Weinstein manifolds. Note that for fixed J any two J-convex functions are homotopic (this is obvious for cobordisms and Proposition 11.22 for manifolds), so in this section we will often omit ϕ from the notation of a Stein cobordism or manifold.

By definition, uniqueness up to deformation equivalence of Stein fillings of M implies uniqueness up to orientation preserving diffeomorphism of Stein fillable contact structures on M. By Corollary 11.21, it also implies uniqueness up to exact symplectomorphism of Weinstein completions of the fillings.

For two Stein structures on the same smooth cobordism W the difference between the notions of homotopy and deformation equivalence lies in the topology of the group $\mathrm{Diff}_+(W)$ of orientation preserving diffeomorphisms of W. For instance, by a theorem of Cerf [**30**] the group $\mathrm{Diff}_+(B^{2n})$ of the closed unit ball B^{2n} is connected for $n > 2$, and hence there is no difference between homotopy and deformation equivalence of Stein structures on the ball B^{2n} if $n > 2$. On the other hand, it is unknown whether the group $\mathrm{Diff}_+(B^4)$ is connected, and consequently we do not know whether deformation equivalent Stein structures on B^4 are homotopic.

In this section we will use the method of filling by holomorphic discs to establish uniqueness up to deformation equivalence of Stein fillings of certain smooth and contact 3-manifolds.

Stein fillings of S^3. The following result, which is Theorem 1.7 from the Introduction, first appeared with a sketch of a proof in [**47**] (for the diffeomorphism part see [**83, 43, 133**]).

THEOREM 16.5. *Let (W,J) be a tame compact complex surface with J-convex boundary diffeomorphic to S^3. Suppose that W is minimal. Then W is diffeomorphic to the* 4*-ball. Moreover, (W,J) admits a J-convex Morse function constant on ∂W with a unique critical point, the minimum.*

PROOF. **Step 1.** Let us first show that the manifold W is diffeomorphic to the ball. Note that it follows from [**83**], [**43**], and [**133**] that W is diffeomorphic to a ball, possibly blown up in a few points. In order to see that it is actually a ball, we will use a theorem of Bogomolov and de Oliveira from [**20**]. Let us pick a collar neighborhood $C = M \times [0,\varepsilon] \subset W$ of the boundary $M \times 0 = M = \partial W$ such that the hypersurfaces $M_r = M \times r$, $r \in [0,\varepsilon]$, are J-convex. After deforming the collar neighborhood near two points on M_ε (using, e.g., Proposition 2.12), we may assume that M_ε satisfies the following conditions:

(i) there exist two points $q_\pm \in M_\varepsilon$ and holomorphic coordinates (z_1,z_2) on neighborhoods $U_\pm \subset W$ of $q_\pm$ in which $q_\pm$ has coordinates $(0,\pm i)$ and $M_\varepsilon \cap U_\pm$ correspond to the following parts of the unit sphere $\{|z|^2 = |z_1|^2 + |z_2|^2 = 1\} \subset \mathbb{C}^2$,

$$M_\varepsilon \cap U_+ = \{|z| = 1, y_2 > 1-\varepsilon\}, \quad M_\varepsilon \cap U_- = \{|z| = 1, y_2 < -1+\varepsilon\};$$

(ii) the hypersurface $M_\varepsilon \setminus (U_+ \cup U_-)$ is real analytic.

Hence, after replacing W by the region bounded by M_ε, we may assume without loss of generality that $M = \partial W$ itself, rather than M_ε, satisfies properties (i) and (ii).

The induced contact structure ξ on $M \cong S^3$ is symplectically fillable and thus tight. By uniqueness of the tight contact structure on S^3 (see [**44**]) it follows that (M, ξ) is diffeomorphic to S^3 with its standard contact structure described in Example 16.2. Hence $M \setminus \{q_+, q_-\}$ can be foliated by a family of 2-spheres S_t, $t \in (-1, 1)$, each having exactly two complex points which are both elliptic. In the above neighborhoods $U_\pm \cap M$, these spheres can be chosen as the intersections of M with the real hyperplanes $y_2 = t$, $t \in (-1, -1+\varepsilon) \cup (1-\varepsilon, 1)$. Moreover, we can arrange that there exists a real analytic diffeomorphism $f : \partial D \times [-1+\varepsilon, 1-\varepsilon] \to M \setminus (U_+ \cup U_-)$ such that $f(\partial D \times t) = S_t$. Here D is the Levi-flat 3-ball defined in (16.4).

By a theorem of Bogomolov and de Oliveira ([**20**], see Theorem 5.64 above) there exists a C^∞-small deformation of J to a complex structure $\widetilde{J}$ which is Stein. In particular, W contains no nonconstant $\widetilde{J}$-holomorphic spheres. So we can apply Theorem 16.3(b) to $\widetilde{J}$. Hence the embedding $f : \partial D \times [-1+\varepsilon, 1-\varepsilon] \hookrightarrow M$ extends to an embedding $F : D \times [-1+\varepsilon, 1-\varepsilon] \hookrightarrow W$ such that the 3-balls $B_t := F(D \times t)$, $t \in [-1+\varepsilon, 1-\varepsilon)$, bounded by the spheres S_t are Levi-flat and foliated by the $\widetilde{J}$-holomorphic discs $\Delta_{t,s} := F\big((D \cap \{x_2 = s\}) \times t\big)$, $s \in (-1, 1)$.

We extend the family B_t to $t \in (-1, 1)$, by defining $B_t := W \cap \{y_2 = t\}$ for $t \in (-1, -1+\varepsilon) \cup (1-\varepsilon, 1)$ using the local coordinates above near the points $q_\pm$. By uniqueness of the holomorphic discs, the B_t fit together smoothly at $t = 1-\varepsilon$ and $t = -1+\varepsilon$, so F extends to a smooth embedding $D \times (-1, 1) \hookrightarrow W$. Denote by $B^4 \subset \mathbb{C}^2$ the unit ball, and let $p_\pm := (0, \pm i)$. Composing F with the inverse of the canonical diffeomorphism

$$D \times (-1, 1) \mapsto B^4 \setminus \{p_+, p_-\}, \qquad ((z_1, x_2), t) \mapsto \Big(\frac{z_1}{\sqrt{1-t^2}}, \frac{x_2}{\sqrt{1-t^2}} + it\Big)$$

yields an embedding $B^4 \setminus \{p_+, p_-\} \hookrightarrow W$, which extends to an embedding $B^4 \hookrightarrow W$ by sending $p_\pm$ to $q_\pm$. Since this embedding induces a diffeomorphism on the boundary and W is connected, it is a diffeomorphism.

Step 2. Now we switch back to the original complex structure J. Since the integral of a taming symplectic form over any nonconstant J-holomorphic sphere is positive, each such sphere must represent a nontrivial second homology class. As W is diffeomorphic to the ball by Step 1, this shows that W contains no nonconstant J-holomorphic spheres. So we can repeat Step 1 with the original complex structure J.

In particular, we assume that $M = \partial W$ satisfies conditions (i) and (ii) in Step 1. So we find a collar neighborhood $C = M \times [0, \varepsilon] \subset W$ of $M = M \times 0$ such that

- the hypersurfaces $M_r = M \times r$, $r \in [0, \varepsilon]$ are J-convex;
- $M_r \cap U_+ = \{|z| = 1-r,\ y_2 > 1-\varepsilon\}$, $M_r \cap U_- = \{|z| = 1-r,\ y_2 < -1+\varepsilon\}$;

see Figure 16.2. Define the smaller collars

$$C' := M \times [0, \frac{\varepsilon}{2}] \subset C'' := M \times [0, \frac{3\varepsilon}{4}] \subset C.$$

Let $F : D \times (-1, 1) \hookrightarrow W$ be the embedding constructed in Step 1. For $\sigma, \tau \in (0, 1)$ set $D_\sigma := D \cap \{|x_2| \leq 1-\sigma\}$ and $W_{\sigma,\tau} := F\big(D_\sigma \times [-1+\tau, 1-\tau]\big)$; see Figure 16.3.

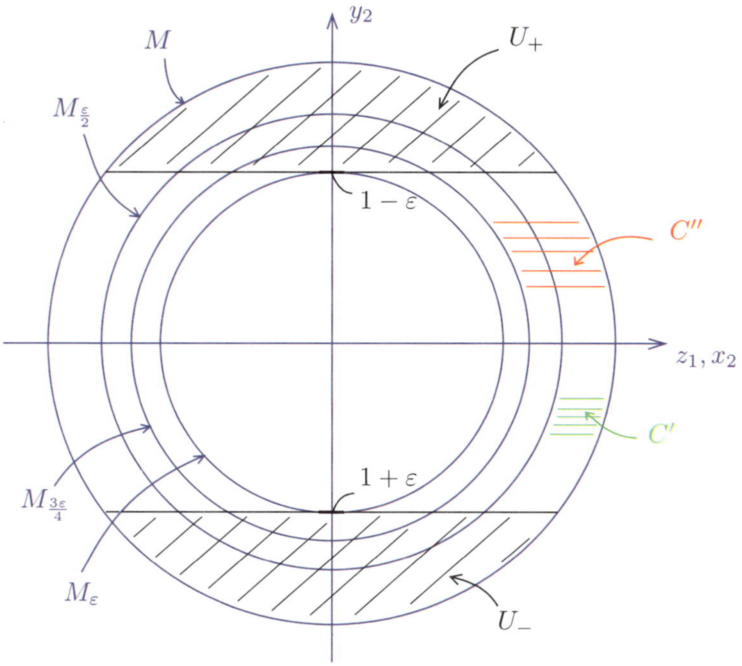

FIGURE 16.2. The different collars of $M = \partial W$.

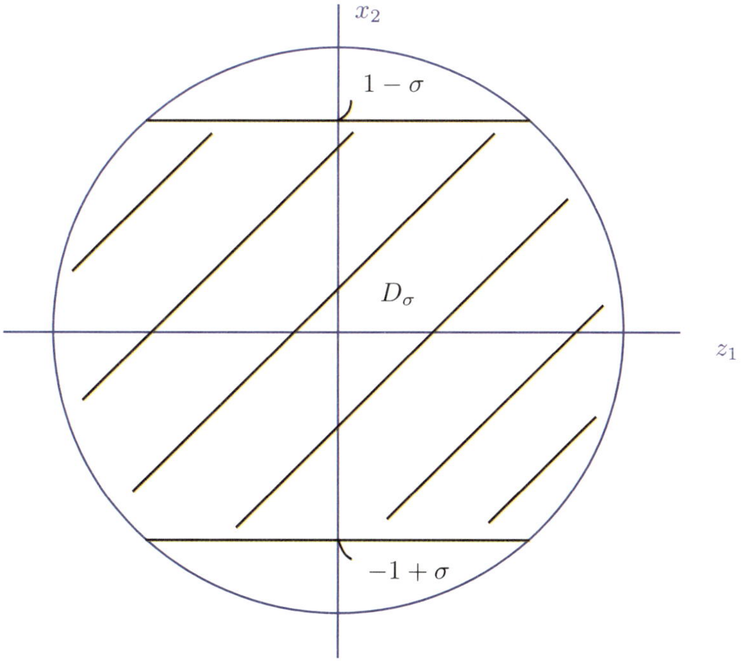

FIGURE 16.3. The truncated Levi-flat 3-ball D_σ.

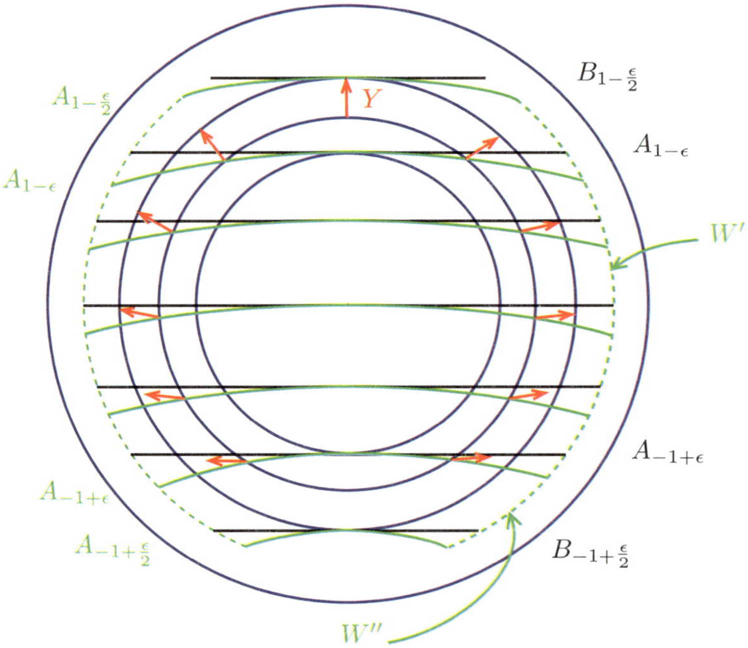

FIGURE 16.4. Deforming the foliation by Levi-flat 3-balls B_t to a foliation by J-convex 3-balls A_t.

Thus

$$W_{\sigma,\tau} = \bigcup_{\substack{|t|\leq 1-\tau \\ |s|\leq 1-\sigma}} \Delta_{t,s}.$$

Let us fix σ, τ so small that $\partial W_{\sigma,\tau} \subset \operatorname{Int} C'$. Note that for each $t \in (-1,1)$ the embedding $F_t : D \hookrightarrow W$, $F_t(z_1, x_2) := F(z_1, x_2, t)$ is real analytic in x_2 and holomorphic in z_1 on the set $\{z_1 \neq 0\}$. Hence by Corollary 5.47, there exists $\delta = \delta(\sigma, \tau) > 0$ such that for any $t \in [-1+\tau, 1-\tau]$ the embedding $F_t|_{D_\sigma}$ extends to a holomorphic embedding

$$\widetilde{F}_t : U_{\sigma,\delta} := \{(z_1, z_2)|\, (z_1, x_2) \in D_\sigma,\ |y_2| < \delta\} \hookrightarrow W.$$

Define the i-convex hypersurface

$$A := \{y_2 = -\delta(|z_1|^2 + x_2^2),\ (z_1, x_2) \in D_\sigma\} \subset U_{\sigma,\delta},$$

and let $A_t := \widetilde{F}_t(A)$; see Figure 16.4. We have $A_t \subset \operatorname{Int} C'$ if $t < -1 + \frac{\varepsilon}{2}$, and if δ is chosen small enough, then $A_t \subset \operatorname{Int} C$ for $t > 1 - \frac{\varepsilon}{2}$. Also if σ is sufficiently small, then $\partial A_t \subset \operatorname{Int} C$ for all $t \in (-1, 1)$.

Note that for sufficiently small δ all the hypersurfaces A_t are transverse to the vector field $X := F_* \frac{\partial}{\partial t}$. Observe also that there exists a vector field Y on C'' which is transverse to the hypersurfaces $M \times r$ for all $r \in [0, \frac{3\varepsilon}{4}]$, and to A_t for $t \geq -1 + \varepsilon$. Set $W' := \bigcup_{|t|\leq 1-\frac{\varepsilon}{2}} A_t$ and $W'' := \bigcup_{t\in[-1+\frac{\varepsilon}{2}, -1+\varepsilon]} A_t \subset W'$. By Proposition 3.25 we find a J-convex function ψ without critical points on W' whose level sets are transverse to X, and on $C'' \setminus W''$ also to Y. We can furthermore assume that its level sets in W'' coincide with the hypersurfaces A_t, $t \in [-1 + \frac{\varepsilon}{2}, -1 + \varepsilon]$. Let ϕ

be a J-convex function on C whose level sets are $M \times r, r \in [0, \varepsilon]$. By a target reparametrization of the function ϕ, we can arrange that on $M \times \frac{3\varepsilon}{4}$ we have $\phi < \psi$ and on $M \times \frac{\varepsilon}{2}$ we have $\phi > \psi$. Hence, according to Corollary 3.20 and Remark 3.24, the function smooth $\max(\phi, \psi)$ on W is J-convex and has a unique nondegenerate critical point, the minimum. □

As a consequence of Theorem 16.5, we obtain the following uniqueness result.

THEOREM 16.6. *Every Stein (or Weinstein) filling of S^3 is deformation equivalent to the standard Stein structure (B^4, i) on the closed unit ball $B^4 \subset \mathbb{C}^2$. In particular, all Stein structures on B^4 are deformation equivalent.*

PROOF. Let (W, J) be a Stein filling of S^3. By Theorem 16.5, W is diffeomorphic to B^4. Moreover, there exists a J-convex Morse function ψ constant on ∂W with a unique critical point, the minimum. Thus (W, J) is deformation equivalent to (B^4, i) by Proposition 11.26. The statement for Weinstein structures follows from that for Stein structures and Theorem 13.5. □

Stein domains with reducible boundary. A 3-manifold M is called *reducible* if it contains an embedded noncontractible 2-sphere $S \subset M$. The following theorem allows us to decompose Stein domains with reducible boundary (see the discussion below).

THEOREM 16.7. *Let (W, J) be a tame compact complex surface with J-convex boundary. Suppose that W contains no nonconstant J-holomorphic spheres. Let $S \subset \partial W$ be an embedded 2-sphere. Then there exists a compact domain $U \subset \operatorname{Int} W$ with smooth J-convex boundary such that the cobordism $W \setminus \operatorname{Int} U$ admits a J-convex Morse function with exactly one critical point of index 1 whose unstable sphere in ∂W is smoothly isotopic to S.*

PROOF. By Theorem 16.4, after a C^0-deformation of S, we may assume that S has exactly two complex points which are both elliptic. Moreover, we can assume that S is real analytic. Hence we can apply Theorem 16.3(a) to construct a Levi-flat ball $B \subset W$ bounded by S and an embedding $F : D \hookrightarrow W$ which is real analytic in the complement of the points $(0, \pm 1) \in D$, and holomorphic along the discs $\Delta_s = \{x_2 = s\} \cap D$, $s \in (-1, 1)$. Let us choose a collar $C = M \times [0, \varepsilon] \subset W$ of $M = M \times 0$ such that each hypersurface $M \times t$, $t \in [0, \varepsilon]$, is J-convex; see Figure 16.5. Set $C' := M \times [0, \frac{\varepsilon}{2}] \subset C$. Fix $\sigma > 0$ so small that $F(\Delta_s) \subset \operatorname{Int} C'$ for $|s| \geq 1 - \sigma$. By Corollary 5.47 there exists a $\delta > 0$ such that the real analytic embedding $F|_{D \cap \{|x_2 \leq 1 - \sigma\}}$ extends to a holomorphic embedding $\widetilde{F} : U_{\sigma,\delta} \hookrightarrow W$, where

$$U_{\sigma,\delta} := \{|z_1|^2 + x_2^2 \leq 1, |x_2| \leq 1 - \sigma, |y_2| \leq \delta\} \subset \mathbb{C}^2.$$

We can assume that $\widetilde{F}(\{U_{\sigma,\delta} \cap \{x_2 = 1 - \sigma\}) \subset C'$. Consider the following vector field on $\mathbb{C}^2$:

$$X := x_1 \frac{\partial}{\partial x_1} + x_2 \frac{\partial}{\partial x_2} + y_1 \frac{\partial}{\partial y_1} - y_2 \frac{\partial}{\partial y_2}.$$

By Lemma 8.47 (with $C = 4/\delta^2$ and $a = 1$) we find an i-convex function $\psi : U_{\sigma,\delta} \to \mathbb{R}$ with the following properties:

(i) ψ has a unique critical point at the origin, of index 1, with stable manifold $\{x_1 = x_2 = y_1 = 0\}$ and unstable manifold $\{y_2 = 0\}$;

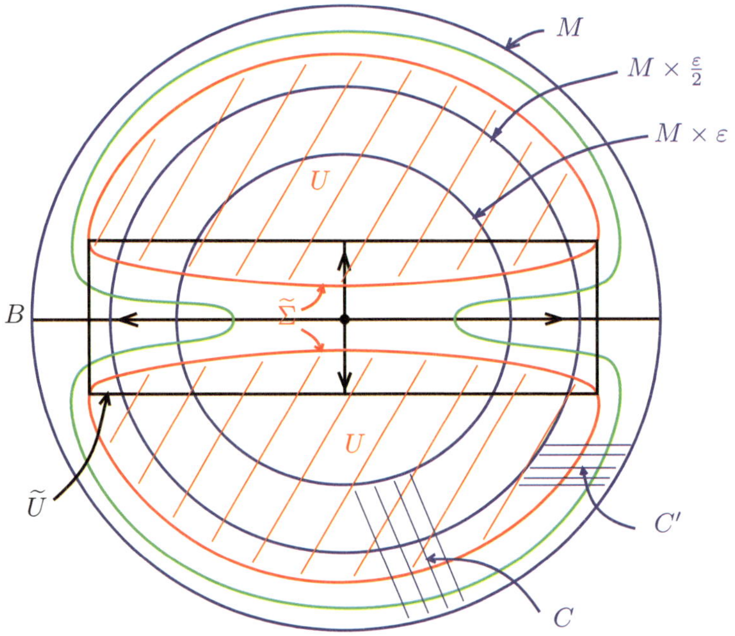

FIGURE 16.5. Decomposing a Stein domain with reducible boundary.

(ii) ψ has the hypersurface $\Sigma := \{y_2^2 = \frac{\delta^2}{4}\left(1 + |z_1|^2 + x_2^2\right)\} \subset U_{\sigma,\delta}$ as one of its level sets;
(iii) $d\psi(X) > 0$ outside the origin;
(iv) $\psi(z_1, z_2) = \psi(z_1, \bar{z}_2)$.

Introduce

$$\widetilde{U} := F(U_{\sigma,\delta}), \quad \widetilde{X} := \widetilde{F}_* X, \quad \widetilde{\Sigma} := \widetilde{F}(\Sigma), \quad \widetilde{\psi} := \psi \circ \widetilde{F}^{-1} : \widetilde{U} \to \mathbb{R}.$$

Let $\phi : C \to \mathbb{R}$ be a J-convex function with regular level sets $M \times t$, $t \in [0, \varepsilon]$. Note that if σ and δ are sufficiently small, then $d\phi(\widetilde{X}) > 0$ in $\widetilde{U} \cap C$.

By a target reparametrization of the function ϕ, we can arrange that $\phi|_{M\times\varepsilon} < \min \widetilde{\psi}$ and $\phi|_{M\times\frac{\varepsilon}{2}} > \max \widetilde{\psi}$. Define a J-convex function $\vartheta : \widetilde{U} \cup C' \to \mathbb{R}$ by

$$\vartheta := \begin{cases} \widetilde{\psi} & \text{on } \widetilde{U} \setminus C, \\ \text{smooth}\max(\widetilde{\psi}, \phi) & \text{on } \widetilde{U} \cap C, \\ \phi & \text{on } C'; \end{cases}$$

see Figure 16.5. Since $\widetilde{X} \cdot \phi > 0$ and $\widetilde{X} \cdot \widetilde{\psi} > 0$ on $\widetilde{U} \cap C$, the function ϑ has a unique index 1 critical point at $\widetilde{F}(0)$ and is constant on M. Set $a := \vartheta|_{\widetilde{\Sigma}\cap(W\setminus C)} = \widetilde{\psi}|_{\widetilde{\Sigma}}$. Then the domain $U := W \setminus \{\vartheta > a\}$ and the function $\vartheta|_{W\setminus \operatorname{Int} U}$ have the required properties. □

Let us discuss the topological implications of Theorem 16.7. Recall that if W is an elementary cobordism of dimension m with a unique index k critical point p, then $\partial_+ W$ is obtained from $\partial_- W$ by *surgery* on the stable sphere $S_p^- \subset \partial_- W$ (see e.g. [**115**]). More abstractly, surgery on an embedded $(k-1)$-sphere S in an $(m-1)$-manifold M with trivialized normal bundle consists of cutting out a tubular neighborhood $S^{k-1} \times D^{n-k}$ of S and gluing in $D^k \times S^{n-k-1}$ via the identity. The sphere corresponding to $0 \times S^{n-k-1}$ in the resulting manifold is called the *belt sphere.*

Thus, in the notation of Theorem 16.7, the boundary $N := \partial U$ is obtained from $M := \partial W$ by surgery on the sphere S, and conversely, M is obtained from N by surgery on the stable sphere in N of the unique critical point in $W \setminus \operatorname{Int} U$. To understand this better, we distinguish two cases.

Case 1. $M \setminus S$ is connected. Then N is the connected manifold obtained by cutting M open along S and gluing 3-balls to the two boundary spheres.

Case 2. $M \setminus S$ has two connected components with closures M_1, M_2. Then $N = N_1 \amalg N_2$ is the disjoint union of the two manifolds obtained by gluing 3-balls to the boundary spheres of M_1, M_2, and M is the connected sum $N_1 \# N_2$. Now there are again two cases.

Case 2a. None of the M_i is diffeomorphic to the 3-ball. Then M is the nontrivial connected sum $N_1 \# N_2$ with none of the N_i diffeomorphic to the 3-sphere.

Case 2b. One of the M_i is diffeomorphic to the 3-ball. Then $N = M \amalg S^3$ and M is the trivial connected sum $M \# S^3$. In this case the domain U in Theorem 16.7 is diffeomorphic to $W \amalg B^4$ and what we see is just the effect of creating a pair of critical points of index 0 and 1 for a J-convex function near the boundary of W.

Note that in Case 2b the sphere $S \subset M$ is *contractible*, i.e., it bounds an embedded 3-ball in M, while in Cases 1 and 2a it does not (so M is reducible).

Combining Theorem 16.7 with the Deformation Theorem 15.14, we obtain

THEOREM 16.8. *Suppose that a closed oriented* 3*-manifold* N *has a unique Stein filling up to deformation equivalence, and* M *is obtained from* N *by surgery on a* 0*-sphere. Then* M *has a unique Stein filling up to deformation equivalence as well.*

PROOF. Let (W, J) and (W', J') be two Stein fillings of M. Applying Theorem 16.7 to the belt sphere $S \subset M$ corresponding to the surgery on N, we find a compact domain $U \subset \operatorname{Int} W$ with smooth J-convex boundary diffeomorphic to N such that the cobordism $W \setminus \operatorname{Int} U$ admits a J-convex Morse function ϕ with exactly one critical point of index 1. Since (W, J) is Stein, by Lemma 5.8 and interpolation, the function ϕ extends (after target reparametrization) to a J-convex function on W. Similarly we find $U' \subset \operatorname{Int} W'$ and $\phi' : W' \to \mathbb{R}$ for (W', J').

By assumption the Stein domains (U, J, ϕ) and (U', J', ϕ') are deformation equivalent. So there exists a diffeomorphism $h : U \to U'$ and a Stein homotopy (J_t, ϕ_t) on U from (J, ϕ) to $(h^*J', h^*\phi')$. After target reparametrization and adjustments near ∂U we may assume that $\phi_t = \phi$ near ∂U for all $t \in [0,1]$. We can extend h over the elementary cobordism $V := W \setminus \operatorname{Int} U \cong W' \setminus \operatorname{Int} U'$ to a diffeomorphism $h : W \to W'$. Moreover, we can arrange that $\phi = \phi' \circ h$ on V. So we obtain two *subcritical* Stein cobordism structures (V, J, ϕ) and $(V, h^*J', h^*\phi' = \phi)$ with the same function which are connected by a Stein homotopy $(J_t, \phi_t = \phi)$ near $\partial_- V = \partial_+ U$. Hence by Theorem 15.14, after target reparametrization of ϕ, the Stein homotopy extends from $\mathcal{O}p\, \partial_- V$ to a Stein homotopy (V, J_t, ϕ) connecting (V, J, ϕ) and $(V, h^*J', h^*\phi' = \phi)$. This homotopy fits together with the homotopy on W to form a Stein homotopy (W, J_t, ϕ) connecting (W, J, ϕ) and $(W, h^*J', h^*\phi')$, thus (W, J) and (W', J') are deformation equivalent. $\square$

Combining Theorems 16.8 and 16.6, we obtain

THEOREM 16.9. *(a) If two closed oriented* 3*-manifolds* M_1, M_2 *have unique Stein fillings up deformation equivalence, then so does* $M_1 \# M_2$.

(b) Any Stein filling of $S^2 \times S^1$ *is deformation equivalent to the canonical (subcritical) Stein structure on* $B^3 \times S^1 = \{(z_1, z_2) \in \mathbb{C}^2 \mid |z_1|^2 + x_2^2 \leq 1\}/y_2 \sim y_2 + 1$ *with* $J = i$ *and* $\phi(z_1, z_2) = |z_1|^2 + x_2^2$.

(c) Any Stein filling of a k*-fold connected sum* $S^2 \times S^1 \# \cdots \# S^2 \times S^1$, $k \geq 1$, *is deformation equivalent to the canonical (subcritical) Stein structure on the* 4*-ball with* k 1*-handles attached.*

PROOF. Part (a) is just a special case of Theorem 16.8. As $S^2 \times S^1$ is obtained from S^3 by surgery on a 0-sphere, part (b) follows from Theorem 16.8 and Theorem 16.6. Part (c) follows from (a) and (b). □

All the above results concerning uniqueness of Stein fillings up to deformation equivalence have Weinstein counterparts as in Theorem 16.6.

Stein fillings of other 3-manifolds. In this section we will use the following terminology. We say that a contact manifold (M, ξ) has a *unique Stein filling up to symplectomorphism* if the following condition is satisfied. Suppose we are given two Stein domains (W_0, J_0, ϕ_0) and (W_1, J_1, ϕ_1) such that the induced contact structures on ∂W_0 and ∂W_1 are isomorphic to (M, ξ). Let $(\widehat{W}_0, \widehat{\omega}_0, \widehat{X}_0, \widehat{\phi}_0)$ and $(\widehat{W}_1, \widehat{\omega}_1, \widehat{X}_1, \widehat{\phi}_1)$ be Weinstein completions of the Weinstein domains $\mathfrak{W}(W_0, J_0, \phi_0)$ and $\mathfrak{W}(W_1, J_1, \phi_1)$. Then there exists a symplectomorphism $h : (\widehat{W}_0, \widehat{\omega}_0) \to (\widehat{W}_0, \widehat{\omega}_0)$ which at infinity sends the Liouville field $\widehat{X}_0$ to the Liouville field $\widehat{X}_1$. In particular, h induces a contactomorphism at infinity. Note that if (M, ξ) has a unique filling up to Stein deformation equivalence, then by Corollary 11.21 it also has a unique filling up to symplectomorphism.

The lens space $L(p, 1)$ admits exactly $(p - 1)$ pairwise non-isotopic Stein-fillable contact structures, see [**43, 67, 101, 95**]. These give rise to $[p/2]$ pairwise non-contactomorphic structures. One of these $[p/2]$ structures is obtained as the quotient of the standard contact structure on $S^3 \subset \mathbb{C}^2$ by the diagonal action $(z_1, z_2) \mapsto (e^{\frac{2\pi i}{p}} z_1, e^{\frac{2\pi i}{p}} z_2)$. We will refer to this structure as *standard*. Thus the standard structure is *universally tight*, i.e., its lift to the universal cover S^3 is tight. One can check that the lifts to S^3 of all other contact structures on $L(p, 1)$ are in homotopy classes of plane fields different from the class of the unique tight structure. Hence, according to [**43**] these lifts are overtwisted. So all nonstandard tight contact structures on $L(p, 1)$ are *virtually overtwisted*, i.e., they lift to overtwisted structures on the universal cover.

The following theorem is a combination of the results of several authors:

THEOREM 16.10 (McDuff [**133**], Plamenevskaya and Van Horn-Morris [**159**], Hind [**95**]). *All tight contact structures on* $L(p, 1)$, $p \geq 2$, *have unique Stein fillings up to symplectomorphism, except for the standard structure on* $L(4, 1)$ *which admits exactly two (non-diffeomorphic) Stein fillings. Moreover, the fillings of the standard structures are unique up to Stein deformation equivalence in each diffeomorphism class.*

The classification up to symplectomorphism of Stein fillings of the standard structures is due to McDuff [**133**], and up to Stein deformation equivalence it is due to Hind [**95**]. The uniqueness result for fillings of the virtually overtwisted structures is proven by Plamenevskaya and Van Horn-Morris [**159**], based on a theorem of Wendl in [**188**].

Lisca [**124**] completely classified Stein fillings up to *diffeomorphism* of all lens spaces $L(p,q)$ endowed with a universally tight contact structure.

REMARK 16.11. All the above results fit into a general program relating the classification of Stein fillings of certain contact manifolds to singularity theory. Namely, if a contact 3-manifold appears as the link of an isolated normal complex surface singularity, then one expects that all Stein fillings are given by the Milnor fibers corresponding to different irreducible components of the so-called miniversal space of deformations of the singularity. For instance, the quotient singularity of $\mathbb{C}^2$ by the diagonal action of $\mathbb{Z}_p$ has irreducible deformation space, except in the case $p = 2$ when there are exactly two irreducible components. This is the source of McDuff's classification result. Némethi and Popescu-Pampu [**146**] have shown that Lisca's Stein fillings of lens spaces correspond exactly to the different smoothings of the associated cyclic quotient singularities of $\mathbb{C}^2$.

The 3-torus T^3 carries infinitely many tight contact structures in the same homotopy class of plane fields [**68, 110**]. By contrast, it was proved in [**46**] that any Stein fillable contact structure on T^3 is contactomorphic to the standard one given by the field of complex tangencies on the boundary of the Stein domain

$$T^2 \times D^2 = \{(z_1, z_2) \in \mathbb{C}^2 |\, y_1^2 + y_2^2 \le 1\}/(x_1 \sim x_1 + 1,\, x_2 \sim x_2 + 1).$$

Wendl has further improved this result and showed

THEOREM 16.12 (Wendl [**188**]). *The standard contact structure on T^3 has a unique Stein filling up to symplectomorphism.*

Not every contact 3-manifold (M, ξ) is Stein fillable. First of all, the contact structure ξ has to be tight (see [**43**]). Moreover, there is a long hierarchy of different degrees of fillability of tight contact 3-manifolds with Stein fillability at the top; see [**45, 54**] for relevant discussions. On the other hand, there are contact 3-manifolds which admit infinitely many non-homeomorphic Stein fillings, see [**156, 174**]. Moreover, there are contact manifolds which admit infinitely many homeomorphic but non-diffeomorphic Stein fillings; see [**6**].

It also turns out that certain 3-manifolds do not admit any Stein fillings at all, regardless of the contact structure they carry. The first such example was obtained by Lisca [**123**] who proved that the Poincaré homology sphere P with one of its orientations admits no positive Stein fillable contact structure. It then follows from Theorem 16.7 that $P\#(-P)$ has no Stein filling with either orientation. Etnyre and Honda [**56**] improved Lisca's result by showing that P admits no positive tight contact structure with the above orientation, and hence $P\#(-P)$ admits no tight contact structure at all. As far as we know there are no known examples of irreducible orientable 3-manifolds which are not Stein fillable with either orientation.

There are no known examples of different Stein domain structures on the same 4-manifold with boundary which are homotopic as almost complex structures but not deformation equivalent (or, more strongly, whose boundaries are not contactomorphic). This is in sharp contrast to the situation in higher dimensions, as we will see in Chapter 17.

16.3. Stein structures on 4-manifolds

In the previous section we proved uniqueness up to deformation equivalence of Stein fillings of certain 3-manifolds. In this section we derive from these results uniqueness up to deformation equivalence of Stein manifolds with certain given ends.

We need some topological preparation. We say that a topological space X is of *finite type* if there exists a compact subset $A \subset X$ such that $X \setminus \operatorname{Int} A$ is homeomorphic to $\partial A \times [0, \infty)$. In this case we call ∂A an *end* of X.

LEMMA 16.13. *Any two ends of a finite type topological space are weakly homotopy equivalent.*

PROOF. Let $B = \partial A$ and $B' = \partial A'$ be two ends of X. Then we find compact intervals $I \subset J \subset [0, \infty)$ and $I' \subset J' \subset [0, \infty)$ such that $B \times I \subset B' \times I' \subset M \times J \subset M' \times J'$ under the homeomorphisms $X \setminus \operatorname{Int} A \approx B \times [0, \infty)$ and $X \setminus \operatorname{Int} A' \approx B' \times [0, \infty)$. Since the induced maps on homotopy groups $\pi_k(B \times I) \to \pi_k(B \times J)$ and $\pi_k(B' \times I') \to \pi_k(B' \times J')$ are isomorphisms, it follows that the map $\pi_k(B' \times I') \to \pi_k(B \times J)$ is an isomorphism as well. Thus for $a \in I'$ the map $B' \approx B' \times a \hookrightarrow B' \times I' \hookrightarrow B \times J \to B$ induced by the obvious inclusions and projections is a weak homotopy equivalence. □

Now we specialize to 4-manifolds. We say that a smooth oriented 4-manifold V is of *finite type* if there exists a compact subset $W \subset V$ with smooth boundary such that $V \setminus \operatorname{Int} W$ is diffeomorphic to $\partial W \times [0, \infty)$. In this case we call the closed oriented 3-manifold ∂W an *end* of V. It follows from Lemma 16.13 that the weak homotopy type of an end of V is determined by the homeomorphism type of V.

Let us say that a closed oriented 3-manifold M is *determined by its homotopy type* if every other closed oriented 3-manifold which is weakly homotopy equivalent to M is actually diffeomorphic to M by an orientation preserving diffeomorphism. Not every closed 3-manifold is determined by its homotopy type, counterexamples being provided by certain lens spaces. On the other hand, Perelman's proof of the geometrization conjecture [**158**] implies

THEOREM 16.14 (Perelman). *The manifolds S^3, $S^2 \times S^1$, $\mathbb{R}P^3$, and connected sums of these are determined up to orientation preserving diffeomorphism by their homotopy type (in fact, by their fundamental group).*

PROOF. Let M be a closed orientable 3-manifold whose fundamental group is a free product of copies of $\mathbb{Z}$ and $\mathbb{Z}_2$. It follows from Perelman's work (see [**141**, Theorem 0.1]) that M is diffeomorphic to a connected sum of copies of $S^2 \times S^1$ and spherical space forms. If $\pi_1(M) = 0$, this implies that M is diffeomorphic to S^3. Otherwise, each spherical space form appearing in the connected sum must have fundamental group $\mathbb{Z}_2$ and hence be diffeomorphic to $\mathbb{R}P^3$, so M is diffeomorphic to a connected sum of copies of $S^2 \times S^1$ and $\mathbb{R}P^3$ (with some orientations). Since all the manifolds S^3, $S^2 \times S^1$ and $\mathbb{R}P^3$ admit orientation reversing diffeomorphisms, we find an orientation preserving diffeomorphism between M and the connected sum of these manifolds with their standard orientations. □

The uniqueness results for Stein domains in Section 16.2 now imply uniqueness results for finite type Stein structures on their interiors:

THEOREM 16.15. *(a) Let W_1, W_2 be compact oriented 4-manifolds with boundary such that $\partial W_1 \# \partial W_2$ is determined by its homotopy type. If finite type Stein structures on the interiors W_1, W_2 are unique up to deformation equivalence, then so are finite type Stein fillings of the interior of the* boundary connected sum $W_1 \#_b W_2$ *(the manifold obtained from $W_1 \amalg W_2$ by attaching a 1-handle connecting W_1 and W_2).*

(b) Let W be a compact 4-manifold bounded by S^3, $S^2 \times S^1$, $\mathbb{R}P^3$, or a connected sum of these. Suppose that Int W *admits a finite type Stein manifold structure. Then* Int W *is diffeomorphic to $\mathbb{R}^4$, $\mathbb{R}^3 \times S^1$, T^*S^2, or the interior of a boundary connected sum of these, respectively, and the finite type Stein manifold structure on* Int W *is unique up to deformation equivalence.*

PROOF. (a) Let (J, ϕ) be a finite type Stein structure on Int $(W_1 \#_b W_2)$. Then for sufficiently large c the manifold $\{\phi < c\}$ is diffeomorphic to Int $(W_1 \#_b W_2)$. Since its end $M_1 \# M_2$ is determined by its homotopy type, the level set $\{\phi = c\}$ is diffeomorphic to $M_1 \# M_2$. Now the claim follows from the corresponding uniqueness result for the Stein domain $\{\phi \leq c\}$ provided by Theorem 16.9(a).

(b) Let (J, ϕ) be a finite type Stein structure on Int W. Then for sufficiently large c the manifold $\{\phi < c\}$ is diffeomorphic to Int W. Since according to Theorem 16.14 its end ∂W is determined by its homotopy type, the level set $\{\phi = c\}$ is diffeomorphic to ∂W. Now the claim follows from the corresponding uniqueness result for the Stein domain $\{\phi \leq c\}$ provided by Theorems 16.6, 16.10, and 16.9(b). □

Combining Theorem 16.15(b) with Corollary 11.27, we obtain the following uniqueness result up to *homotopy* rather than just deformation equivalence.

COROLLARY 16.16. *Any finite type Stein (or Weinstein) manifold structure on $\mathbb{R}^4$ is* homotopic *to the standard structure.* □

As we will see below (Corollary 17.5), $\mathbb{R}^{2n}$ admits infinitely many pairwise non-homotopic (in fact, non-symplectomorphic) finite type Stein structures for any $n \geq 3$.

Nonexistence of Stein structures. An analogue of the Existence Theorem 1.5 fails for 4-manifolds. For example, Theorem 16.15 implies the following nonexistence result.

THEOREM 16.17. *No finite type Stein surface is homeomorphic to $S^2 \times \mathbb{R}^2$.*

PROOF. Suppose that (V, J, ϕ) is a finite type Stein surface homeomorphic to $S^2 \times \mathbb{R}^2$. Then for sufficiently large c the manifold $\{\phi < c\}$ is diffeomorphic to V. Since according to Theorem 16.14 its end $S^2 \times S^1$ is determined by its homotopy type, the level set $\{\phi = c\}$ is diffeomorphic to $S^2 \times S^1$. Hence Theorem 16.15(b) implies that V is diffeomorphic to $\mathbb{R}^3 \times S^1$, which contradicts the hypothesis that V is homeomorphic to $S^2 \times \mathbb{R}^2$. □

REMARK 16.18. In [**125**] Lisca and Matić prove that $S^2 \times \mathbb{R}^2$, with its standard smooth structure, does not admit any (possibly infinite type) Stein manifold structure. Their proof requires the adjunction inequality of Kronheimer and Mrowka [**117**], proven via Seiberg–Witten theory. As Lisca and Matić show in [**125**], this implies that any homologically nontrivial embedded 2-sphere in a Stein surface must have self-intersection index ≤ -2. See [**74, 147, 148**] for further discussion.

In sharp contrast to these nonexistence results, Gompf [**70**] used the technique of Casson handles to prove an analogue of the Existence Theorem 1.5 for 4-manifolds, provided that the smooth structure is allowed to be changed (see Theorem 1.6 in the Introduction):

Every oriented open topological 4-manifold which admits a (possibly infinite) handlebody decomposition without handles of index > 2 is homeomorphic to a Stein surface.

For example, this shows that $S^2 \times \mathbb{R}^2$ is homeomorphic to a Stein surface, which in view of Theorem 16.17 is necessarily of infinite type.

Gompf also proved a topological analogue of the Ambient Existence Theorem 8.16:

THEOREM 16.19 (Gompf [**71**]). *An open subset U of a complex surface V is topologically isotopic to a Stein open subset if and only if it is homeomorphic to the interior of a handlebody without handles of index > 2.*

CHAPTER 17

Exotic Stein Structures

In this chapter we discuss how to distinguish Stein and Weinstein structures up to deformation equivalence. The main tool for this is symplectic homology, which turns out to be an invariant of Liouville manifolds up to Liouville homotopy. By considering their underlying Liouville structures, symplectic homology thus gives rise to a deformation invariant of Weinstein structures. In Section 17.2, we explain constructions of infinitely many pairwise non-deformation equivalent Stein structures on the same manifold that are distinguished by their symplectic homology.

17.1. Symplectic homology

In this section we recall the definition of symplectic homology and some of its properties. For details we refer to [**34, 57, 137, 169, 185**]. We fix a coefficient ring R with unit.

We begin with the completion (V, λ) of a Liouville domain $(W, \lambda|_W)$. Recall that $\lambda = e^r\alpha$ on $V \setminus W \cong \mathbb{R}_+ \times \partial W$, where $\alpha = \lambda|_{\partial W}$. Consider a Hamiltonian function $H : V \to \mathbb{R}$ which outside a compact set is of the form $H(r, x) = h(r)$ for a function $h : \mathbb{R}_+ \to \mathbb{R}$ satisfying $h'' \geq 0$ and $h'(r) \to \infty$ and $r \to \infty$. Define the action functional $\mathcal{A}_H : C^\infty(S^1, V) \to \mathbb{R}$ by

$$\mathcal{A}_H(x) := \int_0^1 x^*\lambda - \int_0^1 H\big(x(t)\big)dt.$$

Its critical points are 1-periodic solutions of the Hamiltonian system $\dot{x} = X_H(x)$, where X_H is the Hamiltonian vector field defined by $i_{X_H} d\lambda = -dH$. Pick an almost complex structure J on V which is compatible with ω in the sense that $g = \omega(\cdot, J\cdot)$ is a Riemannian metric. Moreover, we require that, outside a compact set, J is invariant under translation along $\mathbb{R}_+$, maps ∂_r to the Reeb vector field R_α, and preserves the contact structure $\xi = \ker \alpha$. Gradient flow lines of $\mathcal{A}_H$ with respect to the L^2-metric on the loop space induced by the metric g are maps $u : \mathbb{R} \times S^1 \to V$ satisfying the Floer equation

$$\partial_s u + J(u)\Big(\partial_t u - X_H(u)\Big) = 0.$$

To define Floer homology, we need to pick generic time-dependent perturbations of (H, J). Since the result does not depend on these perturbations, we will suppress them from our discussion. Then let CF_* be the free R-module generated by the critical points on $\mathcal{A}_H$. It is $\mathbb{Z}_2$-graded, and $\mathbb{Z}$-graded if $c_1(V) = 0$, by the Conley–Zehnder indices of 1-periodic orbits. The boundary operator $\partial : CF_* \to CF_{*-1}$ is defined on generators by

$$\partial x := \sum_y \langle x, y\rangle y,$$

where $\langle x, y\rangle$ is the signed count of isolated gradient flow lines from y to x. It satisfies $\partial^2 = 0$ and its homology HF_* does not depend on the choices of (H, J) (within the classes described above). So HF_* is an invariant of the Liouville domain; we denote it by $SH_*(W, \lambda)$ and call it the *symplectic homology* of (W, λ).

Next one observes [**169**] that two Liouville domains with the same completion have isomorphic symplectic homology, so symplectic homology yields an invariant of finite type Liouville manifolds which we still denote by $SH_*(V, \lambda)$. Moreover, symplectic homology is invariant under Liouville isomorphisms, i.e., diffeomorphisms $f : V_0 \to V_1$ between finite type Liouville manifolds (V_0, λ_0) and (V_1, λ_1) such that $f^*\lambda_1 - \lambda_0$ is exact and compactly supported. (This differs slightly from the definition in [**169**] where $f^*\lambda_1 - \lambda_0$ is required to be the differential of a compactly supported function.) Since by Proposition 11.8 homotopies of Liouville domains gives rise to Liouville isomorphisms of their completions, symplectic homology defines a homotopy invariant of Liouville domains.

For a Liouville subdomain W in a finite type Liouville manifold (V, λ), the inclusion $\iota : W \hookrightarrow V$ induces a homomorphism $\iota_\# : SH_*(V, \lambda) \to SH_*(W, \lambda|_W)$. The *transfer map* $\iota_\#$ was introduced by Viterbo [**185**] and it is functorial for nested inclusions. It allows us to extend the definition of symplectic homology to Liouville manifolds (V, λ) that are not of finite type as follows. Pick an exhaustion $V = \bigcup_{k\in\mathbb{N}} V^k$ by Liouville subdomains and use the transfer maps to define the inverse limit

$$SH_*(V, \lambda) := \lim_{\longleftarrow} SH_*(V^k, \lambda|_{V^k}).$$

Functoriality of the transfer map shows that this does not depend on the chosen exhaustion. A similar argument shows that symplectic homology of Liouville manifolds is invariant under exact symplectomorphisms, i.e., diffeomorphisms $f : V_0 \to V_1$ between Liouville manifolds (V_0, λ_0) and (V_1, λ_1) such that $f^*\lambda_1 - \lambda_0$ is exact. Since by Proposition 11.8 homotopies of Liouville manifolds give rise to exact symplectomorphisms, symplectic homology defines a homotopy invariant of Liouville manifolds. Note that for a finite type Liouville manifold, the new definition in terms of the inverse limit is canonically isomorphic to the earlier one. This concludes the definition of symplectic homology, which we summarize in

PROPOSITION 17.1. *Symplectic homology associates to every Liouville manifold (V, λ) a $\mathbb{Z}_2$-graded ($\mathbb{Z}$-graded if $c_1(V) = 0$) R-module $SH_*(V, \lambda)$ which is invariant under exact symplectomorphisms as well as Liouville homotopies. The inclusion $\iota : W \hookrightarrow V$ of a Liouville subdomain W in a Liouville manifold (V, λ) induces a transfer map $\iota_\# : SH_*(V, \lambda) \to SH_*(W, \lambda|_W)$.*

Let us now discuss some further properties of symplectic homology that will be relevant in the sequel. To simplify notation, we will usually drop λ from the notation.

(1) Symplectic homology comes with a canonical map (where $\dim V = 2n$)

$$H^{n-*}(V; R) \to SH_*(V),$$

which behaves naturally with respect to the maps in Proposition 17.1.

(2) The pair-of-pants product on Floer homology induces a product on symplectic homology which makes $SH_{n-*}(V)$ a graded commutative R-algebra with unit. The unit is the image of $1 \in H^0(V; R)$ under the canonical map $H^0(V; R) \to SH_n(V)$. The maps in Proposition 17.1 and in (1) are algebra homomorphisms.

Note that the unit in $SH_*(V)$ is zero if and only if $SH_*(V) = \{0\}$. Since the transfer map sends the unit to the unit, this has the following useful consequence first pointed out by McLean [**137**]: for a Liouville subdomain W in a Liouville manifold V, vanishing of $SH_*(V)$ implies vanishing of $SH_*(W)$.

(3) A somewhat different vanishing result is proved in [**35, 162**]: for a Liouville subdomain W in a Liouville manifold V, Hamiltonian displaceability of W in V implies vanishing of $SH_*(W)$. For example, in the stabilization $(V \times \mathbb{C}, \lambda + \lambda_{\rm st})$ (where $\lambda_{\rm st} = \frac{1}{2}(x\,dy - y\,dx)$ on $\mathbb{C}$) of a Liouville manifold (V, λ), each compact set is displaceable, so any Liouville subdomain of $V \times \mathbb{C}$ has vanishing symplectic homology.

(4) Attaching a Weinstein k-handle along a Legendrian $(k-1)$-sphere $\Lambda \subset \partial W$ to a Liouville domain W yields a new Liouville domain $W \cup_\Lambda H_k$ which contains W as a Liouville subdomain. It has been proved in [**32**] that the transfer map $\iota_\# : SH_*(W \cup_\Lambda H_k) \to SH(W)$ is an isomorphism in the subcritical case $k < n$. For example, this implies that every subcritical Weinstein manifold has vanishing symplectic homology. (Note that this also follows from the vanishing result in (3) and the Splitting Theorem 14.16 above.)

A special case of subcritical handle attaching is the *boundary connected sum* $W_1 \#_b W_2$ of two Liouville domains of dimension $2n \geq 4$. This is the result of attaching a 1-handle to $W_1 \amalg W_2$ connecting two points $p_i \in \partial W_i$. It follows that $SH_*(W_1 \#_b W_2) \cong SH_*(W_1) \oplus SH_*(W_2)$ as an R-algebra.

(5) The behavior of symplectic homology under attaching of a critical handle is more complicated. An answer is given in [**24**] in the form of a *surgery exact sequence*

$$\cdots SH_*(W \cup_\Lambda H_n) \xrightarrow{\iota_\#} SH_*(W) \longrightarrow LH_*^{\rm Ho}(\Lambda) \longrightarrow SH_{*-1}(W \cup_\Lambda H_n) \cdots . \tag{17.1}$$

(At the time of writing this book, the proof of this result is not yet completed; in particular, the compatibility of the maps with the product structures is not yet established.) Here $LH_*^{\rm Ho}(\Lambda)$ is a version of Legendrian contact homology which is in general difficult to compute. We will see in the next section examples where this computation is possible.

(6) The first nontrivial computation of symplectic homology was carried out for cotangent bundles [**185, 165, 1, 2**]: Let M be a closed manifold and denote by $LM = C^0(S^1, M)$ its loop space. Then there is an isomorphism

$$SH_*(T^*M, p\,dq) \cong H_*(LM; R)$$

which relates the pair-of-pants product on $SH_*(T^*M)$ to the Chas–Sullivan loop product on $H_*(LM; R)$. In particular, we have $SH_*(T^*M) \neq 0$. As an application, consider a closed exact (i.e., $\lambda|_L$ is exact) Lagrangian submanifold L in a Liouville manifold (V, λ). Then Weinstein's Lagrangian neighborhood theorem yields an exact symplectic embedding of the unit codisc bundle D^*L into V. So the vanishing results in (2) and (3) imply that, in this situation, $SH_*(V) \neq 0$ and L is not Hamiltonian displaceable in V. (In particular, we recover Gromov's theorem on the nonexistence of closed exact Lagrangian submanifolds in $\mathbb{C}^n$).

17.2. Exotic Stein structures

By an "exotic" Stein structure one means a Stein structure on a manifold such as $\mathbb{C}^n$ which is not deformation equivalent to the standard one. The first examples of exotic Stein structures on $\mathbb{C}^{2m}$, $m \geq 2$, were constructed in 2005 by Seidel and

Smith [**167**]; they were distinguished from the standard structure by the presence of a nondisplaceable Lagrangian torus. In 2009 McLean [**137**] constructed infinitely many pairwise non-deformation equivalent Stein structures on $\mathbb{C}^n$ and T^*M for $n = \dim M \geq 4$, distinguished by their symplectic homologies. Other constructions of exotic Stein structures are given in [**131, 132, 3**]. In particular, Abouzaid and Seidel [**3**] have extended McLean's result to the case $n = 3$.

In this section we will discuss McLean's theorem and explain how, combined with the surgery exact sequence from [**24**], it leads to

THEOREM 17.2. *Let (V, J) be an almost complex manifold of real dimension $2n \geq 6$ which admits an exhausting Morse function with finitely many critical points all of which have index $\leq n$. Then V carries infinitely many finite type Stein structures (J_k, ϕ_k), $k \in \mathbb{N}$, such that the J_k are homotopic to J as almost complex structures and (J_k, ϕ_k), (J_ℓ, ϕ_ℓ) are not deformation equivalent for $k \neq \ell$.*

We begin by recalling McLean's theorem [**137**]. In this section we denote by $SH_*(V)$ symplectic homology (of a Liouville manifold V) with coefficient ring $\mathbb{Z}_2$. Recall from the previous section that $SH_{n-*}(V)$ is a graded commutative ring with unit. Following [**137**], we associate to every Liouville manifold V the quantity

$$i(V) := \text{ number of idempotent elements in } SH_{n-*}(V).$$

The properties of symplectic homology imply the following properties of $i(V)$:

(a) $i(V)$ (which may be infinite) is invariant under exact symplectomorphisms; in particular, it defines a deformation invariant of Stein manifolds.

(b) If $SH_*(V) = \{0\}$, then 0 is the only idempotent and thus $i(V) = 1$; if $SH_*(V) \neq \{0\}$, then the unit and 0 define two different idempotents, and thus $i(V) \geq 2$.

(c) For the *end connected sum* (i.e., the completion of the boundary connected sum of the underlying Liouville domains) of two finite type Liouville manifolds V_1, V_2 of dimension $2n \geq 4$, we have

$$i(V_1 \#_e V_2) = i(V_1) i(V_2).$$

Now we can state McLean's theorem.

THEOREM 17.3 (McLean [**137**], Abouzaid and Seidel [**3**]). *For every $n \geq 3$, there exists a finite type Stein manifold K_n of complex dimension n which is diffeomorphic to $\mathbb{C}^n$ and satisfies $1 < i(K_n) < \infty$.*

By the preceding discussion, this immediately implies

COROLLARY 17.4 ([**137, 3**]). *Let (V, J_0, ϕ_0) be any finite type Stein manifold of complex dimension $n \geq 3$ with $i(V) < \infty$. Then the end connected sums*

$$V_k := V \#_e K_n \#_e \cdots \#_e K_n$$

with $k \geq 0$ copies of K_n define finite type Stein structures (J_k, ϕ_k) on V with the following properties:

- *J_k is homotopic to J_0 as almost complex structures;*
- *$i(V_k) = i(V) i(K_n)^k$ and hence the Stein structures (J_k, ϕ_k), (J_ℓ, ϕ_ℓ) are not deformation equivalent for $k \neq \ell$.*

In particular, by properties (4) and (6) in Section 17.1 the standard Stein structures on $\mathbb{C}^n$ and T^*M satisfy $i(\mathbb{C}^n) = 1$ and $i(T^*M) = 2^c$, where c is the number of connected components of M. So Corollary 17.4 yields

COROLLARY 17.5 ([**137, 3**]). *On* $\mathbb{C}^n$ *and* T^*M, $n = \dim M \geq 3$, *there exist infinitely many finite type Stein structures that are pairwise not deformation equivalent.*

For the proof of Theorem 17.2 we need one more ingredient.

THEOREM 17.6 ([**24**]). *Let* (V, J, ϕ) *be any finite type Stein manifold of complex dimension* $n \geq 3$. *Then there exists a finite type Stein structure* (J', ϕ') *on* V *such that* J' *is homotopic to* J *as almost complex structures and* $SH_*(V, J', \phi') = \{0\}$.

PROOF. (V, J) is obtained from a subcritical Weinstein domain W by attaching $k \geq 0$ n-handles along disjoint Legendrian $(n-1)$-spheres $\Lambda_1, \dots \Lambda_k \subset \partial W$. Since W is subcritical, its symplectic homology vanishes. Now it is explained in [**24**, Section 6.2] that each Λ_i can be modified to a new Legendrian $(n-1)$-sphere $\Lambda_i' \subset \partial W$ with the following properties:

- attaching n-cells to W along $\Lambda_1', \dots, \Lambda_k'$ yields a Stein manifold (V', J', ϕ') diffeomorphic to V;
- the pullback of J' under the diffeomorphism $V \to V'$ is homotopic to J as almost complex structure;
- the Legendrian contact homologies $LH_*^{\mathrm{Ho}}(\Lambda_i')$ vanish.

Hence the surgery exact sequence (17.1) implies vanishing of $SH_*(V', J', \phi')$. □

PROOF OF THEOREM 17.2. By the Existence Theorem 1.5 there exists a finite type Stein structure (J_0, ϕ_0) on V such that J_0 is homotopic to J as almost complex structure. After applying Theorem 17.6, we may assume that $SH_*(V, J_0, \phi_0) = \{0\}$ and thus $i(V, J_0, \phi_0) = 1$. Now Corollary 17.4 yields infinitely many finite type Stein structures (J_k, ϕ_k), $k \in \mathbb{N}$, on V such that the J_k are homotopic to J as almost complex structures and (J_k, ϕ_k), (J_ℓ, ϕ_ℓ) are not deformation equivalent for $k \neq \ell$. □

APPENDIX A

Some Algebraic Topology

In this appendix we collect some standard facts from algebraic topology that are used in the book.

A.1. Serre fibrations

In this section we collect some facts about Serre fibrations that are used in the book; see [**91**] for further discussion. Throughout this section all spaces are topological spaces and all maps are continuous.

Consider a map $\pi : E \to B$. A map $\widetilde{f} : X \to E$ is called a *lift* of $f : X \to B$ if $\pi \circ \widetilde{f} = f$. We say that π has the *lift extension property* for a space pair (X, A) if any map $X \to B$ has a lift $X \to E$ extending any given lift $A \to E$. Let $I := [0, 1]$. The *homotopy lifting property* of π for a space pair (X, A) is the lift extension property for the pair $(I \times X, 0 \times X \cup I \times A)$, i.e., any homotopy $I \times X \to B$ has a lift $I \times X \to E$ extending any given lift $0 \times X \cup I \times A \to E$. The homotopy lifting property for a space X is the homotopy lifting property for the pair $(X, \varnothing)$.

We denote by D^k the closed unit disc in $\mathbb{R}^k$. Note that the homotopy lifting property for D^k implies the homotopy lifting property for all D^ℓ, $\ell \leq k$. Since the pair $(I \times D^k, 0 \times D^k \cup I \times \partial D^k)$ is homeomorphic to $(I \times D^k, 0 \times D^k)$, the homotopy lifting property for D^k implies the homotopy lifting property for the pair $(D^k, \partial D^k)$, and hence for all k-dimensional CW pairs.

The homotopy lifting property for a point is also called the *path lifting property.* It implies surjectivity of π if B is path connected (and of course E is nonempty). The map π is called a *Serre fibration* if it has the homotopy lifting property for all closed k-discs D^k.

Let us fix points $e \in E$ and $b = \pi(e) \in B$ and define the "fiber" $F := \pi^{-1}(b)$. In the following all homotopy groups are taken with base points e (resp. b). We denote by $D^k_{1/2}$ the disc of radius $1/2$.

LEMMA A.1. *Suppose that B is path connected and $\pi : E \to B$ has the homotopy lifting property for D^{k-1}, for some $k \geq 1$. Then the following are equivalent:*

(a) the induced map π_* on homotopy groups is injective on π_{k-1} and surjective on π_k;

(b) $\pi_{k-1} F = 0$;

(c) any map $f : D^k \to B$ with $f|_{D^k_{1/2}} \equiv b$ has a lift $D^k \to E$ extending any given lift $\partial D^k \to E$.

REMARK A.2. Up to the technical condition $f|_{D^k_{1/2}} \equiv b$, part (c) is just the lift extension property for the pair $(D^k, \partial D^k)$. Now any map $f : D^k \to B$ is homotopic rel ∂D^k to one which is constant on $D^k_{1/2}$. Hence, if π has the homotopy lifting property for the pair $(D^k, \partial D^k)$, then part (c) can be replaced by the lift extension property for the pair $(D^k, \partial D^k)$.

PROOF. The homotopy lifting property for D^{k-1} allows us to define a connecting homomorphism $\partial : \pi_k B \to \pi_{k-1} F$ such that we get an exact sequence

$$\pi_k F \to \pi_k E \stackrel{\pi_*}{\to} \pi_k B \stackrel{\partial}{\to} \pi_{k-1} F \to \pi_{k-1} E \stackrel{\pi_*}{\to} \pi_{k-1} B.$$

The equivalence of (a) and (b) follows from this sequence. (b) follows from (c) applied to the map $D^k \to b \in B$ and a lift $\partial D^k \to F$. To show that (b) implies (c), consider a map $f : D^k \to B$ with $f|_{D^k_{1/2}} \equiv b$ and a lift $\widetilde{f} : \partial D^k \to E$. By the homotopy lifting property for S^{k-1}, the map $\widetilde{f}$ extends to a lift $\widetilde{f} : D^k \setminus D^k_{1/2} \to E$. Since $\pi_{k-1} F = 0$, the map $\widetilde{f}|_{\partial D^k_{1/2}} : \partial D^k_{1/2} \to F$ extends to a map $\widetilde{f} : D^k_{1/2} \to F$, so altogether we get the desired lift $\widetilde{f} : D^k \to E$. □

In particular, Lemma A.1 for all $k \geq 1$ together with Remark A.2 yields

COROLLARY A.3. *Suppose that B is path connected and $\pi : E \to B$ is a Serre fibration. Then the following are equivalent:*

(a) π is a weak homotopy equivalence;

(b) the fiber F is weakly contractible;

(c) for each $k \geq 1$, any map $D^k \to B$ has a lift $D^k \to E$ extending any given lift $\partial D^k \to E$.

Recall the following standard construction from topology (see e.g. [**91**]). Given a map $f : X \to Y$ define the space

$$P := \{(x, \gamma) \mid x \in X,\ \gamma : [0,1] \to Y,\ f(x) = \gamma(0)\}.$$

Note that this is just the fiber product over Y of X and the path space of Y. It is easy to see that the inclusion

$$X \hookrightarrow P, \qquad x \mapsto (x, f(x)),$$

where $f(x)$ denotes the constant path at $f(x)$, is a deformation retract and

$$\pi : P \to Y, \qquad (x, \gamma) \mapsto \gamma(1),$$

defines a Serre fibration. Its fiber at $y \in Y$

$$F := \{(x, \gamma) \mid x \in X,\ \gamma : [0,1] \to Y,\ f(x) = \gamma(0),\ \gamma(1) = y\}$$

is called the *homotopy fiber* of f. In particular, we get a homotopy exact sequence

$$\cdots \to \pi_k F \to \pi_k X \stackrel{f_*}{\to} \pi_k Y \to \pi_{k-1} F \to \cdots$$

and Corollary A.3 implies

COROLLARY A.4. *Consider $f : X \to Y$ and define the Serre fibration $\pi : P \to Y$ as above. Suppose that Y is path connected. Then the following are equivalent:*

(a) f is a weak homotopy equivalence;

(b) the homotopy fiber F is weakly contractible;

(c) for each $k \geq 1$, any map $D^k \to Y$ has a lift $D^k \to P$ extending any given lift $\partial D^k \to P$.

More generally, consider a commuting triangle

$$\begin{array}{ccc} X & \xrightarrow{\quad f \quad} & Y \\ & \searrow^{\pi_X} \quad {}^{\pi_Y}\swarrow & \\ & B & \end{array} \tag{A.1}$$

For each $b \in B$ we get an induced map $f_b := f|_{X_b} : X_b = \pi_X^{-1}(b) \to Y_b = \pi_Y^{-1}(b)$. Define the spaces

$$P_b := \{(x, \gamma) \mid x \in X_b,\ \gamma : [0, 1] \to Y_b,\ f(x) = \gamma(0)\},$$
$$P := \{(x, \gamma) \mid x \in X,\ \gamma : [0, 1] \to Y_{\pi_X(x)},\ f(x) = \gamma(0)\} = \bigcup_{b \in B} P_b.$$

The inclusion

$$X \hookrightarrow P, \qquad x \mapsto \big(x, f(x)\big),$$

is again a deformation retract. The projection

$$\pi_P : P \to Y, \qquad (x, \gamma) \mapsto \gamma(1),$$

will in general not be a Serre fibration. Note, however, that the fibers at $y \in Y$ of $\pi_P : P \to Y$ and its restriction $\pi_{P,b} : P_b \to Y_b$, where $b = \pi_Y(y)$, both equal

$$F = \{(x, \gamma) \mid x \in X_b,\ \gamma : [0, 1] \to Y_b,\ f(x) = \gamma(0),\ \gamma(1) = y\}.$$

Hence Lemma A.1 and Remark A.2 imply

COROLLARY A.5. *Consider a commuting triangle as in* (A.1) *and define* $\pi_P : P \to Y$ *and* $\pi_{P,b} : P_b \to Y_b$ *as above. Suppose that* Y *is path connected and* π_P *has the homotopy lifting property for* D^{k-1}, *for some* $k \geq 1$. *Then the following are equivalent:*

(a) the induced map f_* *on homotopy groups is injective on* π_{k-1} *and surjective on* π_k;

(b) $\pi_{k-1} F = 0$;

(c) any map $g : D^k \to Y$ *with* $\pi_Y \circ g|_{D^k_{1/2}} \equiv b$ *has a lift* $D^k \to P$ *extending any given lift* $\partial D^k \to P$;

(d) the induced map $(f_b)_*$ *on homotopy groups is injective on* π_{k-1} *and surjective on* π_k;

(e) any map $g : D^k \to Y_b$ *has a lift* $D^k \to P_b$ *extending any given lift* $\partial D^k \to P_b$.

COROLLARY A.6. *Consider a commuting triangle as in* (A.1) *and define* $\pi_P : P \to Y$ *as above. Suppose that* Y *is path connected and* π_P *is a Serre fibration. Then the following are equivalent:*

(a) $f : X \to Y$ *is a weak homotopy equivalence;*

(b) the homotopy fiber F *is weakly contractible;*

(c) $f_b : X_b \to Y_b$ *is a weak homotopy equivalence.*

A.2. Some homotopy groups

In this section we collect some results on homotopy groups that are used in this book. The following lemma will be useful.

LEMMA A.7. *Let $p : E \to B$ be a Serre fibration with fiber $F = p^{-1}(b)$. Then the map $p : (E, F) \to (B, b)$ induces an isomorphism of homotopy groups*

$$p_* : \pi_i(E, F) \to \pi_i(B).$$

PROOF. The long exact sequences of the pair (E, F) and of the Serre fibration $F \to E \to B$ fit into a diagram

$$\begin{array}{ccccccc} \cdots \pi_i(F) & \longrightarrow & \pi_i(E) & \longrightarrow & \pi_i(E,F) & \xrightarrow{\partial} & \pi_{i-1}(F) \cdots \\ \downarrow \mathrm{id} & & \downarrow \mathrm{id} & & \downarrow p_* & & \downarrow \mathrm{id} \\ \cdots \pi_i(F) & \longrightarrow & \pi_i(E) & \longrightarrow & \pi_i(B) & \xrightarrow{\partial} & \pi_{i-1}(F) \cdots . \end{array}$$

The definition of the boundary maps ∂ shows that this diagram commutes, so by the five-lemma p_* is an isomorphism. □

For $1 \leq k \leq n$ denote by $V_{n,k}$ the *Stiefel manifold* of orthonormal k-frames in $\mathbb{R}^n$, and by $G_{n,k}$ the *Grassmannian* of k-dimensional subspaces in $\mathbb{R}^n$. The obvious projection $p : G_{n,k} \to V_{n,k}$ defines a fibration

$$O(k) \to V_{n,k} \to G_{n,k}$$

with fiber the orthogonal group $O(k)$. For $\ell < k \leq n$ the map $V_{n,k} \to V_{\ell,k}$ that forgets the last $k - \ell$ vectors defines a fibration

$$V_{n-\ell,k-\ell} \to V_{n,k} \to V_{n,\ell}.$$

Here an explicit inclusion $V_{n-\ell,k-\ell} \hookrightarrow V_{n,k}$ is given by adding to a $(k - \ell)$-frame in $\mathbb{R}^{n-\ell} \times \{0\} \subset \mathbb{R}^n$ the last ℓ standard basis vectors. Note that $V_{n,n} \cong O(n)$ and $V_{n,1} \cong S^{n-1}$. Thus the preceding fibration includes the following special cases:

$$V_{n-1,k-1} \to V_{n,k} \to S^{n-1}, \tag{A.2}$$

$$O(n-k) \to O(n) \to V_{n,k}, \tag{A.3}$$

$$O(n-1) \to O(n) \to S^{n-1}. \tag{A.4}$$

Of course, the preceding discussion carries over to the complex case: Just replace everywhere $V_{n,k}$ by the complex Stiefel manifold $V^{\mathbb{C}}_{n,k}$, $G_{n,k}$ by the complex Grassmannian $G^{\mathbb{C}}_{n,k}$, $O(n)$ by the unitary group $U(n)$, and S^{n-1} by S^{2n-1}. Moreover, the fibrations for the real and complex Stiefel manifolds fit into the following commuting diagram, where the vertical maps are the natural inclusions:

$$\begin{array}{ccccc} V^{\mathbb{C}}_{n-1,k-1} & \longrightarrow & V^{\mathbb{C}}_{n,k} & \longrightarrow & S^{2n-1} \\ \downarrow & & \downarrow & & \downarrow \mathrm{id} \\ V_{2n-1,k-1} & \longrightarrow & V_{2n,k} & \longrightarrow & S^{2n-1}. \end{array}$$

Lemma A.7 applied to this diagram yields

COROLLARY A.8. *(a) For $i \leq 2n - 2$ we have*

$$\pi_i(V^{\mathbb{C}}_{n,k}, V^{\mathbb{C}}_{n-1,k-1}) \cong \pi_i(V_{2n,k}, V_{2n-1,k-1}) \cong \pi_i(S^{2n-1}) = 0.$$

(b) For $i = 2n - 1$ the inclusion $(V^{\mathbb{C}}_{n,k}, V^{\mathbb{C}}_{n-1,k-1}) \hookrightarrow (V_{2n,k}, V_{2n-1,k-1})$ induces an isomorphism

$$\mathbb{Z} \cong \pi_{2n-1}(V^{\mathbb{C}}_{n,k}, V^{\mathbb{C}}_{n-1,k-1}) \to \pi_{2n-1}(V_{2n,k}, V_{2n-1,k-1}) \cong \mathbb{Z}.$$

The following lemma gives more information about the homotopy groups of Stiefel manifolds.

LEMMA A.9. *(a) The map* $\pi_i V_{n-1,k-1} \to \pi_i V_{n,k}$ *induced by the inclusion is an isomorphism for* $i < n-2$ *and surjective for* $i = n-2$. *Similarly, the map* $\pi_i V^{\mathbb{C}}_{n-1,k-1} \to \pi_i V^{\mathbb{C}}_{n,k}$ *is an isomorphism for* $i < 2n-2$ *and surjective for* $i = 2n-2$.

(b) $V_{n,k}$ *is* $(n-k-1)$*-connected and* $V^{\mathbb{C}}_{n,k}$ *is* $(2n-2k)$*-connected.*

(c) For $n \geq k+2$, *the group* $\pi_k V_{n,n-k}$ *equals* $\mathbb{Z}$ *if* k *is even or* $k=1$, *and* $\mathbb{Z}_2$ *if* $k > 1$ *is odd.*

PROOF. Part (a) follows directly from the long exact sequence of the fibration (A.2) because S^{n-1} is $(n-2)$-connected. For part (b), let $i < n-k$. Then it follows by induction from part (a) that $\pi_i V_{n,k} = \pi_i V_{n-k+1,1} = \pi_i S^{n-k} = 0$. The complex cases are analogous.

For part (c), let $n \geq k+2$ and $k \geq 2$ (the case $k=1$ is trivial). Then it follows by induction from part (a) that $\pi_k V_{n,n-k} = \pi_k V_{k+2,2}$. Now observe that an element of $V_{k+2,2}$ is a unit vector in $\mathbb{R}^{k+2}$ and a second unit vector orthogonal to the first one. Thus $V_{k+2,2}$ equals the tangent sphere bundle of S^{k+1}, and the fibration (A.2)

$$V_{k+1,1} \cong S^k \to V_{k+2,2} \to S^{k+1}$$

describes this bundle. Now for an oriented sphere bundle $S^k \to E \to B$, the boundary map $\pi_{k+1} B \to \pi_k S^k \cong \mathbb{Z}$ in the long exact sequence is given by evaluation of the Euler class $e(E) \in H^{k+1}(B)$ (this follows directly from the definition of the obstruction cocycle representing the Euler class in [**177**]). Thus the fibration above yields an exact sequence

$$\pi_{k+1} S^{k+1} \cong \mathbb{Z} \stackrel{\cdot\chi(S^{k+1})}{\longrightarrow} \pi_k S^k \cong \mathbb{Z} \to \pi_k V_{k+2,2} \to 0,$$

where the first map is multiplication with the Euler characteristic of S^{k+1}. Since $\chi(S^{k+1})$ is 0 for k even and 2 for k odd, it follows that $\pi_k V_{n,n-k} = \pi_k V_{k+2,2}$ equals $\mathbb{Z}$ for k even and $\mathbb{Z}_2$ for k odd. □

Setting $k = n$ in Lemma A.9(a) we find

COROLLARY A.10. *The map* $\pi_i O(n-1) \to \pi_i O(n)$ *induced by the inclusion is an isomorphism for* $i < n-2$ *and surjective for* $i = n-2$. *Similarly, the map* $\pi_i U(n-1) \to \pi_i U(n)$ *is an isomorphism for* $i < 2n-2$ *and surjective for* $i = 2n-2$.

Define the *stable homotopy groups* $\pi_i O := \pi_i O(n)$ for $i < n-1$ and $\pi_i U := \pi_i U(n)$ for $i < 2n$ (this is independent of n by the preceding corollary). These groups are determined by the celebrated

THEOREM A.11 (Bott periodicity theorem [**21**]). *(a) The stable homotopy group* $\pi_i U$ *equals* 0 *if* i *is even and* $\mathbb{Z}$ *if* i *is odd.*

(b) The stable homotopy group $\pi_i O$ *equals* $\mathbb{Z}_2$ *if* $i \equiv 0$ *or* $1 \pmod 8$, $\mathbb{Z}$ *if* $i \equiv 3$ *or* $7 \pmod 8$, *and* 0 *otherwise.*

We conclude this section with two lemmas from [**143**] that we will need in Appendix B.

LEMMA A.12. *The homomorphism* $i : \pi_{n+1} U(n) \to \pi_{n+1} V_{2n+1,n}$ *is trivial for* $n \neq 2$, *and a surjection* $\mathbb{Z} \to \mathbb{Z}/2$ *for* $n = 2$.

PROOF. If n is odd, the homomorphism $\pi_{n+1}U(n) \to \pi_{n+1}V_{2n+1,n}$ is trivial simply because $\pi_{n+1}U(n) = 0$. Suppose now that n is even. The inclusion map $U(n) \to V_{2n+1,n}$ factors as $U(n) \to O(2n+1) \to V_{2n+1,n}$. Consider the homotopy exact sequence

$$\text{(A.5)} \qquad \pi_{n+1}O(2n+1) \xrightarrow{p} \pi_{n+1}V_{2n+1,n} \xrightarrow{\delta} \pi_n O(n+1) \xrightarrow{j} \pi_n O(2n+1)$$

of the fibration $O(n+1) \to O(2n+1) \to V_{2n+1,n}$. Recall the geometric description of $\ker(j)$ from [**177**, §23]: Gluing two trivial bundles over the $(n+1)$-disc by a map $S^n \to O(m)$ yields a one-to-one correspondence between $\pi_n O(m)$ and isomorphism classes of $O(m)$-bundles over S^{n+1}. Since $\pi_n O(n+2) = \pi_n O(2n+1) = \pi_n O$ by Corollary A.10, the kernel of j classifies stably trivial rank $(n+1)$ oriented vector bundles over S^{n+1}. Next observe that the bundle $O(n+1) \to O(n+2) \to S^{n+1}$ is the bundle of orthonormal frames in the tangent bundle of S^{n+1}. So in the associated exact sequence

$$\pi_{n+1}S^{n+1} \to \pi_n O(n+1) \xrightarrow{j} \pi_n O(n+2) \cong \pi_n O(2n+1),$$

the image of the generator of $\pi_{n+1}S^{n+1}$ (which generates $\ker j$) corresponds to the gluing map of the tangent bundle of S^{n+1}. This shows that $\ker j$ is trivial if and only the tangent bundle of S^{n+1} is trivial, which is the case exactly for $n = 2$ or $n = 6$ (see [**23, 111**]). Since $\pi_{n+1}V_{2n+1,n} = \mathbb{Z}/2$, the sequence (A.5) then implies that for $n \neq 2, 6$ the homomorphism p, and hence the homomorphism $i : \pi_{n+1}U(n) \to \pi_{n+1}V_{2n+1,n}$, is trivial.

It remains to consider the cases $n = 2$ and $n = 6$. In both cases the homotopy groups $\pi_{n+1}U(n)$ and $\pi_{n+1}O(2n+1)$ are in the stable range, so the Bott Periodicity Theorem A.11 yields $\pi_{n+1}U(n) = \pi_{n+1}U = \pi_{n+1}O(2n+1) = \pi_{n+1}O = \mathbb{Z}$. Moreover, it is shown in [**21**] that the quotient O/U is homotopy equivalent to the based loop space ΩO, hence the relative homotopy group $\pi_{n+1}(O/U) = \pi_{n+1}(\Omega O) = \pi_{n+2}O$ vanishes for $n = 2$ and equals $\mathbb{Z}_2$ for $n = 6$. It follows that for $n = 6$ the exact sequence

$$\pi_{n+1}U \to \pi_{n+1}O \to \pi_{n+1}(O/U) \to \pi_n U$$

takes the form $\mathbb{Z} \to \mathbb{Z} \to \mathbb{Z}/2 \to 0$. Hence the homomorphism $\pi_{n+1}U \to \pi_{n+1}O$, and therefore $\pi_{n+1}U(n) \to \pi_{n+1}O(2n+1)$, is multiplication by 2. But then the homomorphism $U(n) \to O(2n+1) \to V_{2n+1,n} = \mathbb{Z}/2$ is trivial. For $n = 2$ we conclude that the map $\pi_{n+1}U(n) \to \pi_{n+1}O(2n+1)$ is an isomorphism. Since $\pi_n O(n+1) = \pi_2 O(3) = 0$, it follows from (A.5) that the homomorphism $p : \pi_{n+1}O(2n+1) \to \pi_{n+1}V_{2n+1,n}$, and hence that map $i : \pi_{n+1}U(n) = \mathbb{Z} \to \pi_{n+1}V_{2n+1,n} = \mathbb{Z}/2$, is a surjection. □

LEMMA A.13. *Suppose n is odd. Let $\delta : \pi_{n+1}V_{2n+1,n} \to \pi_n O(n+1)$ be the boundary homomorphism of the exact sequence* (A.5) *and $k : \pi_n O(n+1) \to \pi_n S^n$ the projection homomorphism of the fibration $O(n) \to O(n+1) \to S^n$. Then $k \circ \delta : \pi_{n+1}V_{2n+1,n} = \mathbb{Z} \to \pi_n(S_n) = \mathbb{Z}$ is multiplication by* 2.

PROOF. The elements of the group $\pi_n O(n+1)$ classify $(n+1)$-dimensional vector bundles over S^{n+1}, and for each $x \in \pi_n O(n+1)$ the element $k(x) \in \pi_n S^n = \mathbb{Z}$ is the Euler number of the bundle corresponding to x. On the other hand, as noted in the proof of Lemma A.12, the image $\delta(\pi_{n+1}V_{2n+1,n})$ classifies stably trivial bundles and is generated by the tangent bundle TS^{n+1}. This implies the claim because $\chi(S^{n+1}) = 2$ for n odd. □

APPENDIX B

Obstructions to Formal Legendrian Isotopies

In this appendix, whose content is partially taken from Murphy's paper [**143**], we study obstructions to formal Legendrian isotopies between genuine Legendrian knots. We mainly restrict ourselves to spherical knots, the case which is most relevant to the content of this book. Namely, we consider the following question: Let (M, ξ) be a $(2n+1)$-dimensional contact manifold, $n \geq 1$. Given a smooth isotopy f_t, $t \in [0,1]$, connecting two Legendrian embeddings $f_0, f_1 : S^n \hookrightarrow (M, \xi)$, what are the obstructions for lifting it to a formal Legendrian isotopy (f_t, F_t^s)?

Recall that a formal Legendrian isotopy is a family $F_t^s : TS^n \to TM$, $s, t \in \mathbb{R}$, of monomorphisms (i.e., injective bundle homomorphisms) covering f_t such that $F_t^0 = df_t$, $F_0^s = df_0$, $F_1^s = df_1$, and F_t^1 are Legendrian monomorphisms $TS^n \to \xi \subset TM$. In view of Theorems 7.1 and 7.9, the problem of lifting f_t to a formal Legendrian isotopy is equivalent to the question whether the isotopy f_t is homotopic to a regular Legendrian homotopy $\widehat{f}_t$ through regular homotopies connecting f_0 and f_1.

To a pair of Legendrian spheres $f_0, f_1 : S^n \hookrightarrow (M^{2n+1}, \xi)$ connected by a smooth isotopy f_t, we will first associate their *relative rotation invariant* $r(f_0, f_1; f_t)$ taking values in $\mathbb{Z}$ if n is odd and vanishing for n even. If $r(f_0, f_1; f_t) = 0$, we will define a secondary invariant, the *self-intersection invariant* $I(f_0, f_1; f_t)$, taking values in $\mathbb{Z}$ if n is odd, in $\mathbb{Z}_2$ if $n > 2$ is even, and vanishing for $n = 2$. Both invariants turn out to depend only on f_0, f_1 and the homotopy class of the isotopy f_t in the space of continuous homotopies connecting f_0 and f_1. The main result of this appendix states that these give complete invariants for Legendrian spheres up to formal Legendrian isotopy.

THEOREM B.1. *A smooth isotopy $f_t : S^n \hookrightarrow M^{2n+1}$, $n \geq 1$, connecting two Legendrian spheres $f_0, f_1 : S^n \hookrightarrow (M, \xi)$ can be lifted to a formal Legendrian isotopy if and only if $r(f_0, f_1; f_t) = 0$ and $I(f_0, f_1; f_t) = 0$.*

Note that for $n = 2$ both invariants vanish, so *any smooth isotopy connecting two Legendrian 2-spheres can be lifted to a formal Legendrian isotopy.*

For n odd one can define another invariant, the *relative Thurston–Bennequin invariant* $tb(f_0, f_1; f_t) \in \mathbb{Z}$. We will show that for Legendrian spheres it is given by $tb(f_0, f_1; f_t) = 2I(f_0, f_1; f_t)$, so for n odd, r and tb are also complete invariants. For homologically trivial Legendrian knots we have $tb(f_0, f_1) = tb(f_0) - tb(f_1)$, i.e., the relative Thurston–Bennequin invariant is the difference of the classical absolute Thurston–Bennequin invariants. We will also discuss the effect of the stabilization construction in Section 7.4 on the invariants and Bennequin's inequality.

A slightly different question from the one discussed so far is the following: When can two Legendrian embeddings $f_0, f_1 : S^n \hookrightarrow (M, \xi)$ be connected by a formal Legendrian isotopy? Clearly, a necessary condition is the existence of a continuous (or equivalently, smooth) homotopy f_t connecting f_0 and f_1. Next we need to deform f_t to a smooth isotopy. If $n > 2$ and M is simply connected, this is always possible by Corollary 7.7, so in this case the question reduces to the one answered by Theorem B.1. Note, however, that the answer may depend on the homotopy class of the chosen homotopy f_t.

Homotopy obstructions to formal Legendrian isotopies.

Let us endow the contact bundle ξ with a compatible complex structure J. Then any Legendrian monomorphism $F : TS^n \to \xi$ can be complexified to a complex isomorphism $F \otimes \mathbb{C} : TS^n \otimes \mathbb{C} \to \xi$.

Fix a point $p \in S^n$. Since $\pi_1 V_{2n+1,n} = 0$ for all $n \geq 1$ (Lemma A.9), we can assume without loss of generality that $d_p f_t : T_p S^n \to \xi$ are Legendrian monomorphisms for all $t \in [0,1]$. Then df_t can be covered by a family of complex bundle isomorphisms $\Phi_t : TS^n \otimes \mathbb{C} \to f_t^* \xi$ such that $\Phi_0 = df_0 \otimes \mathbb{C}$ and $\Phi_t|_{T_p S^n} = d_p f_t \otimes \mathbb{C}$ for all $t \in [0,1]$. We use the Reeb vector field R of a contact form defining ξ to extend the Φ_t to real isomorphisms $\overline{\Phi}_t = \Phi_t \oplus R : (TS^n \otimes \mathbb{C}) \oplus \mathbb{R} \to f_t^* TM$. We will view TS^n as the real subbundle of $TS^n \otimes \mathbb{C}$ and denote by b the inclusion $TS^n \hookrightarrow TS^n \otimes \mathbb{C} \hookrightarrow (TS^n \otimes \mathbb{C}) \oplus \mathbb{R}$.

Thus any homotopy of monomorphisms $F_t : TS^n \to TM$ covering f_t can be equivalently viewed as a homotopy of monomorphisms $\widehat{F}_t := \overline{\Phi}_t^{-1} \circ F_t : TS^n \to (TS^n \otimes \mathbb{C}) \oplus \mathbb{R}$ covering the identity. Let us denote by Mon the space of monomorphisms $TS^n \to (TS^n \otimes \mathbb{C}) \oplus \mathbb{R}$ and by $\mathrm{Mon}_{\mathrm{based}}$ its subspace of *based* monomorphisms F satisfying $F|_{T_p S^n} = b|_{T_p S^n}$. Note that $\widehat{df}_t = \overline{\Phi}_t^{-1} \circ df_t$ is a path in $\mathrm{Mon}_{\mathrm{based}}$ starting at b.

LEMMA B.2. *The inclusion homomorphism* $\pi_1(\mathrm{Mon}, b) \to \pi_1(\mathrm{Mon}_{\mathrm{based}}, b)$ *is an isomorphism.*

PROOF. Restriction to the fibers at p defines a Serre fibration $p : \mathrm{Mon} \to \mathrm{Mon}(T_p S^n, (T_p S^n \otimes \mathbb{C}) \oplus \mathbb{R}) \cong V_{2n+1,n}$ with fiber $\mathrm{Mon}_{\mathrm{based}}$. Consider its homotopy sequence (we drop the basepoint b)

$$\pi_2 \mathrm{Mon} \xrightarrow{p_*} \pi_2 V_{2n+1,n} \to \pi_1 \mathrm{Mon}_{\mathrm{based}} \to \pi_1 \mathrm{Mon} \to \pi_1 V_{2n+1,n} = 0.$$

If $n > 1$, then $\pi_2 V_{2n+1,n} = 0$ by Lemma A.9. If $n = 1$, the map $p : \mathrm{Mon} \simeq \mathrm{Map}(S^1, S^2) \to V_{3,1} \simeq S^2$ is the evaluation map $\gamma \mapsto \gamma(p)$ of a loop at $p \in S^1$ and thus induces (via constant loops) a surjection on π_2. So for each $n \geq 1$ the homomorphism $p_* : \pi_2 \mathrm{Mon} \to \pi_2 V_{2n+1,n}$ is surjective, and the lemma follows. □

Note that the bundle $TS^n \otimes \mathbb{C}$ is trivial because any stably trivial n-dimensional complex bundle over S^n is trivial by Corollary A.10. Let us pick trivializations of the bundles $TS^n \otimes \mathbb{C}$ and $TS^n|_{S^n \setminus p}$. This allows us to identify the homotopy class with fixed endpoints of the path $\widehat{df}_t$, $t \in [0,1]$, in $\mathrm{Mon}_{\mathrm{based}}$ with an element in $\pi_{n+1}(V_{2n+1,n}, U(n))$ which we denote by $[\widehat{df}_t]$. Here we view $U(n) \subset V_{2n+1,n}$ as the subspace of unitary n-frames in $\mathbb{C}^n \subset \mathbb{C}^n \oplus \mathbb{R} = \mathbb{R}^{2n+1}$.

Lemma B.2 ensures that the class $[\widehat{df}_t] \in \pi_{n+1}(V_{2n+1,n}, U(n))$ is independent of the way we make $d_p f_t$ Legendrian. It is also independent of all other choices in the construction. Consider the exact sequence

$$\pi_{n+1} U(n) \xrightarrow{i} \pi_{n+1} V_{2n+1,n} \xrightarrow{j} \pi_{n+1}(V_{2n+1,n}, U(n)) \xrightarrow{\partial} \pi_n U(n). \tag{B.1}$$

We call the image $r(f_0, f_1; f_t) := \partial([\widehat{df}_t]) \in \pi_n U(n)$ the *relative rotation invariant*. Note that $r(f_0, f_1; f_t)$ can be defined for any *homotopy* connecting f_0 and f_1 and depends only on the homotopy class of this homotopy. Indeed, a homotopy between f_0 and f_1 gives an isocontact bundle isomorphism $f_0^*TM \cong f_1^*TM$, so we can write $F_1 = g \cdot F_0$ for a unique map $g : S^n \to U(n)$ whose homotopy class is $r(f_0, f_1; f_t)$. In particular, in a contact manifold M with trivial groups $\pi_1(M)$ and $\pi_n(M)$, the relative rotation invariant $r(f_0, f_1; f_t)$ is independent of f_t and can be denoted by $r(f_0, f_1)$. In this case we can also define the *absolute rotation invariant* $r(f) \in \pi_n U(n)$ of a Legendrian knot f as the relative rotation invariant $r(f, f_{\rm st})$, where $f_{\rm st}$ is a standard Legendrian unknot in a Darboux chart, so that we have $r(f_0, f_1) = r(f_0) - r(f_1)$.

If $r(f_0, f_1; f_t) = 0$, then one can define a *secondary invariant* by lifting the invariant $[\widehat{df}_t] \in \pi_{n+1}(V_{2n+1,n}, U(n))$ to an element

$$s(f_0, f_1; f_t) \in \pi_{n+1} V_{2n+1,n} / i(\pi_{n+1} U(n)).$$

The preceding discussion shows: *vanishing of the rotation invariant $r(f_0, f_1; f_t)$ and the secondary invariant $s(f_0, f_1; f_t)$ is a necessary and sufficient condition for existence of a lifting of the isotopy f_t to a formal Legendrian isotopy.*

For even n the relative rotation invariant $r(f_0, f_1; f_t)$ vanishes because it takes values in $\pi_n U(n) = 0$, and hence the secondary invariant $s(f_0, f_1; f_t)$ is always defined and takes values in $\pi_{n+1} V_{2n+1} / i(\pi_{n+1} U(n)) = \mathbb{Z}_2 / i(\mathbb{Z})$. According to Lemma A.12, the homomorphism i is surjective for $n = 2$ and zero for $n \neq 2$. Thus for even $n \neq 2$ the invariant s is $\mathbb{Z}_2$-valued. For $n = 2$ the invariant s vanishes, so any two Legendrian 2-spheres that are smoothly isotopic are formally Legendrian isotopic.

For odd n we have $\pi_n U(n) \cong \mathbb{Z}$ and $\pi_{n+1} U(n) = 0$, so both the relative rotation invariant $r(f_0, f_1; f_t) \in \pi_n U(n) \cong \mathbb{Z}$ and the secondary invariant

$$s(f_0, f_1; f_t) \in \pi_{n+1} V_{2n+1,n} / i(\pi_{n+1} U(n)) = \pi_{n+1} V_{2n+1,n} \cong \mathbb{Z}$$

are integer valued.

This discussion establishes Theorem B.1 with the invariant $s(f_0, f_1; f_t)$ in place of $I(f_0, f_1; f_t)$. In the following subsection we will define the geometric invariant $I(f_0, f_1; f_t)$ and show that it agrees with $s(f_0, f_1; f_t)$.

The self-intersection invariant.

Let $f_0, f_1 : S^n \hookrightarrow (M, \xi)$ be two Legendrian spheres connected by a smooth isotopy f_t. Suppose that $r(f_0, f_1; f_t) = 0$.

Gromov's h-principle for Legendrian immersions (Theorem 7.9) then implies that the isotopy f_t can be C^0-approximated by a Legendrian regular homotopy $\widetilde{f}_t$ which coincides with f_t at the point p together with its differential (this standardization at p is not really necessary but will be convenient for the following discussion). Let $I_{\{\widetilde{f}_t\}}$ be its self-intersection index as defined in Section 7.1. Recall that it takes values in $\mathbb{Z}$ if n is odd, and in $\mathbb{Z}_2$ if n is even.

If $\widehat{f}_t$ is another such Legendrian regular homotopy connecting f_0 and f_1, then together $\widetilde{f}_t$ and $\widehat{f}_t$ give rise to an element $\Delta \in \pi_{n+1} U(n)$ and the difference $I_{\{\widetilde{f}_t\}} - I_{\{\widehat{f}_t\}}$ is determined by the image of Δ under the map $i : \pi_{n+1} U(n) \to \pi_{n+1} V_{2n+1,n}$. Recall that by Lemma A.12 the map i is surjective for $n = 2$ and vanishes for $n \neq 2$. Since one can always add a new self-intersection point in the interior of a given regular homotopy (see [**191**]), this implies that for $n = 2$ the Legendrian regular homotopy $\widetilde{f}_t$ can be always chosen to have $I_{\{\widetilde{f}_t\}} = 0$. If $n \neq 2$, it follows that $I_{\{\widetilde{f}_t\}}$ does not depend on the choice of the regular homotopy $\widetilde{f}_t$ but only on the homotopy class of the isotopy f_t in the space of homotopies connecting the two knots. Hence in this case we will write $I(f_0, f_1; f_t)$ instead of $I_{\{\widetilde{f}_t\}}$ and call it the *self-intersection invariant*. The following proposition concludes the proof of Theorem B.1.

PROPOSITION B.3. *Suppose that* $r(f_0, f_1; f_t) = 0$. *Then*

$$s(f_0, f_1; f_t) = I(f_0, f_1; f_t)$$

under a suitable isomorphism from $\pi_{n+1}V_{2n+1,n}$ *to* $\mathbb{Z}$ *for* n *odd (resp.* $\mathbb{Z}_2$ *for* $n \neq 2$ *even).*

We first prove this proposition for n odd in the following lemma. Let $\delta : \pi_{n+1}V_{2n+1,n} \to \pi_n O(n+1)$ and $k : \pi_n O(n+1) \to \pi_n S^n$ be the homomorphisms introduced in Lemma A.13.

LEMMA B.4. *Suppose* n *is odd. Let* $f_t : S^n \to M^{2n+1}$ *be a smooth isotopy between two Legendrian spheres* $f_0, f_1 : S^n \to (M, \xi)$. *Suppose that* $r(f_0, f_1; f_t) = 0$. *Then for suitable isomorphisms* $\pi_{n+1}V_{2n+1,n} \cong \pi_n S^n \cong \mathbb{Z}$, *we have*

$$k \circ \delta\big(s(f_0, f_1; f_t)\big) = 2s(f_0, f_1; f_t) = 2I(f_0, f_1, f_t).$$

PROOF. Lemma A.13 implies the equality $k \circ \delta\big(s(f_0, f_1; f_t)\big) = 2s(f_0, f_1; f_t)$. For the second equality, note that the element $k \circ \delta\big(s(f_0, f_1; f_t)\big) \in \pi_n S^n$ has the following geometric interpretation. Pick a trivialization of the normal bundle to $L_0 = f_0(S^n)$ (which exists since L_0 is a Legendrian sphere and hence its normal bundle is isomorphic to the stabilized tangent bundle) and continuously extend this trivialization to the normal bundles of the submanifolds $L_t = f_t(S^n)$. The normalized Reeb vector fields along the Legendrian submanifolds L_0 and L_1 then give us maps $t_j : S^n \to S^n$, $j = 0, 1$. The difference of the homotopy classes $[t_0] - [t_1] \in \pi_n(S^n)$ is equal to $k \circ \delta\big(s(f_0, f_1; f_t)\big)$.

Equivalently, this difference can be interpreted as follows. Let R^t be the Reeb flow for a contact form defining ξ. Consider the singular chain $\overline{C} : S^n \times I \to M \times I$ defined by the map $\overline{C}(x, t) = (f_t(x), t)$, $x \in S^n$, $t \in I = [0, 1]$. For a sufficiently small $\varepsilon > 0$ we denote by $\overline{C}^\varepsilon$ the shifted chain $\overline{C}^\varepsilon(x, t) = (R^\varepsilon \circ f_t(x), t)$. We have $\partial\overline{C} = L_0 \times 0 - L_1 \times 1$ and $\partial\overline{C}^\varepsilon = L_0^\varepsilon \times 0 - L_1^\varepsilon \times 1$. Then the intersection number $\overline{C} \cdot \overline{C}^\varepsilon$ is equal to $[t_0] - [t_1]$.

Note that the intersection number $\overline{C} \cdot \overline{C}^\varepsilon$ is a homological invariant: for any two chains A, A' in $M \times I$ with $\partial A = L_0 \times 0 - L_1 \times 1$ and $\partial A' = L_0^\varepsilon \times 0 - L_1^\varepsilon \times 1$ which belong to the same relative homology classes in $H_{n+1}(M \times I, L_0 \times 0 \cup L_1 \times 1)$ (resp. $H_{n+1}(M \times I, L_0^\varepsilon \times 0 \cup L_1^\varepsilon \times 1)$) as $\overline{C}$ and $\overline{C}^\varepsilon$ we have $A \cdot A' = \overline{C} \cdot \overline{C}^\varepsilon$.

Now given a regular Legendrian homotopy $\widehat{f}_t$ connecting f_0 and f_1, we denote by $\widehat{C}$ and $\widehat{C}^\varepsilon$ the corresponding singular chains $\widehat{C} : S^n \times I \to M \times I$ and $\widehat{C}^\varepsilon : S^n \times I \to M \times I$ given by the formulas

$$\widehat{C}(x, t) = (\widehat{f}_t(x), t), \qquad \widehat{C}^\varepsilon(x, t) = (R^\varepsilon \circ \widehat{f}_t(x), t), \qquad (x, t) \in S^n \times I.$$

By the preceding discussion,

$$\widehat{C} \cdot \widehat{C}^\varepsilon = \overline{C} \cdot \overline{C}^\varepsilon = [t_0] - [t_1].$$

On the other hand, since the Reeb vector field is nowhere tangent to $\widehat{f}_t(S^n)$, the only contributions to $\widehat{C} \cdot \widehat{C}^\varepsilon$ arise from self-intersections of $\widehat{C}$, each self-intersection point contributing two intersections with the same sign since n is odd. Thus

$$\widehat{C} \cdot \widehat{C}^\varepsilon = 2I_{\{\widehat{f}_t\}} = 2I(f_0, f_1; f_t),$$

and hence

$$2s(f_0, f_1; f_t) = [t_0] - [t_1] = 2I(f_0, f_1, f_t). \qquad \square$$

PROOF OF PROPOSITION B.3. For n odd this follows from Lemma B.4. For even $n \neq 2$ one can argue as follows. Denote by $\mathcal{H}$ the space of regular homotopies g_t, $t \in [0,1]$, connecting f_0 and f_1 and homotopic with fixed endpoints to the path f_t, such that each g_t coincides together with its differential with f_t at the point p. In view of the discussion above and the Smale–Hirsch Immersion Theorem 7.1, concatenating g_t with the inverse of the isotopy f_t yields a bijection $S : \pi_0\mathcal{H} \to \pi_{n+1}V_{2n+1,n} \cong \mathbb{Z}_2$. Associating to each g_t its self-intersection index $I_{\{g_t\}}$ also defines a map $I : \pi_0\mathcal{H} \to \mathbb{Z}_2$, which is surjective because one can always add a new self-intersection point in the interior of a given regular homotopy (see [**191**]). Since $S(g_t) = 0$ implies that the path g_t is connected to the isotopy f_t through regular homotopies and thus $I_{\{g_t\}} = 0$, we see that $I = S : \pi_0\mathcal{H} \to \mathbb{Z}_2$. But by definition $s(f_0, f_1; f_t) = S(\widetilde{f}_t)$ and $I(f_0, f_1; f_t) = I_{\{\widetilde{f}_t\}}$ for any Legendrian regular homotopy $\widetilde{f}_t$ in $\mathcal{H}$, so we conclude $s(f_0, f_1; f_t) = I(f_0, f_1; f_t)$. $\square$

The relative Thurston–Bennequin invariant.

Suppose n is odd. As above, we denote by R^t the Reeb flow. Let L_0, L_1, $i = 0, 1$, be two disjoint (not necessarily spherical and even not necessarily diffeomorphic) oriented Legendrian submanifolds which belong to the same homology class in $H_n M$. Pick a singular chain C with $\partial C = L_0 - L_1$. For a sufficiently small $\varepsilon > 0$ we denote by C^ε and L_i^ε the shifted chain $R^\varepsilon(C)$ and shifted Lagrangian submanifolds $R^\varepsilon(L_i)$, $i = 0, 1$.

The *relative Thurston–Bennequin invariant* is defined as the intersection number

$$\mathrm{tb}(L_0, L_1; C) := C \cdot (L_0^\varepsilon + L_1^\varepsilon).$$

If $\widetilde{C}$ is another chain with $\partial\widetilde{C} = L_0 - L_1$, then we have

$$\mathrm{tb}(L_0, L_1; C) - \mathrm{tb}(L_0, L_1; \widetilde{C}) = (C - \widetilde{C}) \cdot (L_0^\varepsilon + L_1^\varepsilon),$$

where $C - \widetilde{C}$ is an $(n+1)$-cycle in M. Hence, if either $H_{n+1}M = 0$ or the manifolds L_0 and L_1 are homologically trivial, the relative invariant $\mathrm{tb}(L_0, L_1; C)$ is independent of the choice of the class C, so in this case we can drop C from the notation.

In particular, for a homologically trivial oriented Legendrian submanifold $L = \partial C$ one can define an *absolute Thurston–Bennequin invariant* $\mathrm{tb}(L) := C \cdot L$, and this definition is independent of the choice of the spanning chain C. For two homologically trivial oriented submanifolds L_0 and L_1, we have $\mathrm{tb}(L_0, L_1) = \mathrm{tb}(L_0) - \mathrm{tb}(L_1)$. Indeed, if $L_i = \partial C_i$ for $i = 0, 1$, then $\partial(C_0 - C_1) = L_0 - L_1$ and hence

$$\begin{aligned}\mathrm{tb}(L_0, L_1) &= (C_0 - C_1) \cdot (L_0^\varepsilon + L_1^\varepsilon) = C_0 \cdot L_0^\varepsilon - C_1 \cdot L_1^\varepsilon + C_0 \cdot L_1^\varepsilon - C_1 \cdot L_0^\varepsilon \\ &= \mathrm{tb}(L_0) - \mathrm{tb}(L_1) + \mathrm{lk}(L_0, L_1^\varepsilon) - \mathrm{lk}(L_1, L_0^\varepsilon) = \mathrm{tb}(L_0) - \mathrm{tb}(L_1).\end{aligned}$$

Here we have used that the linking pairing is symmetric for n odd and thus $\mathrm{lk}(L_0, L_1^\varepsilon) - \mathrm{lk}(L_1, L_0^\varepsilon) = \mathrm{lk}(L_0, L_1) - \mathrm{lk}(L_1, L_0) = 0$. Let us also point out that $\mathrm{tb}(L)$ remains unchanged when one reverses the orientation of L.

If $f_0, f_1 : \Lambda^n \hookrightarrow M^{2n+1}$ are two disjoint parametrized Legendrian embeddings connected by a homotopy f_t, we define

$$\mathrm{tb}(f_0, f_1; f_t) := \mathrm{tb}(L_0, L_1; C),$$

where $L_i := f_i(\Lambda)$, $i = 0, 1$, and $C : \Lambda \times I \to M$ is the singular chain in M realized by the homotopy f_t, i.e., $C(x,t) = f_t(x)$ for $(x,t) \in \Lambda \times I$.

PROPOSITION B.5. *Let $f_t : S^n \hookrightarrow M^{2n+1}$ be a smooth isotopy between two disjoint Legendrian embeddings $f_0, f_1 : S^n \hookrightarrow M$. Suppose that n is odd and $r(f_0, f_1; f_t) = 0$. Then*

$$\mathrm{tb}(f_0, f_1; f_t) = 2I(f_0, f_1; f_t).$$

PROOF. We continue using the notation introduced in the proof of Lemma B.4. We showed there that

$$2I(f_0, f_1; f_t) = \overline{C} \cdot \overline{C}^\varepsilon,$$

where the singular chain $\overline{C} : S^n \times I \to M \times I$ is defined by the map $\overline{C}(x,t) = (f_t(x), t)$ and $\overline{C}^\varepsilon$ is the shifted chain $\overline{C}^\varepsilon(x,t) = (R^\varepsilon \circ f_t(x), t)$. We have $\partial\overline{C} = L_0 \times 0 - L_1 \times 1$ and $\partial\overline{C}^\varepsilon = L_0^\varepsilon \times 0 - L_1^\varepsilon \times 1$.

Let us deform the isotopies f_t and $R^\varepsilon \circ f_t$ to the isotopies $\widetilde{f}_t$ and $\widetilde{f}'_t$ defined by the formulas

$$\widetilde{f}_t := \begin{cases} f_{2t}, & t \in [0, \frac{1}{2}], \\ f_1, & t \in (\frac{1}{2}, 1], \end{cases} \qquad \widetilde{f}'_t := \begin{cases} R^\varepsilon \circ f_0, & t \in [0, \frac{1}{2}], \\ R^\varepsilon f_{2t-1}, & t \in (\frac{1}{2}, 1]. \end{cases}$$

Let $\widetilde{C}$ and $\widetilde{C}'$ denote the corresponding singular chains in $M \times I$:

$$\widetilde{C}(x,t) = (\widetilde{f}_t(x), t), \qquad \widetilde{C}'(x,t)(\widetilde{f}_t(x), t).$$

Since the intersection number depends only on the relative homology classes, we have

$$\overline{C} \cdot \overline{C}^\varepsilon = \widetilde{C} \cdot \widetilde{C}'. \tag{B.2}$$

The latter intersection number can also be computed in a different way. Namely, consider the singular chains $\widetilde{C}_1 := \widetilde{C}|_{S^n \times [0, \frac{1}{2}]}$ and $\widetilde{C}'_1 := \widetilde{C}'|_{S^n \times [0, \frac{1}{2}]}$ in $M \times [0, \frac{1}{2}]$, and the chains $\widetilde{C}_2 := \widetilde{C}|_{S^n \times [\frac{1}{2}, 1]}$ and $\widetilde{C}'_2 := \widetilde{C}'|_{S^n \times [\frac{1}{2}, 1]}$ in $M \times [\frac{1}{2}, 1]$. Then

$$\widetilde{C} \cdot \widetilde{C}' = \widetilde{C}_1 \cdot \widetilde{C}'_1 + \widetilde{C}_2 \cdot \widetilde{C}'_2, \tag{B.3}$$

where the intersections are computed, respectively, in the manifolds $M \times [0, 1]$, $M \times [0, \frac{1}{2}]$ and $M \times [\frac{1}{2}, 1]$. Consider the projections $C := \mathrm{pr}(\widetilde{C}_1)$ and $C' = \mathrm{pr}(\widetilde{C}'_2)$ of the chains $\widetilde{C}_1$ and $\widetilde{C}'_2$, respectively, to the factor M. Then $\partial C = L_0 - L_1$ and $\partial C' = L_0^\varepsilon - L_1^\varepsilon$, hence

$$\widetilde{C}_1 \cdot \widetilde{C}'_1 = C \cdot L_0^\varepsilon \quad \text{and} \quad \widetilde{C}_2 \cdot \widetilde{C}'_2 = L_1 \cdot C' = R^{-\varepsilon}(L_1) \cdot C. \tag{B.4}$$

But for n odd the vector fields $R|_{L_1}$ and $-R|_{L_1}$ are homotopic as sections of the normal bundle to L_1 in M, and therefore

$$R^{-\varepsilon}(L_1) \cdot C = R^\varepsilon(L_1) \cdot C = L_1^\varepsilon \cdot C. \tag{B.5}$$

Combining equations (B.2), (B.3), (B.4), and (B.5), we obtain

$$2I(f_0, f_1; f_t) = \overline{C} \cdot \overline{C}^{\varepsilon} = C \cdot L_0^{\varepsilon} + L_1^{\varepsilon} \cdot C = C \cdot (L_0^{\varepsilon} + L_1^{\varepsilon}) = \mathrm{tb}(L_0, L_1; C). \qquad \square$$

The invariant tb for homologically trivial knots in 3-manifolds was independently defined by Thurston (unpublished) and Bennequin [**16**]. It was generalized to higher dimensions by Tabachnikov [**182**].

The Thurston–Bennequin invariant is most interesting for 3-dimensional contact manifolds. For instance, Legendrian knots in the standard contact $\mathbb{R}^3$ satisfy *Bennequin's inequality* ([**16**])

$$\mathrm{tb}(L) + |r(L)| \leq \chi(\Sigma),$$

where Σ is a Seifert surface for L and $r(L) \in \mathbb{Z}$ is the rotation invariant. In [**44**] this inequality was extended from the standard contact $\mathbb{R}^3$ to all tight contact 3-manifolds. Kronheimer and Mrowka [**117**] and Rudolph [**164**] proved a stronger version of Bennequin's inequality, replacing the 3-dimensional genus of the knot by its 4-dimensional genus. Many other bounds on $\mathrm{tb}(\Lambda)$ have been found; see [**55, 150**] for surveys of this subject.

Finally, let us discuss the effect of the stabilization construction in Section 7.4 on the relative invariants. Recall that this construction (see Proposition 7.12 and Lemma 7.14) associates to a Legendrian knot $f_0 : \Lambda \hookrightarrow (M^{2n+1}, \xi)$ a Legendrian regular homotopy $f_t : \Lambda \to M$ such that f_1 is a Legendrian embedding and $I_{\{f_t\}} = (-1)^{n(n-1)/2}\chi(N)$. Here $N \subset \mathbb{R}^n$ is a compact domain with smooth boundary over which the stabilization is performed. Note that, since the construction is supported in a Darboux chart, it is irrelevant whether Λ is a sphere or not. Since the two knots are connected by a Legendrian regular homotopy, the relative rotation invariant $r(f_0, f_1; f_t)$ is always zero.

For $n = 1$ the Euler characteristic $\chi(N)$ and thus the self-intersection invariant $I(f_0, f_1; f_t)$ can only be positive, so $\mathrm{tb}(f_0, f_1; f_t) = 2I(f_0, f_1; f_t)$ can be any positive even integer, in accordance with Bennequin's inequality (recall that $\mathrm{tb}(L_0, L_1) = \mathrm{tb}(L_0) - \mathrm{tb}(L_1)$ for homologically trivial knots).

For $n = 2$ the self-intersection invariant vanishes, so the two knots f_0 and f_1 are formally Legendrian isotopic. For even $n > 2$ they are formally Legendrian isotopic if and only if $\chi(N)$ is even.

For odd $n > 1$ the two knots are formally Legendrian isotopic if and only if $\chi(N) = 0$ (note that the "if" was also shown in Proposition 7.23). By varying $\chi(N)$, we can arrange the relative Thurston–Bennequin invariant to take any even integer value, which shows that there is no analogue of Bennequin's inequality for Legendrian knots of odd dimensions $n > 1$.

APPENDIX C

Biographical Notes on the Main Characters

In this appendix we sketch biographies of the mathematicians whose work is most relevant to the content of this book. We have grouped them according to their fields (complex analysis, respectively, differential and symplectic topology), and put them in chronological order within each field. The following sources were used in preparation: the internet site Wikipedia; several articles by J. O'Connor and E. Robertson under http://www-history.mcs.st-andrews.ac.uk/Biographies; the article by L. Dell'Aglio on E. Levi under http://www.treccani.it/enciclopedia/eugenio-elia-levi_(Dizionario-Biografico)/ (translated from Italian by A. Gnoatto); the articles [**106**] and [**105**] by A. Huckleberry on K. Stein and H. Grauert; the article [**22**] by R. Bott on M. Morse; the article by J. Zund under http://www.anb.org/articles/13/13-02523.html, and the interview [**193**] with H. Whitney; the book [**14**] by S. Batterson on S. Smale; and the preface to the book [**127**] by J. Marsden and T. Ratiu on A. Weinstein.

C.1. Complex analysis

Friedrich Hartogs (May 20, 1874 – August 18, 1943). Friedrich Hartogs was born in Brussels, Belgium, into the family of a German businessman. Hartogs' family were German Jews and he was brought up in Frankfurt am Main, Germany. He attended the Realgymnasium Wöhlerschule in Frankfurt, graduating from high school in the spring of 1892.

At that time the standard university career for German students involved moving between different institutions and Hartogs followed this route. First he spent a semester at the Technical College at Hannover, followed by a semester at the Technical College in Berlin. He then matriculated at the University of Berlin where he was taught mathematics by, among others, Georg Frobenius, Lazarus Fuchs, and Hermann Schwarz, and he attended physics lectures by Max Planck. Following his studies at the University of Berlin, he went to the University of Munich where he attended courses by Ferdinand von Lindemann and Alfred Pringsheim. In 1901 Pringsheim became a full professor at Munich and he became Hartogs' thesis advisor. In 1903 Hartogs was awarded his doctorate from Ludwig-Maximilians-Universität in Munich, and two years later he received his habilitation.

After that Hartogs became a privatdozent at the University of Munich. In 1909–10 he taught Abraham Fraenkel who, years later, wrote in his memoirs that Hartogs was by nature a consistently shy and rather anxious person. Perhaps for this reason he was promoted only slowly when the outstanding quality of his research would suggest that he might have risen more rapidly through his profession. He became an extraordinary professor in 1912, then ten years later was offered a full professorship at the University of Frankfurt. Hartogs was indeed a very cautious person and he turned down the offer of this chair because, in the difficult financial climate of the time with hyperinflation gripping Germany, he did not feel confident that a privately owned institution, which the University of Frankfurt was, offered the security that he required to support his wife and four children.

In Munich, Hartogs had several outstanding colleagues such as Oskar Perron, Constantin Carathéodory, and Heinrich Tietze. These three professors all made representations to the university arguing that Hartogs should be appointed to a full professorship, and in 1927, five years after turning down the full professorship at Frankfurt, he had at last reached the top of his profession in Munich. Like all Jewish academics, after the Nazi Party came to power in 1933, Hartogs' life became increasingly difficult. In October 1935 he was forced to retire from his professorship, and on 10 November 1938, during the infamous "Kristallnacht", Hartogs was one of those arrested and taken to the Dachau concentration camp. After being held for several weeks during which he was appallingly treated he was, nevertheless, released.

Hartogs' wife was not Jewish, and in 1941 Hartogs and his wife were given advice by a lawyer that in order to protect Hartogs' wife she should divorce him. This was a painful process for Hartogs and the process was deliberately drawn out to be as lengthy as possible. In early 1943 the divorce was finalized but Hartogs continued to live in his wife's house and the authorities turned a blind eye. The indignity and humiliation that Hartogs had suffered for ten years finally became too much for him, and in August 1943 he took his own life.

Hartogs is best known for his discovery of the *Hartogs phenomenon*, contained in his habilitation thesis, that compact singularities of holomorphic functions in $n > 1$ complex variables are always removable (see Section 5.4). This result is in striking contrast to the case of one variable, and marks the beginning of the theory of functions of several complex variables.

Eugenio Elia Levi (October 18, 1883 – October 28, 1917). Eugenio Elia Levi was born in Torino, Italy. His older brother Beppo Levi was also a well-known mathematician. Eugenio Levi graduated in mathematics from the Scuola Normale di Pisa in 1905. From 1906 to 1909 he was assistant of Ulisse Dini in Pisa, then he moved to the University of Genova where he became full professor in 1912. Eugenio Levi was killed in World War I on October 28, 1917, in Cormons, Italy, on the border with today's Slovenia.

In his short life Eugenio Levi wrote 33 papers making fundamental contributions to group theory, the theory of partial differential operators, and the theory of functions of several complex variables. In his work in group theory he discovered what is now called the Levi decomposition, which was conjectured by Wilhelm Killing and proved by Élie Cartan in a special case. In the theory of partial differential operators he discovered the method of the parametrix, which is a way to construct fundamental solutions for elliptic partial differential operators with variable coefficients. The parametrix method is widely used in the theory of pseudo-differential operators.

In the theory of functions of several complex variables Eugenio Levi introduced the *Levi form* and the concept of *(Levi) pseudoconvexity* (called J*-convexity* in this book), which turned out to be one of the key concepts in the theory of functions of several complex variables. The question of whether a bounded domain in $\mathbb{C}^n$ with smooth pseudoconvex boundary is a domain of holomorphy became known as the *Levi problem* and was one of the main driving forces for the development of complex analysis in the first half of the twentieth century. It was only solved in the 1950s by Oka, Bremermann, and Norguet.

Kiyoshi Oka (April 19, 1901 – March 1, 1978). Kiyoshi Oka entered the Imperial University of Kyoto in 1922 to study physics. However, in 1923 he changed subjects to study mathematics, graduating with a degree in mathematics in 1925. In the same year he was appointed as a lecturer in the Faculty of Science at the Imperial University of Kyoto, and in 1929 he was promoted to assistant professor. 1929 was a very significant year for Oka for in that year he took a sabbatical leave and went to the University of Paris, where he met Gaston Julia and became interested in unsolved problems in the theory of functions of several complex variables.

Oka remained on the staff at the Imperial University of Kyoto while he was on leave in Paris, but on his return to Japan in 1932 he accepted a position as assistant professor in the Faculty of Science of Hiroshima University. In 1938 Oka went to Kimitoge in Wakayama to study by himself, and in 1940 he presented his doctoral thesis to the University of Kyoto. After obtaining his doctorate and a short period 1941–42 as research assistant at Hokkaido University, Oka spent the next seven years again at Kimitoge, supported by a scholarship of the Huju-kai Foundation. In 1949, Oka was appointed professor at the Nara University for Women, a post he held until 1964. From 1969 until his death in 1978 he was a professor of mathematics at the Industrial University of Kyoto.

Oka's most famous work was published over the 25-year period 1936–1961, during which he solved a number of important problems in the theory of functions of several complex variables, such as the Cousin problems and the Levi problem. He proved important foundational results such as Oka's coherence theorem (Section 5.6) and the Oka–Weil theorem (Theorem 5.4). *Oka's principle* on holomorphic approximation of continuous sections, introduced by Oka in his work on the Cousin problems and later generalized by Grauert, provided an early example of an h-principle, and it marked one of the points of departure for Gromov's later work on this subject. In the introduction to Oka's collected works [**155**], Henri Cartan describes the way that Oka came into the subject:

> The publication in 1934 of a monograph by Behnke-Thullen marked a crucial stage in the development of the theory of analytic functions of several complex variables. By giving a list of the open problems in the area, this work played an important role in deciding the direction of Oka's research. He set himself the almost super-human task of solving these difficult problems. One could say that he was successful, overcoming one after the other the obstacles he encountered on the way.

Henri Cartan (July 8, 1904 – August 13, 2008). Henri Cartan was born in Nancy, France, and grew up in Paris. His father, Élie Cartan, was a mathematician well known for his work on Lie groups. Henri had a sister and two younger brothers, Jean and Louis, who both died tragically. Jean, a composer, died of tuberculosis at the age of 25 while Louis, a physicist, was a member of the Resistance arrested by the Germans in 1942, deported to Germany in February 1943, and executed after 15 months in captivity.

Cartan studied at the École Normale Supérieure in Paris, where he met and became friends with André Weil who was one year ahead. It was on André Weil's suggestion that Cartan later began working on analytic functions of several complex variables. Among Cartan's teachers at the École Normale were Gaston Julia and his father Élie Cartan. He received his doctorate in 1928 under the supervision of Paul Montel. After positions at the Lycée Caen and the University of Lille, he took up a post at the University of Strasbourg in 1931. When World War II broke out in September 1939, the inhabitants of Strasbourg had to be evacuated and the university was displaced to Clermont-Ferrand. In November 1940 Cartan was appointed professor at the Sorbonne in Paris. He taught in Paris from that time until 1969 (with the exception of two years 1945–46 when he returned to the University of Strasbourg), and then at the Université de Paris-Sud in Orsay from 1970 to until his retirement in 1975.

At the École Normale Supérieure, Cartan started the Séminaire Cartan. Jean-Pierre Serre, who was one of Cartan's doctoral students, suggested that the seminars should be written up for publication and 15 ENS-Seminars written by Cartan were published between 1948 and 1964. These publications played a major role in shaping the modern theory of functions of several complex variables.

Cartan's most important contribution to mathematics is without doubt the introduction of sheaf-theoretical methods into complex analysis and his Theorems A and B for coherent analytic sheaves on Stein manifolds (see Section 5.6). These new techniques allowed him to treat many of the classical problems on several complex variables in a unified manner, thus moving the whole field into a new era. After Cartan had presented his Theorems A and B at the *Colloque sur les fonctions de plusieurs variables* in Brussels in 1953, the German participant Karl Stein commented, "Wir haben Pfeil und Bogen, die Franzosen haben Panzer."[1]

Cartan also made significant contributions to other areas of mathematics such as algebra and topology. His 1956 book *Homological Algebra* with Eilenberg is a classic text which has had a profound influence on the subject over half a century.

An important part of Cartan's mathematical life was taken up with Bourbaki. He was one of the founding members of this group in 1935 together with André Weil, Jean Dieudonné, Szolem Mandelbrojt, Claude Chevalley, René de Possel, and Jean Delsarte.

Cartan was also involved with politics and in particular supporting human rights. In 1974 the Russian authorities placed the mathematician Leonid Plyushch in a special psychiatric hospital. Andrei Sakharov pointed out that this was a political act and Cartan began a strenuous campaign for Plyushch's release. The International Congress of Mathematicians was held in Vancouver in 1974 and this presented an opportunity to gain wide international support for Plyushch with a thousand signatures to a petition for his release. After the Congress Cartan played a major role in setting up the Comité des Mathématiciens to support Plyushch and other dissident mathematicians. In January 1976 the Soviet authorities released Plyushch, which was a major success for Cartan and the Comité des Mathématiciens. But the Comité did not stop after this success. It has supported other mathematicians who have suffered for their political views, such as the Uruguayan mathematician José Luis Massera. For his outstanding work in assisting dissidents Cartan received the Pagels Award from the New York Academy of Sciences.

[1] We have bows and arrows, the French have tanks.

Karl Stein (January 1, 1913 – October 19, 2000). Karl Stein was born in Hamm in Westfalen, Germany. He studied in Münster, where he received his doctorate under the supervision of H. Behnke in 1937. By that time, he had already been exposed to the fascinating developments in the area of complex analysis. The brilliant young Peter Thullen was proving fundamental theorems, Henri Cartan had visited Münster, and Behnke and Thullen had just written their classical book on the subject. The amazing phenomenon of analytic continuation in higher dimensions had already been exemplified more than 20 years before in the works of Hartogs and Levi, while the recent work of Thullen, Cartan, and Behnke had gone much further. It must have been clear to Stein that this was the way to go.

Even though the Third Reich was already invading academia, Behnke kept things going for as long as possible, but this phase of the Münster school of complex analysis could not go on forever. Although Stein was taken into the army, during a brief stay at home, he was able to prepare and submit the paper which contained the results from his Habilitationsarbeit which was accepted in 1940. At a certain point he was sent to the eastern front. Luckily, however, the authorities were informed of his mathematical abilities, and he was called back to Berlin to work until the end of the war in some form of cryptology.

Almost immediately after the war, in a setting of total destruction, Behnke began to rebuild his group, and very soon Stein became the mathematics guru in Münster. At the time there were only two professor positions in pure mathematics, those of Behnke and F. K. Schmidt. Although it must have been very difficult, Behnke somehow found a position for Stein which he held from 1946 to 1955.

In 1955 Stein took a chair of mathematics at the Ludwigs-Maximilians-Universität in Munich, a position he held until his retirement in 1981. There he continued his mathematics and built his own group in complex analysis, one of his best known students being Otto Forster.

Stein made important contributions to many areas of several complex variables. Until the early 1950s his main efforts were directed towards the Cousin problems. In his 1951 paper [**178**] on this subject, he pointed out that most of the results he considered were true under assumptions which now form the definition of a *Stein manifold*; see Section 5.3. The term "varieté de Stein" for these new spaces was introduced by H. Cartan at the *Colloque sur les fonctions de plusieurs variables in Brussels* in 1953. Stein manifolds and their generalizations, Stein spaces, continue to play a central role in complex analysis to this day.

Hans Grauert (February 8, 1930 – September 4, 2011). Hans Grauert was born in Haren-Ems in Niedersachsen (Lower Saxony) in the north of Germany close to the border with the Netherlands. He attended primary and middle school there from 1936 until the end of the war in 1945. He later recalled how he struggled with mathematics as a school boy until a teacher told him it was acceptable to think abstractly, he didn't necessarily need to deal with numbers.

In 1949 he graduated from the Gymnasium in Meppen, Germany, just 12 km from his home-town. He then studied at the University of Münster, where he was awarded his doctorate in 1954 after spending a year in 1953 at the ETH Zürich, where he was influenced by Beno Eckmann. His first paper "Métrique Kaehlérienne et domaines d'holomorphie" was published in French in 1954.

In September 1955 Grauert was appointed as an assistant at the University of Münster, submitting his habititation thesis there in February 1957. His output of published papers was quite remarkable, with 10 major papers published in 1956 and 1957. He spent the year 1957–58 at the Institute for Advanced Study in Princeton, then the spring semester of 1959 at the Institut des Hautes Études Scientifique in Bures-sur-Yvette, France.

In 1959 Grauert was appointed as an ordinary professor at the University of Göttingen to fill the chair which Carl Ludwig Siegel had occupied. He supervised there doctoral studies of 44 students, several of whom collaborated with him on major projects.

Grauert has been the leading mathematician in the theory of several complex variables in his generation. He not only solved several major problems but his work, along with the work of Henri Cartan, very much shaped the development of this field in the second half of 20th century. For example, the following results of Grauert play an important role in this book: Grauert's solution of the Levi problem for complex manifolds and his characterization of Stein manifolds in terms of J-convex functions (Sections 5.2 and 5.3); Grauert's Oka principle (Section 5.5); and his proof that complexifications of real analytic manifolds (*Grauert tubes*) are Stein (Section 5.7).

Grauert also wrote a large number of excellent textbooks, for example the classical books *Theory of Stein Spaces* (1979) and *Coherent Analytic Sheaves* (1984) with R. Remmert.

C.2. Differential and symplectic topology

Marston Morse (March 24, 1892 – June 22, 1977). Marston Morse was born in Waterville, Maine, USA. His mother was Ella Phoebe Marston and his father was Howard Calvin Morse, a farmer and real estate agent. The name "Marston" by which he wanted to be known was therefore his mother's maiden name and not a forename.

Morse received his B.A. from Colby College in Waterville in 1914, and his Ph.D from Harvard in 1917 for his thesis entitled "Certain types of geodesic motion on a surface of negative curvature" under the direction of G. D. Birkhoff. Morse taught briefly at Harvard before entering military service. For the duration of World War I he served as a private in the U.S. Army in France and for his outstanding work in the Ambulance Corps he was awarded the Croix de Guerre with Silver Star. After the war he resumed his academic career. After positions at Harvard (1919–20), Cornell (1920–25), Brown University (1925–26), and again Harvard (1926–35), he moved to the Institute for Advanced Study in Princeton where he remained until his retirement in 1962.

Morse was married twice and had four daughters and three sons.

In 1925 Morse published a paper entitled "Relations between the critical points of a real function of n independent variables" that would shape his mathematical life and that of generations of mathematicians to this day. In this paper he proves the famous *Morse inequalities* for *Morse functions* on a finite dimensional manifold, thus initiating what is now called *Morse theory* (see Chapter 9).

Realizing the power of this theory, Morse devoted a large part of his mathematical life to its extensions and applications. Almost from the beginning he also considered Morse theory on infinite dimensional spaces such as the loop space of a manifold. His groundbreaking work in this direction culminated in his famous book "The Calculus of Variations in the Large" from 1932, where he proved for example the existence of infinitely many geodesics joining any two distinct points for an arbitrary Riemannian metric on a sphere.

Morse also developed topological versions of his theory for very general functions and found applications to other problems, such as the existence of minimal surfaces. Morse theory was not Morse's only contribution to mathematics — in all he wrote about 180 papers and eight books on a whole range of topics — but clearly the most influential one. It was the basis for many spectacular subsequent developments, from Smale's h-cobordism theorem and Bott's periodicity theorem to Floer homology in gauge theory and symplectic topology. Today, Morse theory is an indispensable tool in geometry and topology. Morse functions, and their J-convex analogues, are also the basic objects studied in this book.

Hassler Whitney (March 23, 1907 – May 10, 1989). Hassler Whitney was born in New York City, the son of Edward B. Whitney, a judge, and Josepha Newcomb. His grandfathers were the philologist William D. Whitney and the astronomer Simon Newcomb. Whitney received his first degree from Yale University in 1928, and his Ph.D. in mathematics from Harvard University in 1932 with the dissertation "The coloring of graphs" written under supervision of George D. Birkhoff. After spending the years 1931–33 as a National Research Council Fellow at Harvard and Princeton he returned to Harvard where he was successively promoted until he became full professor in 1946. From 1943 to 1945 he was a member of the Mathematics Panel of the National Defense Research Committee. In 1952 he joined the Institute for Advanced Study at Princeton, where he was professor of mathematics until his retirement in 1977.

Whitney was a keen mountaineer all his life. As an undergraduate in 1929, Whitney and his cousin Bradley Gilman made the first ascent of a 700-foot ridge in New Hampshire which is now known as the Whitney–Gilman ridge. Later climbing partners included the topologists James W. Alexander and Georges de Rham.

Whitney got married three times, the last time in 1986 at the age of 78, and had five children.

Whitney's work covers a wide range of subjects including graph theory, singularity theory, differential and algebraic topology, and geometric integration theory. In his work on graph theory in the early 1930s he made important contributions to the four color problem. In 1936 Whitney introduced the modern definition of a manifold of class C^r. In 1944 studied the self-intersection index of immersions of half dimension and proved the famous *Whitney embedding theorem* that any smooth manifold of dimension $n > 2$ can be embedded in R^{2n} (see Section 7.1). The *Whitney trick* used in this proof was the basis for much later work in differential topology, such as Smale's proof of the h-cobordism theorem. It also underlies all the flexibility results for Stein structures proved in this book.

In the late 1930s Whitney was one of the major developers of algebraic topology, in particular the theory of bundles and characteristic classes. The significance of his work is reflected in a large number of fundamental concepts that now carry his name such as the Whitney sum, Whitney product theorem, and Stiefel–Whitney classes.

In the 1950s Whitney studied the topology of singular spaces and singularities of maps. He introduced the notion of a Whitney stratification which became the basis for the modern theory of stratified spaces. His classification results for singularities of smooth maps (e.g., the Whitney umbrella) led to the new fields of singularity theory and catastrophe theory. Whitney also did foundational work on analytic spaces, as a by-product of which he proved together with Bruhat that every real analytic manifold has a complexification (Theorem 5.41).

In the last two decades of his life Whitney became involved in mathematical education at elementary schools, vigorously opposing calls for more mathematics to be taught earlier in school.

Stephen Smale (born in 1930). Stephen Smale was born in Flint, Michigan, the home base of General Motors. From the age of five he lived on a farm while his father worked in the city for General Motors. Stephen attended an elementary school with only a single classroom about a mile from his farmhouse. In high school his favorite subject was chemistry. His interests had moved to physics by the time he entered the University of Michigan, Ann Arbor, in 1948, but after failing a physics course he turned to mathematics. He was awarded a BS in 1952 and an MS the following year. In 1957 Smale received his Ph.D. from the University of Michigan under the supervision of Raoul Bott. In his thesis he generalized a result proved by Whitney and Graustein in 1937 for curves in the plane to curves in arbitrary manifolds.

After postdoctoral years spent at the University of Chicago (1956–58), the Institute for Advanced Study in Princeton (1958–59), and the Instituto de Mathemática Pura e Aplicada (IMPA) in Rio de Janeiro, Smale was appointed an associate professor of mathematics at the University of California at Berkeley in 1960. After three years at Columbia University, New York, Smale returned in 1964 to a professorship at Berkeley where he remained until his retirement in 1995. After his retirement he took up a professor position at the City University of Hong Kong, a post he held until 2001 and again since 2009. Since 2002 he is also a professor at the Toyota Technological Institute in Chicago.

Smale's mathematical work is impressive both for its depth and its breadth. He made profound contributions to a whole range of subjects including differential topology, dynamical systems, mathematical economics, and theoretical computer science.

In the years after his Ph.D. Smale astounded the mathematical world with a number of breathtaking results in differential topology. In 1957 he found a general classification of immersions of spheres in Euclidean spaces (see Section 7.1), which implied as a special case that the standard 2-sphere in $\mathbb{R}^3$ can be turned inside out by immersions. His thesis advisor R. Bott at first did not believe this result because he could not picture such a *sphere eversion*, but Smale's proof withstood all scrutiny and was finally published in 1959. Only years later did mathematicians succeed in explicitly describing and visualizing a sphere eversion.

In 1961 Smale proved the generalized *Poincaré conjecture* in dimension > 4, followed in 1962 by the *h-cobordism theorem.* His proof, sketched in Section 9.8, is a beautiful application of Morse theory: beginning with an arbitrary Morse function, Smale successively removes critical points as far as the topology allows, crucially applying Whitney's trick in the process. The most startling aspect of these results was that differential topology suddenly looked *simpler* in higher dimensions than in dimensions 3 and 4. Indeed, in the decade following Smale's work many questions were settled for manifolds of higher dimensions (in a new field called *surgery theory*), while the corresponding questions in low dimensions either had negative answers (such as the existence of exotic smooth structures on $\mathbb{R}^4$), were only solved much later (such as the 3-dimensional Poincaré conjecture), or still remain open (such as the 4-dimensional Poincaré conjecture). For his work on the generalized Poincaré conjecture Smale was awarded a Fields Medal at the International Congress of Mathematicians in Moscow in 1966.

In the 1960s Smale's main focus was the theory of dynamical systems where he introduced a number of new concepts such as his famous *horseshoe* and *Morse–Smale systems*, and proved foundational results such as his Ω-stability theorem. In the 1970s Smale applied his ideas on dynamical systems to questions in economics, and since the 1980s he has been mainly interested in theoretical computer science.

In the summer of 1965 Smale played an important role in the early protests against the Vietnam War in Berkeley. He was one of the main organizers of anti-war activities such as the Vietnam Day 1965, attempts to block trains transporting Vietnam troops, and a march to the Oakland Army Terminal. In early August 1966, the House Committee on Un-American Activities in Washington opened an investigation of radical anti-war protests by Smale and others. At that time Smale was in Europe on his way to Moscow for the Fields Medal Ceremony, which led to the following headline in the San Francisco Examiner on August 5, 1966: "UC Prof Dodges Subpoena, Skips U.S. for Moscow."

Mikhail Gromov (born in 1943). Mikhail Leonidovich (Misha) Gromov was born in Boksitogorsk, a town about 200 km east of St Petersburg (or Leningrad as it was then called). Misha did not speak until the war was over, but then began speaking with whole sentences. At the age of 6 he annoyed his first grade teacher by solving a problem given him by mistake and intended for the third graders. The teacher simply refused to believe that Misha solved it by himself. But when Misha was 10 years old the teacher told his mother that Misha will be a math professor, though at that time the future math professor found much more delight in playing with noxious chemicals.

From 1960 to 1969 Gromov studied at Leningrad University, receiving his masters degree in 1965 and the first doctoral ("candidate") degree in 1969 under the direction of V. A. Rokhlin, followed by his second doctoral degree in 1972. During his undergraduate years, he solved several open problems such as a problem of Banach on the characterization of Banach spaces all of whose k-dimensional subspaces are isometric. But his first major achievement was the far-reaching generalization in his Ph.D. dissertation of the Smale–Hirsch immersion theory, which laid the foundation for the area of mathematics that is now known under the name *h-principle* (see Chapter 7). Over the next four years he made several major advances in this theory, culminating in his theory of convex integration inspired by Nash and Kuiper's C^1-isometric embedding theorem.

The h-principle was the subject of Gromov's invited talk at the International Congress of Mathematicians 1970 in Nice (which he was not allowed to attend by the Soviet authorities). This was the first of a series of four invited ICM talks of Gromov, including two plenary addresses.

In 1974 Gromov left Russia and became a professor at the State University of New York in Stony Brook, USA. In 1981 Gromov moved to France and since that time has been a permanent member of the Institute des Hautes Études Scientifique in Bures-sur-Yvette. From 1991 until 1996 he also held a professor position at the University of Maryland, College Park, and since 1997 he has been a professor at New York University.

Gromov made revolutionary contributions to many branches of mathematics. His work transformed several classical areas and led to the creation of entirely new fields. In particular, his work shaped modern Riemannian geometry, and his introduction of new geometric methods into group theory led to the solution of many classical problems and the creation of the theory of hyperbolic groups. His fundamental paper on pseudo-holomorphic curves in symplectic manifolds essentially created the field of symplectic topology.

Alan Weinstein (born in 1943). Alan Weinstein was born in New York City. He received his undergraduate degree from the Massachusetts Institute of Technology, and his Ph.D. from University of California at Berkeley in 1967 under the direction of S.-S. Chern. After postdoctoral years at the Institute des Hautes Études Scientifique in Bures-sur-Yvette, MIT, and the University of Bonn, he joined the faculty at Berkeley in 1969, becoming full professor in 1976. On the occasion of Weinstein's 60th birthday his advisor S.-S. Chern wrote ([**127**]):

> Alan came to me in the early sixties as a graduate student at the University of California at Berkeley. At that time, a prevailing problem in our geometry group, and the geometry community at large, was whether on a Riemannian manifold the cut locus and the conjugate locus of a point can be disjoint. Alan immediately showed that this was possible. The result became a part of his PhD thesis, which was published in the *Annals of Mathematics.* He received his PhD degree in a short period of two years. I introduced him to IHES and the French mathematical community. He stays close with them and with the mathematical ideas of Charles Ehresmann. He is original and often came up with ingenious ideas. An example is his contribution to the solution of the Blaschke conjecture. I am very proud to count him as one of my students.

Weinstein became interested in symplectic geometry and its applications to mechanics already in the early years of his mathematical career. The Marsden–Weinstein reduction continues to play a fundamental role in classical and quantum mechanics and in the study of the geometry of moduli spaces. Weinstein did important work in the theory of periodic orbits of Hamitonian systems. The Weinstein conjecture about periodic orbits of Reeb vector fields, along with Arnold's fixed point conjectures, continues to be one of the driving forces in symplectic topology. Weinstein made fundamental contributions to Poisson geometry, such as the introduction of symplectic groupoids. Intertwined with his work on symplectic geometry and mechanics, Weinstein did extensive work on geometric partial differential equations, eigenvalues, the Schrödinger operator, and geometric quantization.

In [**187**] Weinstein introduced an object which in [**49**] was called a *Weinstein manifold*, and which is one of the main objects studied in this book.

Alan Weinstein is also an inspiring lecturer and a great teacher. Many of the 32 students who obtained a Ph.D. under his direction became themselves well-known mathematicians.

Bibliography

[1] A. Abbondandolo and M. Schwarz, *On the Floer homology of cotangent bundles*, Comm. Pure Appl. Math. **59**, no. 2, 254–316 (2006).

[2] A. Abbondandolo and M. Schwarz, *Floer homology of cotangent bundles and the loop product*, Geom. Topol. **14**, no. 3, 1569–1722 (2010).

[3] M. Abouzaid and P. Seidel, *Altering symplectic manifolds by homologous recombination*, arXiv:1007.3281.

[4] R. Abraham and J. Robbin, *Transversal Mappings and Flows*, with an appendix by A. Kelley, Benjamin, New York-Amsterdam (1967).

[5] T. Akahori, *A new approach to the local embedding theorem of CR-structures for $n \geq 4$ (the local solvability for the operator $\overline{\partial}_b$ in the abstract sense)*, Mem. Amer. Math. Soc. 67, no. 366 (1987).

[6] A. Akhmedov, J. Etnyre, T. Mark and I. Smith, *A note on Stein fillings of contact manifolds*, Math. Res. Lett. **15**, 1127–1132 (2008).

[7] A. Andreotti and T. Frankel, *The Lefschetz theorem on hyperplane sections*, Ann. of Math. **69**, 717–717 (1959).

[8] A. Andreotti and R. Narasimhan, *A topological property of Runge pairs*, Ann. of Math. **76**, 499–509 (1962).

[9] V. I. Arnold, *Ordinary Differential Equations*, MIT Press, Cambridge, Massachusetts (1973).

[10] V.I. Arnold, *Mathematical Methods of Classical Mechanics*, Springer (1978).

[11] V.I. Arnold, *Geometrical Methods in the Theory of Ordinary Differential Equations*, Springer (1983).

[12] W. Ballmann, *Lectures on Kähler manifolds*, ESI Lectures in Mathematics and Physics, European Mathematical Society, Zürich (2006).

[13] A. Banyaga, *Sur la structure du groupe des difféomorphismes qui préservent une forme symplectique*, Comment. Math. Helv. **53**, no. 2, 174–227 (1978).

[14] S. Batterson, *Stephen Smale: The Mathematician Who Broke the Dimension Barrier*, Amer. Math. Soc. (2000).

[15] E. Bedford and B. Gaveau, *Envelopes of holomorphy of certain 2-spheres in $\mathbb{C}^2$*, Amer. J. Math. **105**, no. 4, 975–1009 (1983).

[16] D. Bennequin, *Entrelacements et équations de Pfaff*, Third Schnepfenried geometry conference, Vol. 1 (Schnepfenried 1982), 87-161, Astérisque **107-108**, Soc. Math. France, Paris (1983).

[17] P. Biran, *Lagrangian barriers and symplectic embeddings*, Geom. Funct. Anal. **11**, no. 3, 407–464 (2001).

[18] E. Bishop, *Mappings of partially analytic spaces*, Amer. J. Math. **83**, 209–242 (1961).

[19] E. Bishop, *Differentiable manifolds in complex Euclidean space*, Duke Math. J. **32**, 1–21 (1965).

[20] F. Bogomolov and B. de Oliveira, *Stein small deformations of strictly pseudoconvex surfaces*, Birational algebraic geometry (Baltimore, 1996), 25–41, Contemp. Math. 207, Amer. Math. Soc. (1997).

[21] R. Bott, *The stable homotopy of the classical groups*, Ann. of Math. (2) **70**, 313–337 (1959).

[22] R. Bott, *Marston Morse and his mathematical works*, Bull. Amer. Math. Soc. **3**, no. 3, 907–950 (1980).

[23] R. Bott and J. Milnor, *On the parallelizability of the spheres*, Bull. Amer. Math. Soc. **64**, 87–91 (1958).

[24] F. Bourgeois, T. Ekholm and Y. Eliashberg, *Effect of Legendrian Surgery*, arXiv: 0911.0026.

[25] F. Bruhat and H. Whitney, *Quelques propriétés fondamentales des ensembles analytiques-réels*, Comment. Math. Helv. **33**, 132-160 (1959).

[26] A. Cannas da Silva, *Lectures on Symplectic Geometry*, Springer (2001).

[27] H. Cartan, *Variétés analytiques complexes et cohomologie*, Colloque sur les fonctions de plusieurs variables (tenu à Bruxelles 1953), 41-55, Georges Thone, Liège; Masson & Cie, Paris (1953).

[28] H. Cartan, *Variétés analytiques réelles et variétés analytiques complexes*, Bull. Soc. Math. France **85**, 77–99 (1957).

[29] D. Catlin, *A Newlander–Nirenberg theorem for manifolds with boundary*, Michigan Math. J. **35**, no. 2, 233–240 (1988).

[30] J. Cerf, *La stratification naturelle des espaces de fonctions différentiables réelles et le théorème de la pseudo-isotopie*, Inst. Hautes Études Sci. Publ. Math. **39**, 5–173 (1970).

[31] Y. Chekanov, *Differential algebra of Legendrian links*, Invent. Math. **150**, no. 3, 441–483 (2002).

[32] K. Cieliebak, *Handle attaching in symplectic homology and the chord conjecture*, J. Eur. Math. Soc. (JEMS) **4**, no. 2, 115–142 (2002).

[33] K. Cieliebak, *Subcritical Stein manifolds are split*, preprint 2002.

[34] K. Cieliebak, A. Floer and H. Hofer, *Symplectic homology II: A general construction*, Math. Z. **218**, no. 1, 103–122 (1995).

[35] K. Cieliebak, U. Frauenfelder and A. Oancea, *Rabinowitz Floer homology and symplectic homology*, Ann. Sci. Éc. Norm. Supér. (4) **43**, no. 6, 957–1015 (2010).

[36] J.-P. Demailly, *Complex analytic and differential geometry*, preliminary version available on the author's homepage, http://www-fourier.ujf-grenoble.fr/~demailly/books.html.

[37] P. De Paepe, *Eva Kallin's lemma on polynomial convexity*, Bull. London Math. Soc. **33**, no. 1, 1–10 (2001).

[38] F. Docquier and H. Grauert, *Levisches Problem und Rungescher Satz für Teilgebiete Steinscher Mannigfaltigkeiten*, Math. Ann. **140**, 94–123 (1960).

[39] K. Dymara, *Legendrian knots in overtwisted contact structures on* S^3, Ann. Global Anal. Geom. **19**, no. 3, 293–305 (2001).

[40] T. Ekholm, J. Etnyre and M. Sullivan, *Non-isotopic Legendrian submanifolds in* $\mathbb{R}^{2n+1}$, J. Diff. Geom. **71**, no. 1, 85–128 (2005).

[41] Y. Eliashberg, *Classification of overtwisted contact structures on 3-manifolds*, Invent. Math. **98**, no. 3, 623–637 (1989).

[42] Y. Eliashberg, *Topological characterization of Stein manifolds of dimension* > 2, Internat. J. Math. **1**, no. 1, 29-46 (1990).

[43] Y. Eliashberg, *Filling by holomorphic discs and its applications*, London Math. Soc. Lect. Notes 151, 45–68 (1991).

[44] Y. Eliashberg, *Contact 3-manifolds 20 years since J. Martinet's work*, Ann. Inst. Fourier **42**, 165–192 (1992).

[45] Y. Eliashberg, *A few remarks about symplectic filling*, Geom. Topol. **8**, 277–293 (2004).

[46] Y. Eliashberg, *Unique holomorphically fillable contact structure on the 3-torus*, Internat. Math. Res. Notices 1996, no. 2, 77–82.

[47] Y. Eliashberg, *Symplectic geometry of plurisubharmonic functions*, Notes by M. Abreu, NATO Adv. Sci. Inst. Ser. C Math. Phys. Sci. 488, Gauge theory and symplectic geometry (Montreal, 1995), 49–67, Kluwer Acad. Publ. (1997).

[48] Y. Eliashberg and M. Fraser, *Topologically trivial Legendrian knots*, J. Symp. Geom. **7**, no. 2, 77–127 (2009).

[49] Y. Eliashberg and M. Gromov, *Convex Symplectic Manifolds*, Proceedings of Symposia in Pure Mathematics, vol. 52, Part 2, 135–162 (1991).

[50] Y. Eliashberg and M. Gromov, *Embeddings of Stein manifolds of dimension* n *into the affine space of dimension* $3n/2+1$, Ann. of Math. **136**, 123-135 (1992).

[51] Y. Eliashberg and V. Kharlamov, *On the number of complex points of a real surface in a complex surface*, Proc. Leningrad International Topology Conference, Leningrad, 1982, 143–148 (1984).

[52] Y. Eliashberg and N. Mishachev, *Introduction to the h-Principle*, Graduate Studies in Mathematics 48, Amer. Math. Soc. (2002).

[53] J. Etnyre, *Introductory lectures on contact geometry*, Topology and geometry of manifolds (Athens, GA, 2001), 81–107, Proc. Sympos. Pure Math. 71, Amer. Math. Soc. (2003).

[54] J. Etnyre, *On Symplectic Fillings*, Algebr. Geom. Topol. **4**, 73–80 (2004).
[55] J. Etnyre, *Legendrian and Transversal Knots*, Handbook of Knot Theory, 105–185, Elsevier (2005).
[56] J. Etnyre and K. Honda, *On the nonexistence of tight contact structures*, Ann. of Math. (2) **153**, no. 3, 749–766 (2001).
[57] A. Floer and H. Hofer, *Symplectic homology I: Open sets in C^n*, Math. Z. **215**, no. 1, 37–88 (1994).
[58] G. Folland, *Introduction to Partial Differential Equations*, Princeton Univ. Press (1976).
[59] J. E. Fornaess and B. Stensønes, *Lectures on Counterexamples in Several Complex Variables*, Princeton Univ. Press (1987), reprinted by AMS Chelsea (2007).
[60] F. Forstnerič, *Stein Manifolds and Holomorphic Mappings*, Springer (2011).
[61] F. Forstnerič and F. Lárusson, *Survey of Oka theory*, New York J. Math. **17a**, 1–28 (2011).
[62] F. Forstnerič, E. Løw and N. Øvrelid, *Solving the d- and $\overline{\partial}$-equations in thin tubes and applications to mappings*, Michigan Math. J. **49**, 369–416 (2001).
[63] F. Forstnerič and M. Slapar, *Stein structures and holomorphic mappings*, Math. Z. **256**, no. 3, 615–646 (2007).
[64] H. Geiges, *Symplectic manifolds with disconnected boundary of contact type*, Int. Math. Res. Not. **1994**, no. 1, 23–30.
[65] H. Geiges, *An Introduction to Contact Topology*, Cambridge Univ. Press (2008).
[66] E. Giroux, *Convexité en topologie de contact*, Comment. Math. Helv. **66**, no. 4, 637–677 (1991).
[67] E. Giroux, *Une infinité de structures de contact tendues sur une infinité de variétés*, Invent. Math. **135**, 789–802 (1999).
[68] E. Giroux, *Structures de contact en dimension trois et bifurcations des feuilletages de surfaces*, Invent. Math. **141**, no. 3, 615–689 (2000).
[69] R. Gompf, *A new construction of symplectic manifolds*, Ann. of Math. **142**, 527–595 (1995).
[70] R. Gompf, *Handlebody construction of Stein surfaces*, Ann. of Math. **148**, no. 2, 619–693 (1998).
[71] R. Gompf, *Stein surfaces as open subsets of $\mathbb{C}^2$*, Conference on Symplectic Topology, J. Symp. Geom. **3**, no. 4, 565–587 (2005).
[72] R. Gompf, *Constructing Stein manifolds after Eliashberg*, New perspectives and challenges in symplectic field theory, 229–249, CRM Proc. Lecture Notes 49, Amer. Math. Soc. (2009).
[73] R. Gompf, *Smooth embeddings with Stein surface images*, arXiv:1110.1865.
[74] R. Gompf and A. Stipsicz, *4-Manifolds and Kirby Calculus*, Amer. Math. Soc. (1999).
[75] J. Gray, *Some global properties of contact structures*, Ann. of Math. (2) **69**, 421–450 (1959).
[76] H. Grauert, *Holomorphe Funktionen mit Werten in komplexen Lieschen Gruppen*, Math. Ann. **133**, 450–472 (1957).
[77] H. Grauert, *On Levi's problem and the imbedding of real-analytic manifolds*, Ann. of Math. (2) **68**, 460-472 (1958).
[78] H. Grauert and R. Remmert, *Theory of Stein Spaces*, Springer (1979).
[79] H. Grauert and R. Remmert, *Coherent Analytic Sheaves*, Springer (1984).
[80] P. Griffiths and J. Harris, *Principles of Algebraic Geometry*, John Wiley & Sons, New York (1978).
[81] M. Gromov, *A topological technique for the construction of solutions of differential equations and inequalities*, ICM 1970, Nice, vol. 2, 221-225 (1971).
[82] M. Gromov, *Convex integration of differential relations I*, Izv. Akad. Nauk SSSR Ser. Mat. **37**, 329–343 (1973).
[83] M. Gromov, *Pseudoholomorphic curves in symplectic manifolds*, Invent. Math. **82**, no. 2, 307–347 (1985).
[84] M. Gromov, *Partial Differential Relations*, Ergebnisse der Mathematik und ihrer Grenzgebiete (3) 9, Springer (1986).
[85] M. Gromov, *Oka's principle for holomorphic sections of elliptic bundles*, J. Amer. Math. Soc. **2**, 851–897 (1989).
[86] M. Gromov and Y. Eliashberg, *Removal of singularities of smooth maps*, Izv. Akad. Nauk SSSR Ser. Mat. **35**, 600–626 (1971).
[87] V. Guillemin and A. Pollack, *Differential Topology*, Prentice-Hall, Englewood Cliffs, New Jersey (1974).

[88] R. Gunning, *Introduction to Holomorphic Functions of Several Variables*, Vol. III: Homological Theory, Wadsworth & Brooks/Cole, Belmont (1990).

[89] R. Gunning and H. Rossi, *Analytic Functions of Several Complex Variables*, Prentice-Hall (1965), reprinted by AMS Chelsea (2009).

[90] A. Haefliger, *Plongements différentiables de variétés dans variétés*, Comment. Math. Helv. **36**, 47–82 (1961).

[91] A. Hatcher, *Algebraic Topology*, Cambridge Univ. Press (2002).

[92] A. Hatcher and J. Wagoner, *Pseudo-isotopies of compact manifolds*, Astérisque **6**, Soc. Math. de France (1973).

[93] G. Henkin and J. Leiterer, *Theory of functions on complex manifolds*, Monographs in Mathematics 79, Birkhäuser (1984).

[94] C. D. Hill and M. Nacinovich, *Stein fillability and the realization of contact manifolds*, Proc. Amer. Math. Soc. **133**, no. 6, 1843–1850 (2005).

[95] R. Hind, *Stein fillings of lens spaces*, Commun. Contemp. Math. **5**, no. 6, 967–982 (2003).

[96] H. Hironaka, *Resolution of singularities of an algebraic variety over a field of characteristic zero I, II*, Ann. of Math. **79**, 109–326 (1964).

[97] M. Hirsch, *Immersions of manifolds*, Trans. Amer. Math. Soc. **93**, 242–276 (1959).

[98] M. Hirsch, *Differential Topology*, Springer (1976).

[99] H. Hofer, *Pseudoholomorphic curves in symplectizations with applications to the Weinstein conjecture in dimension three*, Invent. Math. **114**, no. 3, 515–563 (1993).

[100] H. Hofer and E. Zehnder, *Symplectic Invariants and Hamiltonian Dynamics*, Birkhäuser (1994).

[101] K. Honda, *On the classification of tight contact structures I*, Geom. Topol. **4**, 309–368 (2000).

[102] L. Hörmander, *L^2 estimates and existence theorems for the $\overline{\partial}$ operator*, Acta Math. **113**, 89–152 (1965).

[103] L. Hörmander, *An Introduction to Complex Analysis in Several Variables*, D. Van Nostrand Co., Princeton (1966), 3rd edition North-Holland (1990).

[104] L. Hörmander and J. Wermer, *Uniform approximation on compact sets in C^n*, Math. Scand. **23**, 5–23 (1968).

[105] A. Huckleberry, *Hans Grauert: mathematician pur*, Mitt. Deutsche Math.-Verein. **16**, no. 2, 75–77 (2008).

[106] A. Huckleberry, *Karl Stein (1913–2000)*, Jahresber. Deutsch. Math.-Verein. **110**, no. 4, 195–206 (2008).

[107] K. Igusa, *The stability theorem for smooth pseudoisotopies*, K-Theory **2**, no. 1-2 (1988).

[108] H. Jacobowitz, *An Introduction to CR Structures*, Mathematical Surveys and Monographs 32, Amer. Math. Soc. (1990).

[109] E. Kallin, *Fat polynomially convex sets*, Function Algebras (Proc. Internat. Sympos. on Function Algebras, Tulane Univ., 1965), Scott-Foresman, 149–152 (1966).

[110] Y. Kanda, *The classification of tight contact structures on the 3-torus*, Comm. Anal. Geom. **5**, no. 3, 413–438 (1997).

[111] M. Kervaire, *Non-parallelizability of the n-sphere for $n > 7$*, Proc. Nat. Acad. of Sci. USA **44**, 280–283 (1958).

[112] M. Kervaire, *Le théorème de Barden-Mazur-Stallings*, Comment. Math. Helv. **40**, 31–42 (1965).

[113] S. Kobayashi and K. Nomizu, *Foundations of Differential Geometry*, Vol. II, Interscience Tracts in Pure and Applied Mathematics No. 15 Vol. II, John Wiley & Sons, New York (1969).

[114] J. Kohn and H. Rossi, *On the extension of holomorphic functions from the boundary of a complex manifold*, Ann. of Math. **81**, 451–472 (1965).

[115] A. Kosinski, *Differential Manifolds*, Pure and Applied Mathematics 138, Academic Press, Boston (1993).

[116] S. Krantz, *Function Theory of Several Complex Variables*, John Wiley & Sons, New York (1982), 2nd edition reprinted by AMS Chelsea (2001).

[117] P. Kronheimer and T. Mrowka, *The genus of embedded surfaces in the projective plane*, Math. Res. Lett. **1**, no. 6, 797–808 (1994).

[118] M. Kuranishi, *Strongly pseudo-convex CR structures over small balls, Part III*, Ann. of Math. **116**, 249–330 (1982).

[119] H.-F. Lai, *Characteristic classes of real manifolds immersed in complex manifolds*, Trans. Amer. Math. Soc. **172**, 1–33 (1972).
[120] P. Landweber, *Complex structures on open manifolds*, Topology **13**, 69–75 (1974).
[121] S. Lefschetz, *L'Analysis situs et la géométrie algébrique*, Collection de Monographies publiée sous la direction de M. Emile Borel, Gauthier-Villars, Paris (1924).
[122] L. Lempert, *Algebraic approximations in analytic geometry*, Invent. Math. **121**, no. 2, 335–353 (1995).
[123] P. Lisca, *Symplectic fillings and positive scalar curvature*, Geom. Topol. **2**, 103–116 (1998).
[124] P. Lisca, *On symplectic fillings of lens spaces*, Trans. Amer. Math. Soc. **360**, 765–799 (2008).
[125] P. Lisca and G. Matič, *Tight contact structures and Seiberg-Witten invariants*, Invent. Math. **129**, 509–525 (1997).
[126] R. Lutz, *Structures de contact sur les fibrés principaux en cercles de dimension trois*, Ann. Inst. Fourier (Grenoble) **27**, 1–15 (1977).
[127] J. Marsden and T. Ratiu (eds.), *The breadth of symplectic and Poisson geometry*, Birkhäuser (2005).
[128] J. Martinet, *Formes de contact sur les variétés de dimension 3*, Proceedings of Liverpool Singularities Symposium II (1969/1970), 142–163, Lecture Notes in Math. 209, Springer (1971).
[129] J. Martinet, *Singularities of Smooth Functions and Maps*, Cambridge Univ. Press (1982).
[130] P. Massot, K. Niederkrüger and C. Wendl, *Weak and strong fillability of higher dimensional contact manifolds*, arXiv:1111.6008.
[131] M. Maydanskiy, *Exotic symplectic manifolds from Lefschetz fibrations*, arXiv:0906.2224.
[132] M. Maydanskiy and P. Seidel, *Lefschetz fibrations and exotic symplectic structures on cotangent bundles of spheres*, J. Topol. **3**, no. 1, 157–180 (2010).
[133] D. McDuff, *Symplectic manifolds with contact type boundaries*, Invent. Math. **103**, no. 3, 651–671 (1991).
[134] D. McDuff, *Blow ups and symplectic embeddings in dimension* 4, Topology **30**, 409–421 (1991).
[135] D. McDuff, *The local behavior of holomorphic curves in almost complex manifolds*, J. Diff. Geom. **34**, 143–164 (1991).
[136] D. McDuff and D. Salamon, *Introduction to Symplectic Topology*, 2nd edition, Oxford Univ. Press (1998).
[137] M. McLean, *Lefschetz fibrations and symplectic homology*, Geom. Topol. **13**, no. 4, 1877–1944 (2009).
[138] M. Micallef and B. White, *The structure of branch points in minimal surfaces and in pseudoholomorphic curves*, Ann. of Math. **141**, 35–85 (1995).
[139] J. Milnor, *Morse Theory*, Based on lecture notes by M. Spivak and R. Wells, Annals of Mathematics Studies 51, Princeton University Press, Princeton (1963).
[140] J. Milnor, *Lectures on the h-Cobordism Theorem*, Notes by L. Siebenmann and J. Sondow, Princeton Univ. Press, Princeton (1965).
[141] J. Morgan and G. Tian, *Ricci flow and the Poincaré conjecture*, Clay Mathematics Monographs 3, Amer. Math. Soc. (2007).
[142] J. Munkres, *Obstructions to the smoothing of piecewise-differentiable homeomorphisms*, Ann. of Math. **72**, 521–554 (1960).
[143] E. Murphy, *Loose Legendrian embeddings in high dimensional contact manifolds*, arXiv:1201.2245.
[144] R. Narasimhan, *Imbedding of holomorphically complete complex spaces*, Amer. J. Math. **82**, 917–934 (1960).
[145] R. Narasimhan, *A note on Stein spaces and their normalisations*, Ann. Scuola Norm. Sup. Pisa (3) **16**, no. 4, 327–333 (1962).
[146] A. Némethi and P. Popescu-Pampu, *Milnor fibers of cyclic quotient singularities*, arXiv:0805.3449v2.
[147] S. Nemirovski, *Complex analysis and differential topology on complex surfaces*, Russian Math. Surveys **54**, no. 4, 729–752 (1999).
[148] S. Nemirovski, *Adjunction inequality and coverings of Stein surfaces*, Turkish J. Math. **27**, no. 1, 161–172 (2003).
[149] A. Newlander and L. Nirenberg, *Complex analytic coordinates in almost complex manifolds*, Ann. of Math. (2) **65**, 391–404 (1957).

[150] L. Ng, *A Legendrian Thurston–Bennequin bound from Khovanov homology*, Algebr. Geom. Topol. **5**, 1637–1653 (2005).

[151] K. Niederkrüger and O. van Koert, *Every contact manifold can be given a nonfillable contact structure*, Int. Math. Res. Not. IMRN 2007, no. 23.

[152] A. Nijenhuis and W. Wolf, *Some integration problems in almost complex manifolds*, Ann. of Math. (2) **77**, 424–489, (1963).

[153] L. Nirenberg, *Lectures on linear partial differential equations*, Amer. Math. Soc. (1973).

[154] K. Oka, *Sur les fonctions analytiques de plusieurs variables VII: Sur quelques notions arithmétiques*, Bull. Soc. Math. France **78**, 1–27 (1950).

[155] *Kiyoshi Oka: Collected papers*, translated from the French by R. Narasimhan, with commentaries by H. Cartan, edited by R. Remmert, Springer (1984).

[156] B. Ozbagci and A. Stipsicz, *Contact 3-manifolds with infinitely many Stein fillings*, Proc. Amer. Math. Soc. **132**, 1549–1558 (2004).

[157] J. Palis and W. de Melo, *Geometric theory of dynamical systems: An introduction*, Springer (1982).

[158] G. Perelman, *Finite extinction time for the solutions to the Ricci flow on certain three-manifolds*, arXiv:math/0307245.

[159] O. Plamenevskaya and J. Van Horn-Morris, *Planar open books, monodromy factorizations and symplectic fillings*, Geom. Topol. **14**, no. 4, 2077–2101 (2010).

[160] R. M. Range, *Holomorphic Functions and Integral Representations in Several Complex Variables*, Springer (1986).

[161] R. Richberg, *Stetige streng pseudokonvexe Funktionen*, Math. Annalen **175**, 251–286 (1968).

[162] A. Ritter, *Topological quantum field theory structure on symplectic cohomology*, arXiv:1003.1781.

[163] H. Rossi, *Attaching analytic spaces to an analytic space along a pseudoconcave boundary*, Proc. Conf. Complex Analysis (Minneapolis, 1964), 242–256, Springer (1965).

[164] L. Rudolph, *Quasipositivity as an obstruction to sliceness*, Bull. Amer. Math. Soc. **29**, 51–59 (1993).

[165] D. Salamon and J. Weber, *Floer homology and the heat flow*, Geom. Funct. Anal. **16**, no. 5, 1050–1138 (2006).

[166] J. Schürmann, *Embeddings of Stein spaces into affine spaces of minimal dimension*, Math. Ann. **307**, no. 3, 381-399 (1997).

[167] P. Seidel and I. Smith, *The symplectic topology of Ramanujam's surface*, Comment. Math. Helv. **80**, no. 4, 859–881 (2005).

[168] P. Seidel, *Fukaya categories and Picard-Lefschetz theory*, Zurich Lectures in Advanced Mathematics, European Math. Soc. (2008).

[169] P. Seidel, *A biased view of symplectic cohomology*, Current Developments in Mathematics, 2006, 211–253, Int. Press (2008).

[170] J-P. Serre, *Quelques problèmes globaux relatifs aux variétés de Stein*, Colloque sur les fonctions de plusieurs variables (Bruxelles, 1953), 57–68, Georges Thone, Liège; Masson & Cie, Paris (1953).

[171] S. Smale, *A classification of immersions of the two-sphere*, Trans. Amer. Math. Soc. **90**, 281–290 (1958).

[172] S. Smale, *The classification of immersions of spheres in Euclidean spaces*, Ann. of Math. **69**, 327–344 (1959).

[173] S. Smale, *On the structure of manifolds*, Amer. J. Math. **84**, 387–399 (1962).

[174] I. Smith, *Torus fibrations on symplectic four-manifolds*, Turkish J. Math. **25**, 69–95 (2001).

[175] J. Sotomayor, *Generic bifurcations of dynamical systems*, Dynamical Systems (Proc. Sympos. Univ. Bahia, Salvador, 1971), Academic Press, 561–582 (1973).

[176] J. Stallings, *The piecewise-linear structure of Euclidean space*, Proc. Cambridge Philos. Soc. **58**, 481–488 (1962).

[177] N. Steenrod, *The topology of fibre bundles*, Princeton Univ. Press, Princeton (1951).

[178] K. Stein, *Analytische Funktionen mehrerer komplexer Veränderlichen zu vorgegebenen Periodizitätsmoduln und das zweite Cousinsche Problem*, Math. Ann. **123**, 201–222 (1951).

[179] E. Stout, *Polynomial Convexity*, Birkhäuser (2007).

[180] D. Struppa, *The first eighty years of Hartogs' theorem*, Geometry Seminars 1987–1988, Univ. Stud. Bologna, 127–209 (1988).

[181] D. Sullivan, *Cycles for the dynamical study of foliated manifolds and complex manifolds*, Invent. Math. **36**, 225–255 (1976).

[182] S. Tabachnikov, *An invariant of a submanifold that is transversal to a distribution* (Russian), Uspekhi Mat. Nauk **43** (1988), no. 3 (261), 193–194; translation in Russian Math. Surveys **43**, no. 3, 225–226 (1988).

[183] R. Thompson, *Singular values and diagonal elements of complex symmetric matrices*, Linear Algebra Appl. **26**, 65–106 (1979).

[184] W. Thurston, *Some simple examples of symplectic manifolds*, Proc. Amer. Math. Soc. **55**, no. 2, 467–468 (1976).

[185] C. Viterbo, *Functors and computations in Floer homology with applications I*, Geom. Funct. Anal. **9**, no. 5, 985–1033 (1999).

[186] A. Weinstein, *Symplectic manifolds and their Lagrangian submanifolds*, Advances in Math. **6**, 329–346 (1971).

[187] A. Weinstein, *Contact surgery and symplectic handlebodies*, Hokkaido Math. J. **20**, 241–251 (1991).

[188] C. Wendl, *Strongly fillable contact manifolds and J-holomorphic foliations*, Duke Math. J. **151**, no. 3, 337–384 (2010).

[189] H. Whitney, *Analytic extensions of differentiable functions defined in closed sets*, Trans. Amer. Math. Soc. **36**, no. 1, 63–89 (1934).

[190] H. Whitney, *Differentiable manifolds*, Ann. of Math. (2) **37**, no. 3, 645–680 (1936).

[191] H. Whitney, *The self-intersections of a smooth n-manifold in $2n$-space*, Ann. of Math. (2) **45**, 220–246 (1944).

[192] H. Whitney, *On singularities of mappings of Euclidean spaces I. Mappings of the plane into the plane*, Ann. of Math. (2) **62**, no. 3, 374–410 (1955).

[193] H. Whitney, Interview with A. Tucker and W. Aspray, 10 April 1984, The Princeton Mathematics Community in the 1930s, Transcript Number 43 (PMC43), The Trustees of Princeton University (1985).

[194] W.-T. Wu, *On the isotopy of C^r-manifolds of dimension n in euclidean $(2n+1)$-space*, Sci. Record (N.S.) **2** 271–275 (1958).

[195] R. Ye, *Filling by holomorphic curves in symplectic 4-manifolds*, Trans. Amer. Math. Soc. **350**, no. 1, 213–250 (1998).

Index